"What an enjoyable tour of the history of geology, the interaction between geological study and biblical exegesis, and the current field of geology itself! Professors Young and Stearley have shown us why the geological evidence for the age of the earth is persuasive. Their love for the Bible and for the world that God made, their commitment to a biblical worldview, and their patient explanations of scientific principles are exemplary, as is their respect for those Christians with whom they disagree. Thank you, professors, for writing such a helpful book, which I will be glad to have people read."

C. JOHN ("JACK") COLLINS, PROFESSOR OF OLD TESTAMENT, COVENANT THEOLOGICAL SEMINARY

"This book is a masterful, scholarly, engaging and readable defense of the claim that standard geology's dating of the earth is consistent with a high view of holy Scripture and sound Christian doctrine. Its challenge to young-Earth creationism is daunting—not only scientific but biblical, theological, philosophical and apologetical as well. It is the best book I know on this topic."

JOHN W. COOPER, PROFESSOR OF PHILOSOPHICAL THEOLOGY, CALVIN THEOLOGICAL SEMINARY

"*The Bible, Rocks and Time* is a thoughtful book by Christian scientists for anyone interested in reconciling Genesis and geology. Planet Earth and life on our planet have long, complex and deeply interesting histories written in rocks and fossils that will never go away. Here professors Davis A. Young and Ralph F. Stearley lead us through a timely and constructive synthesis of religion and science."

PHILIP D. GINGERICH, PROFESSOR OF GEOLOGY AND PALEONTOLOGY, UNIVERSITY OF MICHIGAN

"The basic geology in *The Bible, Rocks and Time* is "rock solid." Despite my background of over a half century of geologic teaching and research, I could find nothing with which to argue. The book represents a most welcome change, an authoritative and well-documented major work combining solid science and religious history to stand in sharp contrast against the vast religio-scientific literature by those with little sophistication about either or both. This book could and probably should be used as a geology text for Christian colleges, or it might well be the basic text for an innovative 'pro and con' geology course in more traditional colleges and universities."

DONALD U. WISE, PROFESSOR EMERITUS OF GEOLOGY, UNIVERSITY OF MASSACHUSETTS AT AMHERST

"The long-awaited return of Dave Young's *Christianity and the Age of the Earth* has finally come, and in a wonderful new version. Young and coauthor, Ralph Stearley, present us with a unique scholarly treatment of geology's relationship to a key issue of biblical interpretation. They relate well the historical engagement of theology and geoscience with comprehensive support for the establishment of Earth's great antiquity. Every church library, Christian high school, and college seminary and bookstore should have this volume available as an essential reference."

JEFF GREENBERG, PROFESSOR OF GEOLOGY, WHEATON COLLEGE

"The title says it all, as Davis Young has dealt with the issues of geology and the Bible in great depth. Here an internationally recognized historian of geology assesses the questions of geological time and Genesis from the three perspectives of geology, biblical studies and history and does so with great erudition and spiritual sensitivity. The result is excellent. Here is a book which is a must-read—and one you should buy for your pastor or minister."

MICHAEL ROBERTS, AUTHOR OF *EVANGELICALS AND SCIENCE*, AND VICAR OF COCKERHAM, LANCASTER, ENGLAND

THE BIBLE, ROCKS AND TIME

Geological Evidence for the Age of the Earth

Davis A. Young & Ralph F. Stearley

IVP Academic

An imprint of InterVarsity Press
Downers Grove, Illinois

InterVarsity Press
P.O. Box 1400, Downers Grove, IL 60515-1426
World Wide Web: www.ivpress.com
E-mail: email@ivpress.com

InterVarsity Press® is the book-publishing division of InterVarsity Christian Fellowship/USA®, a student movement active on campus at hundreds of universities, colleges and schools of nursing in the United States of America, and a member movement of the International Fellowship of Evangelical Students. For information about local and regional activities, write Public Relations Dept., InterVarsity Christian Fellowship/USA, 6400 Schroeder Rd., P.O. Box 7895, Madison, WI 53707-7895, or visit the IVCF website at <www.intervarsity.org>.

All Scripture quotations, unless otherwise indicated, are taken from the Holy Bible, New International Version®. NIV®. *Copyright ©1973, 1978, 1984 by International Bible Society. Used by permission of Zondervan Publishing House. All rights reserved.*

Aristotle quotation reprinted by permission of the publishers and the Trustees of the Loeb Classical Library from ARISTOTLE: VOLUME VII, Loeb Classical Library® Volume 397, *translated by H. D. P. Lee, pp. 107-9, 119-21, and Fellows of Harvard College. The Loeb Classical Library® is a registered trademark of the President and Fellows of Harvard College.*

Excerpt from The Literal Meaning of Genesis by St. Augustine, *translated and annotated by John Hammond Taylor, SJ, copyright ©1982 by Rev. Johannes Quasten, Rev. Walter Burghardt, SJ, and Thomas Comerford Lawler. Used with permission of Paulist Press, Inc., New York/Mahwah, N.J. <www.paulistpress.com>.*

Quotation taken from Genesis by John H. Walton. Copyright ©2001 by John H. Walton. Used by permission of Zondervan.*

Quotations from PRITCHARD, JAMES; ANCIENT NEAR EASTERN TEXTS RELATING TO THE OLD TESTAMENT-THIRD EDITION WITH SUPPLEMENT. ©1950, 1955, 1969, renewed 1978 by Princeton University Press. Reprinted by permission of Princeton University Press.*

Quotation from The Natural Sciences Know Nothing of Evolution by a. E. Wilder-Smith, ©1981 by The Word for Today. Reprinted by permission of The Word for Today, P.O. Box 8000, Costa Mesa, CA 92628 <www.twft.com>.*

Quotation by A. J. Akridge, copyright 1998 by Creation Science Fellowship, Inc., Pittsburgh, Pennsylvania, USA. Published with permission. All rights reserved.

Design: Cindy Kiple
Images: Howard Kingsnorth/Getty Images

ISBN 978-0-8308-2876-0

Printed in the United States of America ∞

Library of Congress Cataloging-in-Publication Data

Young, Davis A.

*The Bible, rocks, and time: geological evidence for the age of the
earth / by Davis A. Young and Ralph F. Stearley.*
p. cm.
Includes bibliographical references and indexes.
ISBN 978-0-8308-2876-0 (pbk.: alk. paper)
*1. Bible and geology. 2. Earth—Age. 3. Creationism. I. Stearley,
Ralph F. II. Title*
BS657.Y679 2008
231.7'652—dc22

2008017282

P	22	21	20	19	18	17	16	15	14	13	12	11	10	9	8	7	6	5	4	3	2	1
Y	27	26	25	24	23	22	21	20	19	18	17	16	15	14	13	12	11	10	09	08		

To Dottie and Gloria

CONTENTS

PREFACE

THE PROJECT CULMINATING IN THE PUBLICATION OF *The Bible, Rocks and Time* began innocuously enough when Ralph poked his head into Dave's office one day in 2002. He had dropped in for another of the hundreds of informal conversations that we had enjoyed as friends and colleagues in the Department of Geology, Geography, and Environmental Studies at Calvin College. After ten years of teaching geology at New York University and the University of North Carolina at Wilmington, Dave came to Calvin in 1978 to help develop a geology program. As a "hard-rock" geologist, Dave taught mineralogy, igneous and metamorphic petrology, geochemistry and other geology courses. Ralph joined the department in 1992 and lent his expertise to the program by teaching courses in paleontology, stratigraphy, sedimentology and historical geology.

Along with our common love for and commitment to Jesus Christ, the Reformed tradition of Christianity, God's creation and the science of geology, we were both greatly interested in the history of geologic thought as well as issues concerning the integration of Christian faith and the natural sciences, issues that we frequently discussed and were privileged to teach our students at Calvin. Besides these common interests, both of us can attest to the fact that some of our most satisfying and enjoyable experiences at Calvin involved collaborative efforts. On several occasions, we co-taught courses or co-led extended geology field trips outside our regional mid-continent setting.

With his responsibility for teaching historical geology each spring, Ralph annually assigned as required reading a book that Dave had published in

1982, *Christianity and the Age of the Earth.* After many years using the book, Ralph recognized topics that he wished were in the book and also saw places in which the argumentation of the text could be improved or abandoned altogether. Thus, when Ralph showed up in Dave's office that day in 2002 he simply posed the question, "What would you think about collaborating on a revision of *Christianity and the Age of the Earth?"* Dave normally has several writing projects whirling around in his brain, but a revision of this book was not among them. It didn't take long, however, for him to realize the importance of undertaking the project, and the prospect of collaborating with Ralph on the book made the whole idea much more enticing. We agreed to do it.

Given the other teaching, writing and administrative commitments that accompany the academic life, progress at first was slow, limited to fleshing out a broad outline of what we wanted to accomplish and determining the division of labor. The project gathered momentum when Dave retired in June 2004, after thirty-six years of college teaching (twenty-six at Calvin). He and his wife, Dottie, immediately moved to Tucson, Arizona, an incredibly inspiring and stimulating place in which to write. Dave now had ample time to devote to the revision. Ralph was able to devote summers to the project, and during a sabbatical leave in early 2007, he at last had a large uninterrupted block of time. We communicated our progress via email, but midway through that sabbatical Ralph and his wife, Gloria, drove to Tucson for a six-week stay. During this period, we were able to work through and fine-tune the entire manuscript for publication.

The Bible, Rocks and Time is virtually a total rewrite of *Christianity and the Age of the Earth.* Although the theme and format of both books are very similar, they are very different books. The goal of our book is to convince readers, on both biblical and geological grounds, of the vast antiquity of this amazing planet that is our God-given home. Along the way we point out the flaws of so-called young-Earth creationism. Although the issue of Earth's antiquity may seem to be little more than an interesting intellectual exercise that has little immediate bearing on one's life, we point out that this issue can have profound spiritual consequences for the church of Jesus Christ, the individual Christian and the nonbeliever as well.

This book is addressed primarily to Christian pastors, theologians, biblical scholars, students and lay people with some interest in scientific questions, but we extend an open invitation to non-Christians to read the book as well because we not only seek to persuade Christians to abandon any idea that the Bible demands belief in God's creation of the world only a few thousand years ago but also to show non-Christians that acceptance of modern geological conclusions regarding an ancient Earth is by no means incompatible with biblical Christianity. In other words, nobody needs to abandon sound science in order to become a Bible-believing follower of Jesus Christ.

Readers should take note that wherever we make reference to the biblical flood associated with Noah, we have capitalized the words *Flood* and *Deluge*, in distinction from "generic" floods for which we use lowercase. We also refer to a significant component of young-Earth creationism known as "Flood geology," a fundamentally flawed way of looking at Earth history compared with that of standard, mainstream scientific geology. By the same token we have typically capitalized *Fall* in reference to the Fall of Adam because of its unique significance.

The reader will also note that sometimes we capitalize *Earth* and at other times we use lowercase letters. When references to "the earth" occur in the context of the biblical text, we typically use lowercase. On the other hand, we capitalize *Earth* where discussing the planet in a scientific context. To scientists, the planet on which we live is named "Earth" with a capital E just as Jupiter, Venus, Saturn and the other planets of the Solar System are capitalized. Thus, we frequently refer to "Earth." As many readers are not accustomed to that usage, however, we sometimes use the more familiar expression, "the Earth." Terms like "the sun" and "the moon" are treated in similar fashion. Throughout the text, we also normally refer to distance measurements in terms familiar to most English readers such as feet and miles. In places, however, we employ the metric system: centimeters, meters and kilometers. The reader can make a rough conversion between kilometers and miles by remembering that one kilometer is about 0.6 mile.

We are greatly indebted to the assistance of many friends and colleagues in preparation of the book. We have received helpful reviews of historical and biblical chapters from David Rylaarsdam of Calvin Theological Seminary and Pete Enns of Westminster Theological Seminary. Material pertaining to geology was reviewed by several members and friends of the Affiliation of Christian Geologists that include Michael Roberts, Steve Moshier, Kent Ratajeski, Ken Van Dellen, Jeff Greenberg, and our colleague Gerry Van Kooten at Calvin College. The constructive comments of these Christian geologists, all of whom have no doubt that God's Earth is billions of years old, are deeply appreciated. Bruce Rubidge of the Bernard Price Institute, University of Witwatersrand, kindly provided information concerning the vertebrate faunas of the Karoo Formation of South Africa. William Harrison II of Western Michigan University provided historical information on oil and gas exploration in the Michigan Basin.

We also express our gratitude to Jim Bratt, director of the Calvin Center for Christian Scholarship, for his strong support and encouragement. The Center provided a summer stipend for Ralph as well as a generous grant that made it possible for him to spend a significant portion of his sabbatical leave in Tucson so that we could finalize our manuscript. Ralph is also grateful to Calvin College for two Calvin Research Fellowships in support of this work. Ellen Alderink of Calvin's audiovisual department graciously drafted several of the diagrams. Tim and Mary Ann Young generously scanned some of the slides and illustrations.

To the following institutions we acknowledge permission and express gratitude for permission to quote or to use illustrations: Princeton University Press, Zondervan, Continuum International Publishing Group, Harvard University Press, Paulist Press, Creation Science Fellowship, P & R Publishing, Reivew and Herald Publishing Association, Master Books/NewLeaf Publishing Group, Andrews University Press, Wysong Corporation, Institute for Creation Research, The Word for Today, American Association of Petroleum Geologists (AAPG), National Portrait Gallery, Natural History Museum, Geological Society (London), California Institute of Technology, The McGraw-Hill Companies, Geological Society of America, Institute of Medical Anatomy/ University of Copenhagen, National Trust of Scotland/Hugh Miller Museum, Technische Universität Bergakademie Freiberg, International Ocean Drilling Program/Texas A & M University, Dinosaur National Monument, University of Nebraska State Museum, Larry Agenbroad, Walt Brown, Stephen O. Moshier and Jody Brown Zylstra.

Dave is grateful to Dottie for her unflagging support, and Ralph expresses his indebtedness to his parents for their love, support and encouragement over the years and to Gloria and children for their patience during the project.

Davis A. Young   *Ralph F. Stearley*
Tucson, Arizona   *Grand Rapids, Michigan*

List of Figures and Tables

Tables

ABBREVIATIONS

AAPG   American Association of Petroleum Geologists

ANF   Ante-Nicene Fathers, ed. Alexander Roberts and James Donaldson, 10 vols. (Grand Rapids: Eerdmans, 1950-1951)

DSB   *Dictionary of Scientific Biography,* ed. Charles C. Gillispie, 16 vols. (New York: Scribner, 1970-1980)

GSA   Geological Society of America

GSL   Geological Society of London

LCL   Loeb Classical Library

NPNF   Nicene and Post-Nicene Fathers, ed. Philip Schaff and Henry Wace, 14 vols. (Grand Rapids: Eerdmans, 1956)

NPNF2   Nicene and Post-Nicene Fathers, 2nd series, ed. Philip Schaff and Henry Wace, 14 vols. (Grand Rapids: Eerdmans, 1976)

USGS   United States Geological Survey

INTRODUCTION

HUGH MILLER (1802-1856) WAS ONE OF THE eminent figures of nine-teenth-century Scotland (fig. I.1).[1] A native of the coastal town of Cromarty on the Moray Firth, a few miles north of Inverness, Miller lost his father at the age of five and grew up in poverty. As a young man he made a hardscrabble living as a stonemason, but his discovery of the remains of excellently preserved fossil fishes embedded in the blocks of red Devonian sandstones of the quarries in which he labored triggered a deep interest in geology that would alter his life. Fueled by his newfound passion for geology, Miller became an excellent field geologist and paleontologist who acquired considerable knowledge of fossil fishes, thanks to study with Louis Agassiz, the preeminent authority on fossil fishes of his era. Miller's fame, however, rested on his remarkable gift for written expression. Although he acquired a reputation among the Scottish populace for his verse and books about Scottish folklore, it was especially his rare talent for lucid, popular expositions of geology that expanded that reputation. Among his geology books that remain a pleasure to read to this day are

---

[1]On Miller's life see Michael Shortland, ed., *Hugh Miller's Memoir: From Stonemason to Geologist* (Edinburgh: Edinburgh University Press, 1995); and *Hugh Miller and the Controversies of Victorian Science* (Oxford: Clarendon, 1996). Several works by Miller have recently been republished, including *My Schools and Schoolmasters* (Edinburgh: B & W Publishing, 1993); *Legends and Tales of the North of Scotland* (Edinburgh: B & W Publishing, 1994); and his last great work, *The Testimony of the Rocks; or, Geology in Its Bearings on the Two Theologies, Natural and Revealed* (Cambridge: St. Matthew Publishing, 2001). Miller's cottage, now a small museum that is well worth a visit, still stands in Cromarty and contains many displays pertaining to Miller's life and work.

The Old Red Sandstone, The Footprints of the Creator and his greatest work,
The Testimony of the Rocks.[2]

Miller's writings were distinctive, however, because he was also a devout
Presbyterian who was concerned about the spiritual condition of his church,

the national Church of Scotland. Along
with many other Scots, he aligned himself with the evangelical movement led by
Thomas Chalmers that culminated in the
Disruption of 1843 when 450 ministers
walked out of the General Assembly of
the Church of Scotland down the street to
Tanfield Hall where they established the
Free Church of Scotland, a denomination
that recognized that Jesus Christ is the
sole head of the church and that congregations had the right to select their own
ministers. Miller was profoundly influential in the new denomination because, in
view of his literary skill, he was appointed
as editor of *The Witness*, the ecclesiastical
journal of the Free Church of Scotland. In

**Figure I.1. Hugh Miller (1802-1856).
Reproduced by permission of Na-
tional Trust of Scotland/Hugh
Miller Museum.**

its editorial columns, Miller educated members of the Free Church and the
Scottish public not only about ecclesiastical and theological affairs but also
about geology! As both a Christian and an experienced paleontologist, Miller
thought much about the relation of his faith to his geological knowledge, and
he recognized from Scottish geology, with which he had plenty of field experience, that God's Earth was far older than the 6,000 years entertained by
Christians of a bygone era.[3]

In *The Testimony of the Rocks*, the great masterpiece written just before
his tragic suicide in 1856, Miller elegantly laid out a persuasive case on biblical and scientific grounds for an old Earth and a localized Noahic Flood. He

---

[2]Arguably Miller's three most important geological works are *The Old Red Sandstone; or, New
Walks in an Old Field* (Edinburgh: John Johnstone, 1841); *The Footprints of the Creator; or, The
Asterolepis of Stromness* (London: Johnstone and Hunter, 1849); and *The Testimony of the Rocks;
or, Geology in its Bearings on the Two Theologies, Natural and Revealed* (Edinburgh: Thomas
Constable, 1857). This last work was a compilation of public lectures given in Edinburgh and
Glasgow.

[3]Toward the end of his life, Miller was working on a big book summarizing the geology of Scotland. After his death, Miller's wife, Lydia, no mean writer herself, published a set of his lectures
that incorporated material that formed part of the projected magnum opus. See Hugh Miller,
*Popular Geology: A Series of Lectures Read Before the Philosophical Institution of Edinburgh
with Descriptive Sketches from a Geologist's Portfolio* (Boston: Gould and Lincoln, 1859).

predicted that the ideas of a group that he termed "anti-geologists" would soon be as obsolete as those of the astronomers who upheld the geocentric world picture of Ptolemy. Miller's "anti-geology" referred to a belief, based on a very literalistic interpretation of the early chapters of Genesis, in the concept of a very young Earth that had been destroyed by a global Flood. He envisioned a time when it would be plain to everyone that such belief would be a relic of a bygone era. The time was at hand, Miller maintained, when the history of the Earth unfolding through long geological ages would be found "more worthy of its Divine Author than that which would huddle the whole into a few literal days, and convert the incalculably ancient universe which we inhabit into a hastily run-up erection of yesterday."[4]

Miller proved to be a far better geologist than prophet. Little did he suspect that the second half of the twentieth century would witness a stunning and baffling explosion of "anti-geology" in its modern guise of young-Earth creationism (also known variously as scientific creationism, biblical creationism, literal creationism) and its associate, Flood geology. Far from fading away, these convictions about Earth history not only flourished in the evangelical wing of the Christian church but also came to the attention of the general public and attracted a hostile response from the scientific community.

To combat the geological fallacies associated with the modern version of "anti-geology" that infected the Christian community, D. A. Young published *Christianity and the Age of the Earth (CAE)* in 1982.[5] This book was well received by younger Christian natural scientists who had been strongly influenced during their youth by young-Earth creationism and were unsure how to integrate their Christian faith with the scientific knowledge they gained in their formal studies. *CAE* also received favorable notice in a number of recent works on systematic theology. Unfortunately, some people writing about geology in relation to the Bible in recent works on systematic theology could have profited from reading *CAE*. Predictably, leaders in the young-Earth creationist movement failed to render a warm welcome to *CAE*.[6]

Despite the publication of *CAE* and other works by Christians advocating an ancient Earth, young-Earth creationism continues to thrive in the evangelical universe. Miller's prediction seems no closer to fulfillment today than it did a quarter century ago when *CAE* first appeared. Far too many Christian institutions, including colleges, elementary and secondary schools, theological

---

[4]Miller, *The Testimony of the Rocks*, p. 428.
[5]Davis A. Young, *Christianity and the Age of the Earth* (Grand Rapids: Zondervan, 1982).
[6]For example, Henry M. Morris, *Science, Scripture, and the Young Earth* (El Cajon, Calif.: Institute for Creation Research, 1983); and Henry M. Morris and John D. Morris, *Science, Scripture, and the Young Earth*, 2nd ed. (El Cajon, Calif.: Institute for Creation Research, 1989).

seminaries, ecclesiastical denominations, and individual congregations, and far too many individual Christians, including pastors, theologians, educated lay people, leaders and students, along with much of the general population, continue to dwell in appalling gross darkness when it comes to knowledge about the composition, structure, processes and history of the planet on which they live. The sad thing is that this ignorance accompanies the confession of evangelicals that the Earth is a creation of the God they worship and serve. Not infrequently the little "knowledge" that evangelicals possess about Earth's history is fiction rather than fact.

Although public attention has been somewhat diverted from young-Earth creationism and Flood geology in recent years thanks to the rising popularity of the Intelligent Design movement, young-Earth creationism continues to exert pervasive influence throughout the evangelical and fundamentalist world. Within the past half-century, well-meaning but poorly informed Christians have written hundreds of books, pamphlets and articles under the conviction that the globe is only a few thousand years old and that the biblical Deluge was responsible for a vast portion of Earth's layered, fossiliferous sedimentary rocks. Young-Earth creationists with well-honed debating skills take on proponents of strictly materialistic versions of biological evolution on many college and university campuses. Young-Earth creationism and Flood geology are commonly featured on Christian radio programs, and these views appear in curricular materials, especially those employed by the home-schooling movement. In this new age of the Internet, young-Earth creationism has increased its influence through dozens of websites maintained by individuals and organizations such as Answers in Genesis. Of particular significance is the fact that every four years a technical conference on creationism is held in Pittsburgh. The current crop of young-Earth leaders is geologically more savvy than was the case twenty-five years ago, and their geological arguments have become considerably more sophisticated. Fortunately, the newer generation has tried to weed out some of the more egregiously fallacious claims that permeate the movement.

Despite widespread popularity of young-Earth creationism, particularly in evangelical circles, other Christians are convinced, as was Hugh Miller, that a long geological history is both demanded by the scientific evidence and consistent with the inspired account of creation in the book of Genesis.[7] Among these are the authors of this book, along with the vast majority of members of the Affiliation of Christian Geologists. Moreover, the worldwide profes-

---

[7]See, for example, Alan Hayward, *Creation and Evolution: The Facts and Fallacies* (London: Triangle, 1985); Dan Wonderly, *God's Time-Records in Ancient Sediments: Evidences of Long Time Spans in Earth's History* (Flint, Mich.: Crystal Press, 1977); and John L. Wiester, *The Genesis Connection* (Nashville: Thomas Nelson, 1983).

sional geological community, consisting of tens of thousands of geologists, both Christian and non-Christian, is totally convinced of the vast antiquity of our planet. They are also convinced that sedimentary rocks were formed by a wide variety of processes, including deposition by wind, rivers, ocean waves and glaciers along coasts, in the deep ocean, in river valleys, on desert floors, rather than by a single, global, catastrophic Flood. Mainstream geologists are persuaded that young-Earth creationism and Flood geology, despite the façade of scientific sophistication with which they are presented to a geologically naive public, are deeply flawed both scientifically and biblically. Christian geologists are especially concerned that persistent efforts by well-intentioned believers to gain acceptance for young-Earth creationism as a viable alternative to mainstream geological thinking will result in deepening alienation of the scientific community from evangelical Christianity, a trend that does not bode well for evangelization of scientists.

To provide a fresh critique of young-Earth creationism and Flood geology for the benefit of Christians and non-Christians alike, we present *The Bible, Rocks and Time*, a completely revised, rewritten, and updated work based on the theme and format of *CAE*. In support of the vast antiquity of God's Earth, we present a case that has historical, biblical, scientific and philosophical dimensions. We write to convince Bible-believing Christians that the Earth really is extremely old and to show them that acceptance of such a belief need not be in any way a threat to their Christian faith. But we also write to demonstrate to non-Christians who may understandably entertain the false impression that Christianity entails commitments to a young Earth and a global Deluge that such commitments are by no means inherent to Christian faith.

To provide a context for the ongoing controversy about Earth's antiquity, we begin with a survey of the history of ideas about the age of the Earth.[8] While noting the diversity of interpretations of Genesis 1 in the early and medieval church, we point out that the church's scholars took it for granted that the Earth was only a few thousand years old. In the context of the birth and maturing of scientific geology after the Protestant Reformation, we describe the growing awareness than the globe might be older than had traditionally been perceived. Finally, we discuss the varied biblically based responses to the idea of an ancient Earth within the Christian community.

Next, we proceed to the biblical aspects of the discussion about terrestrial

---

[8]For an overview of ideas about the age of the Earth, including the Greeks, see Patrick Wyse Jackson, *The Chronologer's Quest: The Search for the Age of the Earth* (Cambridge: Cambridge University Press, 2006). Some useful older works include Francis Haber, *The Age of the World: Moses to Darwin* (Baltimore, Md.: Johns Hopkins Press, 1959); and Stephen Toulmin and June Goodfield, *The Discovery of Time* (New York: Harper & Row, 1965).

antiquity. Those who argue that the Earth is only a few thousands of years old generally maintain that the six days of creation mentioned in Genesis 1 must be taken as six literal, ordinary, twenty-four-hour days.[9] They also maintain that the Flood of Noah described in Genesis 6—8 covered the entire globe and did a tremendous amount of geological work. Other Christians who believe that the Earth is extremely old have interpreted the creation account of Genesis 1 in a variety of alternative ways, and they also maintain that Genesis does not necessarily require belief that the Flood of Noah covered the entire globe.[10] We present evidence that the biblical text does not demand adherence to the traditional interpretation that God created the world in six successive, twenty-four-hour days only a few thousand years ago. On that basis we claim that the Bible does not demand assent to any specific age for the planet.

In the third section of the book we examine geological evidence pertinent to Earth's antiquity. We show from several angles that geological evidence overwhelmingly indicates that the Earth has had an extremely long, dynamic history. We examine and refute, in light of contemporary geology, many of the geological arguments that are alleged by young-Earth creationists to support a young-Earth view. Some readers without a scientific background may find chapters eleven, fourteen and fifteen daunting, but even without reading those chapters, they should still be able to recognize that geology demands an old Earth.

The book concludes with a look at the philosophical side of the discussion. We demonstrate that, despite their adherence to a philosophical principle of catastrophism, young-Earth creationists, nonetheless, unavoidably often present their case from a "uniformitarian" point of view in much the same way as mainstream geologists do, that is, by appeal to geologic evidence and present-day understanding of natural processes. They simply misinterpret and/or ignore the evidence. We further show that young-Earth creationists completely fail to understand the concept of "uniformitarianism" as employed by contemporary geologists. The final chapter concerns the implications of acceptance of an old Earth for Christian faith and shows that such acceptance need pose no threat to Christianity. In contrast, the cause of Christian faith is damaged by promotion of false belief.

Let's get underway.

---

[9]The second article of the statement of belief of the Creation Research Society says that "All basic types of living things, including man, were made by direct creative acts of God during the Creation Week described in Genesis." Rebuttals of alternative interpretations of Genesis 1 are presented in Henry M. Morris, *Scientific Creationism*, 1st ed. (El Cajon, Calif.: Master Books, 1974).

[10]On the church's response to scientific discoveries pertaining to Noah's flood, see Davis A. Young, *The Biblical Flood: A Case Study of the Church's Response to Extrabiblical Evidence* (Grand Rapids: Eerdmans, 1995).

Part One

HISTORICAL PERSPECTIVES

1

THE AGE OF THE EARTH THROUGH THE SEVENTEENTH CENTURY

THE EARLY CHRISTIAN CHURCH PRODUCED much valuable insight into the creation account of Genesis 1. Even today, Christians benefit from expositions on the *Hexameron* (the six days) by Basil, Ambrose, Chrysostom and others, and they are challenged by the ideas of Origen or Augustine about creation. Although early Christian thinkers dug as deeply into the theological content of Genesis 1 as more recent Christian interpreters have, they lived in a world with very different conceptions about the nature of the cosmos and the place of the Earth within it. As a result, the church fathers could not interpret creation texts in light of the kinds of questions that are raised by modern scientific investigation of the Earth. Nor did they have access to the wealth of archeological material pertaining to the ancient Near East that was discovered during the nineteenth and twentieth centuries.[1] Consequently, the church fathers were not privy to much of the information about the cultural, religious and literary world in which the early Israelites lived and moved and had their being.

---

[1]For a compilation of major documents from the ancient Near East, see James B. Pritchard, *Ancient Near Eastern Texts Relating to the Old Testament* (Princeton, N.J.: Princeton University Press, 1950). See also Alexander Heidel, *The Babylonian Genesis: The Story of Creation*, 2nd ed. (Chicago: University of Chicago Press, 1951) and *The Gilgamesh Epic and Old Testament Parallels*, 2nd ed. (Chicago: University of Chicago Press, 1949).

GREEK VIEWS ON TERRESTRIAL ANTIQUITY

The early church was, however, the beneficiary of several centuries of Greek thought. Although their main scientific interests were astronomy and geometry, some pre-Christian Greeks did inquire into the nature of earthquakes, volcanoes, rivers, fossils and other geologic phenomena. The Greeks even developed a fairly elaborate theory of transgressions of the sea onto land to account for the existence of fossils that are far removed from present coastlines.[2]

The earliest of the Greek writers known to have commented on the existence of fossils was Xenophanes (late sixth to early fifth centuries B.C.), a poet and founder of the Eleatic school of philosophy. Some of his ideas were preserved in the work of the Christian writer Hippolytus (A.D. 170-236), a Roman presbyter and, for a time, an anti-pope. Hippolytus wrote that

> Xenophanes is of opinion that there had been a mixture of the earth with the sea, and that in process of time it was disengaged from the moisture, alleging that he could produce such proofs as the following: that in the midst of earth, and in mountains, shells are discovered; and also in Syracuse he affirms was found in the quarries the print of a fish and of seals, and in Paros an image of a laurel in the bottom of a stone, and in Melita parts of all sorts of marine animals, and he says that these were generated when all things originally were embedded in mud, and that an impression of them was dried in the mud, but that all men had perished when the earth, being precipitated into the sea, was converted into mud; then, again, that it originated generation, and that this overthrow occurred to all worlds.[3]

Xenophanes explained the existence of fossils in terms of flooding of the land by the sea. He may have been influenced in his thinking by the Greek legend of Deucalion's flood. No indication of how much time Xenophanes thought was involved in the process is found in the existing literature from classical times, but the idea of marine transgression became firmly embedded in Greek thought.

Herodotus (fifth century B.C.), the "father" of historians and an astute observer of men, nations and nature, was well aware of the existence of fossils and of various geologic processes. His travels in Egypt led him to speculate that the Nile River valley was originally a gulf much like the Red Sea and that the valley had gradually filled with silt over the course of time.

Suppose, now, that the Nile should change its course and flow into this gulf—

---

[2]Adrian J. Desmond, "The Discovery of Marine Transgression and the Explanation of Fossils in Antiquity," *American Journal of Science* 275 (1975): 692-707. Desmond discussed the views of many of the Greek writers to whom we refer.

[3]Hippolytus *Refutation of All Heresies*, in ANF, 5:17.

the Red Sea—what is to prevent it from being silted up by the stream within say, twenty thousand years? Personally I think even ten thousand would be enough. That being so, surely in the vast stretch of time which has passed before I was born, a much bigger gulf than this could have been turned into dry land by the silt brought down by the Nile—for the Nile is a great river and does, in fact work great changes.[4]

The existence of fossil shells in the hills and salt in the soil supported his idea that the Nile River valley had once been under the sea. There is also some evidence that information about the geography of the Nile delta obtained from Egyptian records played a part in Herodotus's calculations. His estimate appears to be the earliest effort to quantify a geologic process; he believed that probably more than ten thousand years were required for the silting of the Nile delta.

Strabo recorded that Xanthus of Sardis (fifth century B.C.), the historian of Lydia, also taught the former occupancy of land by the sea.

Far from the sea, in Armenia, Matiene, and Lower Phyrgia, he himself had often seen, in many places, stones in the shape of a bivalve, shells of the pectin order, impressions of scallop-shells, and a salt-marsh, and therefore was persuaded that these plains were once sea.[5]

Aristotle (384-322 B.C.), the greatest of the ancient Greek thinkers, was a student of Plato, the tutor of Alexander the Great, and founder of the Lyceum. Consistent with his remarkable breadth of interest, Aristotle contributed more than all other Greeks to knowledge about the Earth. His views on the nature of earthquakes and volcanoes and of different kinds of exhalations from the interior of the Earth, spelled out in considerable detail in *Meteorologica*, exerted tremendous influence on geological thinking in the Western world until well after the Middle Ages. Like his predecessors Xenophanes, Xanthus and Herodotus, Aristotle, too, believed that the sea had once been where land now is:

The same parts of the earth are not always moist or dry, but change their character according to the appearance or failure of rivers. So also mainland and sea change places and one area does not remain earth, another sea, for all time, but sea replaces what was once dry land, and where there is now sea there is at another time land. This process must, however, be supposed to take place in an orderly cycle. . . . But these changes escape our observation because the whole natural process of the earth's growth takes place by slow degrees and over periods of time which are vast compared to the length of our life, and whole peoples are destroyed and perish before they can record the process from beginning to end.[6]

---

[4]Herodotus, *The Histories* (Baltimore, Md.: Penguin Books, 1954), p. 106.
[5]Strabo, *Geography*, Book 1 (London: Heinemann, 1931), pp. 181-83.
[6]Aristotle, *Meteorologica*, LCL (London: Heinemann, 1952), pp. 107-9.

Aristotle believed that all rivers came into being and then eventually disappeared because the universe is eternal and thus plenty of time is available for such continuing changes to occur:

> It is therefore clear that as time is infinite and the universe eternal that neither Tanais nor Nile always flowed but the place whence they flow was once dry: for their action has an end whereas time has none. And the same may be said with truth about other rivers. But if rivers come into being and perish and if the same parts of the earth are not always moist, the sea also must necessarily change correspondingly. And if in places the sea recedes while in others it encroaches, then evidently the same parts of the earth as a whole are not always sea, nor always mainland, but in process of time all change.[7]

Although these and other Greek thinkers saw evidences of geologic change that seemed to extend well into the past, they were unable to develop any convincing means for estimating the amount of time that such changes might have required. Even Herodotus's calculation was only an educated guess. Moreover, no other lines of evidence for the antiquity of the Earth were worked out, so it is little wonder that early Christians were unimpressed by Greek Earth science. Such efforts must have appeared to them as so much speculation.

The chief sources for ideas about Earth's antiquity in early Christian times stemmed primarily from philosophy and the historical records of the Egyptians, Chaldeans and other ancient Near Eastern peoples. The concept of the eternity of the world was a philosophical speculation that was widely held by Greek thinkers. The early Christians opposed the notion of the eternity of the cosmos because of their belief in the creation of the world by God. For them, eternity belonged to God the creator alone. Moreover, historical records of pagan civilizations were regarded by Christians as rather unreliable in comparison with the divinely inspired biblical records. For example, Theophilus (A.D. 115-181), an early bishop of Antioch, condemned

> the empty labor and trifling of these authors, because there have neither been twenty thousand times ten thousand years from the flood to the present time, as Plato said, affirming that there had been so many years; nor yet 15 times 10,375 years, as we have already mentioned Apollonius the Egyptian gave out; nor is the world uncreated.[8]

Then, too, the father of Christian chronography, Sextus Julius Africanus (A.D. c. 160-c. 245) took issue with others as follows:

> The Egyptians, indeed, with their boastful notions of their own antiquity, have put forth a sort of account of it by the hand of their astrologers in cycles and

---

[7]Ibid., pp. 119-21.
[8]Theophilus *To Autolycus*, in ANF, 2:119.

myriads of years; which some of those who have had the repute of studying such subjects profoundly have in a summary way called lunar years; and inclining no less than others to the mythical, they think they fall in with the eight or nine thousands of years which the Egyptian priests in Plato falsely reckon up to Solon. . . . Why should I speak of the three myriad years of the Phoenicians, or of the follies of the Chaldeans, their forty-eight myriads?[9]

CHRISTIAN REACTION TO GREEK SCIENCE

If Greek thinkers did not present convincing evidence for the antiquity of the Earth from their study of the phenomena of nature, they yet must be commended for their attempts to study the Earth. In contrast, so far as we can tell, early Christians showed little interest in investigation of the Earth or, for that matter, in any natural scientific study. To be sure, the opportunity for leisurely investigation of nature was not present during the early centuries of Christian history. Early Christians were repeatedly subjected to intense persecution until the "conversion" of Roman Emperor Constantine in 312. Even after Christianity gained privileged status within the Roman Empire, the intellectual energy of Christian thinkers was devoted to working out the fundamental doctrines of the Christian faith such as the Trinity and the person of Jesus Christ and to defense of the faith against heathen detractors. Christian apologists were often willing to use available scientific knowledge in their defense of the faith, but they also frequently displayed deep ambivalence toward scientific investigation. The study of nature was strongly bound up with Greek philosophy and was therefore suspect in the eyes of some, particularly in the Western church.

For example, Basil (329-379), a bishop of Caesarea, in his sermons on the Hexameron, showed considerable acquaintance with Greek science. Despite his acceptance of the four elements (earth, water, air, fire) as well as the sphericity of the Earth, Basil was still wary because Greek science had not led its practitioners to a true knowledge of God. Moreover, he knew that on any given topic of study there might be a large number of conflicting viewpoints, and so he regarded the primitive scientific enterprise as rather speculative and unfruitful. Because science did not lead to sound knowledge in the way that Scripture did, attempts to investigate the Earth were deemed to be of relatively little value.

> These same thoughts, let us also recommend to ourselves concerning the earth, not to be curious about what its substance is; nor to wear ourselves out by reasoning, seeking its very foundation. . . . Therefore, I urge you to abandon these questions and not to inquire upon what foundation it stands. If you do that, the mind will become dizzy, with the reasoning going on to no definite end.[10]

---

[9]Julius Africanus *Chronography*, in ANF, 6:130-31.
[10]Basil, *On the Hexameron* (Washington, D.C.: Catholic University of America Press, 1963), p. 14.

Other Christian thinkers, like Augustine (354-430), a bishop of Hippo in North Africa and the greatest theologian of the early church, were skeptical of the sphericity of the Earth, a fact that had been established long before the appearance of Christ. Augustine was decidedly hostile to the inference drawn by the Greeks that there might be people living on the other side of the globe. He used reason and Scripture in his criticism of the theory of the antipodes:

> But as to the fable that there are Antipodes, that is to say, men on the opposite side of the earth, where the sun rises when it sets to us, men who walk with their feet opposite ours, that is on no ground credible. And, indeed, it is not affirmed that this has been learned by historical knowledge, but by scientific conjecture, on the ground that the earth is suspended within the concavity of the sky, and that it has as much room on the one side of it as on the other; hence they say that the part which is beneath must also be inhabited. But they do not remark that, although it be supposed or scientifically demonstrated that the world is of a round and spherical form, yet it does not follow that the other side of the earth is bare of water; nor even, though it be bare, does it immediately follow that it is peopled. For Scripture, which proves the truth of its historical statements by the accomplishment of its prophecies, gives no false information; and it is too absurd to say, that some men might have taken ship and traversed the whole wide ocean, and crossed from this side of the world to the other, and that thus even the inhabitants of that distant region are descended from that one first man. Wherefore let us seek if we can find the city of God that sojourns on earth among those human races who are catalogued as having been divided into seventy-two nations and as many languages.[11]

Despite his negative view on the question of the antipodes, Augustine elsewhere evidenced open-mindedness toward science. Other Christians, too, particularly in the Eastern church, clearly appreciated learning and the fruits of learning, including scientific work. For example, Titus Flavius Clemens, generally known as Clement of Alexandria (153-217), head of the Catechetical School in Alexandria and a presbyter in that church, said:

> And by astronomy, again, raised from the earth in his mind, he is elevated along with heaven, and will revolve with its revolution; studying ever divine things, and their harmony with each other; from which Abraham starting, ascended to the knowledge of Him who created them.[12]

And also,

> The same holds also of astronomy. For treating of the description of the celestial

---

[11] Augustine *The City of God*, in NPNF, 2:315-16. Lucius Caecilius Firmianus Lactantius (c. 260-c. 330), a Christian apologist and a tutor of the son of Emperor Constantine, also poked fun at the idea of the antipodes.

[12] Clement of Alexandria *The Miscellanies*, in ANF, 2:498.

objects, about the form of the universe, and the revolution of the heavens, and the motion of the stars, leading the soul nearer to the creative power, it teaches to quickness in perceiving the seasons of the year, the changes of the air, and the appearance of the stars; since also navigation and husbandry derive from this much benefit, as architecture and building from geometry. This branch of learning, too, makes the soul in the highest degree observant, capable of perceiving the true and detecting the false, of discovering correspondences and proportions, so as to hunt out for similarity in things dissimilar; and conducts us to the discovery of length without breadth, and superficial extent without thickness, and an indivisible point, and transports to intellectual objects from those of sense.[13]

As a rule, however, early Christians tended to shy away from the investigation of nature and were skeptical of many conclusions about nature drawn by the Greeks. Moreover, the Greeks had not produced abundant, conclusive, compelling evidence for the antiquity of the world from the phenomena of nature. Thus, early Christian thinkers grappled with the text of the early chapters of Genesis without any significant external pressure from the realm of science. We need not wonder that the early church fathers did not fully work out all the implications of the doctrine of creation, for they had not fully worked out other dogmas either.

The Early Christian View of the Age of the Earth

Despite Herodotus's estimate for the filling of the Nile delta, the virtually unanimous opinion until the time of Augustine among early Christians who wrote about the issue was that human history from the creation of Adam to the birth of Christ had lasted approximately 5,500 years. The fathers probably also regarded the age of the world as the same number of years as that of human history, because their writings generally do not indicate any sharp distinction between the initial creation of the cosmos and the creation of the human race. A widespread conviction that the present world order would last for 6,000 years prevailed. Upon completion of the 6,000 years, Christ would return to establish his kingdom of righteousness and peace for another thousand years. Many early Christian thinkers thus fully expected the millennium to be ushered in around A.D. 500.

These beliefs about the age of the world were based on several lines of evidence. The predominant line of evidence was drawn from the record of God's creation of the world in six days. In general, the church fathers regarded the days of creation as ordinary solar days.[14] Nevertheless, they did not shy away

---

[13]Ibid., ANF, 2:501.
[14]This point has been defended by J. Ligon Duncan and David W. Hall, "The Twenty-Four-Hour

from additionally regarding the days in a more figurative or allegorical sense. Virtually all of the fathers were struck by the statement of Psalm 90:4 that "a thousand years in your sight are like a day that has just gone by, or like a watch in the night" and of 2 Peter 3:8 that "with the Lord a day is like a thousand years, and a thousand years are like a day." They saw no difficulty in connecting the days of creation with 1,000-year periods on the basis of these texts, nor did they detect any problem in viewing the days figuratively as millennia as well as literal ordinary days. For them the concept of day was much broader than an ordinary solar day.

The striking feature of this patristic view, however, is that the equation of days to millennia was applied not to the length of the creation week but rather to the length of human history.[15] There is no evidence that the fathers of the church believed that the work of creation had taken six millennia to complete. They did believe that the totality of human history would occupy 6,000 years, a millennium of history for each of the six days of creation. Why they made this transference from six days of *creation* to 6,000 years of *history* is unclear, although the view may be a holdover from earlier Jewish tradition. No reason for the connection, apart from the statements of Psalm 90:4 and 2 Peter 3:8, was offered by the fathers. The idea of 6,000 years of history was simply assumed and taught. We meet first with this opinion in the *Epistle of Barnabas* and subsequently in the writing of Irenaeus, Hippolytus, Methodius, Lactantius, John of Damascus and several others.

The *Epistle of Barnabas,* whose unknown author was allegedly an Alexandrian Jewish Christian layman who was combating Judaism, probably dates from the early second century. In discussing the true versus the false sabbath, the writer observed that the reference in Genesis 2:2 to God's completion of the work of creation in six days "implies that the Lord will finish all things

---

View" and "The Twenty-Four-Hour Reply," in *The Genesis Debate,* ed. David G. Hagopian (Mission Viejo, Calif.: Crux Press, 2001), pp. 47, 99-102.

[15]In his book *Creation and Time* (Colorado Springs: NavPress, 1994), astrophysicist and Christian apologist Hugh Ross conveyed the impression that there was a wide range of opinions about the nature of the creation days in the early church. He maintained that "writing later in the third century, Lactantius (c. AD 250-325), Victorinus of Pettau and Methodius of Olympus, all concurred with Justin Martyr's and Irenaeus's view of the creation days as thousand-year epochs" (p. 19). He also wrote that "throughout the Dark and Middle Ages, church scholars maintained the tolerant attitude of their forefathers toward differing views and interpretations of the creation time scale" (p. 25). We are unable to agree with Ross's assessment of the views of the early church. The first quotation is incorrect because the church fathers did not believe that the days of creation were 1,000 years long, but they did believe that 1,000 years of human history corresponded to each literal day of creation. The second quotation is somewhat misleading because it conveys the impression that the church fathers might have differed about the time of creation, but we have no evidence that anyone other than Origen thought the world might be older than 5,500 years.

in six thousand years, for a day is with Him a thousand years." After quoting Psalm 90:4, the writer explained that "in six days, that is, in six thousand years, all things will be finished." The reference to God's seventh-day rest meant that the Son would return to "judge the ungodly, and change the sun, and the moon, and the stars." Then God shall "truly rest on the seventh day."[16]

In *Against Heresies,* Irenaeus (120-202), a pupil of the martyr Polycarp, a presbyter and a bishop of Lyons in Gaul and an enemy of the Gnostic heresies that plagued the second-century church, wrote about the apostasy that would take place in the time of the antichrist. Irenaeus noted that, in the Apocalypse, the Apostle John described the great beast whose number is 666 as a way of "summing up of the whole of that apostasy which has taken place during six thousand years." According to him, "for in as many days as this world was made, in so many thousand years shall it be concluded." Appealing to 2 Peter 3:8, he asserted that "the day of the Lord is as a thousand years; and in six days created things were completed: it is evident, therefore, that they will come to an end at the six thousandth year."[17]

Hippolytus discussed the "times" of the world in comments on the visions in the book of Daniel. He asserted that "the first appearance of our Lord in the flesh took place in Bethlehem, under Augustus, in the year 5500." Moreover, he continued, "6000 years must be accomplished, in order that the sabbath may come." Hippolytus mentioned the day of the Lord being as a thousand years and concluded that since "in six days God made all things, it follows that 6000 years must be fulfilled."[18]

In a section on the first and last times of the world in *The Divine Institutes,* Lactantius criticized the philosophers "who enumerate thousands of ages from the beginning of the world." Being "ignorant of the origin of all things," they needed to be set straight that "the six thousandth year is not yet completed." As the basis for his assertion, Lactantius stated that "since all the works of God were completed in six days, the world must continue in its present state through six ages, that is, six thousand years." He repeated the claim that "the great day of God is limited by a circle of a thousand years," citing the Psalm 90:4 text. Just as God labored six days, he said, "His religion and truth must labor during these six thousand years" after which "all wickedness must be abolished from the earth, and righteousness reign for a thousand years."[19]

Victorinus (d. c. 303), a bishop of Pettau in what is now Austria, in writing about the creation of the world, commented on Jesus' "violation" of the

---

[16] *The Epistle of Barnabas,* in ANF, 1:146.
[17] Irenaeus *Against Heresies,* in ANF, 1:557.
[18] Hippolytus *Fragments from Commentaries,* in ANF, 5:179.
[19] Lactantius *Divine Institutes,* in ANF, 7:211.

sabbath described in Matthew 12:5 that the "true and just sabbath should be observed in the seventh millenary of years. Wherefore to those seven days the Lord attributed to each a thousand years." Like the other fathers he referred to Psalm 90:4.[20]

Methodius (c. 260-c. 312) was a bishop of Olympus and one of the first opponents of some of the speculations of Origen. In fragments of documents pertaining to creation, Methodius referred disapprovingly to Origen's suggestions of the eternity of the universe and of the existence of humans prior to Adam. In response he suggested that the antiquity of the world could easily be "reckoned from the creation of the world." After quoting Psalm 90:4, he wrote that, as the creation took six days, so, too, "to our time six days are defined, as those say who are clever arithmeticians. Therefore, they say that an age of six thousand years extends from Adam to our time. For they say that the judgement will come on the seventh day, that is in the seventh thousand years."[21]

Julius Firmicus Maternus, a fourth-century Sicilian Christian who lived during the reigns of emperors Constantine and Constantius, wrote in his book *The Error of the Pagan Religions* that "after long ages, in the last reaches of time, that is, almost at the end of the week of the centuries, the Word of God commingled Itself with human flesh." Although this passage is not quite so straightforward as those quoted above, the reference to the "week of the centuries" is consistent with the idea that the length of human history was tied in with the creation week.[22]

Tychonius lived toward the end of the fourth century and is known primarily for his book of seven rules designed to serve as guidelines in the exegesis of the Bible. The fifth section concerns time. "The world's age," he said, "is six days, i.e., six thousand years. In what is left of the sixth day, i.e. of these 1000 years, the Lord was born, suffered and rose again. Similarly what is left of the 1000 years is called the thousand years of the first resurrection." Later he wrote, in an evident reference to the creation period, that "the first seven days represent 7000 years." Inasmuch as God "made that world in six days, so he makes the spiritual world, which is the church, in the course of six thousand years, and he will stop on the seventh day, which he has blessed and has made eternal."[23]

John of Damascus (c. 676-c. 754), a monk and priest renowned as the most capable defender of the veneration of images, wrote that "seven ages of this

[20]Victorinus *On the Creation of the World*, in ANF, 7:341-43.
[21]Methodius *Extracts from the Work on Things Created*, in ANF, 6:381.
[22]Julius Firmicius Maternus, *The Error of the Pagan Religions* (New York: Newman Press, 1970).
[23]Tychonius, *The Book of Rules*, trans. William S. Babcock (Atlanta: Scholars Press, 1989), pp. 91, 101.

world are spoken of, that is, from the creation of the heaven and earth till the general consummation and resurrection of men." Although the word *age* may have many meanings, he understood that "a period of a thousand years is called an age," suggesting that he understood the seven ages to comprise a span of 7,000 years.[24]

Although many early theologians may have misinterpreted and misapplied Psalm 90:4 and 2 Peter 3:8, it is clear that they had little difficulty making some connection between the six days of creation and much longer periods of time.

The second line of evidence that the world was less than 6,000 years old at the time of Christ was drawn from the presumed chronology of the genealogical accounts of Genesis 5 and 11 and other chronological information in Scripture. Church fathers frequently appealed to such data in an effort to demonstrate the antiquity and greater reliability of the Mosaic revelation as over against Greek or pagan thought in men like Homer, Hesiod or Plato. Justin Martyr (c. 110-165), an administrator of schools of philosophy in Ephesus and Rome, Tatian (110-172), a Christian apologist who eventually turned toward Gnosticism, and Clement of Alexandria frequently accused the Greeks of having plagiarized from the Old Testament.

Appeals to chronology were most fully worked out by Theophilus of Antioch and subsequently by Clement of Alexandria, Julius Africanus, Hippolytus, Origen, Eusebius, Augustine and George Syncellus. In *To Autolycus*, Theophilus calculated that from the creation of the world to the onset of the flood was 2,242 years, to Abraham 3,278 years, to the Babylonian captivity 4,954 years, and to the time of his own writing 5,698 years.[25] He would probably have placed creation at 5529 B.C. In his brief commentary on the Hexameron, Theophilus clearly regarded the days of creation as types.[26] For example, he treated the first three days as a type of the Trinity and the fourth day as a type of man. Nevertheless, he also definitely thought of the creation days as ordinary solar days, for he attempted to date various events and people from the beginning of the creation of the world, not just from the creation of Adam.

Clement of Alexandria freely applied allegorical interpretations to the days of creation.[27] Whether or not he also regarded the days literally is unclear, but he certainly regarded human history as of short duration. His calculations from chronological data in Scripture led him to conclude that 2,148 years and four days had elapsed between the creation of Adam and the onset of the flood

---

[24]John of Damascus *Exposition of the Orthodox Faith*, in NPNF2, 9:19.
[25]Theophilus *To Autolycus*, in ANF, 2:118-20.
[26]Ibid., ANF, 2:100-104.
[27]Clement of Alexandria *The Miscellanies*, pp. 332-33, 513-14.

and that approximately 5,590 years had elapsed between Adam's creation and the birth of Christ.

The most extensive calculations were those of Julius Africanus. In his *Chronography*, Julius elaborately worked out calculations based on both Scripture and secular history and concluded that from Adam to the flood was 2,262 years, and that from Adam to the Advent of the Lord was 5,531 years.[28]

The chronographic tradition was continued by Eusebius (c. 265-340), a bishop of Caesarea who is renowned for his history of the early church and for being chairman of the Council of Nicea in 325. In his *Chronicle*, Eusebius regarded the birth of Abraham as providing the first reliable date and devoted most of his book to the dating of subsequent events. He did offer more speculative dates for individuals going back to Adam, however, and placed the creation of Adam at 5,232 years before the resurrection of Christ.[29]

The chronography of Eusebius set the standard within the Christian community for the next several centuries until the time of George Syncellus, a Palestinian monk of the late eighth and early ninth centuries who became a *synkellos*, that is, a personal secretary to the patriarch Tarasios in Constantinople. He was likely also a bishop in that city. After the death of the patriarch, Syncellus produced a large-scale chronography in which he maintained that Eusebius had departed from the apostolic tradition regarding the time of the appearance of Christ. In his judgment Mary's conception of the Savior had taken place at the beginning of the year 5501. Syncellus also pointed out that Panodoros, an early fifth-century Alexandrian monk, had calculated the birth of Christ 5,493 years after the beginning.[30]

Augustine also maintained that fewer than 6,000 years had passed in the history of the world.[31] In the context of this claim, he was speaking against the vast antiquity claimed by heathen writers for kingdoms like Assyria, Persia and Macedonia. Augustine may have been thinking in terms of human history being less than 6,000 years in duration, because his allegorical ideas about the days of creation could have led him to think that the date of creation was earlier than this. There is, however, no evidence that such was the case, and we need to remember that Augustine's views on creation were not completely consistent. Elsewhere he noted that, according to copies of Scripture available

---

[28] Julius Africanus *Chronography*, in ANF, 6:131-38.

[29] For information on the chronography of Eusebius see Alden A. Mosshammer, *The Chronicle of Eusebius and Greek Chronographic Tradition* (Lewisburg, Penn.: Bucknell University Press, 1979). Much information about the content of the *Chronicle* can also be found in the chronography of George Syncellus. See William Adler and Paul Tuffin, *The Chronography of George Synkellos* (Oxford: Oxford University Press, 2002).

[30] Adler and Tuffin, *Chronography of George Synkellos*.

[31] Augustine *City of God*, in NPNF, 2:233.

to him, 2,262 years had elapsed between the time of Adam and the Deluge, but according to the Hebrew text, 1,656 years.[32]

A third line of argument in support of a short history for the Earth was provided by Hippolytus.[33] When asked how we know that the Lord appeared in the year 5500, Hippolytus did not make the obvious appeal to the genealogies and other chronological data in the Bible, nor did he lean heavily on the purported relation between days of creation and millennia of history. Arguing on symbolic grounds, he maintained that numbers throughout Scripture are not just literal numbers but are also symbols of spiritual realities. For example, he observed that the text of Exodus 25:10 states the dimensions of the ark of the covenant as 2 ½ x 1 ½ x 1 ½ cubits. When added, these three figures yield the sum of 5 ½ cubits. It was plain to Hippolytus that the 5 ½ cubits are a symbol of the 5 ½ millennia of history that had elapsed prior to the coming of Christ.

Hippolytus further appealed to Revelation 17:10 which says that the seven horns on the beast are seven kings of which "five have fallen, one is, the other has not yet come." He completely ignored the reference to kings and regarded the numbers as referring to millennia. From his perspective, five millennia had elapsed, he was living in the middle of the sixth, and the seventh was wholly future. Hippolytus also believed that John 19:14 gives a hint as to the age of the world. The apostle John recorded that, at the trial of Jesus, it was about the sixth hour, that is, noon or halfway through the day. Since one day is as a 1,000 years, half a day equals 500 years. So it was plain to Hippolytus that the end of Christ's life history was in the middle of another millennium.

In a number of his writings, Hilary (c. 318-368), a bishop of Poitiers and a vigorous defender of trinitarianism against Arianism, alluded to his belief that the present world would last 6,000 years, at which time Christ would return. From the parable of the workers in the vineyard presented in Matthew 20, Hilary concluded that the various hours of the day at which the owner of the vineyard hired workers corresponded to epochs in Old Testament history. The eleventh hour, he decided, corresponded to the Lord's birth. Using mathematical argument, Hilary maintained that Christ was born 500 years (at the eleventh hour) into the final millennium on the assumption that the twelfth hour corresponds to the end of the age and that the world is going to last for 6,000 years.[34] Elsewhere he made links between the six days on which the Israelites were allowed to collect manna in the wilderness and 6,000 years of human

---

[32]Ibid., NPNF, 2:300-301.

[33]Hippolytus *Fragments from Commentaries*, in ANF, 5:179.

[34]Hilaire de Poitiers, *Sur Matthieu*, trans. Jean Doignon, vol. 2 (Paris: Les Éditions du Cerf, 1979).

history.[35] Finally, he stated that the six days of marching around the city of Jericho represented the 6,000 years during which the course of generations traces its circle. At the seventh millennium, he said, the church, like Rahab, would be saved.[36]

To arrive at the conclusion that the world was approximately 5,500 years old at the time of Christ, the fathers were compelled to hold to a strictly literal interpretation of the genealogies of Genesis 5 and 11. It did not occur to any of the early Christians that there could be gaps or omissions in those genealogies. As a result, they felt perfectly justified in summing the ages of the patriarchs in Genesis 5, for example, to obtain the number of years between Adam and the Flood. The calculations worked out from these genealogies by Theophilus, Julius and others were based on the Greek Septuagint translation rather than the Hebrew text, so that the sums obtained for the time that had elapsed between Adam and Abraham were greater by several hundred years than they would have been had they followed the Hebrew text.

The fathers also had to reckon the days of creation in Genesis 1 as ordinary days if they were to hold that the Earth was only 5,500 years old at the time of Christ. Absolutely no one in the early church argued that the world is tens of thousands of years old on the grounds that the six days are used figuratively for indefinite periods of time. Not even Origen (185-254), a student of Clement of Alexandria and his successor at the Catechetical School in that city, maintained such a view, although he came closest to holding the great antiquity of the Earth.[37] In fact, Origen argued for a succession of worlds rather than a very ancient existing world. Most of the church fathers plainly regarded the six days as ordinary days. Basil, for example, explicitly spoke of the day as a twenty-four-hour period.[38]

Patristic writers recognized the uniqueness of the six days of creation, however, and so had difficulty in regarding them as literal days *only*. Thus we find allegorical interpretations in profusion within the early church. The *Homilies on Genesis* of Origen are a case in point. In his first homily, after quoting Genesis 1:9, Origen did not discuss what was going on during creation but launched into an allegorical application of the text. "Let us labor, therefore," he exhorted his congregation,

> to gather "the water which is under heaven" and cast it from us that "the dry land," which is our deeds done in the flesh, might appear when this has been

---

[35]Hilaire de Poitiers, *Traité de Mystères*, trans. Jean-Paul Brisson (Paris: Les Éditions du Cerf, 1947), p. 139.

[36]Ibid., p. 157.

[37]Origen *De Principiis*, in ANF, 4:341.

[38]Basil *On the Hexameron*, p. 34.

done so that, of course, "men seeing our good works may glorify our Father who is in heaven." For if we have not separated from us those waters which are under heaven, that is, the sins and vices of our body, our dry land will not be able to appear nor have the courage to advance to the light.[39]

Origen treated each creative act of the Genesis 1 account in similar fashion.

AUGUSTINE'S CONCEPTION OF THE SIX DAYS

Augustine believed that the six days did not refer to a real succession in time because of his conviction that everything had been created simultaneously.[40] In effect, he distinguished between the creative work of God and the literary presentation of that creative work, and held that God created matter in an already formed state rather than creating unformed matter and then superimposing form on that matter at a later time. This idea comes out repeatedly in his work *The Literal Meaning of Genesis.* We read, for example, that "God the Creator did not first make unformed matter and later, as if after further reflection, form it according to the series of works He produced. He created formed matter."[41] More specifically, he affirmed that "God created together both the matter which He formed and the objects into which He formed it." Scripture had to mention both matter and formed objects but could not mention them together. So which should be mentioned first? Obviously, Augustine thought, it was logical to mention raw matter first and then the objects made from it. Thus, "when speaking of matter and form, understand that they exist together, but we must name them separately."[42]

Moreover, because the inspired writer of Genesis was attempting to bring the lofty topic of divine creation "down to the capacity of children," we should understand that God "did not say, 'Let this or that creature be made' as often as the sacred text repeats *And God said.* He begot one Word in whom He said all before the several works were made."[43] In other words, "the account of the stages of the creation of the world" is "not a real progression in time, since God created matter at the very moment he created the world, but a mere progression in the narrative."[44]

[39]Origen, *Homilies on Genesis and Exodus* (Washington, D.C.: Catholic University of America Press, 1982), p. 50.

[40]Augustine borrowed his notion of instantaneous creation from the Apocryphal book Sirach, which says in 18:1 that "He who lives forever created all things at once."

[41]Augustine, *The Literal Meaning of Genesis,* trans. John Hammond Taylor (New York: Newman Press, 1982), p. 36.

[42]Ibid.

[43]Ibid., p. 54.

[44]Ibid., p. 68. Hilary of Poitiers supposedly held a view similar to Augustine, according to Luther, but a close examination of what Hilary said indicates that he was establishing the point that God had a single plan prior to creation rather than that he created all things at once. What

Consistent with his idea that God created all things at once, Augustine departed from the idea that the six days were ordinary days. Even though his views on the subject were the most thoroughly worked out of any of the early Christians, Augustine always approached the nature of the creation days with great caution and humility. "It is," he wrote, "a laborious and difficult task for the powers of our human understanding to see clearly the meaning of the sacred writer in the matter of these six days."[45] He also said that "if we are able to make any effort towards an understanding of the meaning of those days, we ought not to rush forward with an ill-considered opinion, as if no other reasonable and plausible interpretation could be offered."[46] Augustine was the first to propose that the first three days of creation were not ordinary days. He reasoned that these days could not have been ordinary days because they were not marked by the rising and setting of the sun.[47] There must, therefore, be some special meaning for the first three days. But, he went on to argue, that special meaning must not be limited to the first three days but must be applicable to all the creation days.[48] Likewise, the "day" and "night" that God divided and the "light" of the first day need to be interpreted in a different manner from the physical days and nights related to the position of the sun.

Consistent with his neo-Platonic tendencies, Augustine leaned toward the notion that the ideal/spiritual realm is more "real" than the physical/material realm.[49] As a result, the days of creation do not relate to the passage of time in the physical world. The language that he used is unfamiliar to us and not at all easy to grasp. Augustine related the days to spiritual knowledge of the angels contemplating the works of God. God created everything at once, he said, but the one day that God made "recurs in connection with His works not by a material passage of time but by spiritual knowledge, when the blessed company of angels contemplate from the beginning in the Word of God the divine decree to create."[50] "Through this knowledge," he said, "creation was revealed to it as if in six steps called days, and thus was unfolded all that was created; but in reality there was only one day."[51] The light of day one is spiritual light, not material light, but it is, nevertheless, true, literal, not allegorical or metaphorical light. He acknowledged that the "framework of the six days of creation might

---

Hilary denied was that God created the Earth, then thought out the next step, then created the waters, thought out the next step and so on. See Hilary *On the Trinity*, in NPNF2, 9:228.

[45] Augustine, *The Literal Meaning of Genesis*, p. 103.

[46] Ibid., p. 135.

[47] Augustine *City of God*, in NPNF, 2:208-9 and *The Literal Meaning of Genesis*, p. 125.

[48] Augustine, *The Literal Meaning of Genesis*, 134.

[49] Andrew J. Brown, "The Relevance of Augustine's View of Creation Re-evaluated," *Perspectives on Science and Christian Faith* 57 (2005): 134-45.

[50] Augustine, *The Literal Meaning of Genesis*, p. 134.

[51] Ibid., p. 155.

seem to imply intervals of time," but he consistently denied that Genesis 1 told us anything about the amount of time that elapsed during the creation of the world.[52] Indeed, creation occurred in a moment.

Also worthy of note are the facts that Augustine maintained that the events described in the first two verses of the Bible did not belong to the six days and that the seventh day of the creation week still continues. He said that the seventh day has no evening and morning because God had sanctified it to everlasting continuance.[53] Elements of Augustine's unusual interpretation would reappear nearly 1,400 years later in the restitution, day-age and framework interpretations of Genesis 1 that will be discussed in chapters four and five.

The idea that the seventh day of divine rest had no ending and that it continues forever may not have originated with Augustine. Pope Anastasius I (d. 401), who gained fame for his condemnation of the writings of Origen, indicated that Clement of Alexandria, Irenaeus and Justin Martyr thought that the seventh day differed from the first six. According to Anastasius, "the fact that it was not said of the seventh day equally with the other days, 'And there was evening, and there was morning,' is a distinct indication of the consummation which is to take place before it is finished."[54] Whether this consummation referred explicitly to the seventh day at the conclusion of creation or to the seventh millennium corresponding to that day we cannot tell, but the uniqueness of the seventh day of creation week was clearly recognized. Augustine, however, apparently maintained the continuing existence of the seventh day. Despite the idea that the seventh day continues, no serious thought was given to the idea that the other six days might also be long time periods, a fact that is doubly intriguing because the church fathers readily connected creation days to millennia, and although anticipations of internal biblical evidence suggestive of the antiquity of the Earth appear in the creation account, the church fathers failed to carry through the logic of their insights.

Medieval, Reformation and Post-Reformation Views

Churchmen continued to ponder the meaning of Genesis and creation, and numerous studies of creation and of the Hexameron appeared, written by men like the Venerable Bede (673-735), a priest in the monastery at Jarrow who authored the *Ecclesiastical History of the English People*, and Robert Grosseteste (c. 1175-1253), a bishop of Lincoln. From time to time, Augustine's idea that the first two verses of Genesis 1 do not belong to the first day of creation emerged in various guises. Hugh of St. Victor (1097-1141), a scholar and mystic at the

---

[52]Ibid., p. 157.
[53]Augustine *Confessions*, in NPNF, 9:207.
[54]Anastasius *Fragment*, in ANF, 1:301-2.

monastery of St. Victor in Paris and later head of the School of St. Victor, suggested that Scripture does not teach us how long the Earth remained in the state of disorder described in verse 2.[55] Thomas Aquinas (c. 1225-1274), a brilliant Dominican scholar who taught in numerous institutions throughout Europe, particularly Paris, was arguably the greatest of the medieval theologians. He maintained that the creation occurred prior to the six days.[56] Given the prevailing belief that the universe is only a few thousand years old, it is unlikely that these men envisioned the passage of very much time previous to the work of the six days. Medieval works on creation strongly followed the thinking of the church fathers, and little new material of significance regarding the question of the age of the Earth appeared.

During and after the Protestant Reformation, scholars increasingly became interested in deriving the age of the Earth from biblical data. Biblical chronography became a highly refined science whose practitioners eventually sought to determine not only the year in which creation had taken place, but also the season, and in some cases, the month, week, day and hour! These newer determinations of the Earth's age differed significantly from those of the church fathers in that the calculations were based on the Hebrew text rather than the Septuagint translation. Consequently, the typical dates given for the creation of the Earth were approximately 1,500 years more recent than those proposed by the church fathers. Protestants of the Reformation era showed a strong tendency to abandon the more allegorical interpretations of the days of Genesis that had been adopted by some of the church fathers. They advocated a strictly literal view of the days throughout. One important exception, however, was Simon Episcopius (1583-1643), a Dutch theologian who, as a follower of Jacobus Arminius, became a leading exponent of Remonstrant theology in opposition to Calvinism. Episcopius proposed that a period of time elapsed between Genesis 1:1 and 1:2 to account for the fall of the wicked angels.[57]

The great initiator of the German Reformation, Martin Luther (1483-1546), completely rejected Augustine's approach to Genesis 1 and insisted on a strictly literal exegesis of the six days. He taught that the Earth was not yet 6,000 years old in his time and held the opinion that the creation had occurred in the spring.[58] Others in the Lutheran wing of the Reformation who agreed with Luther about creation in the springtime were Philip Melanchthon (1497-1560), the educational reformer, founder of Protestant systematic theology, and au-

---

[55]Hugh of St. Victor, *On the Sacraments of the Christian Faith*, trans. Roy J. Deferrari (Cambridge, Mass.: Mediaeval Academy of America, 1951).

[56]Thomas Aquinas *Sententiarum: A Commentary on the Works of Peter Lombard* 2.13.3.

[57]Simon Episcopius, *Institutiones Theologicae* (Amsterdam: Johan Blaeu, 1650).

[58]Jaroslav Pelikan, *Luther's Works*, vol. 1: *Lectures on Genesis* (St. Louis: Concordia, 1958), p. 3.

thor of the *Augsburg Confession*, and theologian John Gerhard (1582-1637). On the other hand, the German chronologer Sethus Calvisius (1556-1615) calculated that creation had occurred in the autumn of 3944 B.C., whereas Johannes Kepler (1571-1630), the Lutheran astronomer, put creation at 3984 B.C.

The Genevan Reformer John Calvin (1509-1564), in the 1559 edition of his *Institutes of the Christian Religion*, likewise assumed that the world had not yet seen its six thousandth year. For example:

> And indeed, that impious scoff ought not to move us: that it is a wonder how it did not enter God's mind sooner to found heaven and earth, but that he idly permitted an immeasurable time to pass away, since he could have made it very many millenniums earlier, albeit the duration of the world, now declining to its ultimate end, has not yet attained six thousand years.[59]

Other Calvinists, such as Joseph Justus Scaliger (1540-1609), philosopher, textual critic and chronographer at the Academy of Geneva and the University of Leiden, placed creation at 3950 B.C. Jerome Zanchi (Zanchius) (1516-1590), a professor of Old Testament at the College of St. Thomas in Strasbourg and a professor of theology at the University of Heidelberg; Gijsbert Voet (Voetius) (1589-1676), a professor of theology and oriental science at the University of Utrecht and major participant at the Synod of Dort; Samuel des Marets (Maresius) (1599-1673), a professor of theology at the University of Groningen; and Francis Turretin (1623-1687), a professor of theology at the University of Geneva who, years after the death of Galileo, still insisted on biblical grounds that the Earth is fixed in place and that the Sun moves around it, all believed that the creation had taken place in the autumn. In the British Isles, John Lightfoot (1601-1675), an Anglican divine who was a vice chancellor of the University of Cambridge and one of the major architects of the Westminster Confession of Faith, suggested in 1642 that humanity was created at 9 a.m., and two years later he wrote that the creation of the world was initiated on Sunday, September 12, 3928 B.C.[60] In 1650, James Ussher (1581-1656), an archbishop of Armagh, prelate of Ireland, and vice chancellor of Trinity College, Dublin, proposed that the beginning of creation took place at sunset of the night preceding Sunday, October 23, 4004 B.C.[61] All the contributors to the

---

[59]John Calvin, *Institutes of the Christian Religion*, trans. Ford Lewis Battles (Philadelphia: Westminster Press, 1960), p. 160.

[60]John Lightfoot, *A Few and New Observations upon the Booke of Genesis: The Most of them Certaine the Rest Probable All Harmless, Strange, and Rarely Heard of Before* (London: T. Badger, 1642) and *The Harmony of the Foure Evangelists Among Themselves, and with the Old Testament: The First Part, from the Beginning of the Gospels to the Baptisme of our Saviour, with an Explanation of the Chiefest Difficulties Both in Language and Sense* (London: Andrew Crooke, 1644).

[61]James Ussher, *Annales Veteris Testamenti: A Prima Mundi Origine Deducti: Una cum Rerum Asiaticarum et Aegyptiacarum Chronico, a Temporis Historici Principio usque ad Maccabaico-*

Westminster Confession of Faith (1643-1648) who expressed themselves on the subject interpreted the six days of creation as ordinary solar days.[62]

Roman Catholic thinkers of the Counter Reformation also held to a recent creation. Among them were Francisco Suarez (1548-1617), a Jesuit widely regarded as the most prominent scholastic philosopher after Thomas Aquinas, and Tommaso de Vio Gaetani (Thomas Cajetan) (1469-1534), a Dominican cardinal and theologian who is perhaps best known as the one responsible for attempting to keep Martin Luther from abandoning the Catholic faith. Denys Pétau (Dionysius Petavius) (1583-1652), a French Jesuit and professor of divinity at the College de Clermont in Paris, fixed the date of creation at 3984 B.C. Petavius, like his predecessors Peter Lombard (1100-c. 1164) and Aquinas, also followed Augustine in teaching that the creative activity of Genesis 1:1-2 preceded the six days rather being a part of the first day.

With the exception of occasional ideas about Genesis 1 that departed from the rigidly literal interpretation, the almost universal view of the Christian world from its beginnings through the seventeenth century was that the Earth is only a few thousand years old. As naturalist John Ray, whom we will meet in the next chapter, wrote in 1673, the discovery of buried forests that were found in places in Europe that had been part of the sea just five hundred years earlier was "a strange thing considering the novity of the world, the age whereof according to the usual account is not [yet] 5600 years."[63] Not until the development of modern scientific investigation of the Earth itself would this view be seriously called into question within the church.

   rum Initia Producto (Londini: Ex Officina J. Flesher & Prostant apud L. Sadler, 1650).

[62]David W. Hall, "What Was the View of the Westminster Assembly Divines on Creation Days?" in Did God Create in Six Days? ed. Joseph A. Pipa Jr. and David W. Hall (Taylors, S.C.: Southern Presbyterian Press, 1999), pp. 41-52.

[63]John Ray, Observations Topographical, Moral and Physiological, Made in a Journey through Part of the Low-Countries, Germany, Italy and France: With a Catalogue of Plants Not Native of England (London: 1673), p. 8.

2

INVESTIGATION OF THE EARTH IN THE SEVENTEENTH CENTURY

FOR THE MOST PART, SEVENTEENTH-CENTURY SCHOLARS who were interested in studying the Earth did not consider an outside estimate of 6,000 years for Earth history as a restriction on their scientific studies. They assumed the validity of calculations for the age of the Earth obtained by biblical scholars and regarded such calculations as the fruit of considered investigation that was based on the best sacred and secular information and methods available. With few exceptions, they perceived no compelling physical evidence to challenge such age. Most discoveries in the realm of nature could be fitted quite neatly into a short Earth history without any embarrassment. However, foundations were gradually being laid for questioning the accepted opinion about the age of the Earth.[1] Advances in the study of fossils and rock strata were both necessary before such questioning would come about. The Earth would not be perceived as much older than the received opinion until it was recognized that there had been several successions of animal and plant populations through time and that these populations had become preserved in slowly deposited sediments that hardened into rock strata. But before those discoveries were made, first came important advances in understanding the nature of both fossils and rock stratification.

---

[1]On some of the questioning, see Michael B. Roberts, "Genesis Chapter 1 and Geological Time from Hugo Grotius and Marin Mersenne to William Conybeare and Thomas Chalmers (1620 to 1825)," in *Myth and Geology*, GSL Special Publication 273, ed. L. Piccardi and W. B. Masse (London: The Geological Society, 2007), pp. 39-49.

THE NATURE OF FOSSILS AND ROCK STRATA

Today most people characteristically think of fossils as the remains of once-living plants and animals that were entombed in mud, sand or clay that hardened into rock, but there was little agreement on this point until the early decades of the eighteenth century. Some scholars did think of fossils as once-living creatures, but many who accepted that view believed that these objects are the remains of creatures that perished in the Noachian Deluge. Others found it difficult to account for the distribution of fossils in terms of the Flood and spoke about interchanges between land and sea as had the Greeks. Many able scholars were not convinced that fossils are organic at all. Fossils were regarded as *lusus naturae,* that is, sports of nature; as the works of God created in place in the rocks for various inscrutable purposes; as the discards of dry runs at the creation of organic worlds before God, finally satisfied with the outcome, pronounced the creation of existing animals and plants as very good; and as productions of celestial influences, fermenting vapors or plastic virtues in rocks, or subterranean fluid charged with the seminal principles of animals or plants.

Although such explanations seem amusing today, a few centuries ago the very definition of a fossil caused confusion.[2] To naturalists of the sixteenth century, any object dug out of the Earth, including true fossils, minerals, crystals, rocks, concretions and ores, was treated as a "fossil" because the Latin word *fossilium* simply meant "something dug up." Given the state of knowledge, it was not always easy to distinguish among the various objects that came from the ground. Many geological objects *do* superficially resemble organisms without being true fossils. For example, *concretions* can resemble fossilized vegetables, ear lobes, and bananas.[3] Other concretions with mineral coatings look like the shells of turtles. *Dendrites* are branching, needle-shaped crystals of manganese oxide on rock surfaces; their arborescent pattern gives them a striking resemblance to plants. *Geodes* resemble eggs.[4] We need not

---

[2] On ideas about fossils, see Martin J. S. Rudwick, *The Meaning of Fossils: Episodes in the History of Palaeontology,* 2nd ed. (Chicago: University of Chicago Press, 1976).

[3] A concretion is an ellipsoidal to spherical mass contained within sedimentary rocks. Concretions develop during conversion of sediment into rock through chemical processes and commonly grow as concentric shells around a fossil or other inclusions with the sedimentary rock.

[4] A geode is typically a rounded rock, normally limestone, with a cavity that is lined or filled with crystals of various minerals, predominantly either quartz or calcite. In some instances, the cavity is partly or totally filled with banded agate. Geodes form when acidic groundwater percolates through limestone. Chemical reaction between the carbonated water and the calcium carbonate of the limestone produces open passageways, often large enough to be considered caves and cavities. Later, groundwater containing dissolved silica or calcium carbonate may precipitate quartz or calcite along the walls of the fracture or within the cavities. Eventually, the host limestone may weather away entirely except for the material surrounding the crystal-lined

wonder that, if many inorganic rock materials resemble living things, true fossils might be considered simply as another expression of some force in nature that causes rocks to resemble organic materials.

The confusion was compounded because of philosophical predispositions. Neo-Platonism held that there is no sharp distinction between living things and nonliving things. Even the Earth itself was thought to be alive in some way. If so, people would not be surprised to find the Earth studded with objects that look like living things. The nearly alive Earth generated productions that mimicked life. Moreover, many nonliving things display some of the characteristics of living things. For example, *stalactites* and *stalagmites* in caves and mineral crystals exhibit the phenomenon of growth.[5] There was even a medieval belief in male and female stones. In some cases stones were thought to give birth to new stones, as when, for example, a hollow geode was broken open and a much smaller stone was found inside. Thus, powers of reproduction were not uncommonly attributed to rocks. All of nature was seen as replete with all sorts of examples of organic processes and analogies with living creatures. There was no need to rely on any elaborate explanation of fossils in terms of once-living organisms that died, were buried and petrified. The living nature of the Earth was deemed sufficient to produce the shapes of these figured stones.[6]

Another powerful philosophical influence was that of Aristotle. In *Meteorologica,* he stressed the importance of vapors or exhalations in producing terrestrial phenomena.[7] He explained a variety of stones in terms of the actions of wet or dry vapors. Avicenna (980-1037), greatest of the medieval Islamic philosophers, modified and developed Aristotle's theory of exhalations and proposed that rocks form by the twin processes of *congelation* and *conglutination.*[8] At the heart of Avicenna's theory was the congealing *petrifying virtue,* a fluid thought to have the capacity of transforming liquid into solid. The stalactites in caves were obvious examples of the operation of such a petrifying virtue where one could "obviously" see dripping water turning into stone.

---

cavities. Because these cavities often contain quartz or agate, they are much less susceptible to chemical dissolution than the limestone. Consequently, a region that has undergone intense limestone weathering may have numerous geodes lying loose on the ground.

[5]Stalactites are the conical formations that hang, like icicles, from the roofs of caves. They develop by precipitation of dissolved calcium carbonate from groundwater dripping from the ceiling. The conical formations that form on the floor of caves are stalagmites. They develop as calcium carbonate is precipitated where drops of groundwater fall to the floor.

[6]See Frank Dawson Adams, *The Birth and Development of the Geological Sciences* (Baltimore, Md.: Williams and Wilkins, 1938); and Rudwick, *The Meaning of Fossils.*

[7]Aristotle, *Meteorologica,* LCL (London: Heinemann, 1952), pp. 29-35, 205-23.

[8]Edward Grant, ed., *A Source Book in Medieval Science* (Cambridge, Mass.: Harvard University Press, 1974), pp. 616-20.

Hard deposits also formed from hot springs. After Avicenna, scholars believed that petrifying virtues operated throughout nature. Their action in humans produces gallstones, and their action within the atmosphere produces meteoritic stones falling from the sky. Fossils could be explained by the action of petrifying virtue on the "seeds" of animals or plants that had fallen into cracks and become trapped inside rocks where they then grew into petrified shapes resembling mature adults.

However, in contrast to those who held these views that characterized the Middle Ages, some individuals emerged during the Renaissance who believed that fossils are buried organisms without attributing their burial to the Noachian Deluge. The notebooks of the great Renaissance painter, sculptor, architect, engineer and naturalist Leonardo da Vinci (1452-1519) indicate that he did not believe that the Flood was capable of accounting for the distribution of fossil shells.[9] Girolamo Fracastoro (1478-1553), an Italian physician who became an expert in the study of contagious and sexually transmitted diseases, held basically the same idea.[10] The Calvinist ceramist Bernard Palissy (1510-1589) was a self-taught man who relied on his own observations more than on established opinions. As a ceramist, he was very interested in the properties of soils and clays, and while studying natural occurrences of these materials, he became knowledgeable about many geological phenomena, including fossils. His own observations convinced him that fossils are not sports of nature but once-living creatures. He did not, however, believe that the Flood had provided enough time for the observed effects, and he thought that fossils occur too far from the sea and too far above sea level to be accounted for in terms of the Flood. He believed that fossils had once lived in lakes.[11]

Throughout the seventeenth century, discussion and argument over the nature of fossils took on increasing prominence among natural philosophers. Belief in an inorganic origin of fossils persisted on plausible grounds throughout much of the century. One important early seventeenth-century advocate of the position that fossils are *lusus naturae* produced by plastic virtues within the Earth was a professor of mathematics, physics and oriental languages at the Jesuit college in Rome, Athanasius Kircher (1602-1680). While conceding that fossil shells found at low elevations near the sea probably were remnants

---

[9]Leonardo da Vinci, *The Notebooks of Leonardo da Vinci*, trans. Edward MacCurdy, 2 vols. (New York: Reynal and Hitchcock, 1938). On da Vinci, see Charles C. Gillispie, Kenneth D. Keele, Ladislao Reti, Marshall Clagett, Augusto Marinoni and Cecil J. Schneer, "Leonard da Vinci," in *DSB*, 8:192-245.

[10]On Fracastoro, see Bruno Zanobia, "Fracastoro, Girolamo," in *DSB*, 5:104-7.

[11]Bernard Palissy, *The Admirable Discourses of Bernard Palissy*, trans. Auréle La Rocque (Urbana: University of Illinois Press, 1957). On Palissy, see Margaret R. Biswas and Asit K. Biswas, "Palissy, Bernard," in *DSB*, 10:280-81.

of real animals, Kircher disputed the idea that shells found in rocks high in mountains could have the same origin. Instead he attributed them to a variety of causes including the influences of the stars, magnetism and a plastic spirit within the Earth.[12]

Despite Kircher's illustrious reputation as a scholar, a growing number of natural historians found such theories about the origin of fossils too speculative and began to produce detailed studies of the nature of fossils. Robert Hooke (1635-1703), who regarded fossils as organic but rejected the idea that they were remnants of Noah's Flood, provided one of the most comprehensive defenses of this position in the seventeenth century.[13] Hooke was a brilliant inventor, architect, astronomer, physicist and mathematician, in many ways the peer of Isaac Newton, with whom he carried on feuds of varying intensity, allegedly attributable to his own prickly personality. After his education at Oxford, Hooke assisted chemist Robert Boyle (1627-1691) in the construction of an air pump. He became a curator of experiments for the Royal Society of London and a fellow in that scientific society.

Among his astonishing list of achievements, Hooke correlated vibrations to musical notes, invented the balance wheel in a watch, proposed the measurement of gravity by a pendulum, did research on the rotation of planets, attempted to measure the parallax of stars, discovered that the center of gravity of the Earth-Moon system follows an elliptical orbit about the Sun, formulated the proportional relation of stress to strain in elastic behavior now known as Hooke's law, designed buildings, submitted architectural plans for St. Paul's Cathedral after its destruction in the Great Fire of 1666, and became Surveyor of the City of London after the fire!

Along with wide-ranging interests in physics and astronomy, Hooke was fascinated by geological topics and made several very important contributions to geology. A meticulous observer, Hooke was the first to apply the recently invented microscope to the study of geologic materials, and his comparative microscopic investigations of thin transparent slices of fossil woods and modern living woods strongly influenced his attitudes regarding the true character of fossils. In his 1665 book *Micrographia*, he pointed out the cellular structure of petrified wood, thus confirming its identity, and coined the term *cell.*[14]

In 1668, Hooke published *Lectures and Discourses of Earthquakes and Sub-*

---

[12]Athanasius Kircher, *Mundus Subterraneous* (London: J. Darby, 1669). On Kircher, see Hans Kangro, "Kircher, Athanasius," in *DSB*, 7:374-78.

[13]On Hooke, see Margaret 'Espinasse, *Robert Hooke* (Berkeley: University of California Press, 1956); and Ellen T. Drake, *Restless Genius: Robert Hooke and His Earthly Thoughts* (Oxford: Oxford University Press, 1996).

[14]Robert Hooke, *Micrographia* (London: Martyn & Allestry, 1665). See also the discussion on Hooke by Rudwick, *The Meaning of Fossils*.

terraneous Eruptions, a work that began with a thorough description of many fossils and elaborate argumentation concerning how to account for these "Figured Bodies."[15] He thought it very "difficult to imagine that Nature formed all these curious Bodies for no other End, than only to play the Mimick in the Mineral Kingdom, and only to imitate what she had done for some more noble End, and in a greater Perfection in the Vegetable and Animal Kingdoms."[16] He also thought it strange that nature should mimic only certain kinds of creatures.

> If there by the apish Tricks of Nature, Why does it not imitate several other of its own Works? Why do we not dig out of Mines everlasting Vegetables, as Grass for instance, or Roses of the same Substance, Figure, Colour, Smell? Etc. Were it not that the Shells of Fishes are made of a kind of stony Substances which is not apt to corrupt and decay. Whereas, Plants and other animals Substances, even Bones, Horns, Teeth and Claws are more liable to the universal Menstruum of Time. 'Tis probably therefore, that the fixedness of their Substance has preserved them in their pristine Form, and not that a new plastick Principle has newly generated them.[17]

Hooke acknowledged that the two greatest difficulties with the organic view were that fossils are composed of various kinds of stone, clay, marble and so on, and that fossils are buried in places far from the sea. In a series of eleven elaborately developed propositions, Hooke showed how he thought these two main objections could be dispensed with. He argued how buried remains of once-living organisms might gradually be converted into stone through some kind of petrifying process, a discussion demonstrating that he had not yet fully escaped the thinking of Aristotle and Avicenna. He appealed to the stone-making processes observable in caves as an example of how a fluid might transform a fossil into stone, and he also suggested that exhalations or vapors given off by subterraneous earthquakes or eruptions might have caused buried objects to turn into stone.

As for why the fossils are found in their present locations either far above or far below sea level, Hooke again appealed to the action of earthquakes, and reported numerous descriptions of the action of historically known volcanic eruptions and earthquakes in elevating the sea floor, raising mountains, burying animals in volcanic ash, forming islands and so on. As evidence for his great earthquakes, Hooke appealed to the classical poets of Greece and Rome with their stories of the wars of the gods as well as to Plato's account of the sinking of Atlantis, suggesting that these are romantic descriptions of great terrestrial

---

[15]Robert Hooke, *Lectures and Discourses of Earthquakes and Subterraneous Eruptions* (New York: Arno Press, 1978), pp. 280-90.

[16]Ibid., p. 283.

[17]Ibid., pp. 318-19.

upheavals. For Hooke, the fact that no written records of such earthquakes existed provided an indication that the great catastrophes occurred prior to the invention of writing. Nevertheless, he was sure that these catastrophic upheavals occurred after the first creation that, he said, produced a world that was smooth all over and covered with water. Hooke was also greatly impressed by the fact that mountainous areas have the most earthquake activity. Thus he saw the existence of fossiliferous mountains as a result of continued earthquake activity in the past.

Hooke was very skeptical of the claim that the Noachian Deluge could be responsible for all fossiliferous rocks. In his mind, the Flood of Noah had neither the duration nor the power to permit the production of all the shells, not to mention the very great thickness of fossil shell deposits.

Without question, Hooke's younger contemporary, the Dane Niels Stensen (1638-1686), achieved the most perceptive geological thinking of the seventeenth century (fig. 2.1).[18] Stensen began his academic life as a pious Lutheran with interest in theological questions and fascination with things scientific. During university days at the University of Copenhagen he adopted the Latinized name Nicolai Stenonis and thereafter became generally known as Steno. He also developed exceptional skills in anatomy and dissection and wrote a book on the structure of the heart and muscles. He continued his education by taking private medical lessons in Amsterdam, during which period he purchased a sheep's head on which to practice his surgical technique. While probing the blood vessels in the sheep's mouth, he accidentally discovered the duct leading to the parotid gland that provides the mouth with most of its saliva. Later he discovered the glands that produce tears. He received his doctoral degree in medicine from the University of Leiden and wrote a paper on the medicinal properties of hot springs.

In 1665, Steno moved to Florence and became part of a scientific academy that was founded by the pupils of Galileo and sponsored by Grand Duke Ferdinando II of the Medici family. Here Steno met outstanding scientific thinkers, among them Francisco Redi (1626-1697), whose experiments demonstrated that flies and other vermin are not produced by spontaneous generation. Steno thrived in this very supportive environment that was conducive to productive scientific research. The following year, some fishermen caught a great white shark in the Mediterranean Sea that weighed well in excess of one ton. After removing some of the teeth, the fishermen chopped off the massive head and threw the remainder of the rotting body back into the sea. Because Steno was already something of a legend in his own time as a master with the dissect-

---

[18]Alan Cutler, *The Seashell on the Mountaintop: A Story of Science, Sainthood, and the Humble Genius Who Discovered a New History of the Earth* (New York: Dutton, 2003).

ing knife, the shark's head ended up in his dissection theater in Florence. The young anatomist was impressed by the shark's teeth and realized that they strongly resembled the well-known natural, three-sided, sharp-edged stones known as *glossopetrae* ("tongue-stones"). Steno carefully established the identity between *glossopetrae* and shark teeth.[19] He understood, too, that the shark teeth represented a sample case of the larger problem of the presence of remains of marine life that are embedded in rocks far removed from the sea.

As a result, Steno became absorbed in the nature of fossilization. He wanted to develop useful criteria for distinguishing between objects in rock that had been buried with the precursors of the rock material and those that had developed in place within the rock. After a couple of years of careful observation of rocks and fossils, the end result of Steno's research was a brief, tightly reasoned work titled *De Solido intra Solidum naturaliter Contento Dissertationis Prodromus.*[20] *Prodromus* ("Forerunner") was published in 1669, just a year after Hooke's work on earthquakes had appeared.

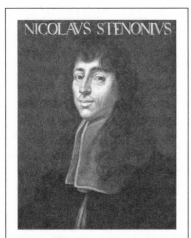

Figure 2.1. Niels Stensen (Steno) (1638-1686). This portrait of Steno was probably painted c. 1667 by Medici court artist Justus Sustermans. Reproduced courtesy of Institute of Medical Anatomy, University of Copenhagen.

Prodromus laid out very careful thinking about the geometrical interrelationships among fossils, crystals, rock fragments and rock strata. From investigation of quartz crystals of all sizes and shapes Steno realized that the angles between two specified adjacent crystal faces are always the same. To this day mineralogists refer to the *law of the constancy of interfacial angles.*

Steno argued persuasively that the objects we now call fossils must be of organic origin. Among the grounds that he offered were the lack of radiating fractures that should have been produced around the fossil had it grown within the rock. In addition, he argued that burial of organic remains under thick accumulations of sediment ought

[19]The recognition that *glossopetrae* are fossilized shark teeth had been made earlier by Guillaume Rondelet (1507-1566), a sixteenth-century authority on fishes. For biographical detail on Rondelet, see Jane M. Oppenheimer, "Guillaume Rondelet," *Bulletin of the History of Medicine* 4 (1936): 817-34.

[20]Niels Stensen, *Prodromus Concerning a Solid Body Enclosed by Process of Nature Within a Solid* (New York: Hafner, 1968).

Figure 2.2. Grand Canyon from Lipan Point, South Rim. The stratigraphy of the canyon illustrates principles first enunciated by Steno. From the *law of superposition,* **geologists infer that layers in the walls of the canyon are successively younger from bottom to top. From the** *law of original horizontality,* **geologists infer that layers in the upper canyon walls remain in their original position whereas gently tilted layers in the bottom of the canyon have been disturbed from their originally horizontal position. The boundary between the upper horizontal layers and the lower tilted layers is an ancient erosion surface termed an** *angular unconformity.* **Photo by D. A. Young.**

to lead to compression of the fossil, an unlikely result had the fossil grown in place.

Through detailed observation and geometrical reasoning, Steno also worked out the fundamental principles of the science of *stratigraphy*. By establishing that sedimentary rock layers were deposited in an essentially horizontal manner on an underlying surface, as is the case with river, lake or beach deposits, Steno formulated the *law of original horizontality*.[21] With the establishment of this principle, it became necessary to explain why many rock layers are severely tilted (fig. 2.2). Steno also demonstrated that strata are not deposited en masse but as a succession of individual layers in which each layer hardened before deposition of the next layer above. Evidence for hardening of a layer was particularly clear where pebbles eroded from the layer that had been incorporated into the next layer above. The successive deposition of strata came to be

[21]Many geologists prefer to use the term *principle* rather than *law* in connection with original horizontality and superposition.

known as the *law of superposition,* which states that, in any undisturbed stack
of sedimentary rock layers, the layers become progressively younger toward
the top of the stack (fig. 2.2).

Steno provided the basis for development of a geologic timescale by distinguishing between primary and secondary rocks. He considered crystalline
rocks in mountains as primary rocks and referred to stratified rocks lying on
top of them as secondary rocks. Moreover, Steno believed that fossils and their
enclosing rock layers were normally produced by flooding rivers, torrents, the
sea or storms. In effect, he perceived that the original environment of sedimentary deposition might be inferred from the nature of the deposits.

Steno noted that, in some rock exposures, horizontal layers lie on top of the
truncated edges of a set of tilted layers. From these geometrical relationships he
inferred that the tilted layers had been deposited, hardened, tilted, eroded and

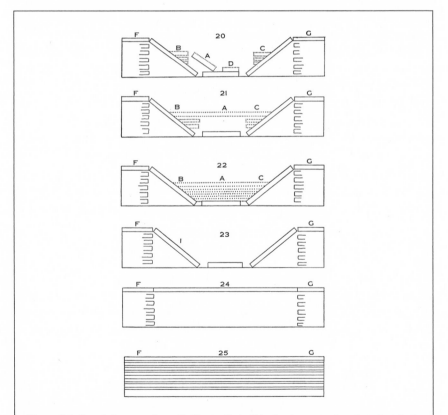

Figure 2.3. Steno's famous schematic illustration of his six-step reconstruction of geologic
events in Tuscany. Diagram 20 represents a cross-section of the existing geology, and diagram 25 represents Steno's conception of the original stratigraphy.

covered by later layers of sediment. By the application of these stratigraphic principles, Steno worked out the relative sequence of geologic events in the Tuscany region of Italy, the first known exercise in applying stratigraphic principles to the rock record (fig. 2.3). He noted that historical records had left no accounts of upheavals of strata that had taken place in the past. As a man of his times, he did not fully follow through on all his ideas but assumed that his conclusions fit into the framework of a 6,000-year Earth history and a universal deluge. Although Steno's contributions were not immediately appreciated, he had firmly established some of the fundamental principles of historical geology and stratigraphy on the basis of which Earth history was later worked out and that would make abundantly clear the vast antiquity of the Earth, a discovery that would have greatly surprised him.

As Steno intended *Prodromus* to serve as a preliminary discussion of the problems of fossilization and stratigraphy, he continued to undertake field investigations in the Alps and elsewhere to accumulate additional support for his thinking. The eagerly awaited sequel, however, never saw the light of day. During his stay in Florence, Steno agonized over his theological and ecclesiastical commitments. In 1667, two years after he arrived in Florence and shortly before he wrote *Prodromus,* Steno converted to Roman Catholicism. In 1675, he was ordained as a Catholic priest, and much to the chagrin of many of his contemporaries and of all geologists ever since, he abandoned his brilliant scientific work, an act that he viewed as a sacrifice to God. Two years later, he was appointed as a bishop with the responsibility of evangelizing the Lutherans in Germany and Scandinavia. Three centuries later, in 1988, Steno was beatified by Pope John Paul II. Apparently only one miracle could be attributed to Steno, one short of the number required for full canonization as a saint.

A year after *Prodromus* appeared, Italian painter Agostino Scilla (1639-1700) published an influential, lavishly illustrated book that argued the case for the organic origin of fossils.[22] To counter the views of Kircher, Scilla made no claim to know how fossils got to be where one finds them, whether by Noah's flood or other means, but he was certain that fossil shells were what they appeared to be, namely, real shells of once-living animals. Unaware of Steno's work, Scilla independently also recognized the true nature of *glossopetrae* as shark teeth.

Despite these cogent defenses of the organic origin of fossils, the idea that the so-called formed stones are the buried remains of once-living organisms faced stringent criticism. Even as Neo-Platonic and Aristotelian influences gradually diminished, the difficulties persisted and found expression in the

---

[22]Agostino Scilla, *La Vana Speculazione Disingannata dal Senso* (Naples: Andrea Colicchia, 1670). On Scilla, see Francesco Rodolico, "Scilla, Agostino," in *DSB,* 12:256-57.

work of some of the ablest natural philosophers. One problem, as Hooke had pointed out, was the fact that many fossils are not composed of the same material as modern animals but are composed of the same material as the surrounding rock. This problem of the *stoniness* of fossils was addressed by Martin Lister (c. 1638-1712), an outstanding collector of fossils, author of several books on fossil shells, and a member of the Royal Society of London.[23] Lister even recognized that only certain kinds of fossils are found in particular rock strata. He was not convinced that they were once alive, however. Lister pointed out that "Iron-stone Cockles are all iron stone; Lime or marble, all limestone or marble; sparre or Christalline-Shells, all sparre, etc., and that they were never any part of an animal."[24] If a fossil shell was at one time a shell then why isn't it composed of the exact material as a modern shell?

Other difficulties with the organic origin of fossils were addressed by John

Figure 2.4. John Ray (1627-1705). Reproduced by permission of National Portrait Gallery, London.

Ray (1627-1705), who generally accepted that position (fig. 2.4).[25] Ray became the premier naturalist of his time, the greatest British naturalist prior to Darwin, and one of the founders of the long tradition of British natural theology. Ray graduated from Cambridge in 1648, was ordained as an Anglican priest, and served as a tutor at the university until 1662 when he resigned because he refused to sign the Act of Uniformity. During his Cambridge years, Ray published a book on plants in the vicinity of Cambridge and became a first-rate botanist. He followed up with thorough field investigation of British plants and undertook a three-year stint of travels throughout Europe with a very close naturalist friend, Francis Willughby. While Ray concentrated on the botany of Europe, Willughby devoted his en-

[23]On Lister, see Jeffrey Carr, "Lister, Martin," in *DSB*, 8:415-17.

[24]Martin Lister, "A Letter of Mr. Martin Lister, Written at York August 25, 1671 Confirming the Observations in No. 74 About Musk-Scented Insects Adding Some Notes upon D. Swammerdam's Book of Insects, and on That of M. Steno Concerning Petrify'd Shells," *Philosophical Transactions of the Royal Society of London* 6 (1671): 2283.

[25]On Ray, see Charles E. Raven, *John Ray, Naturalist: His Life and Works* (Cambridge: Cambridge University Press, 1942).

ergy to the study of European animal life. During an extended stay in Italy, Ray became acquainted with Steno and observed the great anatomist dissect the head of an ox.

After returning to England, and armed with a wealth of specimens and information sufficient to keep him busy for the remainder of his life, Ray produced a steady stream of books, including works on the botany of England, the botany of Europe, the classification of plants and the natural history of plants. In 1671, Ray wrote a lengthy work devoted to cultural and physical observations on his European travels. In this book he devoted space to fossil remains, clearly accepting the idea that they represent once-living animals and plants. A number of his letters to friends also indicate his belief in the organic origin of fossils and his realization that fossil evidence hinted at a world not quite so youthful as almost everyone else believed at the time. Upon the death of Willughby in 1672, Ray diverted his focus from plants so that he could bring the work of his friend on various animals to publication. Thus, there followed a series of large books on birds, fishes, insects and other animals.

Because Ray, a devout Anglican who frequently preached sermons, operated within a thoroughly Christian worldview, he also wrote books in later years dealing with natural theology.[26] Among the works on natural theology were the *Miscellaneous Discourses* of 1692, followed by a second edition published the following year under the revised title *Three Physico-Theological Discourses*, and a third edition written only a year before his death but not published until 1713.[27] In this great work Ray summarized in considerable detail arguments both for and against the organic origin of fossils. Although he clearly believed that most fossils are probably organic, he felt the force of arguments against that hypothesis so keenly that he was inclined to reject the organic origin of at least some fossils that plainly are the remains of animals.

Two major stumbling blocks confronted the naturalists of Ray's era. These obstacles were the *extinction of organisms* and the *position of fossils.* First, Ray was well aware that fossils generally are not identical to modern living organisms even though they bear a strong resemblance. Most fossils are not representatives of living species. If these fossils were once alive, they must have belonged to groups of creatures that are now totally extinct. In Ray's time, however, Christians were not prepared to accept the idea of extinction, because they believed in the concept of the *plenitude of creation.* Seventeenth-century European thinkers could not see how God, having pronounced the

---

[26]John Ray, *The Wisdom of God Manifested in the Works of Creation* (London: Samuel Smith, 1691).

[27]John Ray, *Three Physico-Theological Discourses*, 3rd ed. (1713; reprinted, New York: Arno Press, 1978).

world very good, could permit any of his creatures to vanish from the face of the Earth. Why did God take such extraordinary pains to see that no living creatures completely vanished from the Earth during the Flood? If two of each kind of animal were on the ark, then plainly God did not want any extinctions to occur. So the idea of extinction was not palatable to Ray or his contemporaries. Nonetheless, he suggested that we do not yet fully know all the types of living things on the Earth. There may well be still undiscovered creatures in the depths of the oceans. Maybe some of these unknown beings are the same as what we see embedded in the rocks. If true, fossils could be regarded as once organic creatures, and no problem of extinction would necessarily arise.

Ray's conjecture was certainly valid, because several organisms *have* been discovered to be living today after they were known previously only from the fossil record. In spite of this way out of accounting for the differences between fossils and known living organisms, Ray still could not bring himself to accept the organic origin of the *cornua ammonis*, a type of extinct cephalopod, and some other fossil remains. Although these objects resemble the modern nautilus, they just aren't similar enough, and Ray evidently did not think that modern representatives would be found in the sea. So far, his view on that point has been substantiated.

Ray's second difficulty concerned the location of fossils. Fossils are found not only lying loose on the surface close to shorelines in unconsolidated sediments but also embedded within hard rock deep within quarries and at or near the summits of the highest mountains, far removed both vertically and horizontally from the nearest ocean. How did fossils ever get to be embedded in the rocks high in the mountains far from the sea? One plausible explanation was the Genesis Flood, but Ray and many others did not see how a flood of less than a year's duration could possibly account for the observed distribution of rocks and fossils. If the Flood did not account for the fossils, perhaps great earthquakes had lifted former sea beds to their present elevations as Hooke had proposed. Without question, earthquakes had raised sea floors in historical times but certainly not to the elevations of mountainous regions. In addition, no historical records of any earthquakes of such obviously enormous magnitude existed. Surely earthquakes capable of lifting the sea floor to heights of several thousands of feet would have been recorded by humankind, but no such records were known. As far as Ray was concerned, no known mechanism adequately accounted for the position of the fossils on the assumption that they were once organic remains.

The letters of John Ray reveal that he realized the serious implications of the organic origin of fossils for the Earth's history. He sensed that the Earth's history might have to be extended backward from the commonly accepted

6,000 years to account for the existence of fossil remains in mountain tops, a step he was not ready to take despite the fact that he probably interpreted Genesis 1 as including a reference to a period of chaos that preceded the six days of creation. Michael Roberts has pointed out the widespread acceptance in the seventeenth and eighteenth centuries of the idea that a state of chaos preceded the six days of creation. Although appeal to a period of chaos prior to the six days opened up a potential strategy for expanding the biblical time frame, seventeenth-century proponents of this "chaos" theory like Ray, Burnet and Whiston, as we will see, typically kept well within the traditional limits of around 6,000 years.[28] Ray found it difficult to commit himself regarding the origin of fossils.

By the early eighteenth century, the notion that fossils were formed inorganically faded from acceptance as the study of fossils continued and philosophical objections lost their persuasiveness. The last capable naturalist to advocate an inorganic origin for fossils was Johann Bartholomew Adam Beringer (1667-1740), a professor and dean of the faculty of medicine in the University of Würzburg. Beringer loved to collect and sketch fossils, but he was convinced that they are merely curious shapes produced by nature. Most of his specimens were sent to Beringer by acquaintances, but along with many genuine fossils were also specimens of images of birds sitting in their nests, purely imaginary animals, and Hebrews letters, including the name of Yahweh. Such extremely odd "discoveries" confirmed Beringer's belief that fossils are only sports of nature. Unable to convince Beringer of the spurious nature of some of his "fossils," two colleagues carved and baked fake "fossils," had them brought to Beringer, and then informed him that they had manufactured the latest "finds" after he had assured them that they were genuine. Undeterred, in 1726, Beringer published a book that contains drawings and descriptions of a wide array of the specimens that had been "discovered."[29] After the book was published, Beringer changed his mind about fossils, allegedly because he encountered a slab of rock with his name on it, and spent the rest of his life recalling and destroying all the published copies of his book. This tragic-comic episode provided the deathblow for the concept of the inorganic origin of fossils as a viable option among educated people.

---

[28]R. W. T. Gunther, *Further Correspondence of John Ray* (London: 1928). See letters 151 and 154. On the role of chaos in the interpretation of Genesis 1 and evidence for antiquity, see Michael B. Roberts, "The Genesis of Ray and His Successors," *Evangelical Quarterly* 74 (2002): 143-65; and Roberts, "Genesis Chapter One and Geological Time."

[29]See Georg L. Hüber, Melvin E. Jahn and Daniel J. Woolf, *The Lying Stones of Doctor Johann Bartholomew Adam Beringer, Being His Lithographiae Wirceburgensis* (Berkeley: University of California Press, 1963). Many illustrations of bizarre "creatures" are included.

THE RISE OF DILUVIALISM

Following the lead of Hooke, Steno and Scilla, a growing number of natural philosophers during the late seventeenth and early eighteenth centuries accepted the organic origin of fossils, and some of them believed that they had found a suitable explanation to account for their position in stratified rocks high in mountainous terrains. The Noachian Deluge emerged for a time as the most widely held explanation.[30] If one accepted the Bible as the authoritative Word of God, then it was sensible to assume that this Flood of apparently spectacular proportions was responsible for all the dead plants and animals entombed in the rocks. This general point of view can be traced back to the church fathers. Tertullian (c. 160-c. 225), a Carthaginian presbyter, had spoken of the fossils in mountains as demonstrating a time when the globe was overrun by all waters, although it is not really clear whether he was talking about the Deluge.[31] There are hints that Chrysostom and Augustine thought the Flood was responsible for many fossils, and Martin Luther attributed the origin of fossils to the agency of the Flood.[32]

The most influential expositions of Deluge theories were those that explained the history of the Earth in terms of a combination of biblical information and the physics of the time. French rationalist philosopher, René Descartes (1596-1650), applied laws of mechanics to the development of the Earth.[33] He envisioned the development of a layered spherical Earth by means of the law-bound mechanical interactions of swirling particles. Strongly influenced by the Cartesian view of the origin of the Earth, Thomas Burnet (1635-1715), an Anglican cleric and chaplain to King William III, postulated a history of the planet's development from creation to the final consummation by melding the Bible and Descartes.[34] He elaborated the role of the Flood in great detail.

Burnet's monumental work, titled *The Theory of the Earth Containing an Account of the Origin of the Earth, and of all the General Changes which it hath already Undergone, or is to Undergo till the Consummation of all Things,* first appeared in Latin in 1681.[35] *The Sacred Theory of the Earth,* as it is usually known, is not a scientific work, even though it went beyond the Bible in attempting to paint a picture of the history of the Earth. Burnet's theory was not built up through painstaking observation of the phenomena of nature; rather it was worked out through a combination of reasoning and Scripture.

---

[30]Davis A. Young, *The Biblical Flood: A Case Study of the Church's Response to Extrabiblical Evidence* (Grand Rapids: Eerdmans, 1995).

[31]Tertullian *On the Pallium,* in ANF, 4:6.

[32]Martin Luther, *Lectures on Genesis* (Philadelphia: Fortress, 1967), 1:98.

[33]René Descartes, *Principles of Philosophy* (Lewiston, N.Y.: Edwin Mellen, 1988).

[34]On Burnet, see Suzanne Kelly, "Burnet, Thomas," in *DSB,* 2:612-14.

[35]Thomas Burnet, *The Sacred Theory of the Earth* (London: Centaur, 1965).

This Theory being chiefly Philosophical, Reason is to be our first Guide; and where that falls short, or any other just occasion offers itself, we may receive further light and confirmation from the Sacred writings. Both these are to be lookt upon as of Divine Original, God is the Author of both; He that made the Scripture made also our faculties, and 'twere a reflection upon the Divine Veracity, for the one or the other to be false when right us'd. We might therefore be careful and tender of opposing these to one another, because that is, in effect, to oppose God to himself. As for Antiquity and the Testimonies of the Ancients, we only make general reflections upon them, for illustration rather than proof of what we propose.[36]

There was little recognition of the need for empirical observation, and Burnet even noted that the ancients (many Greek thinkers did try to theorize from geological observation) had never arrived at knowledge of what he was proposing. Only occasionally did he mention some scientific observations when they fell in line with his theory.

Burnet argued from Scripture that there had been a universal Deluge, and he calculated that about eight times the volume of the present ocean would have been required to cover all the land areas on the Earth. Where did all this water come from? It will not do, he said, to argue that God created all that water specifically for the purpose of the Flood and then annihilated it when the Flood was over. We should not rely on a miracle to get us out of a difficulty. Rather, this extra water was built into the planet right from the beginning.

Burnet maintained that it had been "the general opinion and consent of the Learned of all Nations, that the Earth arose from a Chaos."[37] When Moses described the original Earth as "Tohu Bohu, without form and void," he was calling attention to the initial state of Chaos. More specifically, Burnet conceived of the Chaos as a "fluid, dark, confus'd mass, without distinction of Elements . . . without Order, or any determinate Form." This description of Chaos he believed to be so "understood by the general consent of Interpreters, both Hebrew and Christian."[38] Burnet maintained that "all these things arose and had their first existence or production not six thousand years ago."[39]

When the Earth emerged from chaos, it was a perfectly smooth sphere with no irregularities. A differentiation of materials occurred with an oily fluid separating out on top of a watery sphere, which in turn surrounded the Earth's interior. The oily fluid collected earthy particles and became the hardened, smooth crust. Thus, in early times the land rested above a watery subterra-

---

[36]Ibid., p. 26.
[37]Ibid., p. 42.
[38]Ibid., p. 49.
[39]Ibid., p. 42. Michael Roberts suggests that Burnet later expanded his view of Earth time.

nean abyss. That this was so is supported by Scriptures like Psalm 24:2: "for he founded it upon the seas and established it upon the waters." This primitive Earth, suited for paradise, was idyllic. It was perpetually summer. However, the lack of changing seasons caused the Earth to dry out from the heat of the Sun. Not only was the Earth gradually desiccating and shrinking, but the abyss also began to vaporize owing to the heat penetrating the crust. This vaporization created additional pressure on the underside of the crust. At length, in God's providence, the time was ripe for the Flood, a phenomenon that was accomplished by the severe fragmenting of the crust and foundering of the crustal slabs into the abyss at all sorts of angles, thus accounting for mountains and lowlands. The sudden shock of the crustal collapse caused tremendous outward surges of the abyssal waters that inundated the land. In this event, Burnet found the true meaning of the expression in Genesis 7:11: "on that day all the springs of the great deep burst forth."

When the Flood was over, the waters receded into the channels of the sea and the vast subterranean caverns formed by the collapse of the crust. That these subterranean caverns exist was abundantly plain to Burnet because of the existence of smaller, near-surface caves, volcanoes and earthquakes. Moreover, the Mediterranean and Caspian Seas, lacking significant surface outlets, would overflow were they not connected to the oceans by underground passageways.

Although Burnet's book made no mention of fossils, his Flood theory prompted considerable reaction among those who felt that he had taken considerable liberties with the biblical text.[40] His work also stimulated the development of diluvial theories that did account for fossils. The first of such theories was set forth by John Woodward (1665-1722). Woodward was a highly regarded medical practitioner and enthusiastic student of natural history who became a professor of physic (medicine) at Gresham College, London, a position that he held until the end of his life.[41] Woodward put forward his diluvial theory in *An Essay towards a Natural History of the Earth* (1695).[42]

Although Burnet's theory had not been built up inductively from observa-

---

[40]Chief among the critics of Burnet was Erasmus Warren, an Anglican rector of Worlington. Warren accused Burnet of being insufficiently literal in his reading of the biblical text in a 1690 book modestly titled *Geologia; or, A Discourse concerning the Earth before the Deluge, Wherein the Form and Properties Ascribed to It, in a Book Intituled 'The Theory of the Earth,' Are Excepted against and It Is Made Appear, That the Dissolution of that Earth Was Not the Cause of the Universal Flood.*

[41]On Woodward, see Joseph M. Levine, *Dr. Woodward's Shield: History, Science, and Satire in Augustan England* (Berkeley: University of California Press, 1977).

[42]John Woodward, *An Essay toward a Natural History of the Earth* (1695; reprinted, New York: Arno Press, 1978).

tions of the phenomena of the Earth, Woodward, by contrast, was convinced that the only sound way to a true philosophy of the Earth was by means of careful observation. The introduction to his *Essay* repeatedly stressed the importance of observation and pointed out how careful he had been in making observations of the Earth. Woodward's observations led him to realize that England and other places that he had visited were all characterized by stratification of rock layers. Through correspondence with friends and scholars who had traveled widely he learned that England was not unique in regard to the stratification of rocks. Moreover, he observed that English strata are commonly fossiliferous. He became an avid collector of fossils and persuaded travelers and correspondents to send him carefully documented specimens of fossils as well as minerals, rocks and ores.

Woodward accounted for the phenomena of stratification and fossilization by arguing that fossils are truly the remains of organisms. Seeing weaknesses in the idea that the sea and land had changed places, he invoked the agency of a global Deluge. Woodward contended that, during the Flood, the prediluvian world was effectively dissolved in that all rock and soil were completely disaggregated to become a thick slurry. All living things on the surface of the Earth were caught up and destroyed by this particle-charged flood. As the floodwaters diminished and disappeared, the materials in suspension in the waters, including the bodies of plants and animals, gradually settled out in accordance with their specific gravity. Thus, according to Woodward, fossils in certain strata are typically characterized by the same general specific gravity as the rocks that enclose them. Woodward did make some measurements of the specific gravities of fossil shells and rocks, and, to that extent, he based his theory on observation. However, here he began to be rather speculative. Influenced by his knowledge of physics, he failed to base his views on observations, because stratified rock layers were *not* laid down in order of decreasing specific gravity. Relatively low-density rocks commonly occur beneath those with heavier densities, and there is no progressive change in specific gravity from the bottom to the top of a stack of layered sedimentary rocks.

Woodward's theory attracted much attention, in no small part because of an overbearing personality that often succeeded in converting friends into staunch opponents. Edward Lhwyd (1660-1709), a prominent Welsh naturalist in charge of the collections of the Ashmolean Museum at Oxford, was not happy with what he regarded as Woodward's too facile acceptance of the organic origin of fossil remains. Various individuals, particularly John Arbuthnot (1667-1735), a satirist who was sufficiently skilled in mathematics and experimental philosophy to become a member of the Royal Society of London, pointed out that sedimentary rock layers are not arranged in order of specific gravity.[43] Others complained that if rocks were dissolved, the fossils contained in them should also have dissolved.

To answer such criticisms, Woodward promised that he would write a much more through explication of his theory. Although the promised volume never materialized, he responded to some criticisms in letters and in a preliminary work on the natural history of the Earth, published in 1714.[44] In these writings, Woodward contended that the dissolution of materials of the Earth's surface was caused not by the water of the Flood but by a providential suspension of gravity that resulted in the shattering of the granular material of the world into millions of atoms. In contrast, organic materials would retain their integrity and not be subject to the shattering effect because they were more fibrous. With gravity reinstated at the conclusion of the dissolution stage, materials began to settle out in accord with their specific gravities. Because the various settling materials interfered with one another through collisions and obstructions there were deviations from a progressive upward decrease in the specific gravity of layers within a stack of sedimentary rocks. Despite his best efforts to produce a theory of the Flood based on observation and laws of nature, Woodward failed to resist the temptation to invoke a miracle to account for difficulties with the theory.

Woodward was an inveterate and meticulous collector of fossils, minerals, rocks and all manner of antiquities. Because he collected fossil specimens, primarily for their evidential value in support of his diluvial theory, Woodward was concerned that his collection and the beautiful cabinets in which the collection was housed should not be neglected after his death. To that end, he provided in his will for the establishment of a Woodward professorship at Cambridge and bequeathed his collection to the same institution. According to the terms of the will, the professor who held the endowed chair was to be a student of fossils and develop the ideas espoused in Woodward's theory. To this day, a temperature-controlled room in a corner of the Sedgwick Natural History Museum, run by the Department of Earth Sciences at the University of Cambridge, houses the historic collection of specimens.[45] Despite his wishes for a simple burial place, Woodward's remains lie not far from those of Isaac Newton, and his monument is located along the northern nave wall of Westminster Abbey immediately next to the monument to Charles Lyell.

---

[43]John Arbuthnot, *An Examination of Dr. Woodward's Account of the Deluge, etc. with a Comparison between Steno's Philosophy and the Doctor's, in the Case of Marine Bodies Dug out of the Earth* (London: 1697).

[44]John Woodward, *Naturalis Historia Telluris* (London: 1714).

[45]The curatorial staff members at the Sedgwick Museum are proud of the Woodward collection and are delighted to show the collection and the beautiful wood cabinets to interested inquirers.

Subsequent to the appearance of Newton's *Principia Mathematica* in 1687, Burnet's *Sacred Theory* stimulated other writers to blend Scripture with Newtonian, as opposed to Cartesian, physics. William Whiston (1667-1752) produced the premier example of this genre. A very pious young man who planned to enter the Presbyterian ministry, Whiston was educated at Cambridge where he attended Isaac Newton's lectures and demonstrated considerable promise in mathematics.[46] He displayed exceptional gifts in biblical exegesis and, like Newton, was intensely interested in biblical prophecy. Whiston was ordained as an Anglican, and between 1694 and 1698 he served as chaplain to the bishop of Norwich. During this period Whiston published *A New Theory of the Earth* (1696).[47] His theory contained much physics and astronomy of his day and in that sense was more scientific than Burnet's. Nevertheless, Whiston's theory of the Earth was hardly more *geological* than Burnet's, because the theory was not worked out from geological observations to the degree that Woodward's was. The book was essentially a combination of Newtonian physics and Scripture linked to an occasional fact. Whiston had also been influenced by an address of Edmund Halley (1656-1722) before the Royal Society of London in which he appealed to the collision of a comet with the Earth to account for the biblical Flood.[48]

Although Whiston favored Archbishop Ussher's chronology for the Earth, he modified it slightly and concluded that the Earth had been created in 4010 B.C. rather than in 4004 B.C.[49] Like Burnet, he stressed the origin of the Earth from an "ancient Chaos, just before the beginning of the six days of Creation." This chaos was described as a "system of fluids" that was "very dark and caliginous," and a "mixed Compound of all sorts of Corpuscles, in a most uncertain confus'd and disorderly State."[50] Believing that these characteristics of a chaos meshed nicely with those of a comet, Whiston maintained that the original "chaos" of Genesis 1:2 from which the organized planet developed was the atmosphere of a comet. The Earth, then, was originally a comet that was permanently captured by the Sun. At first, the Earth did not rotate on its axis

---

[46]On Whiston, see Maureen Farrell, *William Whiston* (New York: Arno Press, 1981); and James E. Force, *William Whiston: Honest Newtonian* (Cambridge: Cambridge University Press, 1985). Also of importance is Whiston's autobiography, William Whiston, *Memoirs of the Life and Writings of Mr. William Whiston: Containing Memoirs of Several of his Friends also; Written by Himself* (London: 1749).

[47]William Whiston, *A New Theory of the Earth* (London: R. Roberts, 1696; reprinted New York: Arno Press, 1978).

[48]Edmund Halley, "Some Considerations About the Cause of the Universal Deluge," *Philosophical Transactions of the Royal Society of London* 33 (1724): 118-23. Halley delivered a lecture on which the paper was based to the Royal Society of London on 12 December 1694.

[49]Whiston, *A New Theory*, pp. 123-26.

[50]Ibid., pp. 70-71.

daily but had only an annual motion, so that from a given point on the planet's surface there was only one interchange of light and dark in a year. Thus, one of Whiston's novel contentions was that the six days of creation were each one year long, necessitating a date for creation of the cosmos six years earlier than that of Ussher.

At the time of Adam's Fall, the Earth's axis tilted into its present position. The globe began to rotate about its axis and departed from a circular to an elliptical orbit around the Sun. Hence, the repetition of seasons and ordinary solar days began at the time of the Fall of Adam. This state of affairs remained stable until the Deluge when another comet approached. This second comet triggered drenching rains that drowned the globe. Life on the surface was destroyed and various life forms settled out of the floodwaters in accordance with their specific gravity. Like Woodward, Whiston attributed fossil remains in mountains to the action of the Flood.

A striking aspect of Whiston's theory is his notion that the six days of creation were each one year in length because in the beginning there was no daily rotation of the Earth on its axis, only an annual revolution around the Sun. He argued at great length from Scripture in defense of this view and, in so doing, anticipated arguments that reappeared in the day-age theory of the nineteenth century. First, Whiston argued against taking the six days literally by noting that Genesis 2:4 refers to the six days as "the day that the LORD God made the earth and heavens" (NRSV). Second, he posed the equivalence of days and years through such references as 1 Kings 2:11, "the days that David reigned over Israel were forty years" (KJV), or Genesis 5:5, "all the days that Adam lived were nine hundred and thirty years," and particularly in prophetic writings like Daniel 9:24-27 (KJV), where the seventy weeks are weeks of years, not days. Third, he held that the various Levitical sabbaticals suggested a connection between days and years.

Whiston also believed that the events of the six days could not have occurred in twenty-four hours, a point that had just been made a year earlier by Simon Patrick (1626-1707), an Anglican bishop of Chichester and Ely. Whiston felt that runoff of waters from emerging dry lands on day three would require much more than twenty-four hours, especially if subsequent growth of vegetation occurred on that day. Moreover, he maintained that too many events transpired on the sixth day to fit within a twenty-four-hour span.[51] Here we have a catastrophic Flood theory presented in terms of a very recent creation, yet anticipating the day-age view of Genesis 1 that was subsequently developed in defense of the idea that the Earth is extremely old.

---

[51]Ibid., pp. 82-91.

In 1701, Whiston became an assistant to Isaac Newton, and soon thereafter, thanks to Newton's influence, he succeeded the great scientist as Lucasian Professor of Mathematics at Cambridge. He published an edition of Euclid's geometry, Newton's lectures on algebra, and his own lectures on astronomy. He also introduced the first course in experimental physics at Cambridge. Around 1708, his detailed study of the earliest documents of the Christian church led Whiston to doubt the doctrine of the Trinity. He began making his views known, and consequently he was charged with heresy by the authorities at Cambridge who dismissed him from the university in 1710. Nevertheless, he was never defrocked by the Church of England. For the remainder of his life, Whiston published lengthy works defending Arianism, gave public lectures on astronomy and prophecy, wrote books on astronomy, translated the complete works of the Jewish historian Josephus, and received a stipend from Queen Caroline. In the final years of his life, Whiston and Newton fell out of favor with one another.

John Keill (1671-1721), Savilian Professor of Astronomy at the University of Oxford, retained belief in the original chaos entertained by Burnet and Whiston, but he did not appreciate their attempts to provide rational, mechanical explanations for the Deluge. He wanted to restrict the Deluge to the realm of theology. Keill disputed Whiston's claim that a comet passing by the Earth would have an effect on the subterranean abyss, and he doubted that a comet's tail could be converted into rain. He also took issue with Burnet's idea that the Earth originally had a vertical axis of rotation and with his theory of mountain formation. Moreover, he denied that Burnet's solid crust would ever form on the grounds that solid particles would sink into the underlying fluid as soon as they coagulated.[52]

The diluvial hypothesis was not confined to Great Britain. In Germany, diluvialism found an enthusiastic advocate in the naturalist Johannes Jakob Scheuchzer (1672-1733), town physician of Zurich and a professor of mathematics at the University of Zurich.[53] Early in his career, Scheuchzer believed that fossils were sports of nature, but he was strongly influenced by Woodward's *Essay* and carried on extensive correspondence with the English diluvialist. Unlike so many other naturalists of the era, Scheuchzer remained on friendly terms with Woodward, possibly because they never met face to face. After conversion to Woodward's way of thinking, Scheuchzer cleverly attacked

---

[52]John Keill, *An Examination of the Reflections on the Theory of the Earth, Together with a Defence of the Remarks on Mr. Whiston's New Theory* (Oxford, 1699). On Keill, see David Kubrin, "Keill, John," in *DSB*, 7:275-77.

[53]On Scheuchzer, see P. E. Pilet, "Scheuchzer, Johann Jakob," in *DSB*, 12:159; and Robert Felfe, *Naturgeschichte als kunstvolle Synthese: Physikotheologie und Bildpraxis bei Johann Jacob Scheuchzer* (Berlin: Akademie Verlag, 2003).

the inorganic view of fossils in a treatise titled *The Fishes' Complaint and Vindication* (1708).[54] The argument was placed in the mouths of fossil fishes who complained about the unreasonableness of being treated as mere freaks of nature and insisted on being regarded as true fishes that had once lived and been stranded during the recession of the Floodwater. The fishes claimed to have been cosufferers with humans during the Flood and asked for the dignity of being recognized as such. As the climax to his Flood theory, Scheuchzer suggested that some large vertebrae he had encountered were the remains of one of the sinners who had died in the flood. Some time after the publication of his book, Scheuchzer learned that these bones belonged an aquatic reptile now known as *Ichthyosaurus* rather than to a human being.

Despite this mistake, Scheuchzer continued to seek evidence of human remains to prove the occurrence of the Flood. Eighteen years after the appearance of *The Fishes' Complaint,* stone workers uncovered an entire fossilized skeleton and shipped it to Scheuchzer. After its reconstruction, Scheuchzer was persuaded that here at last was the *homo diluvii testis,* the human witness of the Flood, and a giant at that, who had perished in the Noachian Deluge. In 1726, he issued a treatise on the drowned man.[55] Not until 1812, long after Scheuchzer's death, thus sparing him from further embarrassment, was it disclosed by vertebrate anatomist Georges Cuvier, whom we shall meet later, that the "Flood man" was in reality an extinct giant salamander!

By the time the eighteenth century was well under way, the age of the Earth was still generally regarded as about 6,000 years on the basis of literal interpretation of Scripture and available historical records, although some savants had begun to have their doubts. Geologic studies had only begun to hint at evidence contrary to the received opinion, but the stage had been set for the discovery of the much longer antiquity of the planet. By 1750, the true nature of fossils as the remains of buried organisms was almost unanimously accepted. Naturalists were becoming more aware of sequences of stratified rock layers, with their implication of elapsed time. Edward Lhwyd, John Ray, William Whiston, Simon Patrick, Thomas Hobbes and others had begun to wonder if the Earth might be older than was generally believed, and in 1715 Edmund Halley had suggested the possibility of calculating the age of the Earth by determining the amount of salt in the ocean and measuring the rate at which it was added to the ocean. But these were just faint glimmers of things to come. During the eighteenth century, investigators built on the foundation laid by men like Steno, and by the early nineteenth century, the conclusion had become inescapable that the Earth is far older than 6,000 years.

---

[54]Johann Jacob Scheuchzer, *Piscium Querelae et Vindiciae* (Zurich: Gessner, 1708).
[55]Johann Jacob Scheuchzer, *Homo Diluvii Testis* (Zurich: Burkli, 1726).

3

THE EMERGENCE OF
MODERN GEOLOGY

DURING THE EIGHTEENTH CENTURY, the emphasis of Steno and Woodward on stratified rocks coupled with Steno's fundamental stratigraphic principles and methods of reasoning about the geologic history of rocks exerted a growing influence on European naturalists. Several individuals began to describe, sketch and measure successions of stratified rocks. Others recognized the extensive lateral continuity of many strata and noted that stratified rocks typically occur in well-defined vertical successions. These observations on the vertical and regional distribution of rock strata gradually led to development of a geologic timescale and an ever-growing sense of the Earth's great antiquity.

NEW DEVELOPMENTS IN STRATIGRAPHY AND HISTORICAL GEOLOGY

An important conceptual advance took place in the mid-eighteenth century with the expansion of Steno's categorization of primitive and secondary rock deposits by Lehmann, Arduino, Werner and others. Johann Gottlob Lehmann (1719-1767) received a medical degree from the University of Wittenberg in 1741 and practiced medicine in Dresden where close friends introduced him to the world of mining and metallurgy.[1] He eventually became a director of mines and, under the auspices of the Royal Prussian Academy of Science, pursued mining-related geologic studies beginning in 1756. During that year, he

---

[1]On Lehmann, see Bruno von Freyberg, "Lehmann, Johann Gottlob," in *DSB*, 8:146-48.

mapped the stratified rocks in the Harz Mountains and in the Erzgebirge in Saxony in some detail, sketched a geologic cross-section, placed the strata in a historical sequence and developed a classification of mountains.[2]

Lehmann designated the relatively oldest mountains as *Gang-Gebürge* (vein mountains) and found that these steep-sided mountains consist of steeply tilted rocks with abundant ore-bearing veins and very few fossils. He believed that the *Gang-Gebürge* were formed early in the creation period. Lehmann designated materials that are superposed on the flanks of the *Gang-Gebürge* as *Flötz-Gebürge* (stratified mountains containing few ore deposits), comprising thin-bedded, essentially horizontal, water-lain strata containing fossils. He attributed these strata to the action of the universal Deluge and thought that they might have been formed from the debris of older mountains. He also provided detailed descriptions of thirty successive units belonging to the category of *Flötz-Gebürge*. Lehmann also posited a third category consisting of surficial deposits.

Around the same time, Giovanni Arduino (1714-1795) independently established a classification of mountains. Arduino was educated in Verona and worked in mines of Vicenza and Tuscany during his early career.[3] Building on the ideas of earlier Italian workers such as Anton Lazzaro Moro (1687-1764) and Giovanni Targioni Tozzetti (1712-1783) in his early studies, Arduino developed a rough classification of orders of rocks.[4] These orders included various categories of mountains. The first category was *montes primarii* (primary mountains) that were divided into two groups, the lower consisting mainly of schist, possibly some granite and an abundance of ore-bearing veins, and the upper consisting of calcareous and sandy sedimentary rocks. The lower part of Arduino's primary mountains was essentially equivalent to Lehmann's *Gang-Gebürge*. Arduino recognized stacks of layered rocks, predominantly limestone, on the flanks of the primary mountains. These rocks of the *montes secondarii* (secondary mountains) contain fossil remains. On the flanks of the limestone deposits Arduino found poorly consolidated strata of sand, marl and clay that also contain abundant fossil remains, and he designated such terrains as *montes terziarii* (tertiary mountains). He believed that some of these

---

[2]Johann Lehmann, *Versuch einer Geschichte von Flötz-Gebürgen* (Berlin: Gottlieb August Lange, 1756).

[3]On Arduino, see Francesco Rodolico, "Arduino (or Arduini), Giovanni," in *DSB*, 1:233-34.

[4]The classification was introduced in Giovanni Arduino, "Due Lettere . . . Sopra Varie sue Osservazioni Naturali: Al Chiaris. Sig. Cavalier Antonio Vallisnieri Professore di Storia Naturale nell'Università di Padova: Lettera Prima . . . Sopra Varie sue Osservazioni Naturali (Vicenza, 30 Gennaio 1759): Lettera Seconda . . . Sopra Varie sue Osservazioni Fatte in Diverse Parti del Territorio di Vicenza, ed Altrove, Appartenenti alla Teoria Terrestre, ed alla Mineralogia (Vicenza, 30 Marzo 1759)," *Nuova Raccolta di Opuscoli Scientifici e Filiologici (Venezia)* 6 (1760): 99-180.

tertiary strata had been derived by reworking of secondary mountains. Arduino also recognized the existence of volcanic rocks interspersed with, and also lying above, the tertiary deposits. The superposition of these various groups of rocks suggested that primary rocks are the oldest and that tertiary rocks are the youngest. Arduino, however, was not sure that all primary rocks formed at the same time, nor did he attribute secondary rocks to the Flood. He suggested that the strata might be explicable in terms of ordinary processes like erosion and marine deposition of sediment over a long period of time, but he did not speculate on how much time.

After being entrusted with the task of developing agriculture and industry by the Senate of Venice, he worked for a variety of mining companies in the 1770s. About 1775, the Senate employed him to study all the mines in northern Italy. During this time, he undertook investigations of the geology from the Atesine Alps and their foothills to the plains of the Po River. All the while, Arduino continued to refine his classification scheme.[5]

Like Lehmann, Georg Christian Füchsel (1722-1773) was trained in medicine, serving as the town physician of Rudolstadt, but he had plenty of time to pursue his interests in natural history.[6] In describing the stratified rocks of the Thüringer Wald in 1761, Füchsel illustrated general principles of historical geology.[7] He pointed out that the strata are arranged in an orderly sequence, invoked the law of superposition, correlated the layers by means of their characteristic fossils, and produced a geologic map. Füchsel introduced the concept of a *formation*, defining it as consisting of "those strata formed at the same time, of the same material, and in the same manner."[8] He also interpreted strata in terms of modern processes rather than invoking Noah's Flood or other great catastrophes. "We must," he insisted, "take as the norm in our explanation (of the Earth's history) the manner in which nature acts and produces solids at the present time; we know no other way."[9] Only the most recent deposits were considered to be the result of a great Flood.

Similar ideas were entertained by Pyotr Simon Pallas (1741-1811), a member of the St. Petersburg Academy of Sciences and a participant and leader

---

[5]For analysis of the development of Arduino's classification, see Ezio Vaccari, "The 'Classification' of Mountains in Eighteenth Century Italy and the Lithostratigraphic Theory of Giovanni Arduino (1714-1795)," in *The Origins of Geology in Italy*, GSA Special Paper 411, ed. Gian Battista Vai and W. Glen E. Caldwell (Boulder, Colo.: GSA, 2006), pp. 157-77.

[6]On Füchsel, see Bert Hansen, "Füchsel, Georg Christian," in *DSB*, 5:205-6.

[7]Georg Christian Füchsel, "Historia Terrae et Maris, ex Historia Thuringiae per Montium Descriptionem, Eruta," *Acta Academiae Electoralis Moguntiae Scientiarum Utilium, Quae Erfordiae Est* 2 (1761): 44-208.

[8]Ibid., p. 48.

[9]Ibid., pp. 81-82.

in several Russian expeditions devoted to natural history.[10] Pallas considered granite to form the skeleton of mountain systems. On the flanks of masses of granite were secondary limestone deposits containing organic remains. The poorly consolidated rocks of foothills were considered tertiary deposits.[11]

Generally speaking, German and Scandinavian scholars used terminology originated by Lehmann in such classifications, whereas French, Italian and English scholars employed the categories of Arduino. Although early investigations were concerned primarily with rock structure in these classifications, a simple three-part division of rocks and mountains with temporal-historical implications came into being in the middle of the eighteenth century.

The insights of Lehmann were further developed by Abraham Gottlob

Werner (1749-1817), a professor of geognosy in the Bergakademie (Mining Academy) of Freiberg, Saxony, in eastern Germany (fig. 3.1).[12] Werner was an outstanding mineralogist, an expert on the geology of Saxony, author of works on the characteristics of minerals and rock classification, and a brilliant teacher. He was the most prominent adherent of *Neptunism*, the major eighteenth-century school of geologic doctrine. Neptunists regarded water, particularly the ocean rather than Noah's Flood, as the most important geologic agent. In Werner's mature thinking, rocks were deposited from a primeval global ocean that gradually diminished over time. He thought that stratified rocks like sandstone, puddingstone (conglomerate) and shale were deposited by mechanical processes and that crystalline rocks like granite, gneiss, schist and basalt had been precipitated chemically from the primeval ocean, which he conceived of as a highly concentrated aqueous solution.[13] In contrast to Steno, Werner believed

Figure 3.1. Abraham Gottlob Werner (1749-1817). Reproduced by permission of Technische Universität Bergakademie Freiberg/Sachsen Medienzentrum.

[10]On Pallas, see Vasiliy A. Esakov, "Pallas, Pyotr Simon," in *DSB*, 10:283-85.

[11]Pyotr Simon Pallas, "Observations sur la Formation des Montagnes et sur les Changements Arrivés du Globe," *Acta Academiae Scientiarum Imperialis Petropolitanae*, part 1 (1777): 21-64.

[12]On Werner, see Alexander Ospovat, "Werner, Abraham Gottlob," in *DSB*, 14:256-64.

[13]On Werner's ideas about the origin and classification of rocks, see Abraham Gottlob Werner, *Short Classification and Description of the Various Rocks* (New York: Hafner, 1971).

that layered rocks were deposited in much the same orientation in which they occur today. Owing to his remarkable teaching prowess, Werner attracted bright students to the Bergakademie from all over Europe and even America and thereby became the most influential geologist of his era. A host of first-rate geologists helped promulgate Werner's distinctive Neptunist teachings as they took up influential university and government positions in Scotland, France, Germany and the United States.

Werner concluded that rock strata in the vicinity of Freiberg are arrayed in the same unvarying vertical succession wherever they crop out. Although he never studied rocks in other parts of the world, he also believed that the succession of strata found around Freiberg had worldwide validity. His stratigraphic succession began with *Urgebirge* (primeval mountains), that is, crystalline rocks like granite, gneiss and schist, rocks that are now regarded as igneous and metamorphic rocks. He believed that these rocks were chemically precipitated from the long-vanished universal ocean to form a substrate on top of which other rocks were deposited. Above the primeval rocks were the *Übergangsgebirge* (transition mountains), a category that Werner added to his scheme in later years. The transition rocks, a combination of mechanical deposits and chemical precipitates, consisted primarily of greywacke and slate and contained fossils of simple life forms as well as some primitive fishes. Above these rocks were *Flötzgebirge* (stratified mountains), a term that Werner adopted from Lehmann. This series, essentially equivalent to much of the Secondary and Tertiary groups of other geologists, consisted predominantly of well-layered sandstones, limestones, shales and other sedimentary rocks that Werner regarded as mechanically deposited rocks.

Sandwiched among these mechanical deposits he found layers of basalt. Although these layers of basalt are now known to be either lava flows or sills of crystallized magma that had been injected between already existing layers of sedimentary rock, Werner was adamant that basalt is a chemical precipitate having nothing to do with igneous activity. His *Flötz* rocks commonly contain abundant fossils of more advanced life forms, including mammals and trees. Werner held that these materials are overlain by *Aufgeschwemmte* (surficial or alluvial) deposits, roughly

Figure 3.2. William Smith (1769-1839) © Geological Society/NHMPL. Used by permission.

corresponding to Lehmann's third category and consisting of unconsolidated gravels and sands in an essentially horizontal position and containing remains of life forms similar to those in existence today. Werner reserved his fifth stratigraphic category for undeniably volcanic rocks such as lava flows and ash deposits produced by modern volcanoes. He considered volcanic activity to be a strictly recent and accidental geologic phenomenon that was triggered by the burning of subterranean coal deposits.

By the end of the eighteenth century, students of the Earth generally recognized that the positions of individual strata or groups of strata within a stratigraphic sequence demonstrate considerable regularity. There was also growing awareness that different kinds of fossils are found in different strata and in different localities, a fact that Martin Lister had observed as early as the late seventeenth century and that Arduino recognized a century later. It remained for William Smith (1769-1839) (fig. 3.2), a British engineer, to show the intimate connection between types of fossils and the regularity of strata.[14]

Smith was trained as a surveyor who began his career by surveying coal beds of southwestern England and inspecting underground coal mines. He inherited sketches of cross-sections of the individual beds within these coal fields that had been drawn in 1719 and 1725 by John Strachey (1671-1743), an English naturalist.[15] Strachey had sketched horizontal strata resting on steeply inclined beds that included seven coal seams, one of which he successfully used fossils to identify (fig. 3.3). Strachey did not realize the significance of the boundary between the horizontal beds and the underlying tilted beds, and he also believed that the tilted beds projected all the way toward the center of the Earth. Nonetheless, Strachey's drawings assisted Smith's growing perception of the layering, or stratification, of the Earth's surface materials.

In light of the rapidly growing coal industry at the onset of the Industrial Revolution, the need for convenient transportation of coal became apparent to

---

[14]On Smith, see Simon Winchester, *The Map that Changed the World: The Tale of William Smith and the Birth of a Science* (London: Viking, 2001); Hugh S. Torrens, "Timeless Order: William Smith (1769-1839) and the Search for Raw Materials 1800-1820," in *The Age of the Earth: From 4004 B.C. to A.D. 2002*, GSL Special Publication 190, ed. Cherry L. E. Lewis and Simon J. Knell (London: The Geological Society, 2001), pp. 61-83; and Martin J. S. Rudwick, *Bursting the Limits of Time: The Reconstruction of Geohistory in the Age of Revolution* (Chicago: University of Chicago Press, 2003), pp. 431-44.

[15]On Strachey, see Victor A. Eyles, "Strachey, John," in *DSB*, 13:86-87. Strachey's geological contributions may be found in John Strachey, "Observations on the Strata in the Somersetshire Coal Fields," *Philosophical Transactions of the Royal Society of London* 360 (1719): 972-73; "An Account of the Strata in Coal Mines, &c.," *Philosophical Transactions of the Royal Society of London* 33 (1725): 395-98; and *Observations on the Different Strata of Earths and Minerals: More Particularly of Such as Are Found in the Coal-Mines of Great Britain* (London: 1727). See also J. G. C. M. Fuller, "The Industrial Basis of Stratigraphy: John Strachey, 1671-1743, and William Smith, 1769-1839," *Bulletin of the AAPG* 53 (1969): 2256-73.

mine managers. Smith was hired to conduct regional surveys to plot potential canal routes. During the 1790s, he realized that the English countryside is underlain by a thick series of very gently inclined strata that are arranged in unvarying order like a tilted stack of papers. He realized, too, that each group of layers could be traced across the countryside for many miles where the edges of the strata are truncated by the land surface. He also noted that wherever a group of layers crops out, that group occupies the same position relative to other strata. Moreover, Smith, like Lister and Füchsel, discovered that each group of strata in the sequence contains a characteristic suite of fossils that

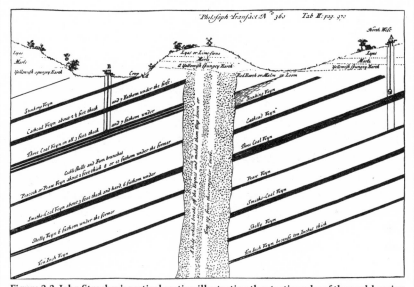

Figure 3.3. John Strachey's vertical section illustrating the stratigraphy of the coal-bearing layered rocks in Somerset, southwest England, 1719.

differs from those in strata above and below. Thus, he came to the profound realization that each group of layers could be identified by its fossil content. Smith eventually developed the ability to predict what rocks and fossils could be expected to occur in distant hills, and he came to know the characteristics of both the rocks and fossils so well that he could amaze his friends by telling them exactly where their fossil specimens had been collected.

By the late 1790s, Smith was being encouraged by close friends, especially the evangelical clergyman Joseph Townsend (1739-1816) to publish his discoveries. Townsend, trained as a physician, was ordained in 1765, but after a few years of ministerial service including a stint as the chaplain to the Duke of Atholl, he turned his attention to his medical practice and writing about economic topics and natural history.[16] He appreciated Smith's work so much that between 1813 and 1815 he published two volumes in which he described some of Smith's discoveries and advocated an Earth older than the traditional 6,000 years.[17]

Eventually, Smith began to publish his observations. In 1816, he produced the first two volumes of *Strata Identified by Organized Fossils*, followed by a third volume in 1817 and a fourth in 1819.[18] These works included illustrations of many of the fossils that he encountered. In 1817, he also published a catalog of his personal collection of nearly 3,000 fossil specimens that had been sold to the British Museum. In this last work, Smith predicted the utility of correlating the stratigraphy of England with that of continental Europe.[19] The formations he had been investigating were mostly within the category of secondary rocks.

Smith's efforts also led to the 1815 publication of an excellent geologic map of much of England and Wales, the first colored geologic map of so large a region.[20] In 1819, he followed up his map by introducing a depth dimension with *Geological Sections*, a series of engraved, hand-colored geologic cross-sections.[21] These graphical works provided the basis for continuing stratigraphic work in England. By the end of his life, Smith's work had been recognized as foundational to the Earth sciences and, in 1831, he was awarded the first Wollaston Medal of the Geological Society of London for his achievements. Upon awarding the medal to Smith, Adam Sedgwick, the president of the Geological Society, called him "the Father of English Geology."

Smith was a practical geologist who did not indulge in theories about whether the strata had been laid down over long periods of time or in a catastrophic Deluge. He apparently viewed the separate layers as separate creations. Other geologists, however, began experiencing difficulty reconciling the extreme regularity of strata over great distances with the action of a global Flood. Even more problematic was the fact that fossil remains are restricted in regular manner to certain strata. A global Flood would have produced a

---

[16]On Townsend, see Frank N. Egerton II, "Townsend, Joseph," in *DSB*, 13:447-49; and A. D. Morris, "The Reverend Joseph Townsend MA, MGS (1739-1816) Physician and Geologist—'Colossus of Roads,'" *Proceedings of the Royal Society of Medicine* 62 (1969): 471-77.

[17]Joseph Townsend, *The Character of Moses Established for Veracity as an Historian, Recording Events from the Creation to the Deluge*, 2 vols. (Bath: 1813-1815).

[18]William Smith, *Strata Identified by Organized Fossils*, 4 vols. (London: Smith, 1816-1819).

[19]William Smith, *Stratigraphical System of Organized Fossils* (London: Williams, 1817).

[20]William Smith, *A Memoir to the Map and Delineation of the Strata of England and Wales* (London: J. Cary, 1815).

[21]William Smith, *Geological Sections* (London: J. Carey, 1819). See also J. G. C. M. Fuller, "'Strata Smith' and His Stratigraphic Cross Sections, 1819" (Tulsa, Okla.: AAPG, 1995), pp. 1-9.

much more chaotic, mixed array of fossils than what was actually observed. Many students of the Earth began to realize that the strata could not have been produced in a one-year Deluge but had to form over a long period of time by deposition in a succession of ancient seas, rivers and floods. They found it hard to envision the regularity of strata and the successions of organic worlds hinted at by the fossils in terms of a 6,000-year-old Earth.

The study of volcanism also played a role in undermining the idea of a very young Earth. In 1751, Jean-Étienne Guettard (1715-1786), a member of the Académie Royale des Sciences de Paris and the Faculté de Médecine de Paris, Adjunct Botanist to the Académie des Sciences, medical botanist for a wealthy French family of nobles, and author of a mineralogical map of France, was studying rocks and landforms in central France and gradually became aware that the features he was looking at were volcanic.[22] To his amazement, he recognized that there were extinct volcanoes in his own native land. The summit craters were pleasant grassy areas in which sheep could safely graze, and villages and houses had been built near the volcanoes with no evident thought of danger. The Romans had even used the local volcanic rock for various construction projects. Clearly the volcanic activity predated the Roman occupation of Gaul, and Guettard could find no mention of any volcanic activity in France among written records. As a result, he wondered if these extinct volcanoes might have been active prior to the advent of people into France. Like Hooke and Steno, Guettard was uncovering evidence of apparently spectacular geological events for which there were no convincing written records. Was there a geological history that preceded human history?

Guettard's investigations were carried farther by Nicholas Desmarest (1735-1815).[23] Not only did Desmarest investigate the volcanic region of central France more thoroughly than Guettard, but he also examined several other active and extinct volcanic terrains throughout Europe. By careful study of several superimposed lava flows, he reached the conclusion that France had experienced a long series of eruptions spaced out over a considerable amount of time. He even recognized that some of the lava flows had stream channels carved into them, and that these channels had subsequently been filled

---

[22]On Guettard, see Rhoda Rappaport, "Guettard, Jean-Étienne," in *DSB*, 5:577-79. The volcanoes of the Auvergne were first described by Jean-Étienne Guettard, "Mémoire sur Quelques Montagnes de la France Qui Ont Été des Volcans," *Mémoires de l'Académie Royale des Sciences pour l'Année 1752* (1756): 27-59. To this day, many of the buildings in the center of Clermont-Ferrand, including the Cathédrale de Notre Dame de l'Assomption, are made of dark lava blocks. Puy-de-Dôme, an extinct volcano, is a famous landmark just to the west of the city.

[23]On Desmarest, see Kenneth L. Taylor, "Desmarest, Nicholas," in *DSB*, 4:70-73. See also Nicholas Desmarest, "Mémoire sur l'Origine et la Nature du Basalte," *Mémoires de l'Académie Royale des Sciences pour l'Année 1771* (1774): 705-75.

in by later flows. This arrangement implied an extensive period of erosion to form the stream channel prior to the eruption of the next lava flow. Elsewhere, beds of sediments were found to lie between lava flows, a feature that also implied a sufficient lapse of time between flow eruptions to allow for intervening episodes of sedimentation. Moreover, the volcanic activity clearly postdated older events represented by the existence of granite and various other rocks. Desmarest suspected from what he saw that Earth history might well be more extensive than was customarily thought to be the case. Moreover, the fact that the volcanic cinder cones are still intact suggested that they either postdated the Flood or else the flood was not global. Presumably the catastrophic Deluge would have destroyed such cones.

More "Theories of the Earth"

The eighteenth century had its share of dedicated men like Guettard and Smith who painstakingly collected geologic data in the field and then left it to men of synthetic genius and vivid imagination to paint the whole canvas and work out the meaning of those data. Among the latter were Benoît de Maillet, the colorful, flamboyant Comte de Buffon, and the colorless, dry Scot James Hutton.

Benoît de Maillet (1656-1738) was a French diplomat who traveled frequently throughout the Middle East and the Mediterranean region.[24] He was also a keen observer of natural phenomena. Evidence of former shorelines above the level of the Mediterranean Sea convinced him that the Earth had experienced a continuous diminution of sea level through time. On this basis he speculated that layered rock formations had been deposited by the shrinking ocean over millions of years. Perceiving that his contention would cause shock and outrage among his contemporaries, de Maillet delayed publication of a book until 1748, some thirty years after its completion, and even then published anonymously. He placed his speculations in the mouth of an imaginary Indian philosopher named Telliamed.[25] Sharp-eyed observers, however, detected that the title of the book was simply the author's name spelled backward! *Telliamed* played no small part in initiating the Neptunist approach to Earth history that characterized much of the eighteenth century and culminated in the work of Werner.

---

[24]On de Maillet, see Albert Carozzi, "Maillet, Benoît de," in *DSB*, 9:26-27.

[25]Albert V. Carozzi, ed., *Telliamed; or, Conversations between an Indian Philosopher and a French Missionary on the Diminution of the Sea* (Urbana: University of Illinois Press, 1968). See also Albert V. Carozzi, "De Maillet's *Telliamed* (1748): An Ultra-Neptunian Theory of the Earth," in *Toward a History of Geology,* ed. Cecil J. Schneer (Cambridge, Mass.: MIT Press, 1969), pp. 80-99.

Georges-Louis Leclerc (1707-1788), commonly known as Comte de Buffon, was born into the home of a wealthy French aristocrat (fig. 3.4).[26] After studying law at Angers University, Buffon's interests turned toward mathematics and natural history. He drew early attention for introducing differential and integral calculus into probability theory and for translating Newton's mathematical work into French. His inheritance of the family fortune as a young man aided his research into natural history. In 1739, Buffon became Keeper of the Jardin du Roi in Paris, and eventually he converted the king's gardens into a center for research. The Muséum d'Histoire Naturelle was eventually constructed on the grounds of these gardens, which came to be known as the Jardin des Plantes.[27] Not only was Buffon an excellent researcher, but he was also an extremely prolific and capable writer. Beginning in 1749, he began issuing volumes of

Figure 3.4. Statue of Georges Louis le Clerc, Comte de Buffon (1707-1788) on the grounds of the Jardin des Plantes, Paris. Photo by D. A. Young.

an encyclopedia titled *Histoire Naturelle*.[28] By the time of his death, thirty-six volumes had been issued and another eight were published posthumously. Buffon attempted to describe and analyze virtually all knowledge of the natural world. An early volume of the encyclopedia was titled *Theory of the Earth*.

In his scientific work, Buffon was influenced by de Maillet's *Telliamed*, Leibniz's *Protogaea* and Newton's *Principia*.[29] Like them, he was interested in developing a theory of Earth history that took advantage of the very latest knowledge of physics. Buffon broke new ground in calculating an age for the Earth that turned out to be significantly greater than 6,000 years.[30] Being acquainted with much of the available geological evidence of the Earth's an-

---

[26]On Buffon, see Jacques Roger, *Buffon: A Life in Natural History*, trans. S. L. Bonnefoi (Ithaca, N.Y.: Cornell University Press, 1997).

[27]A large statue of Buffon is located near the entrance to the Jardin des Plantes in Paris, and Rue Buffon is the street that forms the southeastern boundary of the grounds of the Jardin des Plantes on which the Muséum d'Histoire Naturelle is located (see fig. 3.4).

[28]Georges-Louis Leclerc, Comte de Buffon, *The Natural History of Animals, Vegetables, and Minerals, with the Theory of the Earth in General* (London: n.p., 1976), pp. 358-73.

[29]Gottfried Wilhelm von Leibniz, *Protogaea* (Göttingen: I. G. Schmid, 1749).

[30]For Buffon's forays into "geotheory" see Rudwick, *Bursting the Limits of Time*, pp. 139-50.

tiquity, the Frenchman suspected that the Earth is quite old. He also sought
to work out a theory of the Earth's origin and history in terms of Newtonian
principles. After reading Newton's speculations about the cooling of comets,
Buffon wondered how long it might take for a molten globe the size of Earth
to cool to a temperature at which life could be sustained. In his discussion of
comets, Newton had written that

> a globe of incandescent iron equal to this earth of ours—that is, more or less
> 40,000,000 feet wide—would scarcely cool off in as many days, or about 50,000
> years. Nevertheless I suspect that the duration of heat is increased in a smaller
> ratio than that of the diameter because of some latent causes, and I wish that the
> true ratio might be investigated by experiments.[31]

Stimulated by Newton's open invitation, Buffon decided to conduct a series
of experiments on a group of spheres of varying sizes and materials. He heated
the spheres until they reached white heat and then permitted them to cool. He

Figure 3.5. James Hutton (1726-1797)
© The Natural History Museum, London.
Used by permission.

and his assistants then measured the
amounts of time that elapsed until
the surfaces of the globes cooled to
red heat, until they no longer glowed,
until they could be touched, and until they attained room temperature.
From these experimental data Buffon extrapolated for a body the size
of the Earth.

Preliminary results appeared in
an introductory treatise on the history of minerals.[32] Buffon maintained that a sphere of molten iron
the size of Earth would lose incan-
descence within 42,964 years and
that it would cool to present-day terrestrial temperature in 96,670 years.

Buffon realized, of course, that the Earth is composed of materials other than
molten iron, so he adjusted his numbers for different compositions. He suggested that the Earth consolidated completely in 2,936 years. In 34,270 years,

[31]Isaac Newton, *The Principia: Mathematical Principles of Natural Philosophy*, trans. I. Bernard
Cohen and Anne Whitman (Berkeley: University of California Press, 1999), p. 919. The quotation is from book 3, proposition 41, problem 21.

[32]Georges-Louis Leclerc, Comte de Buffon, "Introduction à l'histoire des minéraux: Partie ex-
périmentale," in *Histoire Naturelle, Générale et Particulière, Supplément*, vol. 1 (Paris: Imprimerie Royale, 1774).

it would be cool enough to touch. Organic life would be possible 1,700 years after that, and within 74,832 years the present temperature of the Earth would be attained. He proposed that with continued heat loss, organic life would no longer be possible in an additional 93,291 years. In his masterwork *Les Époques de la Nature* published in 1778, five years after he had become Count Buffon, the Frenchman laid out in elegant prose a scenario for the history of the Earth in seven successive epochs, consciously styling his reconstruction of Earth history along the lines of the days of creation of Genesis 1.[33]

Privately Buffon believed that the Earth is a great deal older than the 74,832 years he had conceded in print, but, knowing that his conclusions regarding the age of the Earth would create a furor because an earlier work of his had been censored, he attempted to forestall criticism by showing that his conclusions were not out of accord with Genesis at all. He maintained that the days of creation should be taken as long time periods and not as ordinary solar days. "What can we understand by the six days the sacred writer designates so precisely by counting them one after the other," Buffon asked rhetorically, "if not six spaces of time, six intervals of duration?"[34] Along lines reminiscent of Whiston, Buffon suggested that Earth originated when a comet closely approached the Sun and tore off a chunk of solar material. This solar glob was flung out into space where the incandescent mass gradually cooled through successive states until it was habitable. For each epoch, Buffon specified how long it lasted and what sorts of geological and biological events would transpire. Buffon's was one of the earliest efforts to quantify the age of the Earth on scientific grounds.

Very different was the work of Scotland's James Hutton (1726-1797).[35] Although Hutton (fig. 3.5) received a medical degree from the University of Leiden in 1749 with a dissertation on the circulation of blood, his interests lay in chemistry. For a time he joined a friend in running a firm that produced sal ammoniac (ammonium chloride). Thanks to an inheritance and profits from the company, he was able to pursue scientific interests. He inherited two farms, and he visited East Anglia in England to become more proficient in agricultural practice before returning to his farms. In 1767, Hutton moved to Edinburgh permanently where he became a member of a small group of Edinburgh intellectuals that included David Hume, Adam Smith, Thomas Reid and other luminaries of the Scottish Enlightenment. One of his closest friends

---

[33]Georges-Louis Leclerc, Comte de Buffon, *Les Époques de la Nature* (Paris: Éditions du Muséum, 1962).

[34]Ibid., p. 21.

[35]On Hutton, see Dennis R. Dean, *James Hutton and the History of Geology* (Ithaca, N.Y.: Cornell University Press, 1992).

was Joseph Black (1728-1799), a brilliant chemist who studied "fixed air" (carbon dioxide) as well as discovering the phenomena of latent heat of fusion and latent heat of vaporization. The group met regularly to discuss important scientific and philosophical questions. Perhaps stimulated by his observations of river and coastal erosion in East Anglia, Hutton became fascinated by geology. During his Edinburgh years, Hutton became a prolific author, producing several books on scientific and philosophical topics.

Hutton put forward his geological ideas in several preliminary papers, in a lengthy abstract of his theory and finally in *Theory of the Earth* published in 1795.[36] Although Hutton's geological writing lacked the flair of Buffon, it was probably much sounder scientifically than Buffon's effort. Unlike Buffon, Hutton was not interested in applying the principles of physics in a speculative manner to the history of the Earth. Instead, his "geotheory" entailed interpretation of the basic way in which the Earth functions much like a machine.[37] The theory also proposed a fruitful methodology for approaching the study of Earth history on the basis of continued observation of rocks and modern geologic processes over a period of several years. There were speculative elements in Hutton's thinking, but in *Theory of the Earth,* he constructed his theory inductively from the present-day behavior of nature and the characteristics of rocks and strata, unlike Burnet and Whiston, who imposed their theories on geologic phenomena. He applied principles worked out by predecessors like Steno and Arduino more consistently.

Hutton felt that it was unnecessary to invoke processes to explain rock strata and mountains other than ordinary, everyday processes like erosion and sedimentation whose effects we can readily observe. He argued in much more detail than others that rock strata are composed of mineral particles of exactly the same characteristics and constitution as are those found on beaches, sea floors or river beds. It must be, he said, that rock strata were formed on ancient sea beds from layers of loose unconsolidated gravel and sand that were eventually fused into rock and elevated by the internal heat of the Earth. Moreover, everywhere that mountains occur we can see plainly that they are in the process of decay and destruction by the erosive forces of wind, rain and rivers. We can see that the gravels and sands produced by erosion are gradu-

---

[36]See, for example, James Hutton, "The Theory of Rain," *Transactions of the Royal Society of Edinburgh* 1 (1788): 41-86, and *An Investigation of the Principles of Knowledge,* 3 vols. (Edinburgh: 1794). Hutton's geological masterpiece is James Hutton, *Theory of the Earth,* 2 vols. (Edinburgh: Creech, 1795; reprinted by Stechert-Hafner, 1972). Volume 3 was published posthumously: see Archibald Geikie, ed., *Theory of the Earth with Proofs and Illustrations,* vol. 3 (London: The Geological Society, 1899). Volume 3 was recently reprinted: see Dennis R. Dean, ed., *James Hutton in the Field and in the Study* (Delmar, N.Y.: Scholars' Facsimiles and Reprints, 1997).

[37]For an analysis of Hutton's geotheory, see Rudwick, *Bursting the Limits of Time,* pp. 158-72.

ally being carried down to the sea in streams. Therefore, he asked, may we not infer that, just as modern beach and sea-bottom sediments are derived by the erosion of present-day mountains, the strata now composing those mountains were derived from elevated sea-floor deposits that, in turn, had been derived from former mountains in a previous cycle of erosion? Hutton believed that he could see in rock strata the evidence for cycle after cycle of erosion, deposition and uplift to form mountains. He was convinced of the cyclic nature of these processes because, like Steno, he saw and understood the meaning of the

Figure 3.6. Siccar Point on the North Sea coast about 35 miles east of Edinburgh, Scotland. Siccar Point provided a prime illustration of Hutton's contention that Earth history consisted of an ongoing series of cycles of erosion, sedimentation and uplift. Geologist is A. Frazer. Photo by D. A. Young.

numerous unconformities visible in rock outcrops, the most famous of which occurs along the North Sea coast about 35 miles east of Edinburgh at Siccar Point (fig. 3.6). Hutton recognized that unconformities are ancient erosion surfaces that were cut into the beveled edges of tilted strata and subsequently buried underneath new strata.

Hutton was also convinced that the Primary schists and gneisses in the cores of mountains do not date back to creation as Lehmann had suggested but are remnants of very old sedimentary rock layers that had been altered into their present crystalline condition by extreme heat within the depths of Earth. With this proposal, he became the first person to enunciate the concept of *metamorphism of rocks.*[38] In essence, Hutton believed that he saw evidence for several time-consuming cycles of mountain building, erosion and sedimentation that stretched far back into the indefinite past.

Moreover, Hutton espoused the radical idea that large masses of granite are not necessarily Primary rocks. Prior to Hutton, the prevailing opinion was that the granite masses in the cores of mountains were already in place when Secondary layers were deposited on top of them. Hutton's field studies of granite in Glen Tilt in Scotland and on the Isle of Arran in the Hebrides demonstrated that "fingers" of granite emanating from very large granite masses penetrate through the layering of adjacent strata (fig. 3.7).[39] From these field relationships he concluded that the granite solidified from hot melted material that was injected into and across preexisting stratified and massive rocks

Figure 3.7. Outcrop in Glen Tilt, Scotland, where Hutton recognized the intrusive nature of granite. Here small granite dikes cut through metamorphosed schist and limestone. Photo by D. MacIntyre. Used by permission.

---

[38]Rock metamorphism is the process whereby metamorphic rocks are produced from preexisting rocks, commonly sedimentary and volcanic rocks formed on the Earth's surface that have been subjected to burial, recrystallization of their mineral grains, chemical reactions that produced new minerals and development of oriented structure. These processes occur in the solid state under very high pressures and temperatures with or without the aid of fluids.

[39]Hutton's ideas about granite may be found in volume 3 of *Theory of the Earth* and in James Hutton, "Observations on Granite," *Transactions of the Royal Society of Edinburgh* 3 (1794): 77-85.

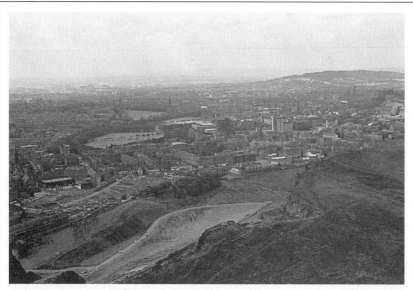

Figure 3.8. View of Edinburgh from summit of Arthur's Seat, an extinct eroded volcano in the southeast corner of the city. The ledge exposed partway down the slope at the lower right is Salisbury Crags. Photo by D. A. Young.

while underground.[40] As a result, the granite bodies must be younger than the overlying Secondary and Tertiary layers. The implication was that, interspersed among cycles of erosion, sedimentation, hardening of strata, tilting and uplift of stratified rocks, there had also been episodes of intrusion of large volumes of granitic melt, a fact that had serious implications for the amount of time involved in these events.

Along similar lines Hutton also took issue with the view of some Neptunists that the rock type basalt is a chemical precipitate from a gradually diminishing universal ocean. Salisbury Crags is a thick layer of dark crystalline basalt-like rock that is well exposed on the northwestern slope of Arthur's Seat, a prominent extinct volcano in southeastern Edinburgh (fig. 3.8). This layer is wedged between layers of sedimentary rocks. Hutton suggested that Salisbury Crags is not an oceanic precipitate but a layered mass of formerly molten igneous liquid that was injected between preexisting layers of rocks while still beneath the surface. As evidence he pointed out places where fragments of the preexisting rocks had been pried up and torn off by invading liquid and floated into the center of the Crags (fig. 3.9). He also showed that the sedimentary rocks ap-

---

[40]Hutton was applying what geologists today term the law (or principle) of cross-cutting relationships.

Figure 3.9. Salisbury Crags, an igneous sill on the flanks of Arthur's Seat. The thin sedimentary layers at the base of the sill and just above the grassy area were injected at considerable depth by viscous magma that tore off slabs of some of the layers. About five feet to the right of geologist, note the broken, upturned layer of sedimentary rock resulting from wedging and prying of magma between layers. At localities such as this Hutton first recognized the intrusive igneous nature of rocks that Wernerian Neptunists had considered precipitates from an allegedly global ocean. Geologist is A. Frazer. Photo by D. A. Young.

peared to have been baked adjacent to the contact with the rock of the Crags. Given his interpretation, the basalt layer would be younger than the layers between which it had been intruded. Hutton's fascination with the role of heat in geologic processes led to his use of the term *plutonism* for his ideas. Hutton and supporters like James Hall (1761-1832), a member of the Edinburgh circle, came to be known as Plutonists, and Plutonists had found evidence hinting at an older world than most people accepted. Hall was the first geologist to melt rocks in a crucible and model igneous phenomena experimentally.

Hutton concluded the first section of *Theory of the Earth* with a controversial statement that he could see in the course of nature "no vestige of a beginning, no prospect of an end."[41] In addition, he had not even bothered to correlate his scheme of geology with the Genesis account. Such by-passing of Scripture was tantamount to blasphemy to traditionalists, and the fact that he saw no vestige of a beginning in the rocks was misinterpreted as an atheistic denial of the doctrine of creation.[42] Hutton was accused of holding

---

[41]Hutton, *Theory of the Earth*, 1:200.
[42]See Playfair's defense of Hutton against Richard Kirwan's charge of atheism in John Playfair,

to eternity of the Earth, but essentially he was acknowledging that Earth processes seemed to reach far back in time and that he was unable to see in the rocks a point at which these processes had not been operative. He could not find the ultimately primeval, first-created rocks. Although Hutton was a deist, his system was not explicitly atheistic, anti-Christian or anti-biblical. He was attempting to explain what he thought the evidence of the rocks indicated.

THE BEGINNINGS OF THEOLOGICAL ADJUSTMENT BY SCIENTISTS TO A VERY ANCIENT EARTH

Hutton's Plutonist views were bitterly attacked by Richard Kirwan (1733-1812), a chemist, mineralogist and Inspector of Mines for Ireland.[43] Kirwan had strong Neptunist sympathies for biblical as well as geological reasons. He maintained that Scripture supported the idea of a gradually diminishing universal ocean, citing Psalm 104:5-9, a text that speaks of the mountains being covered by the waters and the waters then flowing down over the mountains as they arise.[44] Hutton was opposed more gently by Jean-André Deluc (1727-1817), a Swiss Calvinist from Geneva.[45] After holding several important posts in Paris and traveling and collecting specimens throughout the Alps and Jura Mountains, Deluc moved to England in 1773 and became the "intellectual mentor" to Queen Charlotte, the wife of King George III. In 1798, Deluc moved to Germany where he was an honorary professor of philosophy and geology at the University of Göttingen. Throughout his career, Deluc demonstrated great interest in Earth history as an extension of human history and suggested that it might be possible to estimate, from the accumulation of sedimentary deposits in river deltas, the amount of time humans had been on the Earth. The ancient world that preceded humans, he thought, was represented by the fossil-bearing strata and must have been very old.

Although both Deluc and Hutton believed that the world is quite ancient, the Calvinist was nervous about the deist's seemingly open-ended antiquity for the globe. Deluc attempted to refute Hutton on Wernerian grounds in a series of letters and in *An Elementary Treatise on Geology* (1809) by showing that the Neptunist system was the true system of geology because it could be squared

---

Illustrations of the Huttonian Theory of the Earth (Edinburgh: Creech, 1802), p. 120.

[43]On Kirwan, see E. L. Scott, "Kirwan, Richard," in *DSB*, 7:387-90.

[44]Richard Kirwan, "On the Primitive State of the Globe and its Subsequent Catastrophe," *Transactions of the Royal Irish Academy* 6 (1797): 233-308.

[45]On Deluc, see Robert P. Beckinsale, "Deluc, Jean André," in *DSB*, 4:27-29. See also Martin J. S. Rudwick, "Jean-André de Luc and Nature's Chronology," *The Age of the Earth: From 4004 B.C. to A.D. 2002*, GSL Special Publication 190, ed. Cherry L. E. Lewis and Simon J. Knell (London: The Geological Society, 2001), pp. 51-60.

with a basically literal interpretation of Scripture.[46] Genesis 1:2 mentions the "great deep" upon which the Spirit of God moved and which Deluc identified as the Neptunist universal ocean. Even before Smith's detailed paleontological work, Deluc, like Lister and Füchsel before him, had recognized that stratified formations have their own characteristic suites of fossil remains, an arrangement that presumably fit nicely with the successive creations of the six days. Although he was troubled by the vast antiquity intimated by Hutton, Deluc fully recognized that the geologic record requires more time to develop than was allowed by the traditional six literal days of creation. Consequently, Deluc advocated a figurative usage of the six days, commenting that he had spent most of his life in showing "the conformity of geological monuments with the sublime account of that series of operations which took place during the *Six Days*, or periods of time, recorded by the inspired penman."[47] Even some of Hutton's opponents conceded that the Earth is more ancient than a strictly literal interpretation of Genesis 1 allowed.

The leader among Hutton's defenders was John Playfair (1748-1819), a professor of mathematics in the University of Edinburgh and an ordained minister of the Church of Scotland.[48] He both supported Hutton's geology and defended him against the charges of atheism. In *Illustrations of the Huttonian Theory of the Earth* (1802), Playfair maintained that Hutton said that we see "no mark, either of a beginning or an end," not that "the world had no beginning, and will have no end."[49] Likewise, said Playfair, if Hutton were atheistic in his views he would have represented the world as the "result of necessity or chance" rather than continually admiring "the instances of wise and beneficent design manifested in the structure, or economy of the world."[50] Playfair also defended Hutton's ideas respecting Earth's antiquity against those who thought that the high antiquity ascribed by Hutton's theory to the Earth was inconsistent with the scriptural chronology. But, he said,

> this objection would no doubt be of weight, if the high antiquity in question were not restricted merely to the globe of the earth, but were also extended to the human race. That the origin of mankind does not go back beyond six or seven thousand years, is a position so involved in the narrative of the Mosaic books, that any thing inconsistent with it, would no doubt stand in opposition to the testimony of those ancient records. On this subject, however, geology is silent.
>
> On the other hand, the authority of the Sacred Books seems to be but little

---

[46]For a lengthy compilation of Deluc's scientific letters, see Rudwick, *Bursting the Limits of Time*, pp. 670-71.

[47]Jean-André Deluc, *An Elementary Treatise on Geology* (London: 1809), p. 414.

[48]On Playfair, see John Challinor, "Playfair, John," in *DSB*, 11:34-36.

[49]Playfair, *Illustrations of the Huttonian Theory*, p. 120.

[50]Ibid., pp. 121-22.

interested in what regards the mere antiquity of the earth itself; nor does it appear that their language is to be understood literally concerning the age of that body, and more than concerning its figure or its motion. The theory of Dr. Hutton stands here precisely on the same footing with the system of Copernicus.[51]

No specific attempt at harmonization was presented by Playfair who was content simply to deny that Scripture fixed the age of the Earth.

CATASTROPHISM

The figurative interpretation of the six days received further impetus as a result of the catastrophism of Georges Cuvier (1768-1832), a great French comparative anatomist and founder of the discipline of vertebrate paleontology (fig. 3.10).[52] Cuvier was born into a Lutheran family in Montbéliard, a French-speaking region of the German duchy of Württemberg, and he regularly attended church throughout his life. After receiving a high-quality education in Stuttgart, he took on the post of tutor to a Protestant noble family in Normandy. While there, he undertook studies of the marine organisms of the nearby coast. His manifest ability led to his appointment as a professor of anatomy at the Muséum d'Histoire Naturelle in Paris. He remained at this institution from 1795 until his death.

Cuvier was a gifted anatomist who achieved a great deal of fame during his lifetime for his ability to relate the structure of organisms to their biological function. In his emerging role as a global anatomical referee, Cuvier was forwarded many newly collected speci-

Figure 3.10. Statue of Georges Cuvier (1768-1832) at Cuvier's home on the grounds of the Muséum d'Histoire Naturelle, Paris. Decades later physicist Henri Becquerel lived in the home and made his famous discovery of radioactivity. Photo by D. A. Young.

---

[51] Ibid., pp. 125-26.
[52] On Cuvier, see Dorinda Outram, *Georges Cuvier: Vocation, Science, and Authority in Post-Revolutionary France* (Manchester: Manchester University Press, 1984). New translations and fresh interpretations of several of Cuvier's important articles have been published by Martin J. S. Rudwick, *Georges Cuvier, Fossil Bones, and Geological Catastrophes: New Translations & Interpretations of the Primary Texts* (Chicago: University of Chicago Press, 1997).

mens of animals, several of which were fossils. Cuvier accumulated a long list of public honors and appointments. He served for many years as vice president of the Bible Society of Paris. In 1802, Napoleon appointed him one of the six inspectors-general of public education. In 1813, he was appointed to the Council of State, on which he served until his death. King Louis XVIII made him a baron in 1820.

Cuvier began to study the fossils available to him in the Paris region and issued numerous technical papers on a wide range of fossil vertebrates. In one early paper on a fossil opossum, he speculated that the animal might have been buried for "thousands of centuries."[53] In conjunction with his paleontological studies, he worked out the stratigraphy of the Paris basin together with geologist Alexandre Brongniart (1770-1847). A director of the Sevres porcelain factory between 1800 and 1847, Brongniart also taught mineralogy at the University of Paris beginning in 1811, and eventually became a professor of mineralogy at the Muséum d'Histoire Naturelle.[54] Brongniart began to investigate the geology of northern France in the first decade of the 1800s. He adopted the term *formation* to identify a stratigraphic rock unit of uniform composition capable of being mapped, consciously following the terminology conventions of the central German stratigraphic school focused on the Freiberg Mining Academy.

Brongniart and Cuvier collaborated on the mapping and description of the rocks of the Paris region in an early paper of 1808 and in a more comprehensive monograph, *Essai sur la Géographie Minéralogique des Environs de Paris* (1811).[55] Together, they mapped a series of nine superposed formations, the lowest of which was the famous Chalk, exposed in many locales in northwest Europe and on the coast of southern England (fig. 3.11). The strata above the Chalk were assigned to the Tertiary Period. For the 1811 publication, Brongniart also prepared a large colored geologic map. The stratigraphic formations of the Paris Basin, interestingly, alternated between limestones hosting the remains of marine animals and gypsum-bearing mudstones containing the remains of land animals and plants. This plain alternation of marine and terrestrial biotas would greatly influence Cuvier's view of the history of life and of the Earth.

---

[53]Georges Cuvier, "Mémoire sur le Squelette Presque Entier d'un Petit Quadrupède du Genre de Sarigues, Trouvé dans le Pierre à Plâtre des Environs de Paris," *Annales du Muséum d'Histoire Naturelle* 5 (1804): 277-92.

[54]On Brongniart, see Martin J. S. Rudwick, "Brongniart, Alexandre," in *DSB*, 2:493-97.

[55]See Georges Cuvier and Alexandre Brongniart, "Essai sur la Géographie Minéralogique des Environs de Paris," *Annales du Muséum d'Histoire Naturelle* 11 (1808): 293-326; and "Essai sur la Géographie Minéralogique des Environs de Paris," *Mémoires de la Classe des Sciences Mathématiques et Physiques de l'Institut Impérial de France 1810* (1811): 1-278.

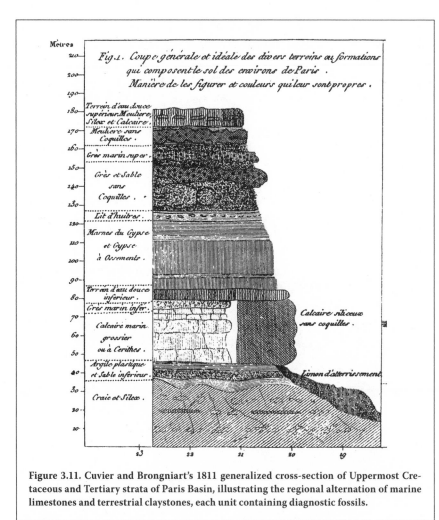

Figure 3.11. Cuvier and Brongniart's 1811 generalized cross-section of Uppermost Cretaceous and Tertiary strata of Paris Basin, illustrating the regional alternation of marine limestones and terrestrial claystones, each unit containing diagnostic fossils.

On the basis of these observations, Cuvier, independent of William Smith in England, discovered that individual groups of strata are characterized by wholly unique faunas. These faunas appeared to have lived for a while before being replaced by wholly new faunas whose fossils are found in successively younger rock strata. Cuvier recognized a whole series of faunal replacements. Unable to reconcile the observed distribution of the fossils with the action of a single universal Flood, he proposed that the fossil evidence was consistent with the idea that the globe had experienced a series of great catastrophes or "revolutions." Cuvier promoted his theory of Earth history in a compilation of his early papers on fossil vertebrates, published in 1812 as *Recherches sur les*

Ossemens Fossiles de Quadrupèdes, a work that firmly established the fact that many types of animals have become extinct.[56] The subtitle of *Recherches* indicated that these quadrupeds appeared to have been destroyed in revolutions of the globe. Cuvier began the first volume of his four-volume collection with a totally new "Discours Préliminaire" in which he spelled out his belief that the fossil evidence could not be fully accounted for by modern-day processes but required a series of catastrophic revolutions. He suggested that the fauna of an area would live for a while, be eradicated in a great catastrophe, and then be followed by the appearance of a wholly new and different fauna, which, in turn, would be catastrophically destroyed and replaced by a new fauna, and so on. Cuvier believed that this succession of revolutions required a long period of time. After maintaining that the documents and legends of numerous civilizations, including the Jews, Greeks and Egyptians, all pointed to a revolution after which the present condition of the human race was established, Cuvier concluded that the final global revolution that brought the continents into their present conditions occurred only five to six thousand years ago.

Figure 3.12. William Buckland (1784-1856) © The Natural History Museum, London. Used by permission.

James Parkinson (1755-1824), following Cuvier, promoted a catastrophist position.[57] Parkinson took over his father's medical practice and gained a wide reputation for his studies of gout and peritonitis as well as for social activism. In 1817, Parkinson wrote *An Essay on the Shaking Palsy*, establishing him as the leading authority on the disease now known as Parkinson's disease. Early in his career, Parkinson also had become interested in geology and paleontology. He published a well-regarded three-volume work on paleontology titled *Organic Remains of a Former World* (1804-1811) in which he presented a synthesis of Cuvier's catastrophic ideas and Genesis.[58] Parkinson

---

[56]Georges Cuvier, *Recherches sur les Ossemens Fossiles de Quadrupèdes, ou l'on Rétablit les Caractères de Plusieurs Espèces d'Animaux que les Révolutions du Globe Paraoissent Avoir Détruites*, 4 vols. (Paris: Deterville, 1812).

[57]On Parkinson, see Patsy A. Gerstner, "Parkinson, James," in *DSB*, 10:321-23.

[58]James Parkinson, *Organic Remains of a Former World* (London: M. A. Nattali, 1803-1811).

treated the days of Genesis as vast periods of time.

Another catastrophist was Oxford's first geologist, William Buckland (1784-1856) (fig. 3.12). Buckland studied at the University of Oxford and remained there as a reader in mineralogy (1813) and then in geology (1819).[59] By all accounts, he was a very charismatic lecturer and mentor. Buckland was ordained as an Anglican clergyman in 1808, and from 1845 until his death he served as Dean of Westminster Abbey, where he instituted educational reforms and physical repairs to the abbey school. Buckland undertook extensive surveys of British regional geology beginning in 1808. By 1811, he was examining layered sedimentary rocks in Ireland and Scotland and, by 1816, continental Europe. We will return to Buckland in chapter four.

Buckland, like Parkinson, adopted a form of Cuvierian catastrophism, but modified it by seeking geologic evidence of a flood that was shorter-lived and more extensive than that of Cuvier.[60] Except for the "Scriptural geologists" whom we note briefly in the next chapter, few geologists in the early nineteenth century attributed the entire stratigraphic record to the Flood. Field geologists were now considering only the most recent, surficial deposits as possible evidences of the Flood. Abundant surficial gravel deposits exist throughout northern Europe, as do large isolated boulders that have obviously been transported great distances from their source areas. Buckland and others attributed such phenomena to the action of great waves of the Noachian Deluge. To such presumed evidence Buckland added evidence from caves and large fissures in which numerous discoveries of large accumulations of fossil mammals had been made. Many of these mammals were similar to modern day animals, and several, such as rhinoceros, hippopotamus and elephant, were no longer living in Europe. In his book *Reliquiae Diluvianae*, Buckland marshaled the evidence for the Flood. But as the Flood was removed from being the sole explanation of geological history to the last in a series of catastrophes, Buckland, like Cuvier, believed that the Earth very likely had a long history prior to the Flood.

THE RISE OF UNIFORMITARIANISM

The catastrophism of Cuvier and Buckland, popular in Europe between 1810 and 1850, was adopted by other prominent geologists such as Adam Sedgwick and George Bellas Greenough (1778-1855). However, during this period a trend developed toward acceptance of a philosophy of Earth history that was foreshadowed by Füchsel, Arduino and Hutton and eventually dubbed

---

[59]On Buckland, see Nicolaas Rupke, *The Great Chain of History: William Buckland and the English School of Geology (1814-1849)* (Oxford: Clarendon, 1983). See also Donald K. Grayson, *The Establishment of Human Antiquity* (New York: Academic Press, 1983).

[60]William Buckland, *Reliquiae Diluvianae* (New York: Arno Press, 1978).

uniformitarianism by English philosopher and historian of science William Whewell (1794-1866).[61] Uniformitarianism was essentially the doctrine that the past history of the Earth could be explained adequately in terms of causes that are presently observed to be in operation on the globe without having recourse to supernatural causes that had ceased to operate. Uniformitarians eschewed the invocation of spectacular one-time catastrophes to account for geologic phenomena when the processes of the present could account for them quite satisfactorily if they operated over a sufficiently long period of time.

Among the more prominent advocates of such an approach was John Fleming (1785-1857), a Scottish biologist and minister who left the Church of Scotland to join the Free Church of Scotland in 1843. Fleming opposed the popular diluvialism of Buckland on the grounds that it was both unscientific and unbiblical.[62] He believed that the gravels and cave deposits described by Buckland could not be explained in terms of one great event. For him, animal extinctions recorded by the fossil bones and the deposition of gravels could be explained only by processes acting gradually over a long time. Fleming also argued that the biblical description of Noah's Deluge failed to match the kind of flood invoked by Buckland.

George Poulett Scrope (1797-1876), an English geologist and member of Parliament, undertook yet another investigation of the volcanic terrain around Clermont-Ferrand in south-central France. From his study of river valleys and volcanic activity in the region, he concluded that currently existing processes acting for very long periods of time could explain the geologic phenomena much better than contemporary catastrophic explanations of the day.[63] He also reasoned that a global Deluge would have obliterated the numerous volcanic cones now exposed.

The uniformitarian viewpoint found its greatest expositor in Charles Lyell (1797-1875) (fig. 3.13).[64] Lyell was born in Kinnordy, Scotland, where his father was a botanist and a very wealthy landowner. Time spent in his youth at a

---

[61]William Whewell, *History of the Inductive Sciences: From the Earliest to the Present Time*, vol. 3 (London: John W. Parker, 1857), pp. 506-20. See chap. 8, "The Two Antagonist Doctrines of Geology."

[62]On Fleming, see Leroy E. Page, "Fleming, John," in *DSB*, 5:31-32. See also Davis A. Young, "Nineteenth Century Christian Geologists and the Doctrine of Scripture," *Christian Scholar's Review* 11 (1982): 212-28.

[63]George Poulett Scrope, *Memoir on the Geology of Central France* (London: Longman, Rees, Orme, Brown and Green, 1827).

[64]On Lyell, see Edward B. Bailey, *Charles Lyell* (Garden City, N.Y.: Doubleday, 1963); Leonard G. Wilson, *Charles Lyell, the Years to 1841: The Revolution in Geology* (New Haven, Conn.: Yale University Press, 1972); and *Lyell in America: Transatlantic Geology 1841-1853* (Baltimore, Md.: Johns Hopkins University Press, 1998).

family home in the country sparked an interest in natural history. As a student at Oxford, he attended Buckland's lectures in geology and fell under his spell. He subsequently moved to London to become a lawyer, but poor eyesight and a growing passion for geology led him to abandon law and devote his full attention to geology, a lifestyle made possible by a generous inheritance and proceeds from his books.

As a student of Buckland, Lyell began his geological career as a catastrophist. He already agreed with Cuvier and Buckland that geologic evidence established that Earth history exceeded 6,000 years. However, during his early excursions in the British Isles, the volcanic regions of the Auvergne in France, Mount Etna and the coastal regions of Italy, he concluded that global catastrophes are unnecessary explanatory agents and that even surficial gravels and boulders might be explained in terms of several local floods. Lyell's *Principles of Geology*, published in three volumes between 1830 and

Figure 3.13. Charles Lyell (1797-1875) © Geological Society/NHMPL. Used by permission.

1833, showed in great detail how geologic phenomena could be explained in terms of modern-day processes such as river and marine erosion and deposition acting at essentially the same rates as now over a long period of time.[65] Lyell's project was spelled out well in the subtitle of the book: *An Attempt to Explain the Former Changes of the Earth's Surface by Reference to Causes Now in Operation*. His masterwork was so successful that it went through twelve editions and remains in print!

Lyell suggested that instead of invoking catastrophes of which we have no experience, we ought rather to invoke processes that we can observe or reasonably infer to have operated from the evidence of the rocks. Being sensitive to religious concerns, Lyell had no intention of denying the existence of Noah's Flood, although he believed that the Flood was an extensive inundation in the Middle East that had relatively little geologic effect. Both he and Fleming suggested that the Flood might have been relatively tranquil. Lyell's uniformitarian approach appeared to require even greater stretches of time than were called for by the proponents of catastrophism.

---

[65]Charles Lyell, *Principles of Geology*, 3 vols. (London: John Murray, 1830-1833).

The most distinctive aspect of Lyell's thinking was the notion that the overall energy and intensity of Earth processes remained essentially constant throughout its history on a global scale. Thus, for example, although the intensity of volcanism remained about the same *on a worldwide scale* throughout Earth history, the intensity of volcanism might have varied from place to place and also throughout time at each specific location. The same was asserted for earthquake activity, glaciation and other geologic processes. Unlike most of his contemporaries, whether catastrophists or uniformitarians, Lyell rejected directionalism in Earth history in the sense that geologic processes like volcanism either decreased or increased in intensity over time *on a global scale.* So committed was he to this steady-state conception of Earth history in his early career that he thought, in spite of appearances of the fossil record to the contrary, that life forms had changed little over time. Despite Lyell's influence, most geologists saw clear signs of direction or progress in Earth history, particularly in regard to the fossil record, and therefore dismissed the steady-state aspect of Lyell's uniformitarianism. We explore the concept of uniformitariansm in more detail in chapter sixteen.

Lyell authored many other books on geology, traveled to North America several times, was knighted by Queen Victoria in 1848, and exerted considerable influence on a young geologist and naturalist named Charles Darwin with whom he ultimately became good friends. The esteem in which Lyell was held by his contemporaries in Great Britain is reflected in the fact that he is buried in Westminster Abbey near John Woodward and not far from Sir Isaac Newton.

Another who did much to overthrow catastrophist theories and to further uniformitarian thinking, Louis Agassiz, did so despite his own catastrophist tendencies. Jean Louis Rodolphe Agassiz (1807-1873) was a native of Switzerland and son of a minister.[66] He received advanced education at the universities of Zurich, Heidelberg, Munich and Erlangen, obtaining a Ph.D. degree from Erlangen in 1829 and a medical degree from Munich in 1830. A collection of fishes from the Amazon River was given to Agassiz for his doctoral study. The result was a book on the fishes of Brazil.[67] In 1831, Agassiz moved to Paris to study comparative anatomy with Cuvier who, observing that his student excelled at work with fossil fishes, entrusted him with all his notes for a book that he had been planning to write on that subject. As a result, Agassiz cemented his growing reputation as a first-rate naturalist with the publica-

---

[66]On Agassiz, see Edward Lurie, *Louis Agassiz: A Life in Science* (Chicago: University of Chicago Press, 1960).

[67]Louis Agassiz, *Selecta Genera et Species Piscium Quos in Itinere per Brasiliam 1817-1820* (Munich: 1829).

tion of a set of five massive volumes titled *Recherches sur les Poissons Fossiles* between 1833 and 1843.[68] During his tenure with Cuvier, Agassiz adopted the catastrophist outlook of his mentor.

In 1832, Agassiz was appointed as a professor of natural history at the Lyceum in Neuchâtel, Switzerland, where he instructed students in paleontology, systematics and eventually glacial studies. Four years later, Agassiz began a careful study of the characteristic deposits such as erratics and moraines that had been left behind by receding glaciers in the Swiss Alps. He noted that glacially produced features like *erratic boulders,* scratches and grooves on rocks, smoothed and polished rock outcrops, and *moraines* existed in regions several miles from the nearest active glaciers.[69] From this distribution of glacial features he postulated that glaciers had been much more extensive in Alpine regions in the past. Agassiz later recognized the same kinds of features all over northern Europe. In 1840, he published a landmark volume titled *Études sur les Glaciers* in which he postulated the existence of a former Ice Age *(Eiszeit)* in which gigantic ice sheets had covered much of northern Europe and northern North America.[70] In that same year, Agassiz and Buckland took a field trip to Scotland where the latter became convinced that many of the allegedly "diluvial" phenomena that he had hitherto attributed to a great deluge had been caused by the action of extinct ice sheets. Lyell, too, at least temporarily, became a convert to the new glacial hypothesis. Thus, the Ice Age replaced a catastrophic flood of nearly global proportions as a causative agent, and, with the disappearance of diluvial catastrophism, it became clearer to active geologists that the Earth had experienced a very long, slow-paced history better explained in terms of observable processes like glaciation, erosion and deposition than in terms of a few very sudden, ancient global catastrophes of which we have no experience.

Agassiz moved to the United States in 1846, wrote another book on glaciation the following year, and in 1848 he became a professor of geology and

---

[68]Louis Agassiz, *Recherches sur les Poissons Fossiles,* 5 vols. (Neuchâtel: Imprimerie de Petit-pierre, 1833-1843).

[69]A moraine is an irregular, somewhat sinuous pile of gravel that is deposited at the melting margin of a glacier. An erratic is a boulder, typically somewhat rounded, that rests on top of soil or bedrock and that consists of rock that has been derived from bedrock at a considerable distance. For example, throughout the lower peninsula of Michigan one occasionally finds small boulders, as well as cobbles and pebbles, of a light-colored conglomerate containing red pebbles of jasper. This rock bears absolutely no resemblance to any bedrock within the lower peninsula, but it is identical to striking outcrops of jasper-pebble conglomerate that form part of the Lorrain Formation along the northern shore of Lake Huron in southern Ontario.

[70]Louis Agassiz, *Études sur les Glaciers* (Neuchâtel: Jent et Gassmann, 1840).

zoology at Harvard College, where he eventually established the Museum of Comparative Zoology. In his later years, the great Swiss scientist remained something of an enigma to his contemporaries because of his advocacy of a polygenetic origin of human beings and resistance to Darwin's theory of evolution.

4

HARMONIZING GEOLOGICAL CHRONOLOGY AND SCRIPTURE IN THE NINETEENTH CENTURY

THE OBSERVATIONS OF WILLIAM SMITH and Georges Cuvier regarding the distinct faunas of successive layers of strata provided the foundation, so to speak, for the nineteenth-century development of biostratigraphy.

THE BEGINNINGS OF BIOSTRATIGRAPHY

Following the work of Smith and Cuvier, European scientists and civil engineers continued to document the fact that fossils are not distributed randomly within the layers of sedimentary rocks that coat large portions of the continents. This nonrandom distribution is manifested in two ways: First, fossils in any given layer form a distinctive suite that can be comprehensively characterized through repeated sampling. Second, these suites of fossils change character upward and downward within the layers of the local rock column. Some fossil organisms have very limited vertical extent, whereas others persist. In some cases, dramatically abrupt vertical changes in the fauna can be documented in the strata, an observation that led to Cuvier's adoption of catastrophism. Another revelation proved even more dramatic. Extensive field observation and collection revealed that the same pattern of vertical succession in fossil suites could be observed over broad regions of the Earth's surface. The pattern can be interrupted by episodes of erosion or nondeposition, but the general order of the fossils remains intact.[1] This phenomenon is sometimes referred to as the *law (or principle) of faunal succession.*

During the first half of the nineteenth century, European and American geologists learned that this pattern could be used as an empirical guide and tool for the correlation of regional stratigraphies. Numerous programs of fossil excavation and collection conducted during the past two hundred years have only repeatedly verified this discovery.

CORRELATION BY FOSSILS AND FURTHER DEVELOPMENT OF THE GEOLOGIC TIMESCALE

Early in his tenure at Oxford, William Buckland began to synthesize data on British stratigraphy for his geological lectures. After an extensive trip to Europe in 1820, he became convinced that similar successions of strata in continental Europe and in Britain demonstrated similarities in their faunas. In 1821, he published his conclusions.[2] Although Buckland's terminology for these stratigraphic groups has passed from general usage, his correlations were sound. British stratigraphic geology during the 1820s and 1830s built on Buckland's and Smith's sequences.

William Daniel Conybeare (1787-1856) was an Anglican country parson who had studied at Oxford, wrote extensively on British regional geology, and ultimately became Dean of Llandaff Cathedral.[3] William Phillips (1775-1828) was a printer and bookseller who wrote geological papers and texts in his spare time.[4] In 1822, Conybeare collaborated with Phillips in updating the latter's *Outline of the Geology of England and Wales.*[5] In the process, Conybeare explicated in great detail the coal-bearing rocks that had prominent surface exposures in north-central England. Conybeare and Phillips christened these layers the Carboniferous System and confirmed Buckland's division of these strata into a lower subdivision, mainly limestones and sandstones, and an upper set of units that contain numerous coal seams.

---

[1]A. O. Woodford, "Correlation by Fossils," in *The Fabric of Geology,* ed. C. C. Albritton (Reading, Mass.: Addison-Wesley, 1963), pp. 75-111.

[2]W. Buckland, "Notice of a Paper Laid Before the Geological Society on the Structure of the Alps and Adjoining Parts of the Continent, and Their Relation to the Secondary and Transition Rocks of England," *Annals of Philosophy* 1 (1821): 450-68; see also Nicolaas Rupke, *The Great Chain of History: William Buckland and the English School of Geology (1814-1849)* (Oxford: Clarendon, 1983).

[3]On Conybeare, see M. J. S. Rudwick, "Conybeare, William Daniel," in *DSB,* 3:395-96.

[4]On Phillips, see Robert P. Beckinsale, "Phillips, William," in *DSB,* 10:585-86.

[5]W. D. Conybeare and W. Phillips, *Outlines of the Geology of England and Wales, with an Introductory Compendium of the General Principles of that Science, and Comparative Views of the Structure of Foreign Countries,* Part 1 (London: 1822).

Meanwhile, across the English Channel, Alexandre Brongniart realized that the fossil sequence that he and Cuvier had established for northern France could be successfully used to relate those rocks to those of other portions of western Europe. In 1821, he compared the fossils contained in the Chalk of the Paris Basin with comparable rocks of England across the Channel and to similar rocks in Poland.[6] He also correlated the Tertiary rocks of the Paris Basin (those strata above the Chalk) with those of northern Italy.[7]

A large number of European geologists followed up on Brongniart's correlation of the French Chalk. In 1822, Jean-Baptiste-Julien d'Omalius d'Halloy (1783-1875), a Belgian statesman, amateur geologist and practicing Catholic Christian, mapped the French Chalk into Belgium, and christened the strata encompassing the chalk the Terrain Cretacé, thus providing the source of the system name, Cretaceous.[8] Omalius would later create a geologic map of the French empire, published in 1823. Others extended the correlations of the Cretaceous System into Denmark, Northern Germany and the Jura Mountains.[9]

During the late 1820s, Charles Lyell began to visit localities around Europe where Tertiary strata could be easily sampled. He was assisted in the classification of the shelled marine invertebrates of these localities by Gerard Deshayes (1797-1875).[10] Deshayes, one of the premier nineteenth-century experts on mollusks, described the Eocene mollusks of the Paris Basin in detail and was eventually appointed to the chair of conchology originally held by Lamarck at the Muséum d'Histoire Naturelle. Lyell adopted a biostratigraphic division of the Tertiary faunas suggested by Deshayes, and in 1833, proposed a subdivision of the Tertiary rocks into three basic groups of strata that he designated the Eocene, Miocene and Pliocene, representing periods of time that he termed "epochs."[11] Lyell would later add the "Post-Pliocene," which included the time of glaciation of Europe prior to human occupation, as well as the brief last stage of geological history, in which humans had made their

---

[6]A. Brongniart, "Sur les Caractères Zoologiques des Formations, avec Application de ces Caractères à la Détermination de Quelques Terrains de Craie," *Annales des Mines* 6 (1821): 537-72.

[7]J. M. Hancock, "The Historic Development of Concepts of Biostratigraphic Correlation," in *Concepts and Methods of Biostratigraphy*, ed. E. G. Kauffman and J. E. Hazel (Stroudsburg, Penn.: Dowden, Hutchinson and Ross, 1977), pp. 3-22.

[8]On d'Omalius d'Halloy, see R. C. Tobey, "Omalius d'Halloy, Jean Baptiste Julien d'," in *DSB*, 10:208-10.

[9]J. B. J. d'Omalius d'Halloy, "Observations sur un Essai de Carte Géologique de la France, des Pays-Bas et des Contrées Voisines," *Annales des Mines* 7 (1822): 353-76; see also W. B. N. Berry, *Growth of a Prehistoric Time Scale, Based on Organic Evolution* (San Francisco: W. H. Freeman, 1968).

[10]On Deshayes, see H. Tobien, "Deshayes, Gerard Paul," in *DSB*, 4:67-68.

[11]C. Lyell, *Principles of Geology*, vol. 3 (London: John Murray, 1833).

appearance.[12] This term would later be replaced by "Pleistocene," Lyell's term for Agassiz's "Ice Age."

Adam Sedgwick (1785-1873), like Buckland and Conybeare, was an ordained Anglican clergyman (fig. 4.1).[13] Sedgwick successfully campaigned for the Woodwardian professorship of geology at Trinity College, Cambridge, in 1818. Although he attained the position having little field experience, with some tutelage by Conybeare he quickly turned himself into one of the best field geologists in England. Sedgwick helped to found the Cambridge Philosophical Society in 1819, and served as the president of the Geological Society of London from 1829 to 1831. His students at Cambridge included the young Charles Darwin. Sedgwick vigorously addressed the regional stratigraphy of southern England, with special emphasis on western England and Wales.

During the early 1830s, Sedgwick teamed up with a retired British army officer, Roderick Murchison (1792-1871), to map the strata of Wales (fig. 4.2).[14] Murchison had attended lectures by Buckland at Oxford, and then continued his geologic education by means of extensive travels in Europe. Sedgwick examined strata in northern Wales while Murchison mapped the strata of southern Wales. The Secondary rock units they examined were well below the strata of the Carboniferous System of England. Sedgwick named the strata of northern Wales the Cambrian System after the Roman name for the province (Cambria); Murchison named his collective system the Silurian System for an ancient Welsh tribe that was antagonistic toward the Roman occupiers of the land.[15] Murchison's strata overlay

Figure 4.1. Adam Sedgwick (1785-1873)
© **The Natural History Museum, London. Used by permission.**

---

[12]C. Lyell, *A Manual of Elementary Geology,* 5th ed. (London: John Murray, 1855); and A. Bowdoin Van Riper, *Men Among the Mammoths* (Chicago: University of Chicago Press, 1993).

[13]On Sedgwick, see M. J. S. Rudwick, "Sedgwick, Adam," in *DSB,* 12:275-79.

[14]On Murchison, see Robert A. Stafford, *Scientist of Empire: Sir Roderick Murchison, Scientific Exploration, and Victorian Imperialism* (Cambridge: Cambridge University Press, 1989).

[15]A. Sedgwick and R. I. Murchison, "On the Silurian and Cambrian Systems, Exhibiting the Order in which the Older Sedimentary Strata Succeed Each Other in England and Wales," *British Association for the Advancement of Science Report,* 5th mtg. (1835): 59-61; R. I. Murchison, "On

Sedgwick's; unfortunately, they could not agree on the boundary between the two systems. Although they originally published joint accounts of the regional stratigraphy, they eventually became estranged. Murchison left England for a time during 1839 and 1840 to study the stratigraphy of the Ural Mountains of eastern Russia. The resolution of the Cambrian/Silurian impasse would not come until 1872, when Charles Lapworth (1842-1920), a professor of geology at the University of Birmingham and an expert on fossil graptolites, created a third system, the Ordovician, to accommodate the strata in the middle of the long succession that Murchison and Sedgwick had described.[16]

Figure 4.2. Roderick Murchison (1792-1871)
© The Natural History Museum, London.
Used by permission.

While yet working together, Sedgwick and Murchison described the rock units directly underlying the Carboniferous System and overlying the Silurian System, which included a prominent formation known as the Old Red Sandstone. They worked on excellent exposures located in Devonshire in southwestern England and named these rocks the Devonian System after the region. Trips to Belgium and Germany confirmed that rocks containing "Devonian" fossils there were also located in beds underlying rocks of the Carboniferous System.[17]

In the Jura Mountains of southern Germany and neighboring parts of

the Silurian System of Rocks," *Philosophical Magazine* 7, ser. 3 (1835): 46-52; R. I. Murchison, *The Silurian System* (London: John Murray, 1839).

[16]C. Lapworth, "On the Tripartite Classification of the Lower Paleozoic Rocks," *Geological Magazine* 6, n. s. (1879): 1-15. On Lapworth, see John Challinor, "Lapworth, Charles," in *DSB*, 8:32-34.

[17]A. Sedgwick and R. I. Murchison, "On the Older Rocks of Devonshire and Cornwall," *GSL Proceedings* 3, no. 63 (1839): 121-23; A. Sedgwick and R. I. Murchison, "On the Distribution and Classification of the Older or Paleozoic Deposits of the North of Germany and Belgium, and Their Comparison with Formations of the Same Age in the British Isles," *Transactions of the GSL* 6, ser. 2 (1842): 221-301; M. J. S. Rudwick, *The Great Devonian Controversy: The Shaping of Scientific Knowledge Among Gentlemanly Specialists* (Chicago: University of Chicago Press, 1985).

France and Switzerland, thick sequences of marine limestones, representing the remnants of old sea floor were exposed during the elevation of the Alps. These rocks, therefore, had been christened the Jurassic well back into the 1700s. Friedrich August Von Quenstedt (1809-1889), a professor of mineralogy and geognosy at the University of Tübingen and an accomplished paleontologist, devoted a considerable portion of his working career to dissecting the stratigraphy of the Juras.[18] His student Albert Oppel (1831-1865) took on the task of expanding Quenstedt's investigation to similar rocks of southern England, which Smith had studied, as well as France, Switzerland and southern Germany.[19] Between 1856 and 1858, he published his masterpiece, *Die Juraformation Englands, Frankreichs, und des Südwestlichen Deutschlands.*[20] Oppel correlated eight separate regional faunas and was able to subdivide the total Jurassic sequence into 33 subunits, termed "zones," based on fossil content. Each zone was identified by 10 to 30 common fossils; each was also named by a typical fossil. Oppel began the practice of naming Mesozoic zones by diagnostic ammonite species, a practice continued to this day. In 1861 he joined the faculty of the University of Munich as professor of paleontology, but, sadly, died of typhoid fever four years later at the age of thirty-four.

Examinations of stratigraphic sequences of fossils soon extended to continents other than Europe. For example, during the late 1830s, James Hall (1811-1898), a young paleontologist employed by the New York State Geological Survey, founded a career by studying intensively the fossiliferous Silurian and Devonian rocks of western New York State.[21] In 1841, he began to collect fossils from the American mid-continent. Extensive collecting demonstrated the regional consistency of his New York biostratigraphic system. In 1843, Hall made explicit comparisons between the Lower Paleozoic fossils of New York and those of England. Between 1855 and 1859, he served as director of the Iowa Geological Survey while still residing in Albany and working as paleontologist for the New York Survey! From 1866 until his death, he was state geologist of New York. Hall was a leader of a movement to establish a separate nomenclature for the North American geologic periods, but by the end of the nineteenth century, his work and that of his students and associates clearly demonstrated the same basic sequence to the fossil record in North America as that of Europe.[22]

---

[18]On Quenstedt, see J. G. Burke, "Quenstedt, Friedrich," in *DSB*, 11:235-36.

[19]On Oppel, see H. Tobien, "Oppel, Albert," in *DSB*, 10:211-12.

[20]A. Oppel, *Die Juraformations Englands, Frankreichs, und des Südwestlichen Deutschlands* (Stuttgart: Ebner and Seubert, 1856-1858).

[21]On James Hall, see John J. Stevenson, "Memoir of James Hall," *Bulletin GSA* 61 (1895): 425-51; and Donald W. Fisher, "Hall, James, Jr.," in *DSB*, 6:56-58.

[22]James Hall, *Geology of New York, Part IV, Comprising a Survey of the Fourth District* (Albany:

By the end of the nineteenth century, the overall nomenclature for the geological time periods from the Cambrian to the Recent was established. The Cambrian through Permian Periods were in turn subsumed under the Paleozoic Era; the Triassic, Jurassic and Cretaceous Periods under the Mesozoic Era; and the Tertiary and Quaternary Periods under the Cenozoic Era. Much later, during the twentieth century, geochronologists would place these three eras under the umbrella of the Phanerozoic Eon (that is, the "time of visible life"). Precambrian time would be divided into the Hadean, Archean and Proterozoic Eons (table 4.1). Scientific attempts to place dates on this timetable are described later in this chapter.

Table 4.1. The Geological Timescale

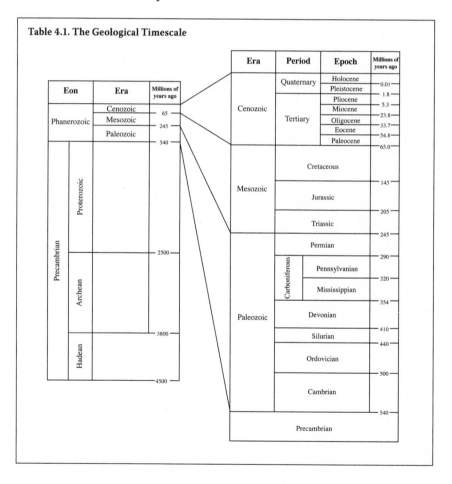

Carroll and Cook, 1843); R. H. Dott Jr., "James Hall's Discovery of the Craton," in *Geologists and Ideas: A History of North American Geology*, Centennial Special Vol. 1, ed. E. T. Drake and W. M. Jordan (Boulder, Colo.: GSA, 1985), pp. 157-67.

ORGANIC EVOLUTION, CATASTROPHISM AND THE EARLY STRATIGRAPHERS

It is extremely important to recognize that hundreds of competent field geologists established the validity and practical utility of the law of faunal succession *prior* to the publication of Darwin's and Wallace's theory of organic evolution. During the early nineteenth century, pre-Darwinian speculations on trends in the history of life were certainly promoted by some authors. The founders of the discipline of biostratigraphy, whether Christian or not, were distinctly nonevolutionists. A fair number of mid-nineteenth-century paleontologists, following Cuvier, viewed the fossil record as containing evidence of radical turnovers of whole faunas. New faunas were then installed by the Creator.

William Smith apparently took no interest in discerning the ages of the rocks he examined. He took at face value the biblical creation and Deluge stories. For example, he thought that the inclined strata of southern England had been created in situ in their present orientation. Smith could have profited by a reading of Steno, who had published his principles a century before Smith's birth!

Cuvier eschewed any idea of organic evolution. As a comparative anatomist, he knew very well the tight connection between the demands of an organism's lifeways and its physical structure. He found the idea of structural change incomprehensible on the grounds that the transformation from one basic body plan to a second would require an intermediate structure that could not serve the functional demands of either of the two lifeways.[23]

Cuvier argued his case vigorously with his fellow Paris Muséum scientists.[24] In the first decade of the nineteenth century, he contended with the Muséum's curator of invertebrate animals, Jean-Baptiste Lamarck (1744-1829).[25] Lamarck had long favored a theory of gradual transformation of organisms through time. They agreed to a test: they opened the bandages of several mummified animals plundered from Egyptian tombs by Napoleon's invading army. The mummies exhibited no significant differences from their modern counterparts. To Cuvier's mind, this test provided clear evidence for the absence of biological change over extended time. During the 1820s, Cuvier argued the concept of biological change with his colleague Geoffroy Saint-Hilaire (1772-1844), curator of vertebrate animals at the Muséum.[26] Their disagreement resulted

---

[23]E. S. Russell, *Form and Function: A Contribution to the History of Animal Morphology* (London: John Murray, n.d.; reprinted by University of Chicago Press, 1982).

[24]Martin J. S. Rudwick, *The Meaning of Fossils: Episodes in the History of Palaeontology*, 2nd ed. (Chicago: University of Chicago Press, 1976); T. A. Appel, *The Cuvier-Geoffroy Debate: French Biology in the Decades Before Darwin* (New York: Oxford University Press, 1987).

[25]On Lamarck, see L. J. Jordanova, *Lamarck* (Oxford: Oxford University Press, 1984).

[26]On Geoffroy Saint-Hilaire, see F. Bourdier, "Geoffroy Saint-Hilaire, Etienne," in *DSB*,

in a prolonged debate before the French Academy of Sciences, extending over several sessions during the spring of 1830.[27] Cuvier evidently felt that a catastrophic view of the history of life provided a framework for fruitful research programs. After all, the succession of rock layers above the Chalk in the Paris Basin provided clear evidence that this history of the Earth must be punctuated by rapid inundations and emergences of crustal provinces. Cuvier could sense that the history he was seeing must have taken much time, thousands of centuries for the Paris Basin strata alone.

Louis Agassiz viewed himself, with some justification, as Cuvier's scientific successor. Agassiz consciously carried on the catastrophic Cuvierian tradition in his fossil researches. He was the leading American antagonist to the Darwinian theory of organic evolution after publication of *The Origin of Species* in 1859. Great numbers of aspiring American naturalists came to study under Agassiz at Harvard, where they were inculcated in catastrophist doctrines. Agassiz clearly saw a progression of life forms in the strata, but viewed this progression as the unfolding of a divine plan.

Alcide d'Orbigny (1802-1857) was perhaps the most accomplished French paleontologist of the generation following Cuvier.[28] He achieved early fame as an explorer of South America (1826-1834), sponsored by the Paris Muséum d'Histoire Naturelle. In 1853, a special government decree established a chair of paleontology for him at the Muséum. He was a good friend and mentor to Albert Oppel during the latter's researches on Jurassic faunas. D'Orbigny worked long and hard detailing the subdivisions of the Cretaceous System in France. In his texts, *Paléontologie Française,* published in several volumes between 1840 and 1846, and *Cours Élémentaire de Paléontologie et de Géologie Stratigraphiques,* published between 1849 and 1852, he subdivided the Cretaceous System into "stages," based on the notion that a typical fauna would be annihilated and replaced by a new fauna.[29] His global ("universal") history of life, outlined in *Prodrome de Paléontologique Stratigraphique Universelle* was likewise catastrophic, divided into 27 stages.[30]

Adam Sedgwick, one of the best field stratigraphers of the nineteenth century, believed that the evidence in the rocks testified to widespread faunal

5:355-58.

[27] Appel, *The Cuvier-Geoffroy Debate.*

[28] On D'Orbigny, see H. Tobien, "Orbigny, Alcide Charles Victor Dessalines D'," in *DSB,* 10:221-22.

[29] A. D. D'Orbigny, *Cours Élémentaire de Paléontologie et de Géologie Stratigraphiques,* 3 vols. (Paris: Victor Mason, 1849-1852).

[30] A. D. D'Orbigny, *Prodrome de Paléontologique Stratigraphique Universelle,* 3 vols. (Paris: 1850-1852).

overturns under Providential guidance.[31] He critiqued his friend Lyell for not allowing past causes of greater magnitude than that of today. While remaining friendly to his former student Charles Darwin, he vigorously opposed Darwin's

Figure 4.3. John Phillips (1800-1874)
© Geological Society/NHMPL.
Used by permission.

notion of gradualistic transformations of life forms. Sedgwick coined the term *Paleozoic* to encompass the life forms (and hence the time interval) of the Cambrian through Permian periods, clearly recognizing the profound magnitude of the end-Permian extinction event.[32]

John Phillips (1800-1874), a committed Anglican, was another accomplished biostratigrapher who opposed the idea of gradualistic biological evolution (fig. 4.3).[33] Phillips was orphaned at an early age and raised by his uncle, none other than William "Strata" Smith. Phillips accompanied Smith on his geological excursions beginning in 1816. Through connection with his uncle, he attained a position

as curator of the museum of the Yorkshire Philosophical Society. From 1834 until 1840, he concurrently held the chair of geology at King's College, London. In 1844, he was appointed a professor of geology at the University of Dublin. In 1853, he was chosen as William Buckland's successor as reader of geology at Oxford. Throughout these decades, he undertook extensive studies of the fossils of Cornwall, Devon and West Somerset. In 1860, Phillips delivered the Rede Lectures at Cambridge, later published as *Life on the Earth: Its Origin and Succession*.[34] Phillips defended the catastrophist view that faunas appeared and

---

[31]A. Sedgwick, "Vestiges of the Natural History of Creation," *Edinburgh Review* 82 (1845): 1-85; Rudwick, *The Meaning of Fossils*; Michael Ruse, *The Darwinian Revolution: Science Red in Tooth and Claw* (Chicago: University of Chicago Press, 1979); James A. Secord, *Victorian Sensation* (Chicago: University of Chicago Press, 2000).

[32]Adam Sedgwick, "A Synopsis of the English Series of Stratified Rocks Inferior to the Old Red Sandstone," *Proceedings of the GSL* 2 (1838): 675-87.

[33]On Phillips, see Jack Morrell, *John Phillips and the Business of Victorian Science* (Burlington, Vt.: Ashgate, 2005); and Anonymous, "Eminent Living Geologists: John Phillips," *The Geological Magazine* 73 (1870): 301-6. See also J. Phillips, *Figures and Descriptions of the Palaeozoic Fossils of Cornwall, Devon and East Somerset* (London: Longman, Brown, Green and Longmans, 1841).

[34]J. Phillips, *Life on the Earth: Its Origin and Succession* (Cambridge: Cambridge University Press,

disappeared from the Earth, requiring periodic creative activity.

Following the introduction of the term *Paleozoic* by Sedgwick, Phillips proposed the terms *Mesozoic* and *Kainozoic* (later changing the spelling of Kainozoic to Cainozoic) for geologic eras.[35] The boundaries of these geologic eras are marked by massive faunal extinctions, plain to the early biostratigraphers. Contemporary biostratigraphers, faced with much more data than were available to their nineteenth-century counterparts, have returned to themes of the significance of mass extinctions in the fossil record.[36]

BIOSTRATIGRAPHY: THE STATE OF THE ART IN 1860

Between 1790 and 1860, a small army of stratigraphers endured long hours in the field amassing, naming and curating thousands of specimens of past organisms. They learned that there is a vertical pattern to the distribution of these organisms. Trilobites and rugose and tabulate corals are restricted to Paleozoic strata. Placoderm fishes are located only in Devonian and lowermost Carboniferous strata. Scleractinean corals, sand dollars and neogastropods are restricted to post-Paleozoic rocks. Ammonites, common faunal constituents of late Paleozoic and Mesozoic strata, were not discovered in post-Mesozoic layers. Plesiosaurs and mosasaurs were found in Mesozoic marine sedimentary rocks but not Paleozoic or Cenozoic sedimentary rocks. Fossil elephants, ground sloths, whales and salmons are restricted to Cenozoic deposits. The vertical pattern was found to be applicable to continents other than Europe, where it had been first delineated.

The founders of the discipline of biostratigraphy were indifferent or even hostile to the concept of organic evolution.[37] Their immediate successors believed that the normal course of nature was interrupted by major or minor catastrophes and that whole biotas were deleted or added to Earth's biological inventory. Many of these scientists were also devout Christians who invoked the Christian God as the purposive populator of the archaic Earth through distinct periods.

By the end of the nineteenth century, the geologic periods had been

---

1860); J. Secord, *Victorian Sensation* (Chicago: University of Chicago Press, 2000).

[35] Phillips, *Figures and Descriptions of the Paleozoic Fossils.*

[36] Norman Newell, "Paleontological Gaps and Geochronology," *Journal of Paleontology* 36 (1962): 592-610; D. V. Ager, *The Nature of the Stratigraphic Record,* 2nd ed. (New York: John Wiley, 1981); S. J. Gould, "The Paradox of the First Tier: An Agenda for Paleobiology," *Paleobiology* 11 (1985): 2-12; D. M. Raup, "Biological Extinction in Earth History," *Science* 231 (1986): 1528-33; D. Erwin, *Extinction: How Life on Earth Nearly Ended 250 Million Years Ago* (Princeton, N.J.: Princeton University Press, 2006).

[37] This historical fact is contrary to the claim made by many twentieth-century Flood geologists to the effect that the biostratigraphic record was created to support some doctrine of organic evolution. This issue is discussed in chap. 8.

sorted out and named. All those stratigraphers who had spent time in the field examining the layered rocks recognized that significant time had elapsed during their formation. The many Christians who took part in the enterprise, like Hugh Miller, evidently believed that an Earth of great antiquity proclaimed God's glory with greater intensity than did a recently created globe.

SCIENTIFIC ATTEMPTS TO DETERMINE EARTH'S AGE

Although throughout the nineteenth century geologists became convinced of the idea of Earth's greatly expanded antiquity, reliable means for figuring out how old it might be were not developed until the latter half. As the geologic timescale took shape, some discussions of the age of the globe included estimates of the *relative* lengths of the various eras and periods. Most geologists reckoned that Earth was probably millions of years old, but imprecise expressions like "inconceivably vast" were commonplace. Throughout the second half of the century, most age calculations were based on estimated rates of erosion and/or sedimentation, estimated rates of salt accumulation in the ocean or geophysical grounds. In the first method, ages were calculated from determinations of total thicknesses of sedimentary rock deposits and estimates of rates of sedimentation and/or erosion. Estimates of the age of the ocean were typically based on measured amounts of dissolved sodium in seawater, measured rates of transport of sodium to the ocean in rivers, estimates of rates of removal of sodium from the ocean and evaluation of other factors that might affect changes in rates of such sources and sinks. Geophysical arguments commonly entailed calculations of rates of heat loss from a cooling Earth or Sun, and the effect of tides on the Earth-Moon system.

The youthful Charles Darwin (1809-1882) undertook important geological investigations in South America and on several oceanic island groups during the voyage of the *H.M.S. Beagle* between the end of 1831 and 1836.[38] He gained recognition as a first-rate geologist long before he became known as the promoter of the theory of evolution by means of natural selection.[39] Darwin's geological experiences impressed upon him the very slow rate at which erosion occurs and induced him to ponder the question of the antiquity of the globe. In *The Origin of Species,* Darwin devoted two full chapters to pertinent

---

[38]For biographical detail on Darwin, see Janet Browne, *Charles Darwin: Voyaging* (New York: Alfred A. Knopf, 1995), and *Charles Darwin: The Power of Place* (Alfred A. Knopf, 2002); and also Adrian J. Desmond and James R. Moore, *Darwin* (New York: Warner Books, 1992).

[39]For example, Darwin wrote on volcanic islands, the geology of South America, and coral islands. Darwin was the first to propose the idea of crystal settling as a mechanism for producing diversity of igneous rock types, on which see Davis A. Young, *Mind over Magma: The Story of Igneous Petrology* (Princeton, N.J.: Princeton University Press, 2003), pp. 127-28.

geologic and paleontologic issues. The question of antiquity was of the utmost importance because, in his judgment, the vast diversity of life forms preserved in the fossil record would be difficult to account for by natural selection unless ample time had been available. Darwin thought that the great thickness of Paleozoic, Secondary and Tertiary strata, which he estimated at a composite 73,000 feet on the authority of geologist Andrew C. Ramsay (1814-1891) of the British Geological Survey, required huge amounts of time to accumulate in light of observed rates of deposition. But it was the slowness of *erosion* that impressed Darwin. Focusing his attention on the eroded, relatively youthful strata that are superimposed on Paleozoic and Secondary strata in the Weald south of London, Darwin, assuming recession of a 500-foot-high cliff at the rate of one inch per century, estimated that a little more than 300 million years had elapsed during the stripping away of the relatively young rocks once covering the Weald.[40]

John Phillips had pondered the question of the Earth's antiquity since the beginning of his career. As early as 1834, he measured the increase of temperature as a function of depth within a large coal mine and concluded that Earth had undergone long-term cooling. He suspected that determination of the rate at which the globe had cooled might help to determine Earth's age. Phillips was also persuaded that an alternative approach would involve measuring thicknesses of sedimentary rock successions as well as the average rate at which sediments are now deposited. Although he had been pondering possible modes of age determination, publication of Darwin's book stirred Phillips into action.[41] His own work convinced him that "the record of marine deposits was reasonably complete and showed no evidence of evolution having occurred."[42] In his 1860 address as President of the Geological Society of London, Phillips accused Darwin of abusing arithmetic in his Weald calculation, charging that his chosen erosion rate of one inch per century was far too low in light of the eighty inches per year he had measured along the Yorkshire coast.[43] In a lecture at Cambridge published later that year, he assumed a general rate of erosion of one inch per year and a total sedimentary rock thickness of 72,000 feet, from which he estimated an age of 96 million years.[44] Thinking that his

---

[40]On erosion of the Weald, see Charles Darwin, *On the Origin of Species by Means of Natural Selection or the Preservation of Favored Races in the Struggle for Life* (New York: Avenel Books, 1979), p. 297.

[41]Jack Morrell, "Genesis and Geochronology," *The Age of the Earth: From 4004 B.C. to A.D. 2002*, GSL Special Publication 190, ed. Cherry L. E. Lewis and Simon J. Knell (London: The Geological Society, 2001), pp. 85-90.

[42]Ibid., p. 88.

[43]John Phillips, "Presidential Address," *Quarterly Journal GSL* 16 (1860): xxvii-lv.

[44]Phillips, *Life on the Earth.*

erosion rate estimate might be even faster, he set a lower limit for Earth's age at 38 million years.

Between 1869 and 1909, other estimates of Earth's age, computed on the basis of erosion and sedimentation rates, were made by biologists Thomas Henry Huxley (1825-1895) and Alfred Russel Wallace (1823-1913), geophysicists James Croll (1821-1890) and Thomas Mellard Reade (1832-1909), and geologists Charles Lyell, Samuel Haughton (1821-1897), Alexander Winchell (1824-1891), Sir Archibald Geikie (1835-1924), Wilber John McGee (1853-1912), Warren Upham (1850-1934), Charles Doolittle Walcott (1850-1927), William J. Sollas (1849-1936), and Jakob Johannes Sederholm (1863-1934).[45] Their calculations yielded wildly disparate results ranging from 3 million (Winchell) to 15 billion years (McGee), although the majority of estimates fell in the range of 70 to 100 million years. It is small wonder that such imprecision characterized the results of these calculations in light of the fact that estimates of the maximum thickness of sedimentary rocks ranged from as little as 72,000 feet to as much as 335,800 feet, whereas guesses at the average rate of sediment accumulation varied from as low as 100 years to as many as 8,616 years required to deposit one foot of sediment. Given the difficulties in making accurate estimates, particularly of rates of sediment deposition, most geologists regarded ages based on sedimentary rock thickness with only a moderate level of confidence.

The problem was compounded because estimates based on sedimentary rock thicknesses typically took into account only those rocks deposited since the onset of the Cambrian Period. Because early Precambrian sedimentary rocks are uncommon, geologists experienced great difficulty estimating the amount of time that had elapsed before the deposition of the fossiliferous sedimentary rocks. Although skeptical of treating any one number as highly accurate, geologists widely accepted the idea that the Earth is on the order of a few tens of millions of years old.

Geologists eagerly pursued other methods in hopes of arriving at a reliable age for planet Earth. One method entailed measurement of the rate at which dissolved salts released into surface waters during chemical weathering of minerals and rocks are transported into the oceans. In the simplest terms, the age of the ocean could be calculated if the total amount of a given chemical element, particularly sodium, in the ocean and the rate at which the chemical element in question is added to the ocean by rivers were known. This general method for calculating Earth's age was first proposed in 1715 by Edmund Halley, but he believed that it was not yet possible to make the calculation unless

---

[45]For further details on these estimates of Earth's age, see Patrick N. Wyse Jackson, *The Chronologers' Quest: The Search for the Age of the Earth* (Cambridge: Cambridge University Press, 2006).

values of the saltiness of the ocean could be measured a long time in the future or be found in documents from ancient Greece and compared with then current values.[46]

In the nineteenth century, geologists better understood the complexities of the method and made various corrections for such matters as recycling of salts in sea spray, deposition of salts in evaporite layers, and variations in the amounts of salts weathered from land areas over time. In 1876, Reade estimated an age of 25 million years on the basis of sulfate weathering and 200 million years on the basis of chloride weathering.[47] The premier exponent of the salt accumulation method, John Joly (1857-1933), a professor of geology and mineralogy at Trinity College in Dublin, Ireland, estimated in 1899 the age of the ocean to be 80 to 90 million years. A year later, he raised his estimate to 100 million years.[48] Joly maintained that sodium was the most reliable chemical element on which to base such calculations.

Other scientists attempted geophysical approaches to figuring out the age of the Earth. Chief among these was William Thomson (1824-1907), otherwise known as Lord Kelvin, arguably the greatest scientist of the nineteenth century (fig. 4.4).[49] Thomson was born in Belfast and moved to Glasgow, Scotland, at the age of six when his father, James Thomson, received an appointment as a professor of mathematics at the University of Glasgow. Young Thomson was a child prodigy, entering the university at the age of ten, the youngest on record. He also studied at Cambridge and Paris, and, in 1846 at the age of twenty-two, became a professor of mathematics and natural philosophy at the University

---

[46]Edmund Halley, "On the Cause of the Saltness of the Ocean, and of the Several Lakes that Emit No Rivers; with a Proposal, by Means Thereof, to Discover the Age of the World," *Philosophical Transactions of the Royal Society of London* 29 (1715): 296-300.

[47]T. Mellard Reade, "President's Address," *Proceedings of the Liverpool Geological Society* 3 (1876): 211-35.

[48]On Joly, see H. H. Dixon, "John Joly 1857-1933," *Obituary Notices of the Royal Society* 3 (1934): 259-86; J. R. Nudds, "The Life and Work of John Joly (1857-1933)," *Irish Journal of Earth Sciences* 8 (1986): 81-94; and Patrick N. Wyse Jackson, "John Joly (1857-1933) and his Determinations of the Age of the Earth," in *The Age of the Earth: From 4004 B.C. to A.D. 2002*, GSL Special Publication 190, ed. Cherry L. E. Lewis and Simon J. Knell (London: The Geological Society, 2001), pp. 107-19. On Joly's estimates of the age of the Earth, see John Joly, "An Estimate of the Geological Age of the Earth," *Scientific Transactions of the Royal Dublin Society* 7 (1899): 23-66; and John Joly, "Geological Age of the Earth," *Geological Magazine* 7, new ser., decade 4 (1900): 220-25.

[49]On Kelvin, see Joe D. Burchfield, *Lord Kelvin and the Age of the Earth* (New York: Science Publications, 1975); Harold I Sharlin and Tiby Sharlin, *Lord Kelvin, the Dynamic Victorian* (University Park: Pennsylvania State University Press, 1979); and D. Lindley, *Degrees Kelvin* (Washington, D.C.: Joseph Henry Press, 2004). Another important work that deals with Kelvin is Stephen G. Brush, *A History of Modern Planetary Physics*, vol. 2: *Transmuted Past: The Age of the Earth and Evolution of the Elements from Lyell to Patterson* (Cambridge: Cambridge University Press, 1996).

of Glasgow, a post he held for the remainder of his life. Throughout his career, Thomson was particularly interested in thermodynamics, the science that investigates the relationship between heat and energy. He defined the absolute temperature scale, established that heat is produced by molecular motion rather than being a fluid, developed the concept of kinetic energy, and produced one formulation of the second law of thermodynamics. In addition, Thomson invented a number of instruments and was heavily involved in the project to lay telegraph cables across the floor of the Atlantic Ocean.

Figure 4.4. William Thomson, Baron Kelvin (1824-1907). Reproduced by permission of National Portrait Gallery, London.

Thomson took his cue from the earlier work of John Phillips in evaluating the age of the Earth on the basis of cooling history. In 1862, Thomson published the first in a series of major papers on the age of the Earth.[50] In this initial paper, he evaluated the thermal history of the Sun. He maintained that the Sun originated from "a coalition of smaller bodies that fell together by mutual gravitation" thereby generating heat. Given a lack of known mechanisms that could compensate for the loss of solar energy by radiation throughout its history, Thomson concluded that the Sun must be cooling. From estimates of the temperature and other thermal properties of the Sun based on its known chemical composition, from measurements of the rate at which heat is radiated from its surface and from estimates of the total amount of the Sun's original heat produced by coalescence of smaller bodies, Thomson calculated that the Sun had not illuminated the earth for 100,000,000 years, and certainly less than 500,000,000 years. Perceptively, Thomson noted that these estimates might be revised if unknown heat sources in the Sun were eventually discovered.

In the same year, Thomson also published a paper on terrestrial cooling after analyzing the distribution of measured temperatures at various depths within Earth's crust.[51] In this paper, he took issue with geologists of the "extreme qui-

---

[50]William Thomson, "On the Age of the Sun's Heat," *Macmillan's Magazine*, March 1862, pp. 388-93.

[51]William Thomson, "On the Secular Cooling of the Earth," *Royal Society of Edinburgh Transactions* 23 (1862): 157-59.

etist, or 'uniformitarian' school," charging that they had overlooked the essential principles of thermodynamics. He considered the geological speculations of those who adopted "paroxysmal hypotheses" involving "greater extremes of heat, more violent storms and floods, more luxuriant vegetation, and hardier and coarser grained plants and animals, in remote antiquity" to be more probable than those of the uniformitarians on the grounds that the Sun had been hotter in the past than it is now. From estimates of the temperature at which rocks melt, limited measurements on thermal conductivity of rocks, and estimates of how thermal properties of rock might change with increased pressure and temperature in the Earth's interior, Thomson calculated from Fourier's heat conduction equation how long it would take for Earth to cool to its present temperature. He concluded that between 20 and 400 million years were required to achieve complete consolidation *(consistentior status)* of the Earth.

In recognition of his brilliant accomplishments, Thomson was knighted in 1866 and was named Lord Kelvin by Queen Victoria. Not one to rest on his laurels, Kelvin returned to the problem of Earth's antiquity in 1871, this time investigating the relation of tidal friction to cooling of the globe. All geological history, he said, "must be limited within some such period of past time as one hundred million years."[52] He again took issue with uniformitarian geology by criticizing John Playfair's Huttonian suggestion of an unending Sun and Earth. If Playfair were correct, he charged, the Sun would, in effect, be a perpetual motion machine in violation of the principles of thermodynamics.

Clarence King (1842-1901), first director of the United States Geological Survey, published an article in 1893 in which he calculated an age of 24 million years on the basis of the tidal stability of the Earth.[53] In 1895, Kelvin reviewed his own calculations and concluded that "I am not led to differ much from [King's] estimate of 24 million years."[54] Two years later, Kelvin gave an address on the topic of the age of the Earth as an abode fitted for life.[55] He reviewed several prior estimates and again challenged the uniformitarian ideas of Lyell and Playfair. He concluded that "we have now good reason for judging that it was more than 20 and less than 40 million years ago; and probably much nearer 20 than 40." He iterated his general agreement with King's calculation

[52]William Thomson, "On Geological Time," *Transactions of the Geological Society of Glasgow* 3 (1871): 1-28. See also an earlier paper, William Thomson, "The 'Doctrine of Uniformity' in Geology Briefly Refuted," *Proceedings of the Royal Society of Edinburgh* 5 (1865): 512-13.

[53]Clarence King, "The Age of the Earth," *American Journal of Science* 45, ser. 3 (1893): 1-20. On King, see T. Wilkins, *Clarence King: A Biography* (Albuquerque: University of New Mexico Press, 1988).

[54]Lord Kelvin, "The Age of the Earth," *Nature* 51 (1895): 438-40.

[55]Lord Kelvin, "The Age of the Earth as an Abode Fitted for Life," *Science* 9 (1899): 665-74, 704-11.

and summed up with the following paragraph:

> Whatever may have been the true history of our atmosphere it seems certain that if sunlight was ready the earth was ready, both for vegetable and animal life, if not within a century, at all events within a few hundred centuries, after the rocky consolidation of its surface. But was the sun ready? The well-founded dynamical theory of the sun's heat carefully worked out and discussed by Helmholtz, Newcomb, and myself, says NO if the consolidation of the earth took place as long as 50 million years; the solid earth must in that case have waited 20 or 50 million years for the sun to be anything nearly as warm as he is at present. If the consolidation of the earth was finished 20 or 25 million years ago the sun was probably ready, though probably not then quite so warm as at present yet warm enough to support some kind of vegetable and animal life on the earth.[56]

Kelvin lived just long enough to see his elaborate calculations and persuasive physical reasoning undermined by the finding that the newly discovered phenomenon of radioactivity provides a source of significant amounts of heat within the Earth and the Sun (chap. 5).

Charles Walcott, Director of the U.S. Geological Survey, summed up much of the thinking about the age of the Earth at the end of the nineteenth century by stating that "geologic time is of great but not infinite duration. I believe that it can be measured by tens of millions, but not by single millions or hundred of millions, of years."[57] Kelvin certainly would have agreed.

REACTIONS TO GEOCHRONOLOGICAL DEVELOPMENTS IN THE NINETEENTH CENTURY

Despite the inexorable accumulation of evidence supporting the idea that the Earth is far in excess of 6,000 years old, there were predictable antagonistic responses to the developing trends. In the early nineteenth century, a torrent of books and pamphlets designed to uphold the traditional point of view on the age and history of the world, including a global Deluge, were published.[58]

---

[56]Ibid., p. 711.

[57]C. D. Walcott, "Geologic Time as Indicated by the Sedimentary Rocks of North America," *Journal of Geology* 1 (1893): 676.

[58]Among the writings of the scriptural geologists were books by Sutcliffe, Penn, Bugg, Young and Fairholme. See, for example, Joseph Sutcliffe, *Short Introduction to the Study of Geology, Comprising a New Theory of the Elevation of the Mountains and the Stratification of the Earth, in Which the Mosaic Account of the Creation and the Deluge Is Vindicated* (London: 1817); Granville Penn, *A Comparative Estimate of the Mineral and Mosaical Geologies* (London: Ogle, Duncan, 1822); George Bugg, *Scriptural Geology* (London: 1826-1827); George Young and J. Bird, *Geological Survey of the Yorkshire Coast, Describing the Strata and Fossils Occurring between the Humber and the Tees, from the German Ocean to the Plain of York* (Whitby: 1828); and George Fairholme, *New and Conclusive Physical Demonstrations Both of the Fact*

The "heretical" and "infidel" tendencies of modern geology were roundly condemned by some churchmen, few of whom had any knowledge of geology, although there were a handful of individuals who had produced acceptable field-based studies of regional geology in Great Britain. These "Scriptural geologists," however, found themselves increasingly marginalized by the vast majority who had extensive working geological knowledge and were now convinced that the Earth is very old.

By the latter half of the century, relatively few hostile denunciations of geology or published defenses of Flood geology appeared. There were exceptions, including Carl Friedrich Keil (1807-1888), a renowned Lutheran commentator and a professor of exegesis and oriental languages at the University of Dorpat, who continued to espouse a six-day creation, the restriction of fossil formation to the period after Adam's Fall, and Flood geology.[59] Also significant were the writings in the 1850s of Eleazar Lord (1788-1871) and his younger brother, Daniel Nevins Lord (1792-1880), the lay sons of Nathan Lord, a Congregationalist minister and president of Dartmouth College.[60] The Lords believed that capitulation to modern geology with its call for long ages undermined biblical authority and the Christian faith. Also significant were the teachings of Ellen Gould White (1827-1915), the founder of Seventh-day Adventism, who claimed to have visions from God about the creation of the world in six literal days as well as of a global Deluge that buried all life and produced the fossils.[61] White's teachings and attitudes toward geology profoundly shaped not only Seventh-day Adventist thought but also the twentieth-century young-Earth creationist movement that is introduced in chapter five.[62]

In the opposite direction, a tide of rationalistic higher critical theories concerning the origin of the biblical text continued to rise, particularly in Germany. Coupled with the rise of the views espoused by scholars like Karl Heinrich Graf (1815-1869) and Julius Wellhausen (1844-1918) were a loss of confidence in Scripture as a divinely inspired book and departure from the doctrines of orthodox Christianity. Rationalistic higher critical scholars generally accepted the discoveries of natural science but normally paid little heed to seeking har-

---

and Period of the Mosaic Deluge (London: James Ridgway, 1837).

[59]C. F. Keil, *The Pentateuch*, vol. 1, 3rd ed. (Leipzig: Dorffling und Franke, 1878); and C. F. Keil, "Die Biblische Schöpfungsgeschichte und die Geologischen Erdbildungtheorien," *Theologische Zeitschrift Dieckhoff und Kliefoth* (1860): 479.

[60]See especially Eleazar Lord, *The Epoch of Creation: The Scripture Doctrine Contrasted with the Geological Theory* (New York: Charles Scribner, 1851); and David N. Lord, *Geognosy; or, The Facts and Principles of Geology against Theories*, 2nd ed. (New York: Franklin Knight, 1857).

[61]Ellen Gould White, *Spiritual Gifts: Important Facts of Faith, in Connection with the History of Holy Men of Old* (Battle Creek, Mich.: Seventh-day Adventist Publishing Association, 1864).

[62]On Ellen Gould White, see Ronald L. Numbers, *The Creationists: The Evolution of Scientific Creationism* (New York: Alfred A. Knopf, 1992), pp. 73-74.

monization between the biblical text and the geological and astronomical evidence because they tended to view early Genesis only as a human document that expressed Israel's concept of beginnings in mythological terms.

A growing number of orthodox evangelical Christian writers, including geologists, preachers, biblical scholars and theologians, accepted and accommodated their thinking to the mounting evidence for terrestrial antiquity. In response, they began to develop a variety of strategies purporting to show how the biblical data are consistent with the findings of geology.[63] Having been encouraged to look afresh at the biblical creation accounts, experts in the original languages became persuaded that there is no conflict between the data of nature and the teaching of Scripture. These individuals continued to insist on the inspiration of the Bible and refused to call Genesis a myth in order to explain difficulties. It was, however, accepted that the traditional exegesis of Genesis 1 was not the only one that adequately satisfied the biblical data.

In particular, several geologists in the nineteenth century were outspoken Christians who were concerned to uphold Scripture. They had no intention of denying the Christian faith. The nineteenth century thus became an age of harmonization, a time when orthodox evangelical theologians and scientists generally adopted either the restitution or the day-age interpretation of Genesis 1.[64]

CONCORDISM: THE RISE OF THE RESTITUTION INTERPRETATION

As we saw in chapters one and two, several writers proposed that the first two verses of Genesis 1 *preceded* the work of the six days.[65] Some said that Scripture did not give any clues as to how much time was involved before the six days began. Others, especially in the seventeenth century, stated that some period of time, during which a state of disorder or chaos existed, had elapsed before the work of the six days got underway. A few, like Episcopius, entertained the idea that a period of time elapsed between the first two verses to accommodate the fall of the angels. However, given that virtually everyone in the Western world until well into the eighteenth century still believed in a cosmos that was only a few thousands of years old, almost no one was prepared to suggest that the work described in the first two verses of Genesis 1 lasted tens of thousands of years or even more prior to the work of the six days.

---

[63]On some harmonistic efforts, see Davis A. Young, "Scripture in the Hands of Geologists, Part II," *Westminster Theological Journal* 49 (1987): 257-304.

[64]For a survey of many of the theories of harmonization, see Bernard J. Ramm, *The Christian View of Science and Scripture* (Grand Rapids: Eerdmans, 1954). For American attempts to integrate geology and the Bible, see Herbert Hovenkamp, *Science and Religion in America* (Philadelphia: University of Pennsylvania Press, 1978).

[65]Michael B. Roberts, "Geology and Genesis Unearthed." *The Churchman* 112 (1998): 225-55.

Among the more adventuresome views was that of Simon Patrick (1626-1707), who wrote:

> How long all things continued in mere confusion after the chaos was created, before light was extracted from it, we are not told. It might have been, for any thing that is here revealed, a great while; and all that time the mighty Spirit was making such motions in it, as prepared, disposed, and ripened every part of it for such productions as were to appear successively in such spaces of time as are here afterwards mentioned by Moses.[66]

Patrick also suggested, like Whiston, that the sixth day could have required a considerable amount of time in order to accomplish everything that is recorded as having occurred on that day. These ideas, however, had no impact on theology in terms of calling into question the idea that the Earth is only about 6,000 years old.

Ideas along these lines developed further throughout the eighteenth century. Johann August Dathe (1731-1791), a professor of Oriental literature at the University of Leipzig, believed that the first two verses preceded the six days. Dathe, however, also proposed that Genesis 1:2 indicates that the Earth had undergone a remarkable change, and he translated the verse to read, in effect, "Afterward, the earth became a waste and desolation."[67]

The growing accumulation of evidence for an ancient Earth eventually led to a modification of these ideas about Genesis 1:1-2 in relation to the six days through the notion that the long periods of time required by geology could adequately be accounted for by assuming that an implied gap between the first two verses of Genesis 1 accommodated a condition that lasted an indeterminate amount of time and that preceded the six days of creation. Allegedly, some sort of devastation befell the Earth at the conclusion of the indeterminate period, rendering the planet "waste and void" as recorded in Genesis 1:2. What was new was the idea that the six days pertained not to the original creation but to a restitution, reconstruction, renovation, restoration or re-creation of the Earth. The days of Genesis 1 were transformed from days of creation to days of restoration, and the view came to be known as the gap, restitution or ruin-reconstruction interpretation.

Arguably the earliest advocate of the restitution theory was Thomas Chalmers (1780-1847), the acknowledged leader of the evangelical party within the Church of Scotland who ultimately led the evangelicals out of the state church

---

[66]Patrick's comments are quoted in Edward Hitchcock, *The Religion of Geology and Its Connected Sciences* (Boston: Phillips, Sampson, 1851), p. 42.

[67]J. August Dathe, *Ex Recensione Textus Hebraei et Versionum Antiquarum Latine Versi, Notisque Philologicis et Criticis Illustrati*, book 6 (Halle: 1791).

to form the Free Church of Scotland in the Disruption of 1843.[68] Raised in a home of pious Presbyterians, Chalmers attended the Universities of St. Andrews and Edinburgh where he studied mathematics under John Playfair. Although his deepest interests were in mathematics and chemistry, Chalmers pursued ordination in spite of lukewarm commitment toward ministry. His biographer noted that "to fill the mathematical chair in one of our universities was the high object of his ambition."[69] Chalmers served as an instructor of mathematics in St. Andrews and devoted less than the appropriate attention to ministerial tasks. His father wrote him "that science had the hold which he wished so much that the gospel of God's redeeming grace should have."[70] Even after his ordination to the parish of Kilmany in 1803, Chalmers continued to give lecture courses in chemistry that included references to geology, a subject that attracted his attention even before Playfair published his *Illustrations of the Huttonian Theory.* Chalmers was sufficiently alarmed by the attitude of many ministers that geology spoke in infidel tones and conflicted with the divine records in Scripture that he decided to take a stand, thus becoming

> the first clergyman in this country who, yielding to the evidence in favor of a much higher antiquity being assigned to the earth than had previously been conceived, suggested the manner in which such a scientific faith could be harmonized with the Mosaic narrative.[71]

Thus, in one of his chemistry lectures given in St. Andrews in 1804, while he was yet in his early twenties, Chalmers announced, after extolling the labors of science, that

> there is a prejudice against the speculations of the geologist which I am anxious to remove. It has been said that they nurture infidel propensities. By referring the origin of the globe to a higher antiquity than is assigned to it by the writings of Moses, it has been said that geology undermines our faith in the inspiration of the Bible, and in all the animating prospects of immortality which it unfolds. This is a false alarm. *The writings of Moses do not fix the antiquity of the globe. If they fix anything at all, it is only the antiquity of the species.*

He went on to warn the church against fanatics, false friends, and bigots who would bring "religion into contempt, by throwing over it the deformity of an illiberal and contracted superstition."[72]

The following year, Chalmers got into hot water when his presbytery ob-

---

[68]On Chalmers, see William Hanna, *Memoirs of the Life and Writings of Thomas Chalmers, D.D., LL.D.,* 3 vols. (New York: Harper & Brothers, 1851).
[69]Ibid., p. 67.
[70]Ibid., p. 77.
[71]Ibid., p. 390.
[72]Ibid., pp. 90-91 (italics original).

jected to his teaching mathematics and chemistry. Despite the rebuff, Chalmers proceeded to apply for professorial positions in natural philosophy at St. Andrews and mathematics at Edinburgh, a chair that Playfair vacated to become professor of natural philosophy at Edinburgh. Turned down by both institutions, however, Chalmers began to undergo a gradual spiritual transformation, brought about in part by the premature death of two of his siblings. By 1810, he had fully committed himself to the service of God rather than his own ambitions. Despite renewed attention to the demands of ministry, however, Chalmers never lost his interest in scientific subjects.

The publication of James Parkinson's alternative day-age harmonization stimulated Chalmers to promote his own harmonization scheme more vigorously, but having decided to commit all his energy to Christian ministry, Chalmers began to focus on the defense of Christianity. In 1813, he published an article on the evidences for Christianity in the Edinburgh Encyclopedia that was reprinted with slight modification as *The Evidences and Authority of the Christian Revelation*. The following year he also penned a review of Cuvier's *Essay on the Theory of the Earth*. In all of these writings, Chalmers developed his thinking about the relationship of geology to Christian faith. In his encyclopedia article on Christianity he posed this question:

> Does Moses ever say that there was not an interval of many ages between the first act of creation, described in the first verse of the book of Genesis, and said to have been performed at the beginning, and those more detailed operations the account of which commences at the second verse? . . . or does he ever make us to understand that the genealogies of man went any further than to fix the antiquity of the species, and, of consequence, that they left the antiquity of the globe a free subject for the speculations of philosophers?[73]

And then in his review of Cuvier, he wrote:

> Should the phenomena compel us to assign a greater antiquity to the globe than to that work of days detailed in the book of Genesis, there is still one way of saving the credit of the literal history. The first creation of the earth and the heavens may have formed no part of that work. This took place at the *beginning*, and is described in the first verse of Genesis. It is not said when the *beginning* was. We know the general impression to be that it was on the earlier part of the first day, and that the first act of creation formed part of the same day's work with the formation of light. We ask our readers to turn to that chapter, and to read the first five verses of it. Is there any forcing in the supposition that the first verse describes the primary act of creation, and leaves us at liberty to place it as far back as we may; that the first half of the second verse describes the state of the earth (which may already have existed for ages, and been the theater of

---

[73]Thomas Chalmers, "Christianity," *Edinburgh Encyclopedia* 6 (1813): 383.

geological revolutions) at the point of time anterior to the detailed operations of
this chapter, and that the motion of the Spirit of God, described in the second
clause of the second verse, was the commencement of these operations? In this
case, the creation of light may have been the great and leading event of the first
day, and Moses may be supposed to give us, not a history of the first formation
of things, but of the formation of the present system.[74]

The key point iterated by Chalmers was the now-familiar claim that the
creation of the heavens and the earth preceded the six days. The creation of
the heavens and the earth was said to have taken place in the beginning, but
when that beginning was is not recorded. Where Chalmers differed with his
predecessors was in maintaining that vast amounts of time might have elapsed
between the work of the initial creation and the onset of the six days' work and
that one could quite nicely fit all the geological activity exposed in the rock
record into a presumably vast stretch of time prior to the six days. The original world, he thought, might have been a suitable home for life, but something
had subjected the world to desolation. Chalmers maintained that the six days
were ordinary days of *recreating* the earth from the chaotic condition that had
befallen the initial creation.

In 1816, Anglican John Bird Sumner (1780-1862) published a book in which
he defended the gap theory.[75] Sumner became a bishop of Chester in 1828 and
the Archbishop of Canterbury twenty years later.

Several prominent Christian geologists gave active support to the restitution
theory. As both an Oxford geologist and an Anglican minister, Buckland was
naturally interested in harmonizing revelation with the discoveries of his science. In his Bridgewater Treatise, he defended both Scripture and geology and
adopted the restitution theory put forward by Chalmers.[76] In Buckland's opinion it was no longer possible to argue that the earth is only 6,000 years old:

> The truth is, that all observers, however various may be their speculation, re-
> specting the secondary causes by which geological phenomena have been brought
> about, are now agreed in admitting the lapse of very long periods of time to have
> been an essential condition to the production of these phenomena.[77]

He rejected other harmonizations, especially the idea that all rock strata

---

[74]Thomas Chalmers, "Review of Cuvier's Essay on the Theory of the Earth," *Christian Instructor*
9 (1814): 273.

[75]John Bird Sumner, *Treatise on the Records of Creation and the Moral Attributes of the Creator*
(London: J. Hatchard, 1816).

[76]William Buckland, *Geology and Mineralogy Considered with Reference to Natural Theology*
(London: William Pickering, 1837). Buckland's work is volume 6 of the Bridgewater Treatises
that were designed to set forth "the power, wisdom, and goodness of God as manifested in the
creation."

[77]Ibid., p. 13.

could be attributed to the Deluge or to the time between the creation of man and the onset of the Deluge.

> Some have attempted to ascribe the formation of all the stratified rocks to the effects of the Mosaic Deluge; an opinion which is irreconcilable with the enormous thickness and almost infinite subdivisions of these strata, and with the numerous and regular successions which they contain of the remains of animals and vegetables, differing more and more widely from existing species, as the strata in which we find them are placed at great depths.[78]

Although Buckland insisted that the days of creation could be taken as long periods of time, he believed that such a position did not yield the needed harmonization inasmuch as the order of geological events did not appear to agree completely with that of Moses, especially as regards the relative time of appearance of vegetation. Instead, for Buckland, the book of Genesis expresses

> an undefined period of time, which was antecedent to the last great change that affected the surface of the earth, and to the creation of its present animal and vegetable inhabitants; during which period of long series of operations and revolutions may have been going on; which, as they are wholly unconnected with the history of the human race, are passed over in silence by the sacred historian, whose only concern with them was barely to state, that the matter of the universe is not eternal and self-existent, but was originally created by the power of the Almighty.[79]

And, moreover,

> this first evening may be considered as the termination of the indefinite time which followed the primeval creation announced in the first verse, and as the commencement of the first of the six succeeding days, in which the earth was to be fitted up, and peopled in a manner fit for the reception of mankind. We have in this second verse, a distinct mention of earth and waters, as already existing, and involved in darkness; their condition also is described as a state of confusion and emptiness, *(tohu bohu),* words which are usually interpreted by the vague and indefinite Greek term, "chaos," and which may be geologically considered as designating the wreck and ruins of a former world.[80]

The restitution theory received one of its ablest treatments in the hands of John Pye Smith (1774-1851), an English divinity tutor at nonconformist Homerton College.[81] Smith, like Chalmers, held that the date of creation is not revealed in the Mosaic account; that the events of the geological record could be applied

---

[78]Ibid., pp. 16-17.
[79]Ibid., p. 19.
[80]Ibid., pp. 23-26.
[81]John Pye Smith, *On the Relation Between the Holy Scriptures and Certain Parts of Geological Science* (New York: D. Appleton, 1840). On Smith, see J. Medway, *Memoirs of the Life and Writings of John Pye Smith* (London: 1833).

to Genesis 1:1; that recently, as suggested by verse 2, the earth was brought into a state of disorganization or ruin, that is, the earth *became* waste and void or formless and empty (not *was* waste and void); and that God subsequently adjusted the surface of the earth to its presently existing condition over a period of six natural days. Smith, however, added an intriguing twist to the restitution theory by reasoning that the condition of chaos and darkness from which the earth was re-created applied only to a region between the Caucasus Mountains and the Caspian Sea south to the Persian Gulf and the Indian Ocean. Outside of this area, over most of the earth, life and sunshine continued to exist and the condition expressed by Genesis 1:2 did not apply. According to Smith, the six days refer to the restoration or re-creation of that localized area in which the first humans were placed at the end of the creation week.

Although concerned more with astronomy than with geology, another important work that sought to harmonize Scripture with scientific discovery was *The Bible and Astronomy* by J. H. Kurtz (1809-1890), a professor of church history in the University of Dorpat in what is now Lithuania.[82] The book first appeared in 1842 and enjoyed success through several editions. Like previous adherents of the restitution theory, Kurtz believed that Genesis 1:1 describes a situation existing prior to the work of the six days. The desolation and waste of verse 2 was possibly a state of ruination of a previously existing creation of God. Kurtz was intrigued by evidence in Scripture for the fall of angels, and he maintained that this rebellion coincided with the production of the earth's ruined state. The following six days were days of restitution. Thus far, he restated the classic restitution theory, but his view of the six days was unique.

Kurtz maintained that Genesis 1 is an account of events that were handed down by tradition and incorporated into the Pentateuch by Moses. The originator of the tradition did not compose it on his own, because he had neither been present at creation nor would he be able to understand the work of creation by intuitive insight. Even Adam in his state of moral and intellectual perfection before the Fall could not have penetrated the nature of the acts of creation by intuition. Instead, Kurtz maintained, the account was given to the first recipient, whether Adam, Enoch or one of the other early patriarchs, by divine revelation in a prophetic vision. God revealed to the original seer a series of successive scenarios of the creation. The days were "prophetic days."

Kurtz did not mean that it took six days for the visions to occur, but rather that the events occurred in six prophetic days. A prophetic day was not necessarily an ordinary day but a day of indeterminate length, he said, and in order to determine just how long these prophetic days were, we would need to consider how the term *day* is used in its immediate context. Kurtz was con-

---

[82]J. H. Kurtz, *The Bible and Astronomy*, 3rd ed. (Philadelphia: Lindsay and Blakiston, 1857).

vinced that these prophetic days of restitution were natural days because they were marked by recurring periods of light and dark. But again, this did not necessarily mean twenty-four hours, since other planets have regular periods of recurring light and dark that differ from the terrestrial twenty-four hours. Despite his adherence to the restitution hypothesis and his insistence on six natural days for the restitution, Kurtz adopted elements of a popular rival view, the day-age hypothesis. Aspects of his view would also reappear in the revelation-day view of the twentieth century.

Kurtz had a potent impact on the thinking of Franz Delitzsch (1813-1890), a Lutheran theologian at the Universities of Rostock, Erlangen and Leipzig, and one of the great biblical exegetes of the latter half of the nineteenth century. After corresponding with Kurtz, Delitzsch, who had previously argued against the restitution idea, eventually adopted it, including the idea of a pre-Adamic rebellion in heaven followed by a visitation of divine judgment upon the earth.[83] Also in Germany, prominent conservative Lutheran commentator Ernst Wilhelm Hengstenberg (1802-1869), a professor of theology at the University of Berlin, held to the restitution theory.

In the United States, the restitution theory was promoted by Edward Hitchcock (1793-1864), President of Amherst College.[84] Hitchcock was not only an excellent geologist who did much of the pioneering field work in New England, but he was also a professional theologian, eminently qualified to attempt the harmonization laid out in his book. Nevertheless, there are no novel aspects to Hitchcock's version of the restitution interpretation.

One of the most influential expositions of the restitution hypothesis was that of G. H. Pember, laid out in *Earth's Earliest Ages*. Pember laid great stress on the fall of Satan and the wicked angels during the primeval history of the earth prior to the six days. This fall of the spirits was connected with a great global catastrophe to which may be attributed the remains of the geologic time periods. Thus, the six days have nothing to do with the geologic record but are seen as days of restoration of the earth from the great catastrophe that occurred at the time of Satan's rebellion. Pember's work, in turn, exerted considerable influence on Cyrus Ingerson Scofield (1843-1921), a pastor of Congregationalist churches in Texas and Massachusetts and the author of the footnotes in the first edition of the dispensationalist Scofield Reference Bible first published in 1909.[85]

---

[83]Delitzsch argued against the restitution hypothesis in *New Commentary on Genesis* (Edinburgh: T & T Clark, 1888) but later adopted that view in *A System of Biblical Psychology* (Edinburgh: T & T Clark, 1899).

[84]Hitchcock, *The Religion of Geology*.

[85]George Pember, *Earth's Earliest Ages* (Old Tappan, N.J.: Fleming H. Revell, n.d.). *Scofield Reference Bible* (New York: Oxford University Press, 1909).

CONCORDISM: THE DAY-AGE INTERPRETATION

The restitution interpretation had plenty of competition from the day-age interpretation of Genesis 1, a view that originated in the harmonizing efforts of Whiston, Buffon, Deluc, Cuvier and Parkinson. The essential point of this latter interpretation is that the six days of creation should not be regarded as ordinary days that were determined by the rotation of the Earth with respect to the Sun or another light source but should be considered as time periods of indeterminate length. If these days were truly of indeterminate length, then we may legitimately equate the vast amounts of time required for the geological history of the Earth with the six days. If, however, the six days were long periods of time, then the question arises as to whether the sequence of events in Genesis 1 matches the sequence of events disclosed by geological inquiry. Advocates of the day-age hypothesis have commonly been enamored of pointing out that there is general agreement between the two sequences. The sequences proposed by different authors, however, were so varied as to suggest that the harmonizations were based less on the text than on the biases of their authors.[86]

In the nineteenth century, geologists provided the most detailed elaborations of the day-age theory. Excellent defenses of the view were presented by Benjamin Silliman (1779-1864) of Yale College in a supplement to an American edition of Robert Bakewell's *Introduction to Geology*; by James Dwight Dana (1813-1895), Silliman's son-in-law and academic successor at Yale; by Arnold Guyot (1807-1884) of the College of New Jersey, later to become Princeton University; and by Sir John William Dawson (1820-1899), the first principal of McGill University.[87] Without question these four men were among the best North American geologists of the nineteenth century. Dana is commonly considered the premier nineteenth-century American geologist, and Dawson the premier nineteenth century Canadian geologist.[88] Among prominent ad-

---

[86]Young, "Scripture in the Hands of Geologists, Part II."

[87]For Silliman's contribution, see Robert Bakewell, *An Introduction to Geology* (New Haven, Conn.: H. Howe, 1833). This work, the 2nd American edition based on the 4th London edition, was reprinted in 1978 by Arno Press, New York. As the American editor, Benjamin Silliman of Yale added a supplement in which he discussed at length "the consistency of geology with sacred history." This discussion includes pp. 389-466 of Bakewell's work. For the others, see James Dwight Dana, "Creation: or, the Biblical Cosmogony in the Light of Modern Science," *Bibliotheca Sacra* 42 (1880): 201-24; Arnold Guyot, *Creation* (New York: Scribner, 1884); and J. William Dawson, *The Origin of the World according to Revelation and Science*, 7th ed. (London: Hodder & Stoughton, 1898).

[88]On Dana, see M. L. Prendergast, *James Dwight Dana: The Life and Thought of an American Scientist*, 2 vols., Ph.D. dissertation, University of California, Los Angeles, 1978. On Dawson, see Susan Sheets-Pyenson, *John William Dawson: Faith, Hope, and Science* (Montreal: McGill-Queen's University Press, 1996).

vocates of the day-age theory in Germany were Friedrich Pfaff (1825-1886), a geologist at the University of Erlangen, and Otto Zöckler (1833-1906), a professor of theology at the University of Greifswald.[89]

Arguably the most eloquent exposition of the day-age view was that of Hugh Miller.[90] Miller, Guyot and Dawson were also influenced by Kurtz in that they regarded the first chapter of Genesis as a prophetic vision in which more phenomenal or optical language was used. They maintained that Genesis 1 does not present absolute physical truth but presents the phenomena as they might have appeared to an imagined observer. For example, the Sun and the Moon are referred to in the language of appearances and not as absolute fact, for the Sun and the Moon in terms of absolute physical truth are far outside of the expanse of the sky. Miller expressed great admiration for the many able spokesmen for the restitution view like Chalmers and Buckland but found himself unable to accept that harmonization. It was evident to him from the geologic record that there had never been a sudden break from an initial pristine world to a desolate chaos plunged in darkness and covered with the deep. To Miller, all the geological evidence indicated continuity between the past and the time of the appearance of humans.

Although numerous geologists defended the day-age interpretation, many Bible expositors and theologians also gave support to the notion that the days of creation were not ordinary days but indeterminate periods. Such German commentators as the early Franz Delitzsch and Johann Peter Lange (1802-1884), a professor of theology in the University of Bonn, favored this view.[91] In the United States an important contribution was made by Tayler Lewis (1802-1877), a theologian of the Reformed Church in America.[92] Lewis provided the most exhaustive nineteenth-century exegetical study of Genesis 1 from the point of view that the creation days cannot possibly be twenty-four hours long. What makes Lewis's work all the more compelling was his desire *not* to be influenced by current scientific findings. In Scotland, Baptist expositor Alexander Maclaren (1826-1910) viewed the six days as long time periods.[93]

Many works on systematic theology also adopted variants of the day-age

---

[89]Friedrich Pfaff, *Schöpfungsgeschichte,* 3rd ed.(Heidelberg: 1881); and Otto Zöckler, *Geschichte der Beziehungen zwischen Theologie und Naturwissenschaften mit besonderes Rücksicht auf die Schöpfungsgeschichte,* 2 vols. (Gütersloh: C. Bertelsmann, 1877-1879).

[90]Hugh Miller, *The Testimony of the Rocks; or, Geology in Its Bearings on the Two Theologies, Natural and Revealed* (Edinburgh: Thomas Constable, 1857).

[91]Johann Peter Lange, *Commentary on the Holy Scriptures: Genesis* (Grand Rapids: Zondervan, n.d.).

[92]Tayler Lewis, *The Six Days of Creation* (Schenectady, N.Y.: Van Debogert, 1855).

[93]Alexander Maclaren, *Expositions of Holy Scripture: Genesis, Exodus, Leviticus and Numbers* (Grand Rapids: Eerdmans, 1932).

theory. Among these were the systematic theologies of the American Presbyterians Charles Hodge (1797-1878), a professor of exegetical, didactic and
polemical theology at Princeton Theological Seminary; William G. T. Shedd
(1820-1894), a professor of systematic theology at Union Theological Seminary
in New York; Augustus A. Strong (1836-1921), a pastor of Baptist churches
in Massachusetts and Ohio before becoming a president of Rochester Theological Seminary where he also served as a professor of biblical theology; and
Arminian theologian John Miley (1813-1895), a pastor in nineteen Methodist
churches and a professor of systematic theology at Drew University.[94]

Several exegetical arguments that the days of creation were protracted
time periods appeared repeatedly in the writings of these men. They maintained that the Hebrew word for "day" in Scripture commonly denotes a long
period of time rather than an ordinary day. In fact, at least once in the creation account itself (Genesis 2:4) the word *day* refers to the entire period of
creation. Further, the word *day* is used in several different senses in Genesis
1, so that it cannot be dogmatically asserted that the six days must be treated
as ordinary days.

Second, it was argued along the lines of Augustine that at least the first
three days cannot be treated as ordinary days inasmuch as the Sun, in relation
to which the Earth's rotation serves as a chronometer, was not yet in existence,
at least in respect to its being a time measurer. Not until day four were the
heavenly bodies made to serve for signs and seasons and for days and years.

Third, several writers noted that the events depicted in the six days could
not all have transpired within a twenty-four-hour span. Day six included the
creation of animals, the creation of Adam, the planting of the garden, the
placement of the man in the garden, his observation and naming of the animals, his deepening loneliness, his deep sleep and the creation of Eve. Many
found it difficult to conceive how all of this might occur in one ordinary day.
Tayler Lewis, in a similar vein, pointed out that the accounts of some of the six
days, as with the vegetation of day three, describe natural growths according
to the nature of the created thing, and that these growths cannot be viewed as
taking only one ordinary day.[95]

The fourth major argument generally put forward by advocates of the day-
age hypothesis was that the seventh day, the day of God's rest, continues to
the present and is, therefore, a long period of time. The fact that the biblical

---

[94]Charles Hodge, *Systematic Theology*, 3 vols. (New York: Scribner, 1872-1873); William G. T.
   Shedd, *Dogmatic Theology*, 3 vols., 2nd ed. (New York: Scribner, 1889); Augustus H. Strong,
   Systematic Theology (Philadelphia: Griffith and Rowland, 1886); John Miley, *Systematic Theol-*
   ogy (New York: Eaton and Mains, 1892).
[95]Lewis, *The Six Days of Creation.*

text does not say of the seventh day, as it does of the other six, that "there was evening and there was morning—the seventh day," was viewed as one clear indication that the seventh day has never terminated. Further, New Testament passages such as Hebrews 4 lend further credence to the notion of the continuing existence of God's sabbath. If the seventh day was a long period of time, then it is also clear, according to the argument, that the preceding six days might also legitimately be treated as long periods of time of indeterminate length.

Some biblical scholars ably summarized the biblical and scientific arguments for and against the various interpretations of Genesis 1 without adopting a specific view. One of the outstanding individuals in this category was Franz Heinrich Reusch (1823-1900), a professor of Catholic theology at the University of Bonn.[96]

BIBLICAL GENEALOGIES

Another major theological advance occurred with the publication of an article on primeval chronology by William Henry Green (1825-1900), a professor of Oriental and Old Testament literature at Princeton Theological Seminary.[97] Green addressed the issue of the antiquity of humanity in light of the genealogies of Genesis 5 and 11. He showed that it is impossible to obtain accurate dates for various biblical events like creation, the creation of Adam or the Flood just by adding up the ages of the patriarchs given in these chapters. From several other genealogies in Scripture, and especially from a comparison of the corresponding genealogies of 1 Chronicles 6 and Ezra 7, he showed conclusively that gaps and omissions are commonplace in biblical genealogy. Inasmuch as this is the case elsewhere in Scripture, we have no way of knowing whether there may have also been deliberate omissions in the genealogies of Genesis 5 and 11. If there have been omissions then we cannot calculate the antiquity of the human race. Humanity may be much older than the 6,000 years traditionally allowed. Green suggested that, in many cases, biblical authors deliberately adjusted their material with a view to obtaining a symmetrical arrangement.

---

[96]F. H. Reusch, *Nature and the Bible: Lectures on the Mosaic History of Creation in Its Relation to Natural Science,* trans. Kathleen Lyttleton, 4th ed. (Edinburgh: T & T Clark, 1886).

[97]W. H. Green, "Primeval Chronology," *Bibliotheca Sacra* 47 (1890): 285-303.

5

ANTIQUITY OF THE EARTH

Twentieth Century to the Present

AS THE NINETEENTH CENTURY PASSED INTO HISTORY, no professional geologist with years of field experience would ever dream of thinking that the planet was only a few thousand years old. In fact, most Christian scholars had made peace with the idea of the Earth's great antiquity, persuading themselves that Genesis 1 can be interpreted in ways that are consistent with the concept that the universe was created in more than six literal days and more than only a few thousand years ago.[1] They also became comfortable with the idea that the human race was considerably more than six thousand years old, although, given the dearth of hominid fossils at that time, doubts persisted as to a long human evolutionary lineage.

Prominent exemplars of turn-of-the-century Christian accommodation to a long terrestrial prehistory included, for example, Benjamin Breckinridge Warfield (1851-1921), a professor of didactic and polemic theory at Princeton Theological Seminary. Warfield authored an article on John Calvin's doctrine

---

[1]For additional background information see David N. Livingstone, *Darwin's Forgotten Defenders: The Encounter Between Evangelical Theology and Evolutionary Thought* (Grand Rapids: Eerdmans, 1987); Ronald L. Numbers, *The Creationists: The Evolution of Scientific Creationism* (New York: Alfred A. Knopf, 1992); and Peter Bowler, *Reconciling Science and Religion: The Debate in Early Twentieth-Century Britain* (Chicago: University of Chicago Press, 2001).

of creation in which he attempted to demonstrate that Calvin adopted an evolutionary view of creation.[2] Warfield thought that if Calvin's doctrine of evolution were to be of use in accounting for "the mode of production of the ordered world, these six days would have to be lengthened out into six periods—six ages of the growth of the world."[3] Although Warfield did not explicitly claim the day-age view as his own, one can assume that he felt comfortable with that view, given his appreciation for the achievements of science and also because the day-age view was that of his mentor, Charles Hodge. Warfield was also comfortable with the possibility that the human lineage might stretch back a few hundred thousand years.[4]

In the Roman Catholic tradition, Warfield's almost exact contemporary, Joseph Pohle (1852-1922), who served as a professor of theology, philosophy, apologetics or dogmatics at several universities in Germany, England and the United States, also adopted a version of the day-age theory.[5] The Catholic Encyclopedia of 1912 included a very informative article on the antiquity of the world, in which the various attempts to determine the age of the Earth on scientific grounds were reviewed.[6] The article already included a reference to the early work on radioactive dating of minerals by Rutherford, Soddy, Strutt and Boltwood, whom we discuss below. The review of the scientific evidence was straightforward and objective, with neither endorsement nor repudiation of the claims. No assessment of Genesis 1 in light of the scientific findings was included.

Of very great significance was the flexibility exhibited by the initiators of the fundamentalist movement with regard to any viewpoint concerning Genesis 1. The several booklets that comprise *The Fundamentals* were published during the 1910s. The editing of the series was accomplished anonymously, the arrangements were made for a print run of 250,000 copies, most of which were sent gratis to pastors, professors at theological seminaries and Sunday-school teachers.[7] James Orr (1844-1913), a Calvinist Presbyterian and a professor of apologetics and systematic theology in the United Free Church College in Glasgow, contributed the article on "The Bible and Science" for *The*

---

[2]B. B. Warfield, *Evolution, Scripture, and Science: Selected Writings*, ed. Mark A. Noll and David N. Livingstone (Grand Rapids: Baker, 2000).

[3]Ibid., p. 309. The original paper is B. B. Warfield, "Calvin's Doctrine of Creation," *Princeton Theological Review* 13 (1915): 190-255.

[4]B. B. Warfield, "On the Antiquity and the Unity of the Human Race," *The Princeton Theological Review* 9 (1911): 1-25.

[5]Joseph Pohle, *God and the Author of Nature and the Supernatural* (St. Louis: Herder, 1934).

[6]Lukas Waagen, "The Antiquity of the World," in *The Catholic Encyclopedia*, vol. 15 (New York: The Catholic Encyclopedia, 1912).

[7]*The Fundamentals: A Testimony*, 12 vols. (Chicago: Testimony Publishing, 1910-1915).

Fundamentals. Orr thought that "there is no violence done to the narrative in substituting in thought 'aeonic' days—vast cosmic periods—for 'days' on our narrower, sun-measured scale. Then the last trace of apparent 'conflict' disappears."[8] Thus, so far as an "official" position of early fundamentalism can be detected (undoubtedly early fundamentalists included a broad spectrum of belief on the interpretation of Genesis 1), this public position was that of a day-age approach compatible with a belief in an Earth of great antiquity.

A fair examination of the state of the interaction of Christian theology with the Earth sciences in the earliest twentieth century shows that, as a rule, Christians abandoned neither Christian faith nor a belief in an infallible Bible as a result of the discoveries of geology regarding the Earth's great antiquity. At the same time, hostile attacks on Christianity by geologists were exceedingly rare.

NEW GEOCHRONOLOGIC METHODS

Although nineteenth-century scientific methods for determining the exact age of the planet provided crude estimates, a momentous turning point came in 1896 when the phenomenon of radioactivity was discovered by Henri Becquerel (1852-1908), a professor of applied physics at the Muséum d'Histoire Naturelle in Paris.[9] Becquerel accidentally discovered that a photographic plate wrapped in opaque black paper showed indications of exposure when placed next to samples of uranium compounds.[10] At the University of Paris, Marie Curie (1867-1934) and her husband Pierre Curie (1859-1906), who coined the term *radioactivity,* showed that the chemical element thorium (Th) possesses the same radioactive property as uranium.[11] In addition, the Curies successfully performed chemical separations of two hitherto unknown radioactive chemical elements, polonium (Po) and radium (Ra), that are intimately associated with uranium in natural samples. The studies of Pierre Curie also indicated that salts of radium continuously give off heat.[12]

Beginning in 1899, Ernest Rutherford (1874-1937), MacDonald Professor of Physics at McGill University in Montreal, conducted several studies of

---

[8]James Orr, "Science and Christian Faith," in *The Fundamentals,* 4:101. See also James Orr, *The Bible Under Trial* (London: Marshall, 1907).

[9]On Becquerel, see Alfred Romer, "Becquerel, [Antoine-] Henri," in *DSB,* 1:558-61.

[10]Henri Becquerel, "Émission de Radiations Nouvelles par l'Uranium Métallique," *Comptes Rendus de l'Académie des Sciences* 122 (1896): 1086-88.

[11]On Marie Curie, see Adrienne R. Weill, "Curie, Marie (Maria Sklodowska)," in *DSB,* 3:497-503; and Susan Quinn, *Marie Curie: A Life* (New York: Simon & Schuster, 1995). On Pierre Curie, see Jean Wyant, "Curie, Pierre," in *DSB,* 3:503-8.

[12]P. Curie and A. Laborde, "Sur la Chaleur Dégagée Spontanément par les Sels de Radium," *Comptes Rendus de l'Académie des Sciences* 136 (1903): 673-75.

radioactivity (fig. 5.1).[13] He pointed out the ionizing character of alpha rays, suggested that the chemical element helium (He) might be produced during radioactive processes and observed the exponential decay of radioactive thorium compounds. In 1903, Rutherford and his associate Frederick Soddy (1877-1956) showed that radioactivity is a decay process in which radioactive elements spontaneously change into different kinds of elements while emitting alpha and beta particles at measurable rates. They presented a simple mathematical formulation of the radioactive decay process that is discussed in more detail in chapter fourteen. They also subjected radioactive materials to varying temperatures and a variety of "chemical and physical agencies" and found that, for a specific radioactive element, there was "no alteration" of the value of the decay constant, the property denoting the rate at which the radioactive element will decay.[14] In 1903, Rutherford and Soddy also measured the amounts of heat released by radioactive decay processes. Becquerel, the Curies, Rutherford and Soddy were all awarded Nobel Prizes for their work on atomic physics and chemistry.

The realization that radioactivity produces heat was revolutionary, because here was a source of heat unknown to Lord Kelvin when he was computing ages for the Sun and Earth. Several scientists immediately realized that Kelvin's estimates for the age of Earth were put in jeopardy by the new discovery and that Earth might be considerably older than Kelvin had surmised. In 1907, Kelvin went to his grave, adjacent to that of Isaac Newton and eventually very close to that of Rutherford, in Westminster Abbey, refusing to concede that radioactivity generated

Figure 5.1. Ernest Rutherford (1874-1937). Reproduced by permission of National Portrait Gallery, London.

a sufficient amount of heat to discredit the validity of his estimates. Despite Kelvin's resistance, most geologists accepted that Earth might be much older

[13]On Rutherford, see Lawrence Badash, "Rutherford, Ernest," in *DSB*, 12:25-36; Arthur Stewart Eve, *Rutherford: Being the Life and Letters of the Rt. Hon. Lord Rutherford, O.M.* (Cambridge: Cambridge University Press, 1939); and John Rowland, *Ernest Rutherford: Atom Pioneer* (London: Laurie, 1955).

[14]Ernest Rutherford and Frederick Soddy, "Radioactive Change," *Philosophical Magazine* 5, ser. 6 (1903): 576-91.

than Kelvin's proposal because of the discovery of radioactive heating.[15] Although the question remained as to exactly how old Earth is, the answer also lay with radioactivity.

In 1905, Rutherford gave the Silliman Lectures at Yale, in which he proposed that radioactivity had the potential to be a geologic timekeeper.[16] He reported age calculations of two specimens that he had analyzed for their abundances of uranium and helium. One specimen of the mineral fergusonite yielded an age of 497 million years, and a specimen of uraninite from Glastonbury, Connecticut, yielded an age of 500 million years. Noting that helium is a noble gas that normally does not chemically combine with other elements and that could readily leak out of a radioactive mineral in which it had been produced, Rutherford pointed out that his calculated ages probably represented minimum ages. Conjecturing that the chemical element lead (Pb) is produced by radioactive decay of uranium and radium, he suggested that dates based on measurement of uranium to lead ratios should be more reliable than those based on uranium to helium ratios.

Robert John Strutt (1875-1947), son of another illustrious Nobel-prize winning physicist, John William Strutt (the third Baron Rayleigh), conducted research in the Cavendish Laboratory at Cambridge beginning in 1899.[17] In 1905, he analyzed radioactive minerals for their contents of uranium, thorium, radium and helium, confirmed that helium is produced by the decay of thorium and showed that the helium content of minerals is a function of their age.[18] He obtained an age of 2.4 billion years for a sample of thorianite from Ceylon (now Sri Lanka) from abundances of radium and helium, but he concluded that the age was invalid because much of the helium had been produced by thorium as well as radium. Strutt also determined an age of 16.5 million years for a sample of pitchblende from Cornwall, England, that contained no thorium.

Intrigued by Rutherford's Silliman Lectures, Bertram Boltwood (1870-1927), a professor of physics at Yale, set out to test Rutherford's suggestion that uranium-lead (U-Pb) dating might be superior to uranium-helium (U-He) dat-

---

[15]It has very recently been shown that the known distribution of radioactivity does not of itself invalidate Kelvin's estimates. On the other hand, in 1895, John Perry demonstrated that Kelvin's estimate of the age of the Earth would be invalidated by the existence of convection in the Earth's interior, but Perry's analysis was overlooked. For details, see Philip England, Peter Molnar and Frank Richter, "John Perry's Neglected Critique of Kelvin's Age for the Earth: A Missed Opportunity in Geodynamics," GSA Today 17 (2007): 4-9.

[16]Ernest Rutherford, Radioactive Transformations (New York: Charles Scribner's Sons, 1906).

[17]On Strutt, see Thaddeus J. Trenn, "Strutt, Robert [Robin] John, Fourth Baron Rayleigh," in DSB, 13:107-8.

[18]R. J. Strutt, "On the Radio-active Minerals," Proceedings of the Royal Society of London 76, ser. A (1905): 88-101.

ing.[19] Discovering that lead is invariably found in uranium-bearing minerals, he proposed that lead is the ultimate end product of uranium decay.[20] He also found that minerals in older rocks, as determined by their stratigraphic position, invariably contain greater quantities of lead and helium than younger minerals, thus confirming the notion that both lead and helium are generated during the radioactive decay of uranium.

In a follow-up study, Boltwood evaluated the Pb/U ratios of 43 mineral samples, mostly uraninite, from localities in the United States, Norway, Sweden and Ceylon.[21] He calculated, for example, an age of 410 million years for uraninite from Glastonbury, Connecticut, compared with Rutherford's 500 million year U-He date. He also obtained an age of 510 million years for uraninite from Spruce Pine, North Carolina, 1,700 million years for uraninite from Anneröd, Norway, and 2,200 million years for thorianite from the Sabaragamuwa Province in Ceylon.

Strutt reported in 1908 on ages calculated from U/He ratios for phosphate minerals in sedimentary rocks of known stratigraphic age.[22] He obtained ages ranging from 0.225 to 141 million years but failed to obtain good correlation with relative stratigraphic ages, concluding that spurious ages resulted from extensive leakage of helium. Strutt moved to Imperial College, London, where he served as a professor of physics. In 1919, he became the fourth Baron Rayleigh. Among Strutt's later achievements was confirmation of the presence of ozone in the atmosphere.

Because so many nineteenth-century age determinations were about 25 to 100 million years and because the calculations of Lord Kelvin were in the same range, a few geologists, notably George Becker of the U.S. Geological Survey and John Joly, were highly skeptical of the results obtained from analysis of radioactive minerals.[23] Becker believed that these minerals could not be hundreds of millions of years old because the values drastically conflicted with the reasonably consistent, much lower age determinations based on other methods.

In 1911, Arthur Holmes (1890-1965), a brilliant British geologist just embarking on his career, joined Strutt at Imperial College to work on uranium-

---

[19]On Boltwood, see Alois F. Kovarik, "Bertram Borden Boltwood, 1870-1927," *National Academy of Sciences, Biographical Memoirs* 14 (1932): 69-96.

[20]Bertram B. Boltwood, "On the Ultimate Disintegration Products of the Radioactive Elements," *American Journal of Science* 20, ser. 4 (1905): 253-67.

[21]Bertram B. Boltwood, "On the Ultimate Disintegration Products of the Radioactive Elements, Part II: The Disintegration Products of Uranium," *American Journal of Science* 23, ser. 4 (1907): 77-88.

[22]R. J. Strutt, "On the Accumulation of the Helium in Geological Time," *Proceedings of the Royal Society of London* 81, ser. A (1908): 272-77.

[23]George Becker, "The Age of the Earth," *Smithsonian Miscellaneous Collections* 56 (1910): 1-28.

bearing minerals from Oslo, Norway.[24] Recalculating Boltwood's data with an improved value for the decay constant for uranium, and adding his own dates from Oslo, Holmes showed that there is a good correlation between dates obtained by uranium-lead dating and the relative stratigraphic age of the mineral.[25] Clearly recognizing the great importance of radioactivity in addressing the problem of the Earth's age, Holmes pursued a life-long quest to solve that problem.

A geologist on the Yale faculty, Joseph Barrell (1869-1919), realized from the work of his physics colleague Boltwood and that of Strutt that there was a correlation between the relative geologic age of minerals and the radiometric ages obtained from them.[26] Barrell joined the Yale faculty in 1903 to teach structural geology, but he proved to be a remarkable generalist and synthesizer, successfully attacking large-scale geologic problems. Toward the end of his career, Barrell produced an enormous paper reviewing the rhythms and the measurement of geologic time.[27] Combining published radiometric dates on minerals of known *relative* geologic age with estimates of the *relative* lengths of the various geologic time periods based on sediment thicknesses, Barrell constructed the first Phanerozoic timescale to incorporate ages obtained by radioactive methods. He placed the beginning of the Paleozoic Era between 550 and 700 million years ago, the beginning of the Mesozoic Era between 135 and 180 million years ago, and the beginning of the Cenozoic Era about 55 to 65 million years ago. Given the primitive state of radiometric dating in the early twentieth century, Barrell's proposals were remarkable in light of currently accepted values of 542 million years for the beginning of the Paleozoic Era, 251 million years for the beginning of the Mesozoic Era, and 66 million years for the beginning of the Cenozoic Era.[28]

Understanding of the nature of these unstable elements increased dramatically throughout the twentieth century. In the 1910s, the discoverer of the electron, Joseph John Thomson (1856-1940), successor to Lord Rayleigh as Cavendish Professor of Experimental Physics at the University of Cambridge from 1884 to 1918, performed experiments demonstrating that many chemical

---

[24]On Arthur Holmes, see Cherry L. E Lewis, *The Dating Game: One Man's Search for the Age of the Earth* (Cambridge: Cambridge University Press, 2000).

[25]Arthur Holmes, "The Association of Lead with Uranium in Rock-Minerals, and Its Application to the Measurement of Geological Time," *Proceedings of the Royal Society of London* 85, ser. A (1911): 248-56.

[26]On Barrell, see Charles Schuchert, "Joseph Barrell, Engineer-Geologist 1869-1919," *National Academy of Sciences, Biographical Memoirs* 12 (1929): 3-40.

[27]Joseph Barrell, "Rhythms and the Measurement of Geologic Time," *Bulletin GSA* 28 (1917): 745-904.

[28]For a very recent refinement of the geologic timescale, see Felix Gradstein, James Ogg and Alan Smith, eds., *A Geologic Time Scale* (Cambridge: Cambridge University Press, 2004).

elements consist of multiple forms of their atoms that came to be known as *isotopes*.[29] The discovery of isotopes permitted scientists to distinguish radioactive atoms from nonradioactive atoms of the same chemical element or to distinguish among different kinds of radioactive atoms for a different element. For example, it was found that uranium-bearing minerals contain two major radioactive forms, the isotopes uranium 238 (^{238}U) and uranium 235 (^{235}U), each of which decays at a different rate and, therefore, has a different decay constant. In addition, these two isotopes decay to different ultimate daughter products, the isotopes lead 206 (^{206}Pb) and lead 207 (^{207}Pb), respectively. As a result, scientists learned that uranium-bearing minerals offered the possibility of at least two independent dating methods, provided that techniques for accurate isotope analysis could be developed. Thomson was yet another Nobel Prize winner in physics and is buried next to Rutherford in Westminster Abbey.

During the 1920s, the *mass spectrograph* was developed by Frederick Aston (1877-1945) to yield accurate and precise analyses of isotopes of uranium, lead and several other chemical elements. Routine mass spectroscopic analysis of radioactive isotopes and their daughter products for dating purposes was pioneered during the 1930s and 1940s by Alfred O. Nier (1911-1994), a professor of physics at the University of Minnesota.[30] His work opened the way for more reliable age determinations than had previously been possible. In subsequent years, physicists have successfully measured the decay constants of a wide range of radioactive isotopes. During the 1940s and 1950s, geochronologists developed different dating methods based on the decay of potassium 40 (^{40}K) into argon 40 (^{40}Ar) and rubidium 87 (^{87}Rb) into strontium 87 (^{87}Sr). During this period, the carbon 14 (^{14}C) method, widely used for dating archeological specimens and materials with ages up to a few thousands of years, was developed at the University of Chicago by Willard Libby (1908-1980), a Nobel-prize winning professor of chemistry at the University of California (Berkeley), Chicago and the University of California at Los Angeles (UCLA).[31]

Throughout the twentieth century, better understanding of the geological processes affecting radioactive isotopes led to improvement in sample selection, methods of age determination and interpretation of the geologic significance of an age. A number of radiometric methods were abandoned because they were too difficult or fraught with too many uncertainties, but several

---

[29]On Thomson, see Edward A. Davis, *J. J. Thomson and the Discovery of the Electron* (London: Taylor and Francis, 1997).

[30]On Nier, see H. Craig, "Introduction of Alfred O. C. Nier for the V. M. Goldschmidt Award 1984," *Geochimica et Cosmochimica Acta* 49 (1985): 1661-65.

[31]On Libby, see David Petechuck, "Willard F. Libby 1908-1980: American Chemist," in *Notable Twentieth-Century Scientists*, ed. Emily J. McMurray (New York: Gale Research, 1995), 3:1245-48.

very reliable, well-understood methods are now routinely used in laboratories around the world. The radioactive decay of samarium 147 (^{147}Sm) to neodymium 143 (^{143}Nd) is now very important in dating extremely old minerals and rocks that contain rare-earth elements. Powerful graphical methods for plotting analytical results include the concordia diagram for interpreting analyses of uranium-bearing minerals, and the isochron diagram used in the interpretation of analyses of minerals containing rubidium, samarium, lanthanum or rhenium. The ^{40}Ar/^{39}Ar method takes advantage of loss of argon from potassium-bearing minerals to elucidate their cooling histories. Most of these newer methods are useful for dating materials with ages of millions to billions of years.

A number of methods, collectively referred to as uranium-disequilibrium series dating, are valuable for determining ages in the range of hundreds of thousands of years. Fission-track dating and an updated version of U-He dating that takes advantage of our knowledge of He diffusion from minerals as a function of temperature, have proved their worth for evaluating the uplift and cooling histories of geologic terranes. As a result of the amazingly successful application of a host of reliable dating methods in dozens of laboratories around the world, thousands of radiometric ages of rocks and minerals are obtained each year by a wide spectrum of methods. These ages are routinely tens of thousands, hundreds of thousands, millions or even a few billions of years depending on the geologic circumstances and history of the specimen being dated. Moreover, several nonradiometric dating methods for dating archeological sites and artifacts appeared in the last half century, including obsidian rim hydration, thermoluminescence (TL) and amino-acid racemization dating.[32]

A handful of investigators have also devised methods for calculating Earth's age. Because attempts to determine the age of the oldest rock on Earth provide only a minimum age for the planet, a method was needed for calculating Earth's age based on a model of its chemical and isotopic composition and history. Especially during the 1940s, E. K. Gerling (1904-1985) of the Radium Institute of the Academy of Sciences of the USSR, Arthur Holmes, who was by now the Regius Professor of Geology and Mineralogy at the University of Edinburgh, and Friedrich Georg Houtermans (1903-1966) independently developed very similar, ingenious and mathematically sophisticated methods for calculating the age of the Earth.

The story of Houtermans is dramatic. He was a brilliant physicist who early distinguished himself by calculations of thermonuclear reactions in stars that

---

[32]On archeological dating methods, see R. E. Taylor and Martin J. Aitken, eds., *Chronometric Dating in Archaeology* (New York: Plenum Press, 1997).

laid the basis for the modern theory of stellar nucleosynthesis. After teaching briefly at the University of Berlin, he fled from the growing Nazi threat in Germany to Great Britain, where he was persuaded by a Russian childhood friend to accept a position teaching physics at the University of Kharkov in the Ukraine. During the Stalinist regime, Houtermans was arrested in 1937, charged with being a Nazi spy, tortured until he confessed and then imprisoned in Moscow. In 1939, he was handed over to the Nazi Gestapo and imprisoned in Berlin the following year. With the collapse of Nazism in 1945, Houtermans took up a position at the University of Göttingen and formulated his method for calculating the age of Earth. He finished his career at the University of Bern in Switzerland.

The Gerling-Holmes-Houtermans model calculates the age of Earth from the values of uranium and lead isotopes in lead ore deposits of known age.[33] A critical component for successful application of the method is the ratio of lead isotopes present in Earth at the time of its formation. Of course, no one was around to measure those ratios. In the 1950s, however, Clair C. Patterson (1922-1995), a professor of geochemistry at California Institute of Technology, devised a clever way around the problem by using the lead isotope ratios of the iron sulfide mineral troilite that occurs in fragments of the Canyon Diablo meteorite, the object that excavated Meteor Crater in northern Arizona, as representative of lead isotope ratios of the early Earth.[34] Patterson (fig. 5.2) justified the procedure on the grounds that troilite contains negligible amounts of uranium and thorium. Therefore, the amounts of lead isotopes, as the end products of uranium and thorium decay, would not have changed throughout the history of the meteorite. Moreover, Patterson reasoned that the lead isotope ratios of meteorites and Earth should have been reasonably similar at the time of formation of the Solar System inasmuch as meteorites are derived from the asteroid belt, a part of the Solar System relatively close to Earth. Patterson reasoned that analyzed samples of average ocean sediment should represent the average lead isotope ratio of Earth *at present* given that these oceanic sediments are derived from a wide variety of rocks of various ages from many places on the continental land masses. From these data he calculated an age of 4.55 billion years. Later attempts to calculate Earth's age

---

[33]For detailed explanation of the Gerling-Holmes-Houtermans method for determining the age of the Earth, see G. Brent Dalrymple, *The Age of the Earth* (Stanford, Calif.: Stanford University Press, 1991).

[34]Clair C. Patterson, "Age of Meteorites and the Earth," *Geochimica et Cosmochimica Acta* 10 (1956): 230-37. Also see Stephen G. Brush, *A History of Modern Planetary Physics*, vol. 2, *Transmuted Past: The Age of the Earth and Evolution of the Elements from Lyell to Patterson* (Cambridge: Cambridge University Press, 1996).

Figure 5.2. Clair Patterson (1922-1995)
Courtesy of the Archives, California Institute of Technology.

have entailed various refinements of Patterson's work.[35]

Radiometric dating has not produced a direct determination of the age of Earth because we do not possess a terrestrial sample that was formed at the time of Earth's origin. But ages have been determined from numerous samples of meteorites, and many lunar rocks have been dated. Calculations have also been made to determine the age of the Moon. The radiometric evidence from Earth, meteorites and the Moon consistently points to ages of about 4.5-4.7 billion years.[36] In recent years, geochronologists have vigorously pursued the oldest rock on Earth.[37] On several continents there are rocks that have yielded radiometric dates in the range of 3.5 to 3.8 billion years. Among the very oldest are the Isua Supracrustal Gneisses of western Greenland, dated at 3.8 billion years, and the Acasta Gneisses near Great Slave Lake in the Northwestern Territories of Canada, dated at 4.03 billion years. In early 2001, a team of geo-

---

[35]F. Tera and R. W. Carlson, "Assessment of the Pb-Pb and U-Pb Chronometry of the Early Solar System," *Geochimica et Cosmochimica Acta* 63 (1999): 1877-89.

[36]On the consistency of ages of meteorites, see Dalrymple, *The Age of the Earth*.

[37]For a summary of research into the oldest terrestrial minerals and rocks, see Balz S. Kamber, Stephen Moorbath and Martin J. Whitehouse, "The Oldest Rocks on Earth: Time Constraints and Geological Controversies," in *The Age of the Earth from 4004 B.C. to A.D. 2002*, GSL Special Publication 190, ed. Cherry L. E. Lewis and Simon J. Knell (London: The Geological Society, 2001), pp. 177-203.

chronologists announced that they had determined the age of a very tiny zircon crystal occurring in sedimentary rocks from the Jack Hills in west-central Australia at 4.4 billion years.[38] That remains the oldest known mineral sample on Earth.

In chapters fourteen and fifteen we explain some representative methods of radiometric dating in more detail and refute criticisms of these methods advanced by advocates of a young Earth.

DEVELOPMENTS IN BIBLICAL STUDIES AND THEOLOGY

Although twentieth-century biblical scholars may be excused for not keeping abreast of the details of the various dating methods, many were sufficiently aware of advances in science to appreciate the force of the evidence for the vast antiquity of the Earth. As in the previous century, radical higher critics with no commitment to an infallible Bible accepted scientific findings without showing much concern for how these findings relate to the Bible, but biblical scholars committed to an authoritative Bible and appreciative of evidence for an old Earth were generally more eager to relate scientific discoveries to relevant biblical texts. Such individuals understandably displayed a broad range of exegetical skills as well as great variability in their grasp of scientific developments of their times. Over the course of the century, the restitution theory largely fell out of favor with most evangelical scholars. The New Scofield Reference Bible of 1967, for example, dropped its allegiance to that view while indicating that Genesis provides no information about the age of the world and hinting that the days of creation might have been long periods of time.[39]

The restitution (gap) theory was, however, promoted by Harry Rimmer (1890-1952), a Presbyterian minister, self-styled authority on science and self-appointed critic of evolution. Probably the most capable twentieth-century defense of the theory was advanced by Arthur C. Custance (1910-1985), Head of the Human Engineering Laboratories of the Defence Research Board of Canada and author of the Doorway Papers, a set of sixty-two monographs dealing with various faith and science issues.[40] The New International Version of the Bible includes a footnote to Genesis 1:2 indicating that the word *was* in the phrase "now the earth was formless and empty" could possibly be translated "became." Inclusion of the footnote was likely a concession to those

---

[38]On the Acasta gneisses, see Samuel A. Bowring and I. S. Williams, "Priscoan (4.00-4.03 Ga) Orthogneisses from Northwestern Canada," *Contributions to Mineralogy and Petrology* 134 (1999): 3-16; and on the oldest known mineral, see S. A. Wilde, John W. Valley, William H. Peck and C. M. Graham, "Evidence from Detrital Zircons for the Existence of Continental Crusts and Oceans on the Earth 4.4 Gyr ago," *Nature* 409 (2001): 175-78.

[39]*The New Scofield Reference Bible* (New York: Oxford University Press, 1967).

[40]Arthur C. Custance, *Without Form and Void* (Brockville, Canada: Custance, 1970).

few members of the translation committee who still held the gap view. Most Hebrew scholars, however, maintain that the text does not support the translation "became."[41] If so, the restitution theory falls to the ground.

During the middle decades of the twentieth century, P. J. Wiseman developed the prophetic-day view.[42] Wiseman proposed that the six days were ordinary days on which God revealed the successive events of the creation to a seer. He did not specify who the recipient of the revelation was. In essence, on one ordinary day, God told the seer about the creation of light. On the following day, God told the recipient of the revelation about the separation of the waters, and so on. Wiseman maintained that the days on which the revelations were given were ordinary twenty-four-hour days in the experience of the seer. On his view, however, Scripture provides no information about how long the creative events themselves actually took. Wiseman's hypothesis gained relatively few adherents.

Throughout the first half of the century, both the traditional twenty-four-hour-day view and variations of the day-age view enjoyed support among most orthodox theologians and biblical scholars from many theological traditions. Among Nazarenes, a prominent day-age advocate was H. Orton Wiley (1877-1961), variously President of Northwest Nazarene University, President of Pasadena College, General Secretary of the Department of Education of the Church of the Nazarene, and editor of *The Herald of Holiness,* the official magazine of the Church of the Nazarene.[43] Episcopalian Francis J. Hall (1856-1932), a professor of dogmatic theology at the Protestant Episcopal Church's General Theological Seminary in New York, was another day-age advocate.[44]

Among the various Lutheran scholars, Friedrich Bettex (1837-1915) held to the day-age interpretation. On the other hand, Francis Pieper (1852-1931), Theodore Graebner (1876-1950) and John Theodore Mueller (1885-1967), all of whom taught at the Missouri Synod Lutheran Church's Concordia Theological Seminary, and Herbert C. Leupold (1892-1972) of Capitol Theological Seminary (now Trinity Lutheran Seminary), all promoted the traditional six-day view.[45] More recently, another professor of theology at Concordia Semi-

[41]Weston W. Fields, *Unformed and Unfilled: The Gap Theory* (Nutley, N.J.: Presbyterian & Reformed, 1976).

[42]P. J. Wiseman, *Creation Revealed in Six Days: The Evidence of Scripture Confirmed by Archaeology* (London: Marshall, Morgan & Scott, 1948).

[43]H. Orton Wiley, *Christian Theology,* vol. 1 (Kansas City, Mo.: Beacon Hill, 1959).

[44]Francis J. Hall, *Creation and Man* (New York: Longmans, Green, 1921).

[45]Friedrich Bettex, *The Six Days of Creation in the Light of Modern Science* (Burlington, Iowa: Lutheran Literary Board, 1924); Francis Pieper, *Christian Dogmatics* (St. Louis: Concordia, 1950); Theodore Graebner, *God and the Cosmos: A Critical Analysis of Atheism* (Grand Rapids: Eerdmans, 1932); John T. Mueller, *Christian Dogmatics* (St. Louis: Concordia, 1934); Herbert C. Leupold, *Exposition of Genesis* (Columbus, Ohio: Wartburg, 1942).

nary, Fred Kramer, has rejected the Ussher chronology and warned theologians against attempting to elicit scientific information from the biblical text.[46]

In the Reformed tradition, several theologians defended the idea that the days of Genesis 1 were normal twenty-four-hour days and that the earth was recently created. Among those in the Dutch Calvinist tradition within the United States who held that view were Herman Hoeksema (1886-1965), president of Protestant Reformed Theological Seminary, and Louis Berkhof (1879-1950), a professor of systematic theology and president of Calvin Theological Seminary.[47] In the Netherlands, Herman Bavinck (1854-1921), a professor of dogmatics at the Theological School at Kampen and at the Free University of Amsterdam, provided a remarkably thorough discussion of the doctrine of creation in the light of outdated and partially misunderstood geologic knowledge.[48] Although he contended that most of the days of creation were not ordinary days, he was skeptical about geologic claims to the great antiquity of the Earth. Valentine Hepp (1879-1950), Bavinck's successor as a professor of theology at the Free University of Amsterdam, delivered the Stone Lectures at Princeton Theological Seminary in 1930 on the topic of *Calvinism and the Philosophy of Nature*, in which he vigorously denounced efforts by Christians to find millions of years in Earth history.[49] Hepp called for a total rethinking of geology in terms of a recent creation and a global flood. Gerrit Charles Aalders (1880-1961), a professor of Old Testament at the Free University of Amsterdam, leaned toward the day-age view.[50]

Gordon J. Spykman (1926-1993) was a professor of theology at Calvin College.[51] He drew a distinction between "creating time" and "creational time," with the seventh day serving as the transition between the two. Creational time is the kind of time in which we live. The instruments that we use for measuring time are not adequate for "measuring" creating time. Because creating time is entirely different from the kind of time in which we live, the "'sixth day' marks a cut-off point for theoretical inquiry." If we want to learn anything at all about creation we are strictly limited to revelation. The "series of 'six-day' beginnings is not an index to spans of time by which to calculate the age of the

[46]Fred Kramer, "A Critical Evaluation of the Chronology of Ussher," in *Rock Strata and the Bible Record*, ed. Paul A. Zimmerman (St. Louis: Concordia, 1970), pp. 57-67; and Fred Kramer, "The Biblical Account of Creation," in *Rock Strata and the Bible Record*, pp. 94-113.

[47]Herman Hoeksema, *Reformed Dogmatics* (Grand Rapids: Reformed Free Publishing Association, 1966); and Louis Berkhof, *Systematic Theology* (Grand Rapids: Eerdmans, 1930).

[48]Herman Bavinck, *Reformed Dogmatics*, vol. 2 (Grand Rapids: Baker, 2004).

[49]Valentine Hepp, *Calvinism and the Philosophy of Nature* (Grand Rapids: Eerdmans, 1930).

[50]Gerrit C. Aalders, *The Book of Genesis*, trans. William Heynen, vol. 1 (Grand Rapids: Zondervan, 1981).

[51]Gordon J. Spykman, *Reformational Theology: A New Paradigm for Doing Dogmatics* (Grand Rapids: Eerdmans, 1992).

earth," but points "to orchestrated sets of developing life relationships within the emerging creation order."[52]

Spykman repeatedly stated that we cannot get back of the boundary between "creating time" and "creational time." He rejected efforts to account for the first few seconds of the big bang as purely speculative. Although he insisted that his view did not undermine the scientific enterprise, his erection of a hard and fast boundary between the six days and subsequent history effectively undermined all historical sciences such as paleontology, geology, astronomy and cosmology, because all of them seek to account for what transpired before human beings came on the scene. Spykman maintained that the author of Genesis was familiar with the weekly rhythm of six days labor and one day of rest and accepted that pattern by revelation as a creation ordinance. According to Spykman, "the writer of Genesis then takes this weekly round of six working days (followed by a day of rest, which marks the inception of creational time) and employs it reflexively as his ordering principle in depicting sequentially the mighty creating acts of God."[53]

To summarize, Spykman did not regard the days of creation as twenty-four-hour days because that would superimpose our own time measurement on the days of the creating time. However, far from being open to an old Earth, Spykman rejected any attempt to consider the days of creating time as long periods of time, again because we would be superimposing our categories of time measurement on days of an entirely different order. Spykman's repudiation of science's attempt to explore what happened before the arrival of humans also ruled out the likelihood of a universe and Earth that are billions of years old.

Within the Presbyterian wing of the Reformed tradition, in addition to James Orr and B. B. Warfield, whom we mentioned at the beginning of the chapter, J. Gresham Machen (1881-1937), a professor of New Testament at Princeton Theological Seminary and later a professor of apologetics at Westminster Theological Seminary, and J. Oliver Buswell (1895-1977), president of Wheaton College, favored some variant of the day-age view.[54] Edward J. Young (1907-1968), a professor of Old Testament at Westminster Theological Seminary, maintained that the days might be interpreted as periods of time of indeterminate length.[55] R. Laird Harris, a professor of Old Testament at Faith and Covenant Theological Seminaries, suggested that the days were not ordi-

---

[52]Ibid., p. 153.

[53]Ibid., p. 164.

[54]J. Gresham Machen, *The Christian View of Man* (New York: Macmillan, 1937), pp. 130-31; and J. Oliver Buswell Jr., *A Systematic Theology of the Christian Religion* (Grand Rapids: Zondervan, 1962).

[55]Edward J. Young, *Studies in Genesis One* (Philadelphia: Presbyterian & Reformed, 1964).

nary days without adopting any particular theory of harmonization.[56]

More recently, Robert Reymond, a professor of systematic theology at Covenant Theological Seminary, accepted the traditional twenty-four-hour-day view, stating that he could "discern no reason, either from Scripture or from the human sciences, for departing from the view that the days of Genesis were ordinary twenty-four-hour days." He dismissed modern geology's conclusions regarding the antiquity of the Earth.[57] "There is no reason to believe," he asserted, "that the universe and the earth in particular are billions of years old either, as many astronomers and geologists insist." He maintained that the world was created "with an appearance of age."[58]

James Montgomery Boice (1938-2000), a pastor of Philadelphia's Tenth Presbyterian Church, was a proponent of the day-age interpretation.[59] C. John Collins, a professor of Old Testament at Covenant Theological Seminary, has espoused the "analogical-day" or "anthropomorphic-day" view in which the human work week is viewed as an analog of the creation week.[60] Our days are images of the divine creative days. The creation account employs metaphor and imagery, and the creation week is a part of that extended metaphor so that the time element cannot be pinned down to twenty-four-hour days. As a result, Collins said that the Bible does not establish the age of the universe. On the other hand, he was convinced that the days record events that took place sequentially. John Currid, a professor of Old Testament at Reformed Theological Seminary, advocated the traditional twenty-four-hour-day view.[61] In contrast, John Jefferson Davis, a professor of systematic theology and Christian ethics at Gordon-Conwell Theological Seminary, accepted the geological indications for the vast antiquity of the Earth as well as the big bang theory of modern cosmology.[62]

Among Baptists, theologian Bernard Ramm (1916-1992), who taught at Biola University, Bethel College and Seminary, and Baylor University, before becoming a professor of systematic theology at American Baptist Seminary

---

[56]R. Laird Harris, *Inspiration and Canonicity of the Bible* (Grand Rapids: Zondervan, 1957).

[57]Robert L. Reymond, *A New Systematic Theology of the Christian Faith* (Nashville: Thomas Nelson, 1998), p. 392.

[58]Ibid., p. 396.

[59]James Montgomery Boice, *Genesis: An Expositional Commentary* (Grand Rapids: Zondervan, 1982).

[60]C. John Collins, *Science & Faith: Friends or Foes?* (Wheaton, Ill.: Crossway Books, 2003), and *Genesis 1—4: A Linguistic, Literary, and Theological Commentary* (Phillipsburg, N.J.: Presbyterian & Reformed, 2006).

[61]John D. Currid, *A Study Commentary on Genesis*, vol. 1 (Webster, N.Y.: Evangelical Press, 2003).

[62]John Jefferson Davis, *The Frontiers of Science & Faith: Examining Questions from the Big Bang to the End of the Universe* (Downers Grove: InterVarsity Press, 2002).

of the West, leaned toward the day-age position in his very influential work *A Christian View of Science and Scripture*, although he also saw a strong topical flavoring to Genesis 1.[63] A host of new works in systematic theology have been written by Baptist theologians in recent years. Carl F. H. Henry (1913-2003) was a professor of theology at Northern, Fuller and Eastern Baptist Theological Seminaries, the first editor of *Christianity Today* and a prolific author. Henry wrote at great length about the different views on the days of creation, occasionally noting weaknesses of the various positions, but in the end he never committed himself to a specific view.[64]

Two theologians at Denver Seminary with roots in the Conservative Baptist tradition, Gordon R. Lewis, a professor of Christian philosophy and theology, and Bruce A. Demarest, a professor of Christian theology and spiritual formation, showed strong concordist tendencies by introducing concepts from modern science into their interpretation of the creation days.[65] For example, they stated that "after relating the initial creative act, the text describes the gas cloud that was proto-earth as 'formless and empty.'"[66] They even claimed that "ultimately, responsible geology must determine the length of the Genesis days, even as science centuries earlier settled the issue of the rotation of the earth about the sun."[67] Lewis and Demarest rejected the traditional view of twenty-four-hour creation days and tentatively concluded that "the six consecutive creative acts were separated by long periods of time." [68]

Wayne A. Grudem, a professor of theology at Trinity Evangelical Divinity School and then Research Professor of Bible and Theology at Phoenix Seminary, tentatively accepted the evidence for the Earth's great antiquity in part on the basis of arguments in the original edition of *Christianity and the Age of the Earth*, and he expressed doubts about the validity of scientific creationism.[69] He was noncommittal regarding the interpretation of Genesis 1, however, but found the day-age interpretation to be the most persuasive if one accepts an old Earth.[70] Unable to bring himself to acknowledge the force of the scientific evidence, Grudem wrote that "it is likely that scientific research in the next ten or twenty years will tip the weight of evidence decisively toward either a

---

[63]Bernard Ramm, *The Christian View of Science and Scripture* (Grand Rapids: Eerdmans, 1955).

[64]Carl F. H. Henry, *God, Revelation, and Authority*, vol. 6: *God Who Stands and Stays* (Waco, Tex.: Word, 1983).

[65]Gordon R. Lewis and Bruce A. Demarest, *Integrative Theology*, vol. 2 (Grand Rapids: Zondervan, 1990).

[66]Ibid., p. 27.

[67]Ibid., p. 29.

[68]Ibid., p. 44.

[69]Wayne A. Grudem, *Systematic Theology: An Introduction to Biblical Doctrine* (Grand Rapids: Zondervan, 1994).

[70]Ibid., p. 297.

young earth or an old earth view, and the weight of Christian scholarly opinion (from both biblical scholars and scientists) will begin to shift decisively in one direction or another."[71] Millard J. Erickson, an evangelical Baptist who has held professorial positions in theology at Bethel Theological Seminary, Truett Seminary at Baylor University and Western Seminary in Portland, after reviewing the strengths and weaknesses of the gap theory, the flood theory, the day-age theory, the ideal-time theory (creation of apparent age) and the literary framework theory, cast his lot with "a variation of the age-day theory."[72]

Turning to Roman Catholicism, the Roman Catholic Catechism does not address the question of the age of the Earth.[73] The 1950 encyclical of Pope Pius XII on the subject of biological evolution, *Humani Generis,* dealt very cautiously with the issue, saying nothing specific about the age of the world or the detailed interpretation of Genesis 1.[74] Of greater interest is the 1951 address of Pius XII to the Pontifical Academy of Sciences concerning the proofs for the existence of God in the light of modern natural science.[75] The address included a section on the beginning of the universe in time in which the pope mentioned the recession of galaxies and implications for the age of the universe. He also discussed the radiometric evidence for the age of minerals and of meteorites based on the decay of uranium and mentioned that the Earth and meteorites are on the order of five billion years old. Although the pope appeared to accept these conclusions from science, he did not address how the creation account ought to be understood.

Theologian Joseph Ratzinger, better known to contemporary readers as Pope Benedict XVI, produced a series of homilies on the doctrine of creation in which he maintained that the biblical creation narrative represents a way of speaking about reality that differs from language about physics and biology. Rather than depicting the process of becoming or the mathematical structure of matter, the creation narrative states in a different way that there is only one

---

[71]Ibid., p. 309. One frequently encounters suggestions like those of Grudem that scientific research will ultimately settle the question of the age of the Earth. While sounding conciliatory and open to science, such comments are far too timid and indicate an unwillingness to acknowledge that the evidence in favor of an ancient Earth has been decisive for decades. Even scientific creationists, some in print and some privately, admit that the evidence as we have it favors an old Earth. No amount of additional evidence to support an old Earth will ever persuade willing acceptance of the conclusion if one does not want to be convinced.

[72]Millard J. Erickson, *Christian Theology,* vol. 1 (Grand Rapids: Baker, 1983).

[73]*Catechism of the Catholic Church* (New York: Doubleday, 1994).

[74]Pius XII, *Humani Generis: Encyclical of Pope Pius XII Concerning Some False Opinions Threatening to Undermine the Foundations of Catholic Doctrine,* August 12, 1950. See www.papalencyclicals.net.

[75]Pius XII, *The Proofs for the Existence of God in the Light of Modern Natural Science: Address to the Pontifical Academy of Sciences,* November 22, 1951.

God and that "the universe is not the scene of a struggle among dark forces but rather the creation of his Word."[76] For him the creation account represents truth in a symbolic manner. For example, he asserted that use of the numbers three, four, seven and ten in Genesis 1 does not reproduce the mathematical structure of the universe but tells something about the idea according to which the universe was constructed.

THE FRAMEWORK INTERPRETATION

One of the most important developments in the latter half of the twentieth century was the rising interest in the framework interpretation of Genesis 1 among conservative Christian scholars. The framework view is compatible with acceptance of the high antiquity of the Earth because time considerations are largely removed from the interpretation of Genesis 1. In the framework interpretation, the use of the seven days serves as a literary device or "framework" for presenting the completed work of creation. The days are not regarded chronologically but topically or symbolically, somewhat like the numbered sequences in the visions of Revelation. The days are not necessarily to be viewed as so many periods of time, either shorter or longer, during which the creative work took place.

Often overlooked by contemporary advocates of the framework view is that John D. Davis (1854-1926), a professor of Old Testament at Princeton Theological Seminary, articulated a framework interpretation of Genesis 1. In an 1892 article on the creation of the universe, Davis, best known as editor of *Davis's Dictionary of the Bible*, spoke of the employment of a framework of six days in which the chronological element had been subordinated to topical emphases, and pointed out the parallels between the first set of three days and the second set of three days in Genesis 1.[77]

Early European advocates of the framework view included Arie Noordtzij (1871-1944), a professor of Old Testament at the University of Utrecht, and Nicholas H. Ridderbos (1910-1981), a professor of Old Testament at the Theological College at Kampen.[78] In the United States, the framework hypothesis has been vigorously promulgated by Meredith G. Kline, who has served as a professor of Old Testament at Westminster Theological Seminary (Phila-

---

[76]Joseph Ratzinger, *In the Beginning . . . : A Catholic Understanding of the Story of Creation and the Fall* (Grand Rapids: Eerdmans, 1995).

[77]John D. Davis, "The Semitic Tradition of Creation," *Presbyterian & Reformed Review* 3 (1892): 448-61. This article was reprinted in John D. Davis, *Genesis and Semitic Tradition* (Grand Rapids: Baker, 1980).

[78]Arie Noordtzij, *Gods Woord en der Eeuwen Getuigenis: Het Oude Testament in het Licht der Oostersche Opgravingen* (Kampen: 1924); and Nicholas H. Ridderbos, *Is There a Conflict Between Genesis 1 and Natural Science?* trans. John Vriend (Grand Rapids: Eerdmans, 1957).

delphia), Gordon-Conwell Divinity School and Westminster Theological Seminary (California).[79] Support for various versions of the framework interpretation has appeared in the work of Old Testament professors like Bruce Waltke of Regent College and Dallas, Westminster and Reformed Theological Seminaries; John H. Stek of Calvin Theological Seminary; and Mark Futato of Reformed Theological Seminary; theologians like Henri Blocher, a systematic theologian at the Faculté Libre de Théologie Évangélique, Vaux-sur-Seine; Kenneth A. Mathews, a professor of divinity at Beeson Divinity School; Conrad Hyers, a professor of religion at Gustavus Adolphus College; and Ronald F. Youngblood of Bethel Seminary.[80] A very detailed and articulate defense of the framework theory was penned by Lee Irons, a Presbyterian minister.[81]

In support of the topical, rather than chronological, treatment of the days, proponents call attention to the parallelism between days one though three and days four through six, a textual feature noted at least as early as Thomas Aquinas. Day one speaks of light, day four speaks of light bearers; day two speaks of waters and firmament, day five speaks of dwellers in the waters and birds that fly across the firmament; day three speaks of land, and day six speaks of dwellers on the land. According to many framework advocates, the first set of three days refers to the establishment of realms or kingdoms—light, water and land. In contrast, the second set of three days refers to the population of those realms with various creatures—Sun, Moon, and stars; birds and fishes; land animals and humans. Kline considered these creatures as rulers or creature-kings in their realms. In addition, the creation of the realms and the filling of those realms with creatures remove the three problematic conditions of the initially uninhabitable earth—darkness, formlessness and emptiness (uninhabitability owing to lack of land), and the vast watery abyss. The elimination of the three problem conditions renders the earth suitable for human habitation. One presumed benefit of the framework view is that it downplays

---

[79]Meredith G. Kline, "Because It Had Not Rained," *Westminster Theological Journal* 20 (1958): 146-57; and "Space and Time in the Genesis Cosmogony," *Perspectives on Science and Christian Faith* 48 (1996): 2-15.

[80]For their expositions see Bruce Waltke with Cathi J. Fredricks, *Genesis: A Commentary* (Grand Rapids: Zondervan, 2001); John H. Stek, "What Says the Scripture?" in Howard J. Van Till, Robert E. Snow, John H. Stek and Davis A. Young, *Portraits of Creation: Biblical and Scientific Perspectives on the World's Formation* (Grand Rapids: Eerdmans, 1990), pp. 203-65; Mark D. Futato, "Because It Had Rained; A Study of Gen. 2:5-7 with Implications for Gen. 2:4-25 and Gen. 1:1—2:3," *Westminster Theological Journal* 60 (1998): 1-21; Henri Blocher, *In the Beginning: The Opening Chapters of Genesis* (Downers Grove: InterVarsity Press, 1984); Conrad Hyers, *The Meaning of Creation: Genesis and Modern Science* (Atlanta: John Knox Press, 1984); Ronald F. Youngblood, *The Book of Genesis: An Introductory Commentary*, 2nd ed. (Grand Rapids: Baker, 1991).

[81]Lee Irons with Meredith Kline, "The Framework View," in *The Genesis Debate*, ed. David G. Hagopian (Mission Viejo, Calif.: Crux Press, 2001), pp. 217-56.

the chronological relevance of the sequence of days and thus virtually removes all possibility of conflict with the sequence of events proposed by geology.

ANCIENT NEAR EASTERN CONTEXT OF GENESIS 1

Another extremely important twentieth-century development in evangelical circles is the realization by a growing number of biblical scholars that it is essential to read the creation account in the light of ancient Near Eastern literature and culture. Closely linked with this approach is the concept that Genesis 1 is fundamentally a polemic against the pagan cosmogonies of Israel's neighbors, or at least includes such polemic elements. Sometimes, though not always, proponents of this approach end up advocating something akin to the framework idea. The increasing popularity of this approach has been stimulated by the remarkable wealth of text-documentary discoveries from Near Eastern archeological sites during the past century and a half, which include striking parallels to the biblical accounts of creation and the Flood.

Commentators who have adopted this approach have typically implicitly rejected the entire harmonistic enterprise whereby the meaning of the biblical text has frequently been distorted in the interests of gaining some degree of agreement with contemporary scientific findings. Without explicitly saying as much, proponents of the view under consideration would reject the gap, day-age, intermittent-day and revelation-day interpretations out of hand because they attempt to import modern scientific questions into textual interpretation. Such questions are, in their judgment, alien to the intent of the author of Genesis 1. Likewise, the traditional approach of six sequential twenty-four-hour days as the first six days of Earth history would be rejected to the extent that the approach is employed as a solution to scientific questions such as the age of the Earth or biological evolution.

Proponents include John Walton, a professor of Old Testament at Wheaton College, and Gordon J. Wenham, a professor of Old Testament at Cheltenham and Gloucester College of Higher Education in England. Walton, for example, has written that "since God chose to use language to communicate, he also bound himself to a culture," a condescension that necessarily required accommodation to his audience on his part.[82] Because of this cultural link, Walton maintained that there is a danger that "we may unconsciously impose our ideas, culture, or worldview on the text." As a result, "successful interpreters must try to understand the cultural background of the ancient New East just as successful missionaries must learn the culture, language, and worldview of

---

[82]John H. Walton, *The NIV Application Commentary: Genesis* (Grand Rapids: Zondervan, 2001), p. 22. On the idea of accommodation, see Davis A. Young, *John Calvin and the Natural World* (Lanham, Md.: University Press of America, 2007).

the people they are trying to reach."[83] The geographic and physical conceptions of the world by the people of the ancient Near East, for example, were
considerably different from our own and must be taken into account in our
interpretation of the text. Thus, the firmament of Genesis 1:6-9 is a solid body
as far as the text is concerned. Walton argued throughout his discussion that
the text is talking about the divine establishment or assignment of functions,
rather than the manufacture of structures. The important issue for the Israelites was the function or purpose of things. "The very fact," Walton claimed,

> that the Bible's ability to use Israelite modes of thinking poses such a problem
> for us demonstrates how significantly we have been influenced by certain as-
> pects of our culture. We have been persuaded to believe that truth about origins
> can only be packaged in scientific terms, that the only cosmological reality is a
> scientifically informed reality, that if a cosmological text operates outside of the
> scientific realm, it ceases to be truth.[84]

Because of our obsession with science, Walton maintained that all too
often "we are asking the wrong questions. It only distorts the biblical text
to try to read science between the lines, as if the text were constructed to
accommodate modern scientific understanding. Nevertheless it also undermines the text to reduce it to a harmless variation of primitive mythological
misconceptions."[85]

In a similar manner, Wenham urged "that Genesis be read on its own terms,
not on ours."[86] He maintained that "Gen. 1 again affirms the unity of God over
against the polytheisms current everywhere else in the ancient Near East."[87]
He envisioned Genesis 1—11 as "an inspired retelling of ancient oriental traditions about the origins of the world with a view to presenting the nature of the
true God as one, omnipotent, omniscient, and good, as opposed to the fallible,
capricious, weak deities who populated the rest of the ancient world." He believed that Genesis is also "concerned to show that humanity is central in the
divine plan, not an afterthought" and that "man's plight is the product of his
own disobedience and indeed is bound to worsen without divine intervention."
If these things are indeed true, then "Gen 1—11 is setting out a picture of the
world that is at odds both with the polytheistic optimism of ancient Mesopotamia and the humanistic secularism of the modern world." Given that these
are the overriding concerns of Genesis, we must be careful in examining the

---

[83]Walton, *Genesis*, p. 23.
[84]Ibid., p. 89.
[85]Ibid., p. 94.
[86]Gordon J. Wenham, *Genesis 1—15*, Word Biblical Commentary 1 (Waco, Tex.: Word, 1987),
   p. liii.
[87]Ibid., p. xlix.

details of the text. "Though historical and scientific questions may be upper-most in our minds as we approach the text," Wenham warned, "it is doubtful whether they were in the writer's mind, and we should therefore be cautious about looking for answers to questions he was not concerned with." Wenham was concerned lest we, diverted by our own interests from the central thrust, would "miss what the LORD, our creator and redeemer, is saying to us."[88]

So, advocates of this newer approach to Genesis 1 stress that we must not expect to find in the text of Genesis 1 the answers to questions that are posed by modern scientific discovery about the antiquity of the Earth on the grounds that such questions were not the immediate interest of the biblical authors, divine or human. To do so would be to import an alien concern into our inter-pretation and would end up distorting and obscuring what the text is saying.

Some of these scholars also take pains to point out, following the lead of Gerhard Hasel, that the text is full of indicators that Genesis 1 carries on a sustained polemic against the pagan cosmogonies and "theologies" of Israel's neighbors.[89] The thrust of the text, in stark contrast to the pagan cosmogonies, is that there is only one living and true Creator God who has brought every-thing into existence by the word of his power. Everything that was worshiped by the Mesopotamians, the Egyptians or the Canaanites, such as stars, the sun, animals, trees, rivers and the like are shown by Genesis to be creatures of the living God, effortlessly brought into being. These created beings offer no resistance to God as he makes them. The sun and moon are not even men-tioned by name in the text because their Hebrew names resemble the names of Babylonian or Canaanite deities. The plants and animals reproduce after their own kind, emphasizing that they always remain plants and animals and do not become deities. The sea monsters that often played a role in pagan mythologies are at the behest of the all-powerful God of the Israelites. Rather than being a nuisance to the gods on the one hand or divine on the other hand, as in the case of the Pharaohs, human beings are the image of God, God's rep-resentatives on the earth. And rather than a dualistic world in which good and evil are equally ultimate, the world as created by God is a good world.

What then of the seven days on this approach? Many scholars would argue simply that the text, to stress God's orderly production of the world, took ad-vantage of the existing notion of the perfection of the number seven and the observance of a human work week of seven days by using the idea of a week as an analog. At the same time, God used the seven-day pattern to establish the observance of his sabbath. Advocates of this view claim that the text is

---

[88]Ibid., p. liii.
[89]Gerhard F. Hasel, "The Polemic Nature of the Genesis Cosmology," *The Evangelical Quarterly* 46 (1974): 81-102.

not concerned to tell us when God created the world, how he did it, and not necessarily even in what order he did it. Adherents of the view are also likely to stress the prolific use of symbolic numbers in Genesis 1. In particular, they note the recurrent patterns of three, seven and ten.

Although Wenham, for example, maintained that "there can be little doubt that here 'day' has its basic sense of a twenty-four-hour period," he is not a traditionalist, for he claimed that "it is perilous to try to correlate scientific theory and biblical revelation by appeal to such texts."[90] For Wenham, Genesis 1 "is a polemic against the mythico-religious concepts of the ancient Orient."[91] There is no hint that he thought of the six days as a literal 144 hours of creation. Instead,

> the six-day schema is but one of several means employed in this chapter to stress the system and order that has been built into creation. Other devices include the use of repeating formulae, the tendency to group words and phrases into tens and sevens, literary techniques such as chiasm and inclusio, the arrangement of creative acts into matching groups, and so on.[92]

"By speaking of six days of work followed by one day's rest," Wenham noted, "Gen. 1 draws attention to the correspondence between God's work and man's and God's rest and a model for the sabbath, but that does not necessarily imply that the six days of creation are the same as human days."[93] In effect, Wenham envisioned an analogy between the divine work week and the human week.

THE HIGHER CRITICAL TRADITION

Scholars who are committed to the results of source criticism that attributes Genesis 1 to a hypothetical editor (P) of a priestly school existing at the time of the exile typically downplay scientific considerations and view the days of creation as ordinary days, although they are more inclined to view the biblical language as a reflection of Israel's own thinking rather than as a divine revelation couched in language understandable to the Israelites. Gerhard von Rad (1901-1971), who taught Old Testament at the Universities of Leipzig, Jena, Göttingen and Heidelberg, maintained that Genesis 1 is concentrated priestly doctrine, and that it is "not concerned at all with things that interest the paleontologists only and that we could ignore."[94]

Claus Westermann, a professor of Old Testament exegesis at the University of Heidelberg, considered the question whether the world originated exactly

---

[90]Wenham, *Genesis 1—15*, p. 19.
[91]Ibid., p. 37.
[92]Ibid., p. 39.
[93]Ibid., p. 40.
[94]Gerhard von Rad, *Genesis: A Commentary* (Philadelphia: Westminster Press, 1961).

as the Bible said or as the natural sciences suggest to represent a complete misunderstanding of the Bible's reflection on creation by both sides. Creation, he said, must be understood in the language of myth and ritual, but myth must not be opposed to history. Instead, "myth must be regarded as a reflection on reality, as a presentation of what has actually happened" as it "accorded with man's understanding of existence and of the world in the early period."[95] Westermann maintained that P (i.e., the presumed priestly author of Genesis 1) presented "a whole, an articulated chronological unity, which is a whole because of its goal. It is not a question of seven times twenty-four hours." He saw the seven days as something like a parable in which P wanted to say that "time properly ordered and directed in carefully regulated periods towards its God-given goal, began with creation."[96] Neither von Rad nor Westermann touched on any relation between the creation account and the findings of modern science, nor did they address any of the concordistic views.

Langdon Gilkey, of the University of Chicago Divinity School, also stressed that Genesis 1 is not a literal historical document that is interested in scientific questions. Rather, in speaking of the act of creation, an act that transcends and begins history but is not incorporated into history, it employs the analogical and metaphorical language of myth.[97]

CHRISTIANS IN THE NATURAL SCIENCES

Christians in the natural sciences, particularly Christian professional geologists, are overwhelmingly in agreement that the Earth is extremely old, and they accept the current determination of approximately 4.55 billion years as the most reliable value for the age of Earth. Only a handful of members of Christian scientific organizations such as the American Scientific Affiliation and the Affiliation of Christian Geologists in the United States or Christians in Science in Great Britain, all composed predominantly of evangelical Christians, adhere to the idea of a young Earth that is only a few thousand years old. Proponents of a young Earth who are scientists are far more likely to be located in the Creation Research Society, of which very few members are practicing geologists. Although the scientific elite of the twentieth and twenty-first centuries have generally not written about Bible-science issues as did Hitchcock, Dana, Guyot and Dawson in the nineteenth century, there have been a number of important works written by professional Christian scientists in the second half of the twentieth century. Among Christian geologists who

---

[95]Claus Westermann, *Creation* (Philadelphia: Fortress, 1974).

[96]Claus Westermann, *Genesis 1—11: A Commentary* (Minneapolis: Augsburg, 1984).

[97]Langdon Gilkey, *Maker of Heaven and Earth: A Study of the Christian Doctrine of Creation* (Garden City, N.Y.: Doubleday, 1959).

adopted the day-age interpretation at one time or another are John Wiester, Donald Daae, Daniel E. Wonderly and Davis A. Young, who later abandoned that view.[98] Mormon geologist William Lee Stokes of the University of Utah also advocated the day-age view.[99] The framework view was favored by Michael R. Johnson of the Geological Survey of South Africa.[100]

Among astronomers and physicists, Howard Van Till of Calvin College leaned toward the framework view.[101] Astrophysicist-theologian Robert Newman of Biblical Theological Seminary in Pennsylvania and his physicist colleague Herman Eckelmann adopted an unusual variant of the day-age view in which the six days of creation were interpreted as ordinary days. These six days of divine fiat, however, were said to be separated by vast stretches of time, perhaps millions of years long, in which the fulfillment of the fiats was carried out.[102] British physicist Alan Hayward, astrophysicist Hugh Ross of Reasons to Believe and David Snoke, a professor of physics at the University of Pittsburgh, all adopted the day-age view.[103]

Among chemists, Charles Hummel of Barrington College advocated the framework theory in *The Galileo Connection*, whereas Russell Maatman of Dordt College preferred the day-age view.[104]

REACTIONARY DEVELOPMENTS

An astonishing and perplexing aspect of the twentieth-century scene was the

---

[98]John L. Wiester, *The Genesis Connection* (Nashville, Tenn.: Thomas Nelson, 1983); Dan Wonderly, *God's Time-Records in Ancient Sediments: Evidences of Long Time Spans in Earth's History* (Flint, Mich.: Crystal Press, 1977); Donald Daae, *Bridging the Gap: The First Six Days* (Calgary, Alberta: Genesis International Research Publishers, 1989); and Davis A. Young, *Creation and the Flood: An Alternative to Theistic Evolution and Flood Geology* (Grand Rapids: Baker, 1977).

[99]William Lee Stokes, *The Genesis Answer: A Scientist's Testament for Divine Creation* (Englewood Cliffs, N.J.: Prentice-Hall, 1984); and *The Creation Scriptures: A Witness for God in the Scientific Age* (Salt Lake City, Utah: Starstone Publishing, 1979). The latter work is explicitly Mormon, referring to verses not only from the Bible, but also the Book of Mormon, and the Doctrine and the Covenants.

[100]Michael R. Johnson, Genesis, *Geology, and Catastrophism: A Critique of Creationist Science and Biblical Literalism* (Exeter: Paternoster, 1988).

[101]Howard Van Till, *The Fourth Day* (Grand Rapids: Eerdmans, 1986).

[102]Robert C. Newman and Herman J. Eckelmann Jr., *Genesis One and the Origin of the Earth* (Downers Grove: InterVarsity Press, 1977).

[103]Alan Hayward, *Creation and Evolution: The Facts and Fallacies* (London: Triangle, 1985); Hugh Ross, *The Fingerprint of God: Recent Scientific Discoveries Reveal the Unmistakable Identity of the Creator*, 2nd ed. (Orange, Calif.: Promise Publishing, 1991); and David Snoke, *A Biblical Case for an Old Earth* (Grand Rapids: Baker, 2006).

[104]Charles E. Hummel, *The Galileo Connection: Resolving Conflicts Between Science & the Bible* (Downers Grove: InterVarsity Press, 1986); and Russell W. Maatman, *The Bible, Natural Science, and Evolution* (Grand Rapids: Reformed Fellowship, 1970).

remarkable, unabated resurgence of belief among many Christians in the sciences in the crucial geologic role of the biblical Flood and in the idea that the Earth is only a few thousands of years old—and this in the face of increasing geologic and astronomical evidence for the vast antiquity of the Earth and the universe.

The Flood geology movement in America has steadily gathered momentum throughout the twentieth century. Ronald Numbers, a historian of science at the University of Wisconsin and a former Seventh-day Adventist, has shown that much of the impetus for the resurgence of Flood geology can be traced to the influence of Seventh-day Adventist founder Ellen Gould White (1827-1915).[105] The major spokesman for Flood geology in the early decades of the century was self-taught Seventh-day Adventist "geologist" George McCready Price (1870-1963), who authored several books that defended catastrophic geology and attacked standard geologic theory.[106] Price rejected faunal succession, the stratigraphic column and *overthrust faults*.[107] In 1946, Seventh-day Adventist Harold W. Clark (1891-1986), a protégé of Price, published *The New Diluvialism*. Although Clark endorsed Flood geology, he succeeded in arousing Price's ire because he conceded the reality of both the geologic column and great overthrusts.[108] In subsequent years, several Seventh-day Adventist scientists, like Ariel Roth of Andrews University, Leonard Brand of the Geoscience Research Institute, and Arthur Chadwick and Elaine Kennedy of Loma Linda University, have expressed openness to the concept of an ancient Earth, but they are still committed to the idea of a very recent creation of life and a global Deluge. Seventh-day Adventists have also been very much in the thick of the search for remains of Noah's ark.[109]

Devotion to a young Earth and Flood geology spilled over from Seventh-day Adventism across a wide spectrum of denominations as well as Mormons and

---

[105]On E. G. White, see Ronald L. Numbers, *The Creationists: The Evolution of Scientific Creationism* (New York: Alfred A. Knopf, 1992), pp. 73-75.

[106]On Price, see Numbers, *The Creationists*, pp. 72-101. Among Price's more important works are George McCready Price, *The New Geology* (Mountain View, Calif.: Pacific Press, 1923), *Illogical Geology: The Weakest Point in the Evolution Theory* (Los Angeles: Modern Heretic Co., 1906), and *The Modern Flood Theory of Geology* (New York: Fleming H. Revell, 1935).

[107]Overthrust faults are typically gently dipping planar surfaces along which plates of rock (commonly stratified and up to thousands of feet thick) have been pushed up and over other masses of rock. Instances are known where rock masses have been transported more than 100 kilometers along thrust surfaces. Overthrusts invariably occur in association with strongly folded rocks in portions of Earth's crust that have been subjected to intense compression.

[108]On Clark, see Numbers, *The Creationists*, pp. 123-29. Clark's major work was Harold W. Clark, *The New Diluvialism* (Angwin, Calif.: Science Publications, 1946).

[109]On the search for Noah's ark, see Davis A. Young, *The Biblical Flood: A Case Study of the Church's Response to Extrabiblical Evidence* (Grand Rapids: Eerdmans, 1995), pp. 314-19. Several other sources are noted in this review of "arkeology."

Muslims. In recent decades, a host of biologists, physicists, chemists, geographers and engineers, but *extremely few geologists and astronomers*, have been insisting on a return to belief in creation in six twenty-four-hour days only a few thousand years ago; an abandonment of all theories of harmonization of modern geology with Scripture; and wholehearted acceptance of a catastrophic global Deluge that produced most of the stratigraphic and paleontological record.

In contemporary America there now exists a vigorous movement within the evangelical wing of the church that favors recent creation and Flood geology. This movement has very strong transdenominational support among Christians who are not engaged in scientific endeavor. Outside of Seventh-day Adventist circles, young-Earth and Flood geology has especially shown considerable popularity among conservative Lutherans. In 1931, a Norwegian Lutheran pastor, Byron Nelson (1893-1972), wrote an interesting and very sympathetic history of the flood theory.[110] At the Missouri Synod Lutheran Church's Concordia Theological Seminary in St. Louis, professor of Old Testament Alfred Rehwinkel (1887-1979) published *The Flood*.[111] The movement received its strongest impetus, however, with the publication in 1961 of *The Genesis Flood* by John C. Whitcomb, a professor of Old Testament at Grace Theological Seminary, and Henry M. Morris (1917-2006), then a professor of hydraulic engineering at Virginia Polytechnic Institute and later the founder and president of the Institute for Creation Research in El Cajon, California.[112] There followed a long list of writings devoted to the Flood and to the young-Earth theory, including numerous books by Henry Morris, including *Science, Scripture, and the Young Earth*, a critique of the first edition of *Christianity and the Age of the Earth*.[113]

Also of great significance was the formation of creation-oriented organizations. The landscape has been littered with essentially one-man young-Earth creationist operations, such as Kent Hovind's Creation Science Evangelism, Walt Brown's Center for Scientific Creation and Carl Baugh's Creation Evidence Museum. Of greater impact are organizations like Answers in Genesis (AiG) and the Institute for Creation Research (ICR). With origins in Australia, AiG has featured writers and speakers like Ken Ham and Andrew Snelling and

---

[110]Byron C. Nelson, *The Deluge Story in Stone: A History of the Flood Theory of Geology* (Minneapolis: Augsburg, 1931).

[111]Alfred M. Rehwinkel, *The Flood in the Light of the Bible, Geology, and Archaeology* (St. Louis: Concordia, 1951).

[112]John C. Whitcomb Jr. and Henry M. Morris, *The Genesis Flood: The Biblical Record and its Scientific Implications* (Philadelphia: Presbyterian & Reformed, 1961).

[113]Henry M. Morris, *Science, Scripture, and the Young Earth* (El Cajon, Calif.: Institute for Creation Research, 1983).

maintains a very active website. ICR was founded in 1970 by Henry Morris as the research arm of Christian Heritage College, but since 1981 it has been an autonomous graduate school devoted to the discovery of scientific support for a young Earth, a global Flood and other catastrophic interpretations of geology, and nonevolutionary explanations of the origins of organisms. ICR has regularly published a series, titled Acts and Facts, devoted to challenging standard geology and espousing young-Earth claims.

The Creation Research Society (CRS), founded in 1963, originally had a strong Missouri Synod Lutheran influence. Ronald Numbers has pointed out that one-third of the original charter membership came from that denomination, including biologists Walter Lammerts, Wilbert Rusch, John Klotz and Paul Zimmerman. Since 1964, the Creation Research Society has published a journal devoted to the espousal of catastrophism, Flood geology and a young Earth.

The young-Earth creationist movement has made inroads into the home-schooling movement, and curricular materials commonly endorse young-Earth claims. Young-Earth creationism is frequently featured on Christian radio and television programs. Repeated efforts, almost inevitably unsuccessful, have been made by advocates of a young Earth to introduce their brand of creationism into public school instruction. In 1981, Arkansas passed a law mandating public schools to give "balanced treatment" to both creation science and evolution in science courses, but the law was declared unconstitutional by U.S. District Judge Overton in the case of *McClean v. Arkansas.* A similar act favorable to creationism was adopted in Louisiana and then struck down in 1987 by the U.S. Supreme Court in the case of *Edwards v. Aguillard.*[114]

A significant development within the creationist movement was the establishment of the International Conference on Creationism held every fourth year in Pittsburgh, Pennsylvania.[115] Unlike most of the early creationist literature, many of the papers in the technical proceedings volumes published by the conference are marked by considerable scientific and mathematical sophistication. The movement has also been encouraged by the entrance of a handful of younger scholars with doctoral degrees in geology, a serious lack in the early days of creationism. The presence within the Flood geology movement of people with doctoral degrees like Leonard Brand, Arthur Chadwick, Stephen Austin, Kurt Wise, Andrew Snelling, Elaine Kennedy and Marcus Ross has not only lent greater scientific sophistication to the Flood geology movement but also has served as an internal check on some of the more egregious geologic

---

[114]On creationism in the courts, see Edward J. Larson, *Trial and Error: The American Controversy over Creation and Evolution,* 3rd ed. (New York: Oxford University Press, 2003).

[115]Every four years, proceedings volumes of the International Conference on Creationism are published by Creation Science Fellowship, Inc. of Pittsburgh, Penn.

errors in the writing of enthusiastic but ill-informed creationists. The publications of some of the more recent advocates of Flood geology such as Ariel Roth and Leonard Brand also have a much more irenic and moderate tone that provides a welcome contrast to the sarcastic, sometimes disrespectful tone and unwarrantedly dogmatic pronouncements of earlier creationists. Young-Earth creationists have, of course, continued to issue books and articles designed to convince people of the truth of a young Earth, and many of these will be noted in succeeding chapters.

Despite the facts that young-Earth creationism has become considerably more sophisticated and that some of its proponents are much more geologically knowledgeable than were earlier advocates like Price or Morris, the claims advanced in favor of a young Earth or of Flood geology remain unacceptable to the scientific community. Thus their claims should also be unacceptable within the church, which, of all places, ought to be committed to truth and reality—for the simple reason that the young-Earth creationist claims lack scientific credibility. They neither discredit evidence for an old Earth nor compel acceptance of a young Earth or a global Flood. Some of their claims are examined in chapters eight through fifteen.

A factor contributing to the remarkably widespread acceptance of young-Earth creationism since the nineteenth century is the strong link geology has acquired with the theory of biological evolution by natural selection, extending not only to lower animals and plants but to the human race as well.[116] The scientific, strictly biological conception of evolution, unfortunately, has on occasion been transformed into an antireligious and anti-Christian philosophy by scientists and philosophers who are committed to or lean toward materialism.[117] The materialistic philosophies of these writers take human beings out of the realm of creatures who are accountable to the creator God and place them into a realm where they are subject only to blind, mechanical forces and inherited instincts. As a result of such popularizations, the scientific theory of biological development has become so closely identified with a materialistic worldview in the minds of many Christians, including young-Earth creation-

---

[116]The close linkage of young-Earth creationism, Flood geology and anti-evolutionism has been amply demonstrated by Numbers, *The Creationists*.

[117]For a small, representative sampling of materialism presented as natural science, see Richard Dawkins, *The Blind Watchmaker* (Harlow, Essex: Longman Scientific and Technical, 1986); Daniel Dennett, *Darwin's Dangerous Idea: Evolution and the Meanings of Life* (New York: Simon & Schuster, 1995); Jacques Monod, *Chance and Necessity: An Essay on the Natural Philosophy of Modern Biology* (New York: Alfred A. Knopf, 1971); Peter W. Atkins, *The Creation* (San Francisco: W. H. Freeman, 1981); Francis Crick, *Life Itself: Its Origin and Nature* (New York: Simon & Schuster, 1981); and Richard Dawkins, *The God Delusion* (Boston: Houghton Mifflin, 2006).

ists, that they throw out the baby with the bath water, calling not merely for the repudiation of materialism but also for the rejection of evolution as a legitimate scientific theory.[118]

Without question, any purely materialistic philosophy is hostile to Christianity and ought to be opposed by Christians. Christians should not, however, attempt to disprove a materialistic evolutionary theory by discrediting the antiquity of the Earth. *Evolutionary materialism and the antiquity of the Earth are two distinct issues.* If the vast antiquity of the Earth is amply demonstrated, one must still evaluate the data and theory of evolution on their own *scientific* merits.[119]

To summarize, in the face of the facts that the scientific community was virtually unanimous in accepting the vast antiquity of the Earth throughout the twentieth century right up to the present, and that geologic (and astronomical) evidence for such antiquity has continued to accumulate, significant segments of the Christian church have regressed by welcoming scientifically discredited ideas that include Flood geology and a very young Earth. For whatever reasons, acceptance of scientific findings prevails among academic theologians and the vast majority of Christian geologists, whereas acceptance of young-Earth creationism, anti-evolutionism and Flood geology prevails among pastors and lay Christians. Moreover, an astonishing number of Christians in sciences other than geology and astronomy, along with numerous Christian physicians and engineers, are quite enthusiastic in their support of these discredited theories.

This state of affairs, in our judgment, reflects lack of appropriate geologic knowledge coupled with existence of tendencies to read Scripture in overly literalistic ways that fail to take into account the ancient Near Eastern cultural background of much of Scripture and the primary pedagogical concern of the Bible. In the rest of the book, our aim is to challenge readers, whether Christian or not, to see the weaknesses in a literalistic interpretation of Genesis 1 and to appreciate the force of the divinely established geologic evidence for an extremely ancient Earth.

---

[118]The failure to give adequate recognition to the distinction between the scientific theory of biological evolution and the philosophy of materialistic evolutionism characterizes such works as Nancy Pearcey, *Total Truth: Liberating Christianity from Its Cultural Captivity* (Wheaton, Ill.: Crossway Books, 2004); Charles Colson and Nancy Pearcey, *How Now Shall We Live?* (Wheaton, Ill.: Tyndale House, 1999); and Phillip E. Johnson, *Darwin on Trial* (Downers Grove: InterVarsity Press, 1991).

[119]By way of example, Hugh Ross, an astrophysicist, and C. John Collins, an Old Testament theologian, are both enthusiastic advocates of an old Earth who, nevertheless, ardently oppose biological evolution.

Part Two

BIBLICAL PERSPECTIVES

6

THE BIBLE AND THE
ANTIQUITY OF THE EARTH

Part One

AS NOTED IN THE PREVIOUS CHAPTER, some Christian groups challenge the notion, widely accepted throughout the twentieth century by Christian and non-Christian scientists, that the Earth is billions of years old. That challenge is made on biblical, scientific and philosophical grounds. First and foremost, the challenge is made in terms of biblical interpretation. Because the biblical commitments of young-Earth creationists are determinative in their thinking, we tackle some of the biblical issues in this and the next chapter before we deal with scientific and philosophical questions. Second, their challenge to the antiquity of the Earth is presented in scientific terms. In chapters eight through fifteen we refute several purported scientific arguments for a young Earth and present some of the evidence favoring the great antiquity of the Earth. Third, the challenge to the Earth's high antiquity is made philosophically. We discuss the philosophical issues in chapters sixteen and seventeen.

So, let's begin our look at the biblical teaching about the age of the Earth. For some readers, the teaching of the Bible on this issue seems so *obvious* that they suspect anyone who disputes what they perceive as the obvious teaching is guilty of deliberate rejection of Scripture. To dispute their understanding is equivalent to questioning God himself! But let's be cautious. Whether we are aware of it or not, every reader of the Bible unavoidably interprets all biblical texts. Each of us brings to the biblical text assumptions and expectations of which we are not always aware. A challenge for all Christians is to become hermeneutically self-conscious by familiarizing themselves with those assumptions and expectations that influence the way in which they might understand a particular passage. We call attention to some of the assumptions that people bring to Genesis 1. All Christians need to do a better job of distinguishing a properly based interpretation that recovers the actual teaching of the Scripture from a flawed interpretation that is unknowingly imposed on the text and that distorts the intended message. Because the interpretation of Genesis 1 is particularly challenging, we encourage believers to study that crucial chapter with humility and openness to new insights that they may not have encountered before.

Caution is definitely in order when dealing with the beginning chapters of the Bible. The great church father Augustine spent a lifetime thinking about the proper meaning of Genesis 1. He wrote about early Genesis in at least four of his books. His most mature reflection on Genesis is *The Literal Meaning of Genesis,* a commentary on the first three chapters that was written when he was in his fifties.[1] In book 1, chapter 20, we find Augustine's startling comment:

> [I had] worked out and presented the statements of the Book of Genesis in a variety of ways according to my ability; and, in interpreting words that have been written obscurely for the purpose of stimulating our thought, I have not rashly taken my stand on one side against a rival interpretation which might possibly be better. I have thought that each one, in keeping with his powers of understanding, should choose the interpretation that he can grasp.[2]

He also said in book 1, chapter 18 that

> in matters that are obscure and far beyond our vision, even in such as we may find treated in Holy Scripture, different interpretations are sometimes possible without prejudice to the faith we have received. In such a case, we should not rush in headlong and so firmly take our stand on one side that, if further progress in the search of truth justly undermines this position, we too fall with it. That would be to battle not for the teaching of Holy Scripture but for our own, wishing its teaching to conform to ours, whereas we ought to wish ours to conform to that of Sacred Scripture.[3]

---

[1]Augustine, *The Literal Meaning of Genesis,* trans. John Hammond Taylor (New York: Newman Press, 1982).
[2]Ibid., pp. 43-44.
[3]Ibid., p. 41.

Augustine cautioned the church not to be dogmatic about specific interpretations of Genesis 1—3. Readers of the Bible would do well to heed his advice, especially in view of the host of interpretations of Genesis 1 available. Maybe readers can learn something of value from each point of view.

Given the variety of interpretations of Genesis 1 that have been put forward, what should the Christian layperson believe? Advocates of various positions have persuasively set forth their cases. In this and the following chapter we do not provide a definitive and comprehensive solution to the interpretation of Genesis 1. In a spirit of openness, however, we make several observations that must not be ignored in arriving at an understanding of what the Bible does or does not teach about the age of the Earth, particularly in light of Genesis 1. Readers who want to pursue further details of the discussion about Genesis 1 may wish to consult *The Genesis Debate,* a book that provides detailed arguments for three of the major positions, responses to each by advocates of the two opposing positions, and final replies by the advocates of the position being presented. The three viewpoints debated in that book are the traditional twenty-four-hour-day view, the day-age view and the framework view. The absence of a presentation of the gap theory is one indication of that theory's loss of popularity in recent decades.[4]

THE TRADITIONAL VIEW

No appeal to scientific evidence will ever convince many young-Earth creationists of the great antiquity of the Earth because they are persuaded that the only reliable and definitive source of information about the age of the Earth is the Bible. Inasmuch as millions of Christians are led by their pastors to believe that the Bible teaches that the world was created in six literal days only a few thousand years ago, they have been conditioned to believe that the Bible unequivocally does teach that view. They may also commonly receive intimations from the pulpit that the findings of scientists about an old Earth and related issues are to be viewed with great suspicion, because they perceive

---

[4]D. G. Hagopian, ed., *The Genesis Debate: Three Views on the Days of Creation* (Mission Viejo, Calif.: Crux Press, 2001). Unfortunately, Hagopian's book has omitted some very important alternative points of view. For example, a defense of the "analogical" day view is given in Collins, *Science and Faith: Friends or Foes?* (Wheaton, Ill.: Crossway Books, 2003). Readers should also digest John Stek's very important article titled "What Says the Scripture?" in Howard J. Van Till, Robert E. Snow, John H. Stek and Davis A. Young, *Portraits of Creation: Biblical and Scientific Perspectives on the World's Formation* (Grand Rapids: Eerdmans, 1990), pp. 203-65. Stek offered a very detailed exegetical study of Genesis 1 and of the creation texts in Scripture. Another very stimulating interpretation of Genesis 1 is found in John H. Walton, *NIV Application Commentary: Genesis* (Grand Rapids: Zondervan, 2001). Walton makes an interesting case that the creation account of Genesis 1 primarily has in view God's establishment of the *functions* of created things more than their physical origination or "manufacture."

that scientists are hostile to Christian faith and because they suspect that the idea of an old Earth and the much-feared theory of biological evolution go hand in glove. For them it is a matter of believing, pure and simple, what the omnipotent, omniscient, truth-telling God has said rather than what fallible, sinful and sometimes truth-distorting scientists say.

There are two major Bible-based reasons why so many Christians are taught that the Earth is only a few thousand years old. The first reason is that the genealogies throughout the Bible provide numbers suggesting the ages of characters like Adam, Noah or Joseph. There are also numbers that concern the amount of time elapsed between events, such as the 480 years between the exodus and the building of the Solomonic temple (1 Kings 6:1). Throughout church history, scholars have been unable to resist adding up the various numbers as we saw in chapter one. When such numbers are added, one obtains a history that is only a few thousands of years long. Assuming that these calculations are correct, however, we end up with the age of humanity, not the age of the universe or of the Earth.

Many biblical scholars starting with William Henry Green (1825-1900) have pointed out that the genealogies were not intended to tell us how long humans have been on the Earth.[5] The genealogies are not continuous; they contain gaps, some of them quite large.[6] Commentators have also pointed out that many of the genealogies, such as the one of Jesus in Matthew 1, have been deliberately structured to achieve numerical symmetry. Matthew's genealogy of Jesus Christ is divided into three groups of fourteen names apiece. But in order to achieve the desired fourteen names in one of the groups, Matthew omitted the names of three of the kings of Judah: Ahaziah, Joash and Amaziah. Such considerations call into question the propriety of using the genealogies to calculate the age of humanity. Although we believe that the human race is far older than 4004 B.C. or 3928 B.C. or some similar number, the antiquity of humanity is not the focus of our book, so we will not pursue this issue further.

The second reason that Christians adopt a young-Earth stance is because Genesis 1 seems to teach that the six days of creation were only twenty-four hours long. They appear to be normal, twenty-four-hour solar days deter-

---

[5]For discussions of the genealogies of Genesis 5 and 11, see William Henry Green, "Primeval Chronology," *Bibliotheca Sacra* 47 (1890): 285-303; Umberto Cassuto, *A Commentary on the Book of Genesis*, Part 1, trans. Israel Abrahams (Jerusalem: Magnes Press, 1972); R. R. Wilson, "The Old Testament Genealogies in Recent Research," *Journal of Biblical Literature* 94 (1975): 169-89; Carol A. Hill, "Making Sense of the Numbers of Genesis," *Perspectives on Science and Christian Faith* 55 (2003): 239-51; and Kenneth A. Kitchen, *The Reliability of the Old Testament* (Grand Rapids: Eerdmans, 2003).

[6]Green, for example, called attention to the contrast between the genealogy of Ezra given in Ezra 7:1-5 and that contained in 1 Chronicles 6:3-14.

mined by the spin of the earth on its axis. The six days of creation also are said to have occurred in temporal sequence. They refer to the amount of time during which specific creative events were accomplished on the earth or in the heavens. If we add six ordinary days to whatever date we arrive at for the age of humanity on the basis of biblical genealogies, we end up with virtually the same age for the beginning of creation. If we are strict literalists in regard to the genealogies and decide that Adam was created in 4000 B.C., and we tack on six solar days, then obviously creation still occurred in 4000 B.C. or very late 4001 B.C. If we allow for some flexibility in the genealogies and decide that they allow for the creation of Adam in 20,000 B.C., we still end up with a universe created in 20,000 B.C. if we merely add six more days to our total.

Adherence to the traditional six sequential, normal, twenty-four-hour days on which the whole of the work of creation was accomplished, therefore, virtually locks one into belief in a young earth that is only a few thousand years old, because no traditionalist is prepared to state that the biblical genealogies are so elastic as to allow for, say, millions of years of human history. From a biblical perspective, the issue of the antiquity of the earth boils down to the interpretation of Genesis 1. Moreover, the traditional interpretation is the only one of which we are aware that locks us in to a very young earth. Other interpretations like the gap, day-age, framework, revelation-day, intermittent-day or analogical-day views can be compatible with an acceptance of either a young or an old earth and universe. Because the traditional view of Genesis 1 is the only one that virtually *requires* commitment to a few-thousand-year-old Earth, our discussion in chapters six and seven interacts primarily with that interpretation.

If it can be shown that the Bible does not demand adherence to the traditional interpretation of Genesis 1, then it means that Scripture is noncommittal in regard to the age of the Earth. And if the Bible does not specify the age of the earth, we are free to evaluate geologic clues to find out its age. Moreover, if we can show that the traditional view does not compel acceptance, we are not required to provide another ironclad interpretation to take its place. We can follow Augustine's advice to adopt tentatively the best interpretation that one can come up with. Our plan, then, is to make a case that Scripture does not demand acceptance of the traditional interpretation of Genesis 1 and, with it, the idea of a young earth.

THE ROLE OF TRADITION IN THE TRADITIONAL VIEW

We first respond to various nonexegetical arguments that have been advanced in favor of the traditional view or against rival interpretations. Then we look at some of the exegetical questions. The traditional twenty-four-hour-day view has been ably defended, for example, in *The Genesis Debate* by J. Ligon Duncan and David W. Hall, both of whom are pastors in the Presbyterian Church in America.[7] In both their positive statement in support of the traditional view and in replies to criticisms, they repeatedly hammered home their contention that the twenty-four-hour interpretation of Genesis 1 is the historic view of the church. Although conceding that Augustine and Origen held to more figurative interpretations of the days of creation, they argued vigorously that virtually all the other church fathers who wrote about the issue, such as Basil of Caesarea, assumed or explicitly asserted that the days of creation were twenty-four hours long. This view, they maintained, persisted through the time of the Reformers and began to be challenged only with the advent of the scientific worldview. Duncan and Hall went to great lengths to demonstrate that none of the framers of the Westminster Confession of Faith, to which they, as Presbyterian ministers, subscribe, adopted any view of the days of creation other than the traditional twenty-four-hour-day interpretation.[8] According to them, the framers of the Confession, following Calvin, opposed Augustine's ideas that God created the entire world in a moment and that the creation days were figurative. Duncan and Hall chastised proponents of alternative views for dismissing the great minds of the church's past and intimated that there is something suspicious afoot if, in hundreds of years of church history, no theologian ever came up with an alternative interpretation on exegetical grounds until the rise of science.

We agree that church history ought to play an important role in our interpretation of the biblical text. Christians should not lightly cast aside widely held interpretations of the past, and they need to respect the views of the church fathers. The church fathers, however, were no more infallible than are modern exegetes, and, in addition, they did not have access to important information about the ancient Near East that we do that bears on the interpretation of Genesis 1. As a result, they could not be expected to have addressed some of the matters we consider today. In many ways we are in a better position than our forebears to interpret the text properly. Moreover, it is Scripture and not tradition, however honorable, that is the final court of appeal in establishing an appropriate interpretation of Scripture. We all need to remind ourselves that the church fathers virtually unanimously held other biblical interpretations that have subsequently been rejected. One example would be the inter-

---

[7]J. Ligon Duncan and David W. Hall, "The Twenty-Four-Hour View," in Hagopian, *The Genesis Debate*, pp. 21-66. See also "The Twenty-Four-Hour Reply," pp. 95-119, and "The Twenty-Four-Hour Response," pp. 165-77, 257-68.

[8]David W. Hall, "What Was the View of the Westminster Assembly Divines on Creation Days?" in *Did God Create in Six Days?* ed. Joseph A. Pipa Jr. and David W. Hall (Taylors, S.C.: Southern Presbyterian Press, 1999), pp. 41-52.

pretation that the six days of creation correspond to six millennial epochs of history and that the seventh day corresponds to a final thousand-year reign of Christ. Eventually the church realized that these views, even though they were dominant in the first few centuries of the church, should not be maintained. As another example with which we are more familiar, for seventeen centuries, scholars interpreted biblical texts such as Psalm 93:1 or Joshua 10:12-14 in accord with the universal belief that the Earth is stationary. That interpretation must be incorrect if the Earth moves around the Sun.

So, as much as we may admire the great tradition of adherence to the view that creation occurred in six twenty-four-hour days, we must finally turn for our understanding not to tradition but to Scripture read within, not in isolation from, the context of the world in which it was written. The correctness of the traditional view is not guaranteed by the mere fact that it is traditional. As long as Scripture is the foundation of Christian theology, the time *never* comes when the church can say that exegetes may no longer alter or fine tune the interpretation of a specific text in light of new information and insights. Scripture, after all, is an inexhaustible mine. Even discoveries from the natural world, which is, after all, God's world, can serve as an impetus for renewed examination of the text and the development of better interpretations.

SHOULD BIBLICAL INTERPRETATION MAKE CONCESSIONS TO SCIENCE AND CULTURAL ISSUES?

But some scholars aren't so sure about the claim we made in that last sentence. Closely related to the argument that tradition favors the twenty-four-hour view is the common criticism, also voiced by Duncan and Hall against interpretations like the gap, day-age and framework interpretations, that they are novel and that nobody would ever have thought of them had it not been for the rise of modern science. These interpretations are, therefore, viewed as concessions to science, more beholden to science than to the text of the Bible. As a result, such interpretations, in their judgment, weaken the authority of Scripture and place natural science on a par with, or even above, Scripture.

Moreover, references to modern natural science by adherents of the traditional view often carry a negative tone of suspicion. Natural science is frequently referred to pejoratively as "secular" science and may be linked to non-Christian worldviews. One favorite charge is that the influence of Darwinian evolution was a major factor in the rush to adopt these alternative interpretations. Why, it is asked, should we overthrow nearly two millennia of "purely" biblical exegesis by the spiritual, godly intellects of old who were not misled by the fallible attempts of humans to interpret nature?

Several things need to be said in response to such suggestions. First, it is unwise to pit natural science as a fallible human enterprise that is always adjusting to new data over against an infallible Scripture. If we do that, we are comparing two things that should not be directly compared. The interpretation of the Bible is also always done by fallible human beings, the Roman Catholic belief in an infallible pope or an infallible teaching magisterium of the church notwithstanding, despite the fact that the Holy Spirit leads his church in the enterprise of biblical interpretation. If Spirit-led biblical interpretation were a foolproof procedure yielding infallible results, the church obviously would not possess very much, if any, variation of interpretation of a wide range of texts by a host of godly people. The church certainly would not be splintered into so many denominations, each of which is fully convinced that its beliefs are the most faithful to Scripture. If biblical exegetes were infallible, they would all agree on the interpretation of every text.

Because God also reveals himself in nature, it is unwise, therefore, for exegetes or Christians in general to disparage the fruits of the scientific investigation of God's handiwork by dismissing scientific findings without even understanding what it is they are dismissing. Moreover, godly exegetes have changed their minds because of continued engagement with Scripture just as much as scientists have changed their views in response to new discoveries from the natural world.

The charge that modern natural science is "secular" or based on a non-Christian worldview is overstated and misleading. The scientific enterprise is conducted by hundreds of thousands of individuals around the globe. A significant proportion of practicing scientists are Christians who operate with Christian presuppositions. Christian astronomers, biologists, chemists, cosmologists, geologists, paleontologists, physical anthropologists and physicists are not unwitting dupes who go along blindly with whatever an allegedly atheistic scientific establishment dictates. Although it *is* the case is that some *individual scientists* are indeed "secular" and operate with non-Christian worldviews, that charge cannot properly be laid at the feet of the entire enterprise. Moreover, although there were other important historical factors, modern natural science emerged out of a Christian milieu. Many significant early scientists like Gesner, Rondelet, Kepler, Galileo, Gassendi, Pascal, Boyle, Ray and Linnaeus were Christians. Even antitrinitarians like Newton and Whiston devoted as much of their time to biblical studies as they did to scientific work. Many of the founders of geology discussed earlier were Christians: Steno, Woodward, Deluc, Buckland, Fleming, Sedgwick and Conybeare, for example. They did their scientific work for the glory of God.

Christians who do not like the conclusions that natural science has drawn about the antiquity of the Earth are on shaky ground if they attempt to discredit scientific work because the conclusions or theories of science change from time to time. It is precisely this openness and changeability of science that has led to solid and reliable knowledge in instance after instance. There are certain ideas about which science is extremely unlikely to change because the evidential support accumulated over the years for those ideas has achieved overwhelming proportions. Physics is not going to change in regard to the applicability of Newton's laws of motion to the mechanics of the solar system. Science is not going to abandon the cell theory of biology, the existence of the electromagnetic spectrum, the data about the rotation and revolution periods of Mars, or the basic principles of electronics. These matters are firmly established on an abundance of critical evidence. The antiquity of the Earth falls into that category.

One cannot legitimately dismiss the idea of an old Earth on the basis that science changes its mind. What might conceivably change is our understanding of the exact age of the Earth, although the currently accepted age of 4.5 to 4.6 billion years has much in its favor.[9] So, scientists may continue to refine the exact chronology of the Earth, but the evidence for the general idea of the vast antiquity of the Earth is so overwhelming and credible that geology will *never* return to the idea that our globe is only a few thousands of years old. That would be a case of "been there, done that." As we saw in chapter two, mainstream science formerly accepted a young Earth, but, in light of the accumulating *God-created* evidence in the rocks, abandoned that idea. Science would no more return to the idea of a young Earth than it would return to acceptance of the idea that the Sun and stars revolve around the Earth.

Instead of viewing modern natural science as a threat or a competitor to Scripture, Christians ought to consider this enterprise as something potentially good and positive in relation to the Bible. We challenge readers to consider that the God who has created this world and who reveals himself in the created order has, in his infinite wisdom, provided modern natural science not just for the technological creature comforts it can provide us but also as an extremely valuable intellectual tool to aid the church in weeding out questionable interpretations of some Bible passages. Although we believe that natural science does not and cannot provide a positive interpretation of what a biblical text says, science certainly can raise questions about the validity of traditional interpretations, thus encouraging us to rethink more thoroughly what the text is really saying. The discoveries of Copernicus and Galileo were providential gifts from God that forced Christian scholars to rethink the meaning of texts

---

[9]G. Brent Dalrymple, "The Age of the Earth in the Twentieth Century: A Problem (Mostly) Solved," in *The Age of the Earth: 4004 B.C. to A.D. 2002*, GSL Special Publication 190, ed. Cherry L. E. Lewis and Simon J. Knell (London: The Geological Society, 2001), pp. 205-21.

like Psalm 93:1 and Joshua 10:12-14. Reinterpretation of these texts did *not* come about because of strictly exegetical examination of the text in complete isolation from all other factors. The scientific discovery that the Earth spins on its axis and revolves about the Sun drove exegetes to reconsider the meaning of the text.

More specifically, Duncan and Hall, as well as many others, have charged that allegiance to day-age and gap interpretations and the accompanying departure from the traditional interpretation can be traced to the evolutionary theory of Darwin. Such an assertion, however, is historically incorrect and misleading. As we saw in chapters three and four, *the day-age and gap theories were in vogue long before Darwin.* Chalmers was advocating the gap theory before Darwin was born. Many geological advocates of these interpretations, such as Deluc, Cuvier, Parkinson, Buckland, Sedgwick, Hitchcock and Miller, were completely opposed to evolutionary interpretations of the fossil record. Some of the leading advocates of the day-age interpretation who lived after Darwin announced his theory, such as Dana, Guyot and especially Dawson, rejected Darwinian evolution. So, advocates of the traditional view cannot legitimately argue support for their view on the grounds that Darwinism has inspired alternative interpretations.

Finally, the exegetes of the scientific era are not the only ones who have been influenced by their cultural milieu. It is impossible to exegete Scripture in a cultural vacuum. Every biblical interpreter throughout the history of the church, including the church fathers, has been unavoidably shaped and influenced by the cultural context in which he or she lived. To be sure, the Christians of today read the Bible through the lens of the knowledge that the Earth is a globular planet that orbits the Sun. The exegetes of old were not influenced by those findings of science. They were, however, inhabitants of a culture that was shaped by a different set of scientific ideas. Their world was dominated by the geocentric hypothesis, and the world as they conceived it had no American continents or inhabitants! For many of them the sky was solid, and almost everyone assumed that the world was only a few thousands of years old. They were also much indebted to Platonic philosophy. Early exegetes would inevitably have interpreted biblical texts in light of these assumptions about their world, because they were just as subject to external influences as we are today. Contemporary believers need to be conscious about cultural influences as they approach the Bible, and instead of pouring modern culture into their reading of the text, they need to learn to read the text in light of the culture in which it was originally addressed. After Christians understand what the text meant originally, they will be in a position to apply it to their own situation. We return to this point later.

In summary, the traditional twenty-four-hour-day view is not validated simply because rival interpretations may have been conceived after the rise of modern science.

Historicity

Now let's consider the issue of the historicity of Genesis 1. One of the claims frequently made by proponents of the traditional twenty-four-hour-day view is that alternative interpretations of Genesis 1, particularly the framework view, threaten the historicity of Scripture by threatening the historicity of the creation account. Charges are sometimes made that at least some nontraditional views relegate Genesis 1 to the realm of fiction, leaving us simply with a theological statement devoid of historical content.

In response, let's establish a few ground rules. We agree with proponents of the traditional view that Scripture is not simply a record of the religious beliefs of ancient Israel. Nor is it a collection of purely fictional myths, legends and manufactured history. The Bible is not a book of human religious opinion. We believe that the Bible is the infallible Word of God, and is, therefore, normative for our faith and life. Scripture teaches what we are to believe concerning God and what duty God requires of us. As such, Christians must take the utmost care in dealing with the text.

We also agree with proponents of the twenty-four-hour-day view that the Bible is a historical book. The Bible is fundamentally the story that recounts God's mighty acts in history with the patriarchs, with the people of Israel, in the advent of Jesus Christ, in the establishment of the church and in the spread of the gospel to the Gentiles during the apostolic era. The Bible tells us about God's interaction with real human beings who lived in earthly times and earthly places and who had distinctive personalities and roles in life like Abraham, Moses, Deborah, David and Isaiah. A large portion of the Bible consists of narratives—like Genesis, parts of Exodus, Joshua, Judges, Ruth, Samuel, Kings, Daniel, Chronicles, Ezra, Nehemiah, Esther, the Gospels and Acts—that purport to be historical and that give the impression that they took place in real space and time.

Moreover, despite the claim of some critics that many of the major personages of the Bible, like Abraham or David, or the events portrayed in the Bible, like the Israelite conquest of the land of Canaan, are fictitious, Egyptologist Kenneth A. Kitchen of the University of Liverpool has demonstrated in compelling fashion that the historical elements of the Old Testament are strikingly consistent with abundant archeological evidence coming from throughout the Near East. He has roundly taken rationalistic theologians to task for spinning theological fantasies out of thin air without regard for the realities of the archeological record and has convincingly shown that archeology lends support to the historical reliability of the Old Testament narratives.[10] F. F. Bruce of the University of Manchester and many other scholars have also made compelling cases for the reliability of the New Testament documents. James D. G. Dunn, for example, has effectively shown that it is extremely unlikely that the early church invented the portraits of Jesus given to us in the Gospels.[11] Instead, they give us a reliable picture of Jesus as he was remembered by those who had contact with him.

To say that the Bible is a historical book, however, does not mean that every biblical passage is historical in the same way or even that every biblical passage intends to convey historical information. Some of the poems of the Psalms, the oracular utterances of the prophets, portions of the apostolic letters, and the apocalyptic vision of Revelation, so richly laden with imagery and symbolism, obviously are not historical narratives. Of course, these were all written in definite historical contexts and addressed the conditions of the times in which they were written. Even so, many of the Psalms are timeless hymns of praise. Proverbs and Ecclesiastes are examples of timeless wisdom literature. So there is much in the Bible that is not expressly history writing.

In making the claim that the Bible is a historical book, too, it is extremely important to make a distinction between the event that is being reported and the manner in which the event is being reported. There is a great difference between what actually transpired in space and time and the manner in which the historical events are reported or portrayed. Not all historical events in the Bible are presented as straightforward narratives. For example, we have already seen that biblical genealogies are commonly interrupted by omissions of generations. In some cases, a historical event may be celebrated in poetic fashion as is the case with Moses' celebration of divine deliverance from Pharaoh and the Egyptians (Exodus 15:1-18) or Deborah's celebration of divine deliverance from Sisera at the hands of the opportunistic Jael and her death-inflicting tent spike (Judges 5). Although these poems have a historical referent, that is, they poetically recount historical events, they also contain metaphorical elements, dramatic hyperbole and so on. In other words, we do not rely on these passages

---

[10]Kitchen, *The Reliability of the Old Testament*.

[11]On the reliability of the New Testament, see, for example, F. F. Bruce, *The New Testament Documents: Are They Reliable?* (Grand Rapids: Baker, 2003); James D. G. Dunn, *Jesus Remembered* (Grand Rapids: Eerdmans, 2003); Craig L. Blomberg, *The Historical Reliability of the Gospels*, 2nd ed. (Downers Grove: InterVarsity Press, 2007); Scot McKnight, *Jesus and His Death: Historiography, the Historical Jesus, and Atonement Theory* (Waco, Tex.: Baylor University Press, 2005); N. T. Wright, *The Challenge of Jesus: Rediscovering Who Jesus Was and Is* (Downers Grove: InterVarsity Press, 1999); M. J. Wilkins and J. P. Moreland, eds., *Jesus Under Fire* (Grand Rapids: Zondervan, 1995); and Paul Rhodes Eddy and Gregory A. Boyd, *The Jesus Legend: A Case for the Historical Reliability of the Synoptic Jesus Tradition* (Grand Rapids: Baker, 2007).

to provide specific details of the events because that is not their intent. Thus, it is unlikely that the stars literally fought against Sisera or that the Egyptian army literally sank like lead into the sea.

Consider too the parable that Nathan the prophet told to King David subsequent to his sin with Bathsheba. We know that the parable was a parable because the surrounding narrative in 2 Samuel 11—12 has given us the real, historical, sordid story of David's moral failure. Curiously, David did not realize that Nathan was telling him a parable. After Nathan told the story of the rich man, the poor man and his ewe lamb, David obviously did not interpret the story as parable; he interpreted it in a strictly literal fashion as an actual historical occurrence in which some rich fellow stole a poor neighbor's sheep to entertain a guest. David was so incensed against the rich man that he demanded his death. Nathan was telling the historical truth, but he was telling it as a parable, a literary form that many of us learned in our Sunday school days is an "earthly story with a heavenly meaning." To his deep embarrassment and humiliation David found out that *he* was the rich man who stole the ewe lamb, not literally but figuratively. The parable is historical in that it refers to an actual event, but we don't turn to the parable itself to glean specific historical details of the event. That is not Nathan's intent in telling the parable. Our point is that some biblical passages that may be "historical" in a loose sense are not necessarily historical in the sense of providing lots of factual content. In other words, historical content may be transmitted by means other than strictly narrative prose.

Consider for a moment Psalm 78 and Psalm 105, two psalms that review the events of the exodus in considerable detail. These psalms are historical in that they refer to a foundational event in the history of Israel. But look at Psalm 78:44-51, verses that review the plagues. The psalmist, Asaph, has reversed the order of the plagues of flies and frogs and of the locusts and hail from the sequence given in Exodus 7—11. Moreover, he omitted the plagues of gnats, boils and darkness and mentions only livestock in connection with hail. Clearly this very historical psalm is not intended for the reader to come away with sequential details or even all the details of the plagues. The same can be said for Psalm 105:28-36, in which the plagues of Egypt also come into view. Here the psalmist switched the order of the plagues of gnats and flies, left out the plagues of boils and livestock, and moved the plague of darkness all the way to the beginning of the list. Again, in this historical psalm, the psalmist obviously had no interest in worrying about the correct chronological order or completeness of reporting. The differences in description of the same event in some parallel passages in Kings and Chronicles suggests the same. Historicity does not necessarily entail strict chronology, completeness or even specificity of detail. It all depends on what literary genre is employed.

If we assume that, unlike Proverbs, Genesis 1 is indeed concerned with history, then we must ask what it means to say that Genesis 1 is historical. What kind of history writing is Genesis 1? In what way does it present historical events? Genesis 1 clearly is not typical Hebrew poetry, nor is it a parable or an allegory. A lot of writers have said that Genesis 1 is a narrative, but, if so, it is a narrative that is utterly unlike any other narrative in the Bible or any other narrative ever written. For one thing, no human eyewitness was present at creation who could then later compose such a narrative. We believe that those writers are on target who suggest that Genesis 1 is a literary genre unlike any other. As John Stek has put it, Genesis 1 is sui generis, a genre in a class by itself, a genre of which we have no other examples.[12]

But if Genesis 1 is not a typical narrative, does that mean it is not historical? Well, not necessarily. We think that Genesis 1 *is* historical. But how is it historical? Obviously it is not a descriptive human eyewitness report of creation, but is it a descriptive report that God, the only witness to the act of creation because he was the creator, showed in a vision or a dream to Moses or Adam, or implanted into the reflective thinking of Moses, or revealed face to face by the angel of the Lord to Adam in the garden or to Moses on the mountain of God? These are conceivable possibilities, but we're not convinced of the validity of any of them, and the Bible doesn't tell us. Even though Moses may have composed Genesis 1, we just don't know how God revealed what we find in the creation account.

The fact that Genesis 1 is so laden with symbolic numbers, repetitive structure, anthropomorphic and metaphorical elements, allusions to ancient Near Eastern concepts, and the literary convention of seven days (about which we will say more in the next chapter) leads us to believe that it is not the intent of Genesis 1 to provide the kind of detailed factual information that one would find in a straightforward narrative report, nor the kind of detailed factual information that would be useful in developing a scientific reconstruction of the historical unfolding of the universe.

Genesis 1, however, is decidedly historical in the sense that God *did* make the universe and *did* make all the various entities within it. God did make the earth. God did make light. God did make the seas. God did make plants. God did make the sun. God did make the moon. God did make the stars. God did make the sea creatures. God did make the birds. God did make the various land animals. God did make human beings. These were all events that took

---

[12]John H. Stek, "What Says the Scripture?" in Howard J. Van Till, Robert E. Snow, John H. Stek and Davis A. Young, *Portraits of Creation: Biblical and Scientific Perspectives on the World's Formation* (Grand Rapids: Eerdmans, 1990).

place in the real world of space and time, not in some idealized mental world. How much more historical can we get than that? It does not matter one iota whether God created all these different things instantaneously or over long periods of time. The point of Genesis 1 is that God made them. Created things did not make themselves, and the "gods" worshiped by all the rest of the world did not make them either. Instead, divine creation happened. Nor does it matter whether the various events are presented in a sequential fashion or not. No matter what the sequence of events may have been, they were all events. They all happened. God brought them about. The world actually did come into existence in response to God's effective word. The universe came to be as a result of God's initiative.

So, the charge that the day-age, framework or the analogical-day view or views other than the traditional view undermine the historicity of the creation account is unfounded. These views are just as historical as the traditional view. As regards historicity, they differ from the traditional view only in terms of the mode of presentation of the events of creation. We explore the character of that mode of presentation in the next chapter.

THE PERSPICUITY OF THE BIBLE

Duncan and Hall stated that the traditional view of Genesis 1 is the obvious, face-value teaching of the text and that adherence to nontraditional interpretations, in particular the framework view, violates the perspicuity of Scripture. Regarding the framework view, they state that "the perspicuity of Scripture is jeopardized by this eccentric hermeneutical approach."[13] They also say that "the framework theory seriously undermines our confidence in the clarity of Scripture."[14] They think that God reveals himself so that even children and nonliterary specialists can understand his mind, at least in substance, but the framework view is "too foggy" for children to grasp.[15]

Without defending the framework interpretation, we suggest that the claims of Duncan and Hall regarding the perspicuity, that is, the clarity of the Bible, misrepresent that doctrine. The doctrine of biblical perspicuity does not teach that every individual text in the Bible is equally clear or necessarily even clear at all. If that were the case, we would not invoke the principle that more difficult and obscure texts should be interpreted in light of clearer texts. Peter

---

[13]Duncan and Hall, "The Twenty-Four-Hour Response," in Hagopian, *The Genesis Debate*, p. 263.

[14]Ibid., p. 265.

[15]Arguably, Duncan and Hall are correct that God might reveal himself so that children could understand, but perhaps the best way for God to have done this for children and everyday people in the ancient Near Eastern world was through the vehicles of story and imagery, not through scientific description.

would not have said that there are things in the writings of Paul that are difficult to understand (2 Peter 3:16). Jesus would not have said that he taught in parables lest those who heard him might understand and believe. Jesus wanted to close the eyes and ears of the hard of heart lest they understand and believe! A puzzling assertion indeed!

Article 1 of the Westminster Confession of Faith asserts that not all parts of Scripture are equally clear. What it particularly affirms is that the message of salvation through faith in the person and work of Jesus Christ *is* clear. The message of salvation is plainly taught in the Bible so that anyone can understand the message, provided that the requisite hard work of study and consultation of the appropriate helps takes place.[16] So the perspicuity of Scripture focuses on the central thrust of the Bible, the gospel of salvation. But even the message of salvation is not obviously and immediately clear to everyone upon the first reading of the text. Most Christians understand that Isaiah 53 is an uncannily detailed prophecy of the suffering of Jesus Christ, the one who would come as Messiah. But Isaiah 53 was not particularly clear to the Ethiopian eunuch who asked the help of Philip the evangelist in understanding the text. Who was Isaiah talking about anyway? The eunuch wanted to know.

If we followed the reasoning of Duncan and Hall, we might argue that children should be able to get the gist of the book of Revelation or the Song of Songs or the prophetic utterances of Ezekiel. But after two millennia of grappling with these and other peculiar portions of Scripture, few if any biblical scholars are prepared to state that they now have those passages figured out to their satisfaction. Luther was sufficiently puzzled by the New Testament book of James that he didn't think it belonged in the Bible. Even that great commentator John Calvin admitted to being baffled by Revelation. Despite all the qualifiers, we still adhere to the perspicuity of Scripture.

From another angle, however, Genesis 1 is remarkable for the fact that even a child *can* grasp the substance of the chapter no matter how the details play out, no matter whether the traditional, framework, day-age, gap, analogical-day, intermittent-day, revelation-day or any other view is the right one. Every child comes away from Genesis 1 understanding that God made the whole world! Isn't that the substance of Genesis 1? Isn't that what the chapter is really driving at?

---

[16]The Westminster Confession of Faith 1.7 says, "All things in Scripture are not alike plain in themselves, nor alike clear unto all; yet those things which are necessary to be known, believed, and observed, for salvation, are so clearly propounded and opened in some place of Scripture or other, that not only the learned, but the unlearned, *in a due use of the ordinary means,* may attain unto a sufficient understanding of them."

Finally, we remind those who think that the traditional view of Genesis 1 is obvious and clear that Augustine, the greatest theologian of the early church, the theologian who studied Genesis in more depth than anyone else in the first millennium of church history, concluded in his most mature study, *The Literal Meaning of Genesis,* that the early chapters of Genesis are *obscure* and, because of that obscurity, humility and lack of dogmatism in interpretation of the creation account is called for! These are wise and humbling words from the great Spirit-filled scholar of the early church. Those who are so sure about the correctness of the traditional view or of any alternative views would do well to back up just a little bit. Deep study and wrestling with the text should, in the end, prove to be more worthwhile than a superficial reading. The doctrine of the perspicuity of the Bible may not be used as an excuse for accepting superficial readings of the text or as a means for determining the proper outcome of exegesis. Specifically, the doctrine of the Bible's perspicuity does not automatically confirm the traditional view of Genesis 1.

THE INERRANCY OF THE BIBLE

Some Christians are probably fearful that adoption of any view of Genesis 1 other than the traditional view poses a threat to the inerrancy of the Bible. Adherents of alternative interpretations may be accused of rejecting "the literal truth" of Genesis 1 and, therewith, biblical inerrancy. And yet proponents of alternative views such as the day-age or framework views routinely profess their commitment to biblical inerrancy. We too accept biblical inerrancy, provided that the concept is properly understood as affirming the errorless character of matters that the Bible is actually teaching. Inerrancy does not apply to words or sentences taken in isolation from their broader contexts or in disregard for the literary genre in which they are employed.

The notion of biblical inerrancy does not prejudge the proper interpretation of any specific passage. Christians are not committed by acceptance of biblical inerrancy to a particular interpretation of a text before they have done the hard work of exegesis. Precisely because the Bible is a Word from God, the God of all truth, the God who cannot lie, Scripture does not err either. But the notion of inerrancy also implies that interpreters should be extremely careful to understand exactly what is being said in any given text. One cannot claim inerrancy for a quick, simple face-value interpretation of a text in an English translation that has ignored the nuances of the original language, the literary genre, the context of the passage or even the original cultural setting in which the text was written, if that is known.

Furthermore, interpreters will go astray if they insist that Scripture records all details in a manner that is consistent with contemporary standards of historical and scientific accuracy. For example, all too often Christians have assumed the scientific truthfulness of material that occurs in Hebrew poetry. Matthew Fontaine Maury (1806-1873), the founder of modern oceanography, allegedly saw the allusion to the "paths of the seas" in Psalm 8:8 as providing hints to the existence of oceanic currents. But to seek such information in the Bible is a misuse of Scripture. The best exponents of biblical inerrancy have consistently pointed out that the Bible does not speak in the precise, technical language of modern science. John Calvin even suggested that, in some instances, the writers of Scripture accommodated themselves to erroneous beliefs about the natural world.[17] He also urged that those who wanted to learn astronomy should not try to find it in the Bible.

It is probable too that the establishment of the firmament recorded for the second day of creation entails divine accommodation to the universally held ancient Near Eastern belief that the stars are attached to a solid dome or vault that covers the Earth.[18] The citizens of the ancient Near Eastern world understood *raqia'* to be a solid. Even many of the church fathers understood the firmament to be a solid vault in the sky.[19] The Bible asserts that God made a *raqia'* a solid dome, but our belief in inerrancy surely does not demand that we regard the sky as a solid just because the Bible says so, especially since we know that the sky isn't solid! Any doctrine of inerrancy needs to acknowledge the role played by accommodation to the ancient Near Eastern context in which the Israelites lived.

Scripture is a record of the deeds of God in history. Its message concerns God's redemptive plan and acts. Christians must, therefore, understand the intent of biblical passages. It is the message intended by the original author that is utterly trustworthy, not our interpretation shaped by contemporary science. The Bible was not given as a handbook of facts of ancient history or of natural science but as the story of God's saving work in a world full of fallen, sinful humans. As a result, as readers approach Genesis 1 they need to ask what God and the human author intended to teach, and that means that both the redemptive thrust of the Bible and the ancient Near Eastern context in which it is embedded must be taken into account.

God did not give us the Bible simply to convey all sorts of bits of isolated

---

[17]Although Calvin seemed to think that the biblical writers knowingly accommodated their teaching to the level of their audience, perhaps it makes more sense to say that God accommodated himself to his audience through the human authors who were just as much children of their times as their readers.

[18]Paul H. Seely, "The Firmament and the Water Above, Part I: The Meaning of *raqia'* in Gen. 1:6-8," *Westminster Theological Journal* 53 (1991): 227-40.

[19]Adherents of the notion that the firmament was solid included Origen, Basil, Ambrose, Augustine and possibly John Chrysostom.

information to us so that we would be more knowledgeable. The Bible may contain allusions to the natural world, but that does not mean that it is a text of scientific data. Nor does it mean that the Bible was given to us specifically to teach small details about the natural world. If we grasp the intent of the writer of Genesis 1 and recognize that he was not trying to teach us a thumbnail sketch of contemporary natural science, we will be on the way to discovering the inerrant teaching of the text.

7

The Bible and the
Antiquity of the Earth

Part Two

Now let's start evaluating in detail some of the textual arguments that are frequently invoked in favor of the traditional twenty-four-hour-day view.

The Analogy with Exodus 20

One of the most widely used reasons for adopting the traditional view is the version of the fourth commandment given in Exodus 20:11. In this commandment, the Israelites were told to remember the sabbath day, to do all their labor in six days, and to cease labor on the seventh day. They were to do this because God made the heavens, the earth, the sea and all that is in them in six days and then rested the seventh day. Because the days of human labor and rest are the ordinary, twenty-four-hour, solar days that we experience, so, it is claimed, the six days of God's creation and the one day of his rest obviously must also have been ordinary, solar days.

Interpretations of parallels or analogies between God and humanity are fraught with dangers, however. Consider another aspect of the fourth commandment, namely, rest. One could just as well argue that because our need to cease from labor on the sabbath results from physical fatigue, therefore, the rest of God is of exactly the same sort, namely, a rest necessitated by fatigue. But we know that God does not wear out, tire or need rest as we do. God's rest is not identical to our rest. Human rest is a creaturely model or image of God's rest. The relation here is one of analogy, not identity.

Or consider the demand of Jesus in the Sermon on the Mount (Matthew 5:48) that his disciples are to be perfect as the heavenly father is perfect. But for us to be perfect is impossible if the perfection required is identical to the divine perfection. What is required can only be a creaturely perfection or wholeness that is a finite analog of God's perfection. The same may be said of love. Just a few verses earlier Jesus told his disciples to love their enemies so that they might be sons of their father in heaven, who shows his love for the evil and unrighteous by giving them sun and rain (Matthew 5:43-45). Certainly human love is analogous, not identical, to divine love.

Think too of Moses who parted the waters of the Sea of Reeds with an outstretched hand (Exodus 14:16, 21-27). But then in Moses' song of deliverance he sings of God's outstretched hand (Exodus 15:6, 12). Clearly we are not to understand that because the outstretched hand of Moses was part of his physical, biological being, the outstretched hand of God is likewise a biological appendage. Again, the relation is one of analogy and the hand of God is used as a metaphor for God's power.

Several theologians, such as W. G. T. Shedd and C. John Collins, have maintained that the days of divine creation and the day of God's rest are not identical to the days of human experience.[1] They are *divine* "days" in which God accomplished his creative work. Our days are scale models or analogs of his "days," and the divine days are to be understood as a metaphor. If the days of creation are "divine days" of which human days are creaturely analogs, then the divine days transcend human experience, and we are forced to conclude that the amount of time that elapsed during the creative work "week" of the Creator is not in view in Genesis 1.

THE MEANING OF *CREATE*

Many Christians assume that God's creation of the world entailed a series of instantaneous, purely miraculous events that resulted in a full-blown, fully functioning, ready-to-go world. The moment that God spoke, light appeared; the next moment he spoke, the waters divided; the next moment plants and full-grown trees with fruit on them appeared; on day four heavenly bodies

---

[1]William G. T. Shedd, *Dogmatic Theology*, 3 vols., 2nd ed. (New York: Scribner, 1889); and C. John Collins, *Science and Faith: Friends or Foes?* (Wheaton, Ill.: Crossway Books, 2003).

suddenly appeared in the sky; an empty sky was suddenly filled with flying birds and empty seas were suddenly full of schools of fish on day five; full-grown animals were suddenly walking around on Earth's surface on day six. In short, many Christians maintain that the creation account is full of the supernatural, miraculous action of God. Robert Reymond, for example, stated in regard to the early chapters of Genesis that "the problem in these chapters for many scholars, simply put, is the distinctly *supernatural* character of the events which they report."[2]

Duncan and Hall maintained that "the supernatural, not the natural, saturates the whole creation narrative."[3] Concerning creation-oriented texts such as Isaiah 65:17-18, they wrote that "these verses contain no hint that creation was—or will be—anything other than quick, instantaneous declaration (by God's speaking), and independent of gradually unfolding naturalistic processes."[4] They criticized the day-age interpretation for being "rooted in a rationalistic approach that leaves little room for miracle and faith . . . their model leaves little, if any, room for the miraculous."[5] According to Duncan and Hall, the day-age model of Hugh Ross and Gleason Archer is suspect because "creation involved a providential implanting" after which "nature took its course over millions or billions of years."[6] Not to be "out-miracled" by Duncan and Hall, Ross and Archer retorted that "the difference between our view and Duncan and Hall's view on this point is as follows: we acknowledge hundreds of millions of miracles over millions, even billions of years, while they acknowledge hundreds or thousands of miracles compressed into a 144-hour sequence."[7] They argued that they "believe in more dramatic and more frequent miracles" than do Duncan and Hall.[8]

In our haste to convince fellow Christians that we believe in miracles, however, we should not forget to ask the question, what does the text of Genesis 1 (and other creation texts) really say? We suspect that most Christians would

---

[2]Robert L. Reymond, *A New Systematic Theology of the Christian Faith* (Nashville: Thomas Nelson, 1998), p. 384.

[3]J. Ligon Duncan and David W. Hall, "The Twenty-Four-Hour Reply," in *The Genesis Debate: Three Views on the Days of Creation*, ed. David G. Hagopian (Mission Viejo, Calif.: Crux Press, 2001), p. 111.

[4]J. Ligon Duncan and David W. Hall, "The Twenty-Four-Hour View," in *The Genesis Debate: Three Views on the Days of Creation*, ed. David G. Hagopian (Mission Viejo, Calif.: Crux Press, 2001), p. 42.

[5]J. Ligon Duncan and David W. Hall, "The Twenty-Four-Hour Response," in *The Genesis Debate: Three Views on the Days of Creation*, ed. David G. Hagopian (Mission Viejo, Calif.: Crux Press, 2001), p. 172.

[6]Ibid.

[7]Hugh Ross and Gleason L. Archer, "The Day-Age View," in *The Genesis Debate: Three Views on the Days of Creation*, ed. David G. Hagopian (Mission Viejo, Calif.: Crux Press, 2001), p. 196.

[8]Ibid., p. 200.

agree, as we would too, with the historic Christian doctrine that the initial act of creation that brought into existence the material from which God formed the habitable, orderly cosmos, was an *ex nihilo* creative act, a sheer, totally supernatural miracle that could not have entailed God's use of any secondary means or natural processes whatsoever because secondary means and natural processes had not yet been created and did not yet exist. The beginning of God's work of creation had to be a miracle of the purest kind. The question before us, however, is whether the subsequent creative work of the six days mentioned in Genesis 1 involved purely supernatural, miraculous acts. We are not here asking whether God could have performed purely supernatural miracles during the six days. Of course he could. Everything that occurred during the six days could have been brought into existence via purely supernatural acts. But that's not in question. We are asking what the text says. Does the text of Genesis 1 unequivocally insist on purely supernatural acts during the six days?

Let's look at some of the reasons that people think there were miracles during creation week. First of all, many Christians are under the impression that the word *create*, that is, the Hebrew term *bara*, includes within itself the very concept of instantaneousness and the exclusion of all use of secondary causes.[9] For them, the word *create* is identical "to create out of nothing miraculously and instantaneously." They think that the word *create* means "immediate" creation. But that is not the case. The verb *bara*, generally translated "create," refers to an action that can be performed only by God and that brings into existence something that is new and unanticipated. People cannot create. Forces cannot create. Processes cannot create. Only God can and does create. But the verb itself tells us nothing about how God brought the new thing into existence. It does not tell us whether the act was "immediate" or "mediate," that is, whether God used means or not. Nor does it tell us whether God brought into existence the new thing out of preexistent material. To create does not always entail creation out of nothing. Nor does it tell us whether the act was effected instantaneously or over time.

It is true that where *bara* is used in the Hebrew Old Testament, means and materials are never mentioned. From the fact that we never read an expression such as "God created *(bara)* new thing x from material y by means of method z," might induce us to infer that creation must be "immediate" and ex nihilo. But that would be incorrect. There is no question that some creative acts did entail preexistent material. Although the Bible does not say that "God created

---

[9]John H. Stek, "What Says the Scripture?" in H. J. Van Till, R. E. Snow, J. H. Stek and D. A. Young, *Portraits of Creation: Biblical and Scientific Perspectives on the World's Formation* (Grand Rapids: Eerdmans, 1990), pp. 203-65.

man in his own image out of the dust of the ground," it does say that "God created man," and it also says that God "formed the man from the dust of the ground." Is it not a reasonable inference that man was created from already existing matter in light of all the textual evidence?

Then consider the fact that the prophet Isaiah spoke about the creation of the nation of Israel (e.g., Isaiah 43:1). The nation of Israel was something new and unanticipated, brought into being by the extraordinary power of God, but to maintain that this creation of Israel entailed no means is absurd on the face of it. God used all sorts of means in creating Israel. He clearly used ordinary processes like birth to bring succeeding generations into being. He also performed extraordinary acts along the way. God used an extraordinary act to bring Isaac into being, but even then he used the womb of Sarah. There is no indication that the twelve sons of Jacob/Israel were conceived and born in anything other than the usual way. Both methods were employed in the creation of Israel. The deliverance of Israel from Egypt, also rather significant in the creation of Israel, was effected by a series of ten plagues. The plagues represented a series of remarkably timed divine acts that were designed to achieve the purposes of God, but God also used ordinary events, albeit on a grand scale and amazingly timed, like swarms of locusts and great hailstorms. So, even though the term *create* is not used with reference to means and materials, the biblical concept of creation clearly allows room for both. God is free to create without means and free to create with means. He is free to create out of nothing. He is free to create from existing material.

Many Christians also think that to create something inherently entails the concept of "instantaneous production." A careful examination of the usage of *bara,* however, indicates that *bara* does not require instantaneousness of fulfillment. God, for example, created the people of Israel, but he did so over the course of centuries, not instantaneously, although they were constituted as a nation at the exodus. He still creates animals (Psalm 104:30), but does anyone think that animals appear out of nowhere? He is said to create people (Psalm 102:18), but does anyone believe that people come into the world without a mother? God creates darkness (Isaiah 45:7), and he creates the wind (Amos 4:13). All of these examples refer to the creation of things that obviously took or take time to bring to fruition.

The term *create* does not inherently entail the notion of instantaneous fulfillment. Surely when Psalm 102:18 says that God creates people it does not mean to teach that God creates each individual instantaneously out of nothing without using any biological and chemical processes. So, the verb *create* in and of itself excludes neither time, process nor secondary causes. God is the only agent who creates, and when he creates, he creates something new. But he is free to create instantaneously or over time; he is free to create directly or by means of processes; and he is free to create out of nothing or out of previously existing material. John Stek has amply demonstrated that the Old Testament routinely interchanges other words like *fashion, establish, form* and *make* as parallel substitutes for the word *create*.[10]

VERBAL FIAT CREATION

Many Christians also believe that the verbal nature of divine creation indicates that creation was purely supernatural. God, by divine fiat or command, simply speaks things into existence instantaneously, without the need for any time-consuming process. For example, John Currid, in his comments on Genesis 1:12, wrote that the language of this verse "underscores the fact that the directive is instantly fulfilled and completed. Theories which argue that God spoke the commands of creation at this point, but that they were not fulfilled until subsequent ages, do great injustice to the text." He asserted that "a clear sense of the spontaneous and instantaneous cloaks the account." In his opinion, "to deny the immediacy of creation's completion is to reduce or diminish the power of God that is so greatly invested in the account."[11] Duncan and Hall claimed that Genesis 1—2, the rest of the Pentateuch and all of Scripture affirm creation "by the unmediated word of God, as opposed to natural providence over long periods of time."[12] They maintained that the tokens of God's love, namely, the sun, moon and stars, are nowhere described in the Bible "as coming about in any fashion other than by His understanding or by His unmediated word."[13]

Appeal is very frequently made to Psalm 33:6-9 in support of this notion of instantaneousness. "By the word of the LORD were the heavens made," we read, "their starry host by the breath of his mouth." As the Psalm continues we read that God "spoke, and it came to be; he commanded, and it stood firm." Surely, many insist, this text demands that we adhere to instantaneous immediate origination of the universe. All of this, of course, is said to be consistent with Genesis 1, which says repeatedly, "And God said," along with its countersign, "and it was so." Coupled with such textual statements is the fact that the stupendous work of creation is described in essentially one page at the beginning of the Bible. The extreme brevity and conciseness of the entire creation account unavoidably convey an impression of quickness to many readers.

---

[10]Ibid.

[11]John D. Currid, *A Study Commentary on Genesis,* vol. 1 (Webster, N.Y.: Evangelical Press, 2003), pp. 71-72.

[12]Duncan and Hall, "The Twenty-Four-Hour View," p. 36.

[13]Ibid., p. 40.

Because we can read the entire account of creation in one minute, it becomes very easy to develop the sense that God, who is, after all, omnipotent, made the whole thing in a minute or less. Moreover, the extreme brevity of the description of each individual creative act conveys the impression that God said let there be something and it was so and it was good and it's over and done with in a flash. It surely sounds instantaneous. Nor is this view just a modern one. A good many great theological thinkers of the past more or less assumed that such was the case.

These impressions, however, are misleading. On the surface Psalm 33:6-9 sounds like instantaneous fulfillment. After all, who but God can speak a universe into being? But the Psalm speaks of the creation of the starry hosts which, according to Genesis 1, did not appear until at least three days after creation of the heavens. But, more significant, there are a host of texts in Scripture in which we read about God's issuance of his word or speaking or commanding that we should consider. Think about Psalm 147:15-18, a text that tells about God sending forth his word. As a result of his word, the snow falls. God sends his word again, and the snow melts. No interpreter maintains that snow is instantaneously formed or instantaneously melted simply by virtue of God's speaking. Why then do we insist that God's *creative* acts must have been accomplished instantaneously simply because God spoke? Having lived for many years in western Michigan, an area with seemingly endless winters and large accumulations of snow, we have often wished that the snow in our yards would have melted instantaneously, but it never did. We all know that snowfall and snowmelt are God-ordained and God-guided, natural processes that take time and that can be described and modeled mathematically in terms of scientific principles. Here God's word is accomplished through process acting over time.

Consider a couple of other verses in the Psalms. Psalm 105:34-35 tells about God speaking and sending the locusts and grasshoppers to devour the plants of Egypt. But then when we turn to Exodus 10:12-15 we are told that the Lord told Moses to stretch out his hand over the land of Egypt, and when he did, the Lord sent a strong east wind that blew all night and all day. Finally the wind brought in the locusts to do their damage. Even though God spoke, the locusts did not appear instantaneously. A time-consuming process involving a natural phenomenon, wind, was employed in carrying out God's command. One could in effect say, "And God said, 'Let there be locusts to devour the land. And it was so." Well, wasn't it so? Much the same can be said about the parting of the Sea of Reeds portrayed poetically in Psalm 106:9 and described more literally in Exodus 14:21-22. The opening of the Sea of Reeds did not occur instantaneously. The Lord drove the waters back by means of a strong east wind that blew all night.

In the New Testament, Hebrews 1:3 asserts that "the Son is the radiance of God's glory and the exact representation of his being, sustaining all things by his powerful word." But if the word of God ipso facto means that God is doing something miraculous, then we have a major problem because this text is talking about what we would call ordinary providence. It is talking about the upholding of the entire cosmos, including its history, by the divine word. But who can dispute that such verbal upholding of the cosmos takes place through the instrumentality of secondary processes? No, the verbal upholding of the cosmos does not rule out the occurrence of mediated divine action, for we all recognize that the ordinary course of things entails natural processes, "laws of nature," behavioral patterns and the like. The utterance of the divine word hardly excludes God's use of processes that can be investigated by scientific means.

The passages that refer to divine fiats or God's speaking really emphasize the inevitability, not the instantaneousness, of divine accomplishment, not to mention the monotheism of the creative act in contrast to the ancient Near Eastern parallels. In every instance the emphasis is on the fact that what God has ordained, commanded or spoken gets done. That does not mean that it is immediately accomplished in its completeness, but it is *inexorably* accomplished. Isaiah 55:8-11 makes that point beautifully. Like the rain, God's word goes forth from his mouth and accomplishes what he pleases. But the metaphor should be carefully noted. The *word* of God is likened to the rain that descends from heaven, waters the earth, causes seed to be brought forth, and ultimately provides bread for the eater and seed for the sower. The rain accomplishes its job, but that accomplishment occurs through a natural process that takes time. If it is like the rain, cannot the word of God be inevitably, irresistibly effective but still take time for the final accomplishment?

Think about all the things that God has ordained that we know have taken time, a lot of time: all of human history, the call of Israel, the coming of the Messiah, all of church history, our personal sanctification, the return of Christ. In all of these events God's will and word are inevitably being worked out, and his will is being and will be accomplished in its entirety at the appointed time. What he wants done, gets done, and nothing can stop it. But God normally takes a lot of time to accomplish his purposes. But if he takes time to accomplish things during human history and our personal lives, why do Christians have such a problem with God's taking time in his creating the completed universe?

So, when Christians look at the creation account and read that God said, "Let there be thus and so," it means that, sooner or later, quickly or over eons, with or without the use of secondary means, what God called into being was eventually completely accomplished. And because it was accomplished, we read that "it was so." But the text of Genesis 1 tells us absolutely nothing about how long it took for God to complete what he called for, and it tells us absolutely nothing about what procedure he took to accomplish it.[14] The point is that God's creative will was accomplished, a point that stands out in contrast to ancient Near Eastern parallels. God called for light, the skies, the waters, the plants, the sun, the moon, the stars, the fish, the birds, the land animals and human beings. Well, look around. They are all here. What God wanted, happened. The creative word was powerful and effective. If we understand creation in this way, we will see that there is no textual evidence that *demands* the execution of hundreds of millions of miracles or even a few miracles in the sense of immediate, direct, instantaneous, processless, secondary-cause-less actions of the Creator except for the initial ex nihilo act. God was free to use means over time, and he may well have done it that way. To believe that this is so in no way violates the text of Genesis 1.

Of course, every reader is probably still wondering about those seven days! Whatever happened could not have taken more than seven days, many may say. Well, be patient. We're getting to that—right now!

THE SEVEN-DAY STRUCTURE OF GENESIS 1

One of the arguments used by traditionalists in favor of the twenty-four-hour-day view is that wherever we encounter a sequence of numbered days in Scripture, the text invariably refers to ordinary, solar days. As examples, consider Numbers 7:11-83 and Numbers 29:12-38. In these passages ordinary days obviously are in view. Another argument advanced by supporters of the traditional view is that the use of evenings and mornings establishes a context that makes it difficult to say anything other than that ordinary, solar days are in view. These two arguments, plus the fact that the primary meaning of the Hebrew noun *yom* is an ordinary, twenty-four-hour day, would seem to clinch the matter. The text mentions six ordinary days of creation and one ordinary day of divine rest. Believe it or not, we agree that that is the case. So what's the issue? Problem solved. Case closed. If these are ordinary days, creation obviously took 144 hours and the Earth is young. Let's go home.

But again—not so fast. There *is* a problem, because other textual evidence indicates that at least some of these days were *not* twenty-four hours long. As we saw in chapter one, some of the church fathers, including Augustine, picked up on the fact that the textual reference to the seventh day lacks the expression

---

[14]In other words, Genesis 1 is not interested in answering our scientific questions but in stressing God's power and sovereignty as he constructs his cosmic kingdom.

"and the evening and the morning were the seventh day." From that omission, some of the fathers drew the inference that the seventh day is ongoing. A host of theologians and exegetes since then have also drawn the same inference. In addition, Hebrews 4:1-6 refers to the ongoing rest of God as a rest that can be entered into by his people. The allusion to the creation account in this passage has led many commentators to see a link between God's rest that is available to believers today and the seventh-day rest. The text appears to imply that the seventh day has continued throughout all of human history. But how can that be if the seventh day is an ordinary day?

Many interpreters have also noted the presumed improbability that all the events mentioned in the text in regard to the sixth day could have taken place within an ordinary twenty-four-hour period. It is unlikely, they say, that in only twenty-four hours God would have made all the livestock, creatures that crawl along the ground and wild animals by having the land produce them, then created Adam, placed Adam in the garden, given Adam a charge to look out for the garden, given Adam the commandment not to eat from the tree of the knowledge of good and evil, provided Adam with the opportunity to become acquainted with all the livestock, beasts of the field and birds of the air, had Adam give names to all the animals, allowed a deep sense of loneliness and incompleteness to well up within Adam, put Adam into a deep sleep (apparently not a twenty-minute afternoon catnap), made the woman out of Adam's rib, closed up Adam's flesh and then presented Eve to Adam.

Two points in particular call for closer attention. The act of naming the animals is much more significant than it might appear on the surface. For us naming a person or a thing is not always a very serious business. For our child we may pick a name that sounds pretty, or maybe we name a child after a beloved relative or someone that we admire. In the ancient Near East, however, naming someone or something was a more serious matter than that. The act of naming entailed an intimate understanding of the nature and character of the thing that is named. The comments of theologian Karl Barth are to the point here. Barth wrote that

> in the Bible the name of a person or thing is not an accidental appendix or a sign of recognition, but is something that designates the nature and function of the person or thing in question, thus corresponding to it. Israel is not called Israel without reason. Jesus is not called Jesus without reason. Judas is not called Judas without reason. Every person or thing is what its name implies. Namelessness and anonymity mean unreality. For this reason the naming of a thing is never an incidental act in the Bible. It is always a decisive act, as is presupposed even where it is not expressly mentioned. To give a thing a name is thus an act of lordship, and originally and properly an act of divine worship. When man names a

thing (as when he has the task of naming the animals in the second creation story in Genesis 2:19), he does so in some sense as the delegate and plenipotentiary of God and not on his own authority.[15]

The second point concerns Adam's reaction upon seeing Eve for the first time. Adam breaks out in jubilation: "This is now bone of my bone and flesh of my flesh." The term *now* carries the effect of exclaiming, "At last! It's about time! Now there is somebody for me!" The sense of relief from his loneliness strongly suggests that Adam had been alone for quite some time. May we not also assume that God provided them with some time to get acquainted on that sixth day, the day of their creation? Then after all that, God pronounced his blessing on the first pair, charged them with their task and reviewed their diet with them. But again, we have a problem. How can the sixth day be an ordinary, solar day if it is unlikely that all the events described would have occurred in twenty-four hours?

So what is the solution to the problem? Let's remember that no human observer was present at the creation. There was no bystander taking notes or making mental images of everything that was transpiring as God performed his creative acts. No one was recording the event with a camcorder. No one was recording sound. No one was taking photographs. No one was painting pictures. We cannot have ordinary history writing here. So, rather than creation being described in the form of a typical narrative, it is described in a unique literary form that has no exact parallel in any other writing in the world. It is clear that the author, in mulling over the creative work of God that had been revealed to him (how, we do not know), chose to present creation in a very highly structured, very artistic format that entailed the seven days. Why did he choose the seven-day structure?

Before one leaps to the conclusion that the author of Genesis 1 chose the seven-day structure because God actually took six ordinary, solar days to create the finished cosmos and one additional, ordinary, solar day to rest and then revealed that fact to Moses, let's note that seven-day literary structures were common in the ancient Near East.[16] At least a half dozen different examples of the use of a

---

[15]Karl Barth, *Church Dogmatics*, vol. 3: *The Doctrine of Creation, Part One* (Edinburgh: T & T Clark, 1958), p. 124.

[16]Consider that Genesis was composed after many of the ancient Near Eastern sagas and myths that employ the seven-day structure and that were making use of a common literary structure. It is questionable whether the widespread use of seven days in the ancient Near Eastern examples simply reflect an ancient memory that has been deformed or parodied from an original seven-day creation tradition. For one thing, this theory presupposes that God revealed creation in seven days to Adam or Noah or someone in the deep past, but that is unsupportable conjecture. Besides, the number seven is not restricted to creation myths (in fact it is more common outside of creation myths). It is much more likely that the number seven is derived from the

seven-day literary structure in portions of epic tales have been preserved. These include Akkadian and Ugaritic examples, so use of the seven-day structure was not confined to one group of people. The examples we have all bear considerable similarities. Because these seven-day literary structures are so significant for our understanding of Genesis 1, we reproduce below the known examples.[17]

From the Ugaritic Poems about Baal and Anath (e.II A B [vi] lines 17-38) we have:

> [*As for Baal*] his house is built,
>> [*As for Hadd*] his palace is raised.
> They [. . .] from Lebanon and its trees,
>> From [Siri]on its precious cedars.
> . . . [. . . Le]banon and its trees,
>> Si[r]ion its precious cedars.
> Fire is set to the house,
>> Flame to the palace.
> Lo, a [d]ay and a second,
>> Fire feeds on the house,
>> Flame upon the palace:
> A third, a fourth day,
>> [Fi]re feeds on the house,
>> Flam[e] upon the palace.
> A fifth, a s[ix]th day,
>> Fire feeds [on] the house,
> Flame u[pon] the palace.
> There, on the seventh d[ay]
>> The fire *dies down* in the house,
>> The f[la]me in the palace.
> The silver turns into blocks,
>> The gold is turned into bricks.
> Puissant Baal exults:
> "My h(ouse) have I builded of silver;
>> My palace, indeed, of gold."

From the Ugaritic Tale of Aqhat (Aqht A [i] lines 1-19), we have:

> [. . . Straightway Daniel the Raph]a-man,
>> Forthwith [Ghazir the Harnamiyy-man],
> Gives oblation to the gods to eat,

---

number of days in each of the four phases of the moon. In that sense a seven-day structure is built into our experience via the astronomical order that God created.

[17]The following excerpts from Poems About Baal and Anath, The Tale of Aqhat, The Legend of King Keret, and The Epic of Gilgamesh are from James Pritchard, ed., *Ancient Near Eastern Texts Relating to the Old Testament* (Princeton, N.J.: Princeton University Press, 1950).

Gives oblation to drink to the holy ones.
A couch of sackcloth he mounts and lies,
     A couch of [loincloth] and passes the night.
Behold a day and a second,
     Oblation to the gods gives Daniel,
Oblation to the gods to eat,
     Oblation to drink to the holy ones.
A third, a fourth day,
     Oblation to the gods gives Daniel,
Oblation to the gods to eat,
     Oblation to drink to the holy ones.
A fifth, a sixth, a seventh day,
     Oblation to the gods gives Daniel,
Oblation to the gods to eat,
     Oblation to drink to the holy ones.
A sackcloth couch doth Daniel,
     A sackcloth couch mount and lie,
     A couch of loincloth and pass the night.
But lo, on the seventh day,
     Baal approaches with his plea:
"Unhappy is Daniel the Rapha-man,
     A-sighing is Ghazir the Harnamiyy-man;

And a few lines later from the same tale ([ii] lines 28-43), we read:

Straighway Daniel the Rapha-man,
     Forthwith Ghazir the Harnamiyy-man,
Prepares an ox for the skillful ones,
     Gives food to the [ski]llful ones and gives drink
     To the daughters of joy[ful noise], the *swallows.*
Behold a day and a second,
     He give[s f]ood to the skillful ones and dr[in]k
     To the daughters of joyful noise, the *swallows;*
A third, a fo[urth] day,
     He gives food to the skillful ones and drink
     To the daughters of joyful noise, the *swallows;*
A fifth, a sixth day,
     He gives food to the skill[ful] ones and d[rink
     To the d]aughters of joyful noise, the *swallows.*
Lo, on the seventh day,
     Away from his house go the skillful ones,
     The daughters of joyful noise, the *swallows.*—
[. . .] the fairness of the bed [*of conception*],
     The beauty of the bed of *childbirth.*

From the Ugaritic Legend of King Keret (KRT A [iv] lines 104-123), we have three examples:

> March a day and a second;
>> A third, a fourth day;
>> A fifth, a sixth day—
> Lo! At the sun on the seventh:
>> Thou arrivest at Udum the Great,
>> Even at Udum the Grand.
> —Now do thou *attack* the villages,
>> Harass the towns.
> *Sweep* from the fields the wood-cutting wives,
>> From the threshing floors the straw-picking ones;
> *Sweep* from the spring the women that draw,
>> From the fountain those that fill.
> Tarry a day and a second;
>> A third, a fourth day;
>> A fifth, a sixth day.
> Thine arrows shoot *not* into the city,
>> (*Nor*) thy hand-stones *flung headlong.*
> And behold, at the sun on the seventh,
>> King Pabel will sleep
> Till the noise of the neighing of his stallion,
>> Till the sound of the braying of his he-ass,
> Until the lowing of the plow ox,
>> (Until) the howling of the watchdog.

And a few lines later (KRT A [v] lines 216-226) we read:

> He *swept* from the spring the women that drew,
>> And from the fountain those that filled.
> He tarr[ied] a day and a second,
>> A thi[rd, a fou]rth day;
>> A fifth, a sixth day.
> And behold, at the sun on the seventh,
>> King Pabel slept
> Till the noise of the neighing of his stallion,
>> Till the sound of the braying of his he-ass,
> Until the lowing of the plow ox,
>> [(Until) the how]ling of the [wa]tchdog.

Last, we have an example from the flood episode in the Akkadian Epic of Gilgamesh (Tablet XI, lines 138-149):

> I looked about for coast lines in the expanse of the sea:
> In each of fourteen (regions)

There emerged a region (-mountain)
On Mount Nisir the ship came to a halt.
Mount Nisir held the ship fast,
     Allowing no motion.
One day, a second day, Mount Nisir held the ship fast,
     Allowing no motion.
A third, a fourth day, Mount Nisir held the ship fast,
     Allowing no motion.
A fifth, and a sixth (day), Mount Nisir held the ship fast,
     Allowing no motion.
When the seventh day arrived,
I sent forth and set free a dove.
The dove went forth, but came back;
Since no resting-place for it was visible, she turned round.
Then I sent forth and set free a swallow.

Now let's note some important characteristics of these usages of the seven-day literary structure. There are five features to highlight. First, the sequences of seven days in these examples all begin with the use of "one day" or "a day" or "day one." The definite article is lacking, as is the ordinal number. We do not read "the first day." Second, the definite article is lacking for days two through six, although ordinal numbers are used, as in "a second day," "a fifth day," and so on. Third, the definite article is used only with "the seventh day." Fourth, there is a lumping of days in pairs as in "a third, a fourth day." Fifth, the "action" of the first six days is characterized by monotonous repetition and differs from that of the seventh day on which is described a climax or resolution to the action.

The fact that this highly stylized structure was frequently used is an indication that it was a literary convention designed to indicate a *completed* action in which the repetitive action of the first six days created a tension that built up to a definitive resolution or climax on day seven. This notion of completion is totally consistent with the fact that the number seven *repeatedly* shows up in much ancient Near Eastern literature in a manner that is patently symbolic. A survey of the epic literary corpus of Sumer, Akkad and Ugarit discloses not only references to seven days but also to seven ordinances, seven gates of the nether world and seven judges (Inanna's Descent to the Nether World, Akkadian); seven gods of destiny (*Enuma Elish*, Akkadian); seven sages, seven goblets, seven cloaks, seven pits, seven leagues, seven wafers (Epic of Gilgamesh, Akkadian); seven pieces right, seven pieces left, seven wombs, seven males, seven females (Creation of Man, Akkadian); seven gates (Descent of Ishtar to the Nether World, Akkadian); seven ill winds (Myth of Zu, Akkadian); seven-head Shalyat, seven chambers, seven years (Baal and Anath; Ugaritic); seven years, seven sons, seven speeches (Keret, Ugaritic); and seven years (Tale of Aqhat, Ugaritic).[18]

Clearly, the number seven was used in the ancient Near East symbolically for the concept of completeness or perfection. By no means was literalism necessarily intended. Consider the fact that in the Epic of Gilgamesh Utnapishtim's gigantic ship, a perfect cube 120 cubits on a side, with a displacement about five times that of Noah's ark, was completed on the seventh day! Not likely in seven literal days! The number seven, in all likelihood, took on its symbolic significance because the seven-day week roughly corresponds to the length of each of the four phases of the Moon.

We also see the same thing in the Bible. In the Old Testament, the following examples should make the point. If the Israelites disobeyed the Lord, God said that he would afflict them for their sins seven times over (Leviticus 26:18, 21, 24, 28). If the Israelites obeyed God, their enemies would come at them in one direction, but flee from them in seven (Deuteronomy 28:7), but if the Israelites disobeyed, then they would come at their enemies in one direction but flee from those enemies in seven (Deuteronomy 28:25). Samson told Delilah that he would lose his strength if he were bound with seven fresh green thongs or if the seven braids of his hair were woven into the fabric of a loom (Judges 16:7, 13). When Ruth bore her first son, the women of Bethlehem said to Naomi that her daughter-in-law, Ruth, loved her and was better to her than seven sons (Ruth 4:15). When Nebuchadnezzar was angry with Shadrach, Meschach and Abednego, he ordered that the furnace into which they were thrown be heated seven times hotter than usual (Daniel 3:19). In Zechariah's vision, he saw a stone in front of Joshua, the high priest, with seven eyes as well as a golden lampstand with seven lamps on it (Zechariah 3:9; 4:2).

In the New Testament, Jesus told Peter to forgive the one who sinned against him seventy times seven (Matthew 18:22). And, of course, the use of the number seven is pervasive in the book of Revelation: seven spirits, seven candlesticks, seven stars, seven seals, seven horns, seven eyes, seven angels, seven trumpets, seven thunders, seven plagues, seven bowls.

What does all this have to do with Genesis 1? Precisely this. As John Stek has said, "As regards the seven-day structure, any other temporal order would appear to have been unfitting in that ancient world. Throughout the ancient Near East the number seven had long served as the primary numerical symbol of fullness/completeness/perfection, and the seven-day cycle was an old and well-established convention."[19] The seven days of the creation account of Genesis 1 are enumerated in almost the same way as they are in the ancient Near

---

[18]Ibid.
[19]Stek, "What Says the Scripture?" p. 239.

Eastern examples given above. The Genesis account closely resembles these examples in regard to the five features we noted above. This point becomes clearer if it is understood that the New American Standard Bible has correctly translated the Hebrew, whereas the King James Version and the New International Version have obscured the fine points of the Hebrew text. Thus, with regard to the first feature of the convention, the first day in Genesis 1 is "day one" or "one day," *not* the "first day," and the definite article is lacking. Regarding the second feature, the definite article is also *not* used for days two through five. Thus, the New American Standard Bible correctly translated as "a second day" or "a fifth day" rather than "the second day" or "the fifth day" as the New International Version did. As to the third feature, the definite article is used for day seven in Genesis 2:1-3 and probably also for the sixth day.

The fact that Genesis 1 shares these first three features with the ancient Near Eastern examples leads to the very important observation that the sequencing of days in Genesis 1 is *unlike* that of every other occurrence of numbered days in the Bible. David Sterchi has pointed out that there are fifty-five numbered sequences in the Old Testament (other than Genesis 1) in which a noun is followed by, for example, the number two (or second). He said that these sequences all share one characteristic that "the noun is always determined by either the definite article or a pronominal suffix, and the number is always determined by the definite article." Genesis 1 lacks the definite article. Sterchi also stated that "even if we consider an individual day of a month part of a sequence, each numbered day always used cardinal numbers and never ordinals in the calendar formulas. Since the Genesis sequence is a cardinally numbered day followed by ordinally numbered days, there is no real analogy with calendar formulas."[20] Thus Genesis 1 is not like the numbered days of Numbers 7 or 29; it is, however, like the numbered days of the literary seven-day conventions in ancient Near Eastern epic literature. Sterchi also argued that the pattern of Genesis 1 is not used in any other sequence in Scripture.[21]

But there is more. As to the fourth feature, the pairing of days typical of the Near Eastern examples such as "day one, a second day," is missing in Genesis 1. There may be, however, a unique variation of the pairing of days based on their content. Many writers have noted a correlation in content between days one and four, focusing as they do on light and light-bearers; between days two and five, focusing on the firmament and the waters and the animals flying across the face of the firmament and dwelling in the waters; and between days three and six, focusing on the dry land and animals that lived on the land.

---

[20]David A. Sterchi, "Does Genesis 1 Provide a Chronological Sequence?" *Journal of the Evangelical Theological Society* 39 (1996): 529-36.
[21]Ibid., p. 532.

The paired structure is particularly evident when we examine the ordering of blocks of material in Genesis 1 concerning the word-commands. Throughout the chapter there are eight sections that begin with divine command speeches prefaced by "and God said" distributed across the six creative days. Day one (Let there be light) and day two (Let there be an expanse between the waters to separate water from water) contain one command speech apiece, but day three has two separate command speeches (Let the water under the sky be gathered to one place, and let dry ground appear; and also, Let the land produce vegetation: seed-bearing plants and trees on the land that bear fruit with seed in it, according to their various kinds). Similarly day four (Let there be lights in the expanse of the sky to separate the day from the night, and let them serve as signs to mark seasons and days and years, and let there be lights in the expanse of the sky to give light on the earth) and day five (Let the water teem with living creatures, and let birds fly above the earth across the expanse of the sky) include one command speech each. But day six, like day three, has two separate command speeches (Let the land produce living creatures according to their kinds: livestock, creatures that move along the ground, and wild animals, each according to its kind; and also, Let us make man in our image, in our likeness, and let them rule over the fish of the sea, and the birds of the air, over the livestock, over all the earth, and over all the creatures that move along the ground).

Thus, there is a parallelism of structure between days one through three and days four through six. In effect, there are in Genesis 1 three pairs of days just as there are in the ancient Near Eastern examples. The pairing scheme, however, is different. Perhaps the inspired author felt freedom to develop the standard literary formula in much the same way that the sonata allegro form of the first movement of the classical symphony as developed by Haydn, Mozart, Beethoven and Schubert underwent dramatic transformation in the hands of much later symphonists like Mahler or Shostakovich.

Finally, in regard to the fifth feature, we have the repetitious character of the six creative days of Genesis 1 followed by the climactic establishment and blessing of the sabbath on the seventh day. Although the repetitions of Genesis 1 mercifully rise far above the sterile monotony of the Near Eastern examples, the repetitions are nonetheless striking. Each of the six days of creation is characterized by a stylized block of statements (with minor variations): (1) And God said, "Let there be"; (2) And it was so; (3) And God saw that it was good; and (4) The evening and the morning were the nth day. The seventh day completely abandons this repetitious phraseology just as the ancient Near Eastern examples do.

This textual evidence suggests that the author of Genesis 1 employed a standard literary device that was widely used in the ancient Near East in describing creation. In so doing he did two things. He emphasized the completeness of the work of creation, the point made in Genesis 2:1, and he emphasized the establishment of the sabbath on the climactic seventh day. So, supposing that the author used this literary device, what are the implications? The author was trying to convey the completeness of creation in the heightening of suspense as he worked toward the climactic establishment of the sabbath. From the way Genesis 1 is constructed, we can affirm that God did create the cosmos and that he did establish the sabbath rest at the conclusion of his creative work, but the author conveyed that teaching in terms of the seven-day week that everyone in the ancient Near East observed.

As a result, the author could use a literary device built around the ordinary twenty-four-hour days of human experience while at the same time mentioning events that lasted more than twenty-four hours. He did that because he was not concerned to convey information about how long God actually took to create the universe. The author was not trying to convey chronological information because that is not how the device functioned in ancient Near Eastern literature. Had he wanted to do that he would have used a pattern more like that of Numbers 7 where ordinary days are obviously in view.

At the same time, the seven-day literary structure is clearly based on the universally held human week, and one can legitimately suggest that, as a matter of divine condescension, God employed the figure of a human work week as a metaphor or symbol or image of divine creation. Thus, there is merit to the contention of Shedd and Collins that the days may be considered as divine "days" so that human days are analogical to divine "days." But readers of Genesis 1 would still be missing the point of the use of the literary structure if they attempted to derive chronological information such as the actual sequence of the creative events from the text. As Sterchi suggested, "the text is not implying a chronological sequence of seven days. Instead it is simply presenting a list of seven days."[22]

THE ANCIENT NEAR EASTERN CONTEXT OF GENESIS 1

We have called attention to the importance of the relationship between the seven-day structure of Genesis 1 and ancient Near Eastern literary parallels. But the link between Genesis and the ancient Near East goes far beyond the seven-day structure, and if we are to understand Genesis 1 aright, we need to recognize this linkage. So let's explore this linkage further.

Many sincere and devout Christians are oblivious to the fact that, if Mo-

---

[22]Ibid., p. 533.

ses was the author, Genesis 1 was composed in the late second millennium before Christ. Even if one accepted the suspect hypothesis that Genesis 1 was composed during the Babylonian exile by a so-called priestly author, P, as the documentary hypothesis maintains, its composition would still date to the mid-first millennium before Christ.[23] In either case, Genesis 1 was originally addressed to an ancient Israelite audience for whom the times, questions, issues and historical context were extremely different from our own.

If contemporary Christians are to understand Genesis 1 properly it is imperative that they reject the idea that Genesis 1 is speaking directly to them in the language and concepts of the twenty-first century. Thanks be to God, even if Christians adopt that improper mindset they still come away from Genesis 1 grasping the teaching that God is the creator of all things. That, of course, is a great blessing, because that teaching is the major thrust of the chapter. But if Christians want to fathom the insights of Genesis 1 further they need to read that chapter in the context of the ancient Near East of which the early Israelites were a part. This project will entail some major adjustments. Christians will need to immerse themselves in the thought patterns, images, literary styles and conventions, and conceptions of the physical universe of that ancient world of at least three millennia ago. They will need to imagine what Genesis 1 would have meant to an ancient Israelite. In God's providence, all of us are blessed with living in a time when far more is known about the culture and worldview of the ancient Near East than was likely even in the time of Christ because in the last couple of centuries we have been the beneficiaries of a plethora of archeological and linguistic work that has opened up that world.

One major point needs stating: The Israelites did not have the scientific conception of the world that characterizes the early twenty-first century. They would not have understood Genesis 1 as talking about such contemporary scientific issues as the age of the universe, the validity of big bang cosmology,

---

[23]For critiques of the documentary hypothesis with its division of the Pentateuch into various hypothetical source documents labeled J, E, P, D, etc., see Kenneth A. Kitchen, *On the Reliability of the Old Testament* (Grand Rapids: Eerdmans, 2003), pp. 492-94; Kenneth A. Kitchen, *Ancient Orient and Old Testament* (Chicago: InterVarsity Press, 1966), pp. 112-29; and Oswald T. Allis, *The Five Books of Moses*, 2nd ed. (Philadelphia: Presbyterian & Reformed, 1949). See also Claus Westermann, *Genesis 1—11: A Commentary* (Minneapolis: Augsburg, 1984), pp. 574-84. Even Westermann, a devotee of higher critical methods in the study of the Old Testament, has acknowledged that many of the traditional arguments to support the idea of different sources of material in the Pentateuch rest on shaky grounds. Of course, the assumption of generally Mosaic authorship of Genesis and the rest of the Pentateuch and rejection of the classic documentary hypothesis by no means rules out the possibility of subsequent modification and final assembly of the Mosaic texts of the Torah at a later time such as the exile. After all, Deuteronomy concludes with a report of Moses' death. One suspects that Moses did not write that account!

plate tectonics, nuclear physics or biological evolution. An Israelite hearing or reading that the seed-bearing trees or beasts of the field reproduced after their own kind would not imagine that the text had anything to do one way or the other with a theory of biological evolution because such theories were not part of their thought world. They might, however, have understood the text to be teaching that God created an orderly, stable world in which animals are not transformed into deities.

No doubt the ancients were curious about how things worked and where they came from, but they, including the Israelites, were especially very much concerned with the relations among the various competing deities and nature. The divine author of Genesis 1 was certainly concerned that the Israelites understood the true connection between deity and nature. Yahweh, the God of Israel, wanted to make sure that the Israelites realized that he was not simply another local, tribal deity, albeit different from the gods of other nations, but that he is the only true God, the God of the entire cosmos and all that it contains, that he is sovereign over everything, and that he made everything. As descendants of Abraham the Mesopotamian, the Israelites emerged from a pagan, polytheistic culture. Abraham's progeny clearly did not immediately shake off every last vestige of polytheism as soon as Abraham closed the gate of Ur behind him. Recall that Rachel stole her family's household gods, hid them from her suspicious father in her camel's saddle, and sat on them in the tent after she and Jacob had left Paddan-Aram and the service of Laban the Aramean (Genesis 31:19, 30-35).

In time the descendants of Jacob sojourned for several hundred years in Egypt, hardly a bastion of monotheism. Egypt was full of gross polytheism; cats, ibises, snakes, beetles, crocodiles and other animals were revered as gods. It is small wonder that Genesis 1 stressed the point that animals remain their own kind. The Sun was a god to the Egyptians. So was the river Nile. Surely this polytheistic environment powerfully affected the children of Israel. How quickly they slipped into the idolatry of worshiping the golden calf after they had been in the wilderness only a short time (Exodus 32)! Even Aaron the high priest was hardly a shining example of theological and ethical purity, forging the calf while Moses was on the mountain receiving the commandment not to make graven images! And upon entry into the promised land Joshua had to challenge the people not to serve the gods that their forefathers had worshiped but to worship Yahweh alone (Joshua 24: 2, 14-15, 23). They were also warned not to take wives from among the Canaanites and the other surrounding nations lest they be tempted to worship their gods, which, of course, they proceeded to do throughout the course of their history as the books of Judges, the Kings and the prophets (Amos 2:4) emphasize. Even Solomon, credited as the wisest of Israel's kings, failed to heed the warnings and quickly accommodated to the pagan practices of his wives.

Given the pervasive influence and pressure of the polytheistic paganism of Israel's neighbors, Israel desperately needed religious instruction, not answers to twenty-first-century scientific questions. As a result, Genesis 1 is a striking repudiation of all the gods of Mesopotamia, Egypt, Canaan and all of Israel's neighbors. Genesis 1 makes it perfectly clear that *anything* that anyone else worshiped, whether sun, moon, stars, trees, waters, animals, is strictly a creature brought effortlessly and inexorably into being by the one and only living God who simply speaks things into existence. In one short page, all crude cosmogonies of the surrounding nations are demolished. That animals and trees reproduce after their kind means that they remain animals or trees—they do not become gods. The mention of stars on the fourth day almost as an afterthought—oh, by the way, the stars also—downplays their importance (Genesis 1:16). The Sun and Moon are not mentioned by name because the Hebrew words resemble the names of heathen divinities. They are simply lights of differing size and brightness. The great sea creatures that commonly played a role in heathen mythologies as opponents of deity offer no resistance whatever to the creative word of God. Genesis 1 stands in stark theological contrast to the heathen cosmogonic myths with which ancient Israelites must have had some familiarity. As has been proposed by many current Old Testament scholars, Genesis 1 exercises a strongly polemic function.[24]

Of course, after it is understood that the main purpose of Genesis 1 is to distinguish the living God who created heaven and earth from all the pagan divinities, readers are in a position to appropriate its teaching to themselves. Although the reader may no longer worship trees, animals or the sun, he or she is warned in Scripture that the worship of anything that is created, and that means *anything* that isn't God, is idolatry. Whether it is a spouse, children, parents, house, career, hobby, vacation, sports, music, art or anything else, those who place any of those things at the center of their lives are guilty of transforming the good things that God has created into idols and of worshiping the creature rather than the Creator, who is blessed forever.

DIVINE ACCOMMODATION IN SCRIPTURE

The Israelites, as part of the ancient world, had a very different conception of the physical character of the world from ours. As a result, some of the words in Genesis 1 did not convey the same impressions to them as they might to

---

[24]Gerhard F. Hasel, "The Polemic Nature of the Genesis Cosmology," *The Evangelical Quarterly* 46 (1974): 81-102.

us. Thus, when the Israelites read about the creation of the *earth* they did not visualize a spherical globe that spins on its axis and revolves annually around the Sun. Such concepts were totally foreign to the ancient world, to Israel and even to the inspired author of Genesis 1. As a child of his time, not even he knew of the Earth as a planetary globe, despite the fact that he, assuming it was Moses, was educated in the wisdom of Egypt. Nor is there any reason why God should have revealed such information to him. It was the intent of neither the human nor the divine author to convey to the readers of Genesis 1 the idea that the earth God created is a planetary globe. If Moses intended to convey the concept that the term *earth (eretz)* in Genesis 1 refers to a spinning globe that orbits the Sun, he would certainly have lost credibility because *everyone* could plainly see that all of the heavenly bodies revolve around a stationary earth. To the ancient Israelites, the term *earth,* as Paul Seely has shown, was understood to be not a globe, but one continuous landmass that included what we regard as southern Europe into Asia and northern Africa. Moreover, this landmass was part of a flat disk.[25]

In turn, the sea was regarded as an ocean that encircled and also undergirded that single landmass: hence, the references in the Old Testament to the waters under the earth (Exodus 20:4; Deuteronomy 5:8; Psalm 136:6). Again, as Seely has pointed out, that was the common conception among the ancients. In other words, when Israelites heard or read Genesis 1 they would have envisioned creation of a flat disk containing a single landmass surrounded by a world-encircling sea that also undergirded the land mass.

When the Israelites read about God's making of the sun, they would have visualized a bright object in the daytime sky, nowhere near as large as we now know the sun to be, that orbits the stationary earth-disk once a day. To them the sun looked like it went around the earth every day because they took it for granted that the sun actually does orbit the earth every day. Each morning the sun appeared through a kind of slot at the far horizon and disappeared into another slot on the opposite horizon at the end of the day.

But then comes the real kicker. The ancient world universally believed that the dome-like vault of the sky is a glassy, crystalline *solid.*[26] This belief persisted widely in the early church and into the Middle Ages. So accustomed are moderns to the idea that the gaseous sky continues uninterruptedly into distant space that it is difficult to imagine why the ancients considered the sky to be solid. But they did. In fact, at least the ancient Mesopotamians, who lived under

---

[25]Paul H. Seely, "The Geographical Meaning of 'Earth' and 'Seas' in Gen. 1:10," *Westminster Theological Journal* 50 (1997): 231-55.

[26]Paul H. Seely, "The Firmament and the Water Above, Part I: The Meaning of *raqia'* in Gen. 1:6-8," *Westminster Theological Journal* 53 (1991): 227-40.

clear blue skies day after day, envisioned the dome of the sky acting as a dam or barrier to hold back a watery ocean, preventing it from collapsing onto the earth. The blue sky does somewhat resemble a big blue ocean above us. Genesis 1, of course, makes reference to the establishment of a *firmament (raqia')* that separates waters above it from waters below it. Many modern commentators refer to this *raqia'* as the sky. But an ancient Israelite upon hearing or reading about the *raqia'* would immediately have understood it as a reference to the solid sky-dome that rests on the earth-disk as a transparent covering that kept out the waters beyond. The events that transpired on day two of creation, as recorded in Genesis 1, therefore, refer to the placement of a solid dome over the earth. Thus, we read about the stars being *in* the firmament and birds flying *across the face* of the firmament (Genesis 1:17, 20). We read in Job 37:18 a reference to the firmament being hammered out as hard as a mirror. And in Ezekiel 1:22, 25-26 we read of God's throne resting on the firmament-pavement.

But then we are confronted with a problem. How can Scripture assert that God established a solid vault in the sky when everyone knows full well that there is no such solid vault up there? Isn't that a violation of the infallibility and inerrancy of the Bible? Some commentators attempt to avoid the force of the statement by claiming that Scripture is using phenomenal language, the language of appearance. But that's our problem. The Israelites would not have seen it that way. The sky didn't just look solid to them; they believed it to *be* a solid. It helps to keep in mind that God was addressing his ancient audience in terms with which they were familiar, consistent with their understanding of their world. It would have done no good for God to talk about creating a completely open sky because everyone knew that the high sky was solid. It would have done no good to talk about the earth orbiting the sun or the earth as a globe or the earth being billions of years old because those ideas would have been utterly alien to them. Moreover, being incidental to the theological-religious-ethical message of the Bible, such assertions or information would have impeded rather than enhanced the message.

As it was, the master teacher, the God of the universe, employed the time-tested pedagogical method of accommodating himself to the level of his hearers by stooping to their level, addressing them in their language in the context of their world picture. John Calvin, for example, repeatedly emphasized the role that divine accommodation plays in Scripture far more than contemporary commentators do.[27] Perhaps we need to recover his insight.

Where the Israelites needed to be challenged and changed was in their understanding of their relation to God. It clearly would have been counterpro-

---

[27]On Calvin's doctrine of accommodation, see Davis A. Young, *John Calvin and the Natural World* (Lanham, Md.: University Press of America, 2007), pp. 161-90, 210-30.

ductive for God to accommodate himself to the erroneous religious conceptions entertained by the Israelites because those conceptions were precisely what he wanted to change. On the other hand, for the sake of getting across the religious message of Scripture, God could readily accommodate himself to the primitive scientific conceptions of the Israelites. Placing a radically different theology before them would be challenge enough. Why introduce additional hurdles? They didn't need access to a better scientific world-picture. That could come with time, and the human race could figure out the better science on their own. They didn't need special revelation for that.

Why do we mention all of this business about the ancient world? Simply to make the point more forcefully that Genesis 1 was not given in twenty-first-century terms to teach twenty-first-century science or to provide answers to contemporary scientific problems. God has given humanity both the freedom and the intellectual and observational tools to do that on our own. Does the Earth go around the Sun or does the Sun go around the Earth? God has chosen not to tell us that in the Bible. He expects his image-bearers to figure that out if they find themselves curious about the matter. Is the Earth 6,000 years old or 4.5 billion years old or somewhere in between? God has chosen not to mention that in the Bible. In other words, it is not the intent of Genesis 1 to tell us when God created the world or how long it took him to create it. Humans can figure that out using the tools of modern science, another of the marvelous gifts granted by the Creator.

USE OF METAPHOR

There is one final point worth mentioning before we bring this section to a close. This point also ties in closely with the linkage of Genesis to the world of the ancient Near East. Some Old Testament scholars have pointed out the pervasive role of metaphor in Genesis 1. They have said that the actions of God are portrayed in anthropomorphic terms. Because, according to Scripture, God is incomprehensible to his creatures, we can understand who he is and what he is like only in terms of the creaturely language of analogy, metaphor and imagery. John Walton, for example, has suggested that temple imagery plays a role in the creation account. The sabbath rest represents God's enthronement is his newly constructed sanctuary with his throne in heaven and his footstool on earth.[28] Meredith Kline and John Stek have likened the description of creation to an account of God as the great king establishing his kingdom. Kline, for example, noted that the creative fiats are "sovereign decrees" that "clearly evoke the throne of the King of Glory." The first three

---

[28]John H. Walton, *The NIV Application Commentary: Genesis* (Grand Rapids: Zondervan, 2001).

days entail formation of "creation kingdoms" that mirror or copy the heavenly kingdom. The last three days involve creation of "creature kings" that exercise dominion over the creation kingdoms, thus reflecting the "royal rule of the Creator enthroned above."[29] Kline also noted that royal terminology is explicitly used of the heavenly bodies.

Stek especially has drawn attention to a variety of features suggestive of the thoroughgoing employment of a royal-political metaphor. He noted that the application of this metaphor to the realm of the gods in the Near East took place as early as the third millennium B.C. Thus, it would hardly have been a surprise to the Hebrews that the story of creation was presented in terms of a well-established royal-political metaphor. For Stek, the work of creation may be seen as the royal acts of the great king, and the ensuing creation is God's kingdom and the human race is God's royal steward.[30]

As indicators of the functioning of the royal metaphor in Genesis 1, Stek noted several textual elements. First, the divine fiats function like the decrees of the absolute rulers of the ancient world. When an ancient king decreed that something should be done, it got done. When God decrees light, vegetation, humans and so on, it gets done: it is so. As absolute overlords, kings in the ancient world exercised the prerogative of naming. We see several examples of that in the Old Testament. Joseph was renamed Zaphenath-Paaneah by the Egyptian Pharaoh (Genesis 41:45). Daniel and his three friends, Hananiah, Mishael and Azariah were all renamed by Nebuchadnezzar's chief official as Belteshazzar, Shadrach, Meshach and Abednego (Daniel 1:7). And, of course, we have examples of God changing the names of individuals: Abram to Abraham (Genesis 17:5), and Jacob to Israel (Genesis 35:10). In Genesis 1, God exercises his lordship by naming the day, the night, the sky, the land and the seas.

Ancient kings also assigned the diet of those who sat at their table. Joseph's brothers were served with food from Joseph's table (Genesis 43:34). Mephibosheth sat at the table of King David (2 Samuel 9:7, 13). Jehoiakim sat at the table of Nebuchadnezzar (2 Kings 25:29). Daniel and his friends had their food appointed for them by Nebuchadnezzar, although they refused to partake of the kings' fare (Daniel 1:5, 16). God, on the sixth day, assigned the food for the animals and for human beings.

Ancient kings frequently erected images of themselves in territories of their empires as indicators or emblems of their sovereign dominion. They served as reminder of who was in charge. So, too, God set up humanity as his image, indicating by our presence God's ownership and sovereignty over creation. We

---

[29]Meredith G. Kline, "Space and Time in the Genesis Cosmogony," *Perspectives on Science and Christian Faith* 48 (1996): 6.

[30]Stek, "What Says the Scripture?" p. 232-35.

210 THE BIBLE, ROCKS AND TIME

humans are his appointed representatives to the rest of creation.

Even the seven-day structure reinforces the royal motif, for, as Stek pointed out, the manner of narration in Genesis 1 approximates the chronicling of royal acts. According to Stek,

> Consistent with its theme, however, its form (an unadorned "objective" account of "dated" events, using formulaic language) suggests that of a "record," recounting what transpired in God's royal council chambers. It reads like a daybook kept by a recorder of royal executive actions.[31]

All in all, Genesis 1 provides a portrayal of God, the great King, sovereignly building and establishing the parameters of his kingdom by decree. If, in fact, the author of Genesis 1 consciously utilized a royal metaphor, then it would be out of keeping to press the text too rigidly for literal historical details. It would, for example, be a mistake to transfer the metaphor of seven divine days directly to seven literal days of the kind that human beings experience. If royal metaphorical language dominates Genesis 1, then it would be misguided to try to work out a literal chronology of God's creative work that must serve as a guide for scientific research.

CONCLUSION

As promised, we have not attempted to provide a comprehensive interpretation of Genesis 1. Nor have we tried to defend the gap theory, the day-age theory, the revelation-day theory, the intermittent-day theory, the analogical-day theory or the framework theory. Clearly we have sympathy for aspects of the last two views, and readers will also detect that we think it is very important to view Genesis 1 in light of its ancient Near Eastern context. For the reader to think, however, that we are advocating a specific interpretation of Genesis 1 would be to miss our point. We have drawn insights from the work of several biblical scholars to show that Genesis 1 is saturated with features that render it highly unlikely that the author was concerned with interacting with the scientific questions of our day or that he was even concerned about the specific question of the age of the Earth. We have concluded from our survey of the biblical material that the Bible does not teach that the Earth is only a few thousands of years old. We have also concluded that the Bible is not concerned about the age of the Earth at all. The Bible leaves it up to humans to try to figure out how old the Earth is, if that is a question that interests us. For that, one may properly turn to the insights of the natural sciences. To the science of geology, then, we turn in the following chapters to find out what geology can tell us about the age of the Earth.

---

[31] Ibid., p. 241.

Part Three

GEOLOGICAL PERSPECTIVES

8

THE NATURE AND NURTURE OF
THE STRATIGRAPHIC RECORD

DURING THE LATE NINETEENTH AND THROUGHOUT the twenti-
eth centuries, the discipline of stratigraphy continued to accumulate data.
New sources of stratigraphic data became available as techniques for drilling
into the deeper continental crust and the oceanic crust were developed. New
subdisciplines of stratigraphy emerged, such as seismic stratigraphy, based
on images obtained from reflections of seismic waves off subsurface layer
boundaries. Geochronologic techniques, largely based on radiometric age determinations (see chaps. 14 and 15), finally helped to bracket absolute time
periods in the geo-historic record. Stratigraphy was spurred and financed by
many economic applications, especially those related to hydrocarbon exploration and development.

This expansion in scope and intensity of investigation is the first sense of
the word *nurture* in our title for this chapter. Throughout the twentieth century to the present, stratigraphy retained a strong industrial association. For
example, the world's major organization of sedimentary geologists, founded
in 1927, was designated the Society of Economic Paleontologists and Mineralogists (SEPM). The commercial applications of stratigraphic principles became highly successful and included targeting oil- and gas-hosting rock layers submerged beneath hundreds to thousands of feet of water by directed
drilling from billion-dollar offshore structures. The great blessings provided
to the global economy is the second sense of the word *nurture* in that indus-

trial stratigraphy has helped to nurture our present energy-intensive civilization. Anyone reading this book owes some gratitude to the work of industrial stratigraphers.

Stratigraphers have been assisted in their efforts by sedimentologists, the students of sedimentation processes. During the twentieth century, sedimentology achieved great sophistication through applications of basic physics and chemistry to processes of rock weathering, transport of sedimentary particles, precipitation of chemical sediments and deposition of laminated rocks. This effort was tied into the study of modern environments in which sedimentary

Figure 8.1. Grand Canyon of the Colorado River, Arizona, viewed from the South Rim. Photo by R. F. Stearley.

rocks are forming, such as the Mississippi River delta or the salty, limey lagoons of the Persian Gulf. The study of modern analogs of ancient "environments of deposition" was subsidized by hydrocarbon exploration companies, which provided scholarships to graduate students and short courses for their employees. As in the case of industrial stratigraphers, the efforts of practical sedimentologists linking ancient sedimentary settings to modern analogs has yielded benefits to millions, including readers of this book.

In this chapter, we look at the general nature of the stratigraphic record and begin to address the alternative picture of that record painted by Flood geologists. We first outline some basics of stratigraphic practice as well as the magnitude and utility of data accumulated to date. We next review some of the basic principles of Flood geology, as well as some of its predictions concerning sedimentology and the stratigraphic record. We address two of Flood geology's most touted "critiques" of mainstream stratigraphy: the criticism that mainstream geology will not and cannot admit that prehistoric catastrophes are recorded in the rocks and the criticism that the purported ascending order to fossils in the stratigraphic record is an artificial construct of geologists designed to mislead the public.

In chapter nine, we pay particular attention to fossils and examine claims by Flood geologists that many fossiliferous layers in the rock record could have been produced only by a planetary catastrophe. We show that such claims are exaggerated at best and often misleading. In chapter ten, we look more closely at individual sedimentary rock units to examine claims for rapid or slow sedimentation rates. In particular, we look at categories of evidence that sedimentary geologists employ to deduce various environments of deposition, many of which are decidedly noncatastrophic.

STRATIGRAPHY: BASIC OBSERVATIONS AND TERMINOLOGY

At this point we introduce some terminology used to describe large-scale aspects of layered rocks. A convenient locale to begin our discussion is the Grand Canyon of the Colorado River in Northern Arizona (fig. 8.1). The canyon is in some places more than a mile deep, affording a wonderful view into the spatial extent and geometry of layered rocks. Because Flood geologists such as Steven Austin often refer to these strata, we will mention this locale at various points during the remainder of the book.

At first glance, one fact jumps out at the casual observer: the layered rocks in the walls of the Grand Canyon do *not* occur as one great layer of uniform color and consistency. Rather, the vertical section is divided into a series of layers, each of which maintains chemical consistency, color, uniform internal texture and fossil content. Such units are termed *formations* and are named by a set of rules analogous to the rules by which biological species are named.[1] Figure 8.2 illustrates two of the canyon's prominent formations: the Coconino Formation and the Hermit Formation. A hiker along one of the canyon's many trails can easily verify that the Coconino Formation is composed almost entirely of very pale sand grains of a uniform size, whereas the Hermit Formation, mostly shale, is composed of clay minerals stained to a deep red by a small amount of the iron oxide mineral hematite. The

---

[1] The North American Commission on Stratigraphic Nomenclature, "North American Stratigraphic Code," *AAPG Bulletin* 89 (2005): 1547-91; and Amos Salvador, ed., *International Stratigraphic Guide—A Guide to Stratigraphic Classification, Terminology, and Procedure*, 2nd ed. (Boulder, Colo.: The International Union of Geological Sciences and GSA, 1994).

grains in the Coconino Formation are well cemented, and the rock holds up a stout vertical cliff, whereas the grains in the Hermit Formation are poorly cemented, and the rock erodes to form a slope.

Upper and lower boundaries to particular formations are typically distinct (fig. 8.2). These *stratal boundaries* are generally abrupt transitions from one rock type to another. Such abrupt boundaries represent pauses or *hiatuses* between one episode of sedimentation and the next. In many outcrops, visual inspection plainly reveals that erosion occurred during the time elapsed between successive depositional events. Erosional boundaries are termed *un-*

Figure 8.2. Coconino Formation (above) and Hermit Formation (below). Photo by R. F. Stearley.

conformities. Unconformities typically demonstrate clearly that the lower unit was lithified (changed to rock) prior to the erosional episode. For example, intact lithified pebbles from the lower unit may be incorporated into the upper unit. In some cases a lower unit may be lithified and then tilted or folded prior to erosion and subsequent deposition. Such a boundary is termed an *angular unconformity.* In the Grand Canyon, a major angular unconformity is visible from the east end of the south rim (see fig. 2.2).

EXTENT, THICKNESS AND CHARACTER OF STRATIFIED ROCKS: OVERVIEW

One of the earliest arguments for the antiquity of the Earth stemmed from the evidence contained within accumulations of sedimentary rock. Thick piles of layered, fossil-bearing sedimentary rocks, such as sandstone, shale and limestone, cover large portions of the continental land masses. Vast areas of Kansas, Nebraska, Iowa and other Midwestern states are underlain by a thickness of a mile or more of sedimentary rock that lies on top of older crystalline basement rocks.[2] Oil and gas wells drilled in Michigan reveal a stack of sedimentary rocks that reaches a collective thickness in excess of 15,000 feet (see chap. 12). A thickness of four to five thousand feet of relatively undeformed layered sedimentary rocks is exposed in the walls of the Grand Canyon (fig. 8.1). Careful examination of the canyon's deep innermost gorge reveals an older, thicker stack of layered sediments beneath this upper stack (fig. 2.2). The underlying stack is approximately 12,000 feet thick. The lower stack was lithified, then tilted and substantially eroded prior to deposition of the upper stack. Thus, even if one ignores the absent rock which correlates to the major erosional interval following tilting of the lower sequence, the Grand Canyon exposes a pile of strata more than three miles thick.[3] As another case, in driving up the canyons east of Salt Lake City, Utah, one works upward through a large stack of tilted sedimentary layers (fig. 8.3). Over northern Utah and southern Idaho, the collective thickness of these layered rocks runs well in excess of 30,000 feet and may range locally to at least 45,000 feet.[4]

Since the eighteenth century, it has become more clearly recognized that

---

[2]B. J. Bunker, B. J. Witzke, W. L. Watney and G. A. Ludvigson, "Phanerozoic History of the Central Mid-Continent, United States," in *Sedimentary Cover—North American Craton, U.S., The Geology of North America,* vol. D-2, ed. L. L. Sloss (Boulder, Colo.: GSA, 1988), pp. 243-60.

[3]Stanley S. Beus and Michael Morales, eds., *Grand Canyon Geology,* 2nd ed. (New York: Oxford University Press, 2003).

[4]William Lee Stokes, *Geology of Utah* (Salt Lake City: Utah Geological and Mineral Survey, 1986). See also Paul K. Link and others, "Middle and Late Proterozoic Stratified Rocks of the Western U.S. Cordillera, Colorado Plateau, and Basin and Range Province," in *Precambrian: Coterminous U.S., The Geology of North America,* vol. C-2, ed. R. C. Reed et al. (Boulder, Colo.: GSA, 1993), pp. 463-595.

Figure 8.3. Tilted strata, Big Cottonwood Canyon, Utah. The road descends toward the west; Proterozoic strata tilt down toward the east. Photo by R. F. Stearley.

sedimentation occurs not only on the sea floor but in a large variety of environments such as river and stream valleys, floodplains, deltas, lakes and desert basins. Sediments are transported by running water, wind and glacial ice. Sediments produced in these environments have distinctive characteristics that relate to the physical and chemical contexts in which the sediments formed. For example, windblown sand grains deposited in a desert are readily distinguishable from particles produced, transported and deposited by glaciers or from fine clay particles that have settled out in the deep ocean. During the past 150 years, geologists have become increasingly adept in recognizing the products of various environments in the sedimentary rock record. The nature of this evidence is elaborated in chapter ten.

This strong evidence of development of thick piles of layered sediments in ancient basins with identifiable contexts such as deserts, lakes, rivers, deltas, shores and open oceans indicates that long time spans must have elapsed during the formation of local stacks of sediment, inasmuch as the formation of deltas, glaciers, lakes and so on involves measurable processes that take considerable time. By comparing modern processes of sedimentation with the evidences in the sedimentary rock record, geologists have concluded that Earth must be far older than was assumed three hundred years ago. The physical evidence contained within sedimentary rocks provides a powerful argument that Earth is much older than just a few thousand years. In fact, as discussed in chapters three and four, many geologists in the eighteenth century and most geologists in the nineteenth century surmised that Earth is quite old on this very basis.[5]

ADVANCES IN STRATIGRAPHIC DATA ACQUISITION

Since the mid-nineteenth century, the accessibility of the upper portion of Earth's crust and the volume of stratigraphic data have increased exponentially. Following on the heels of the original bedrock mappers of the late eighteenth and early nineteenth centuries, thousands of field stratigraphers have measured successions of layers in outcrops. A single vertical sequence exposed on a cliff or a hillslope is termed a *stratigraphic section*. Part of the toolkit of any field geologist is the ability to measure a "strat" section. Measuring a section includes taking photographs in the field, collecting representative samples, noting features observed in outcrops and writing detailed descriptions of the rocks. Field measurements are typically supplemented by examination of the microscopic structure of the rocks with the petrographic microscope, a method first applied during the 1850s by Henry Clifton Sorby.[6] All the observed data are summarized on a stratigraphic chart (fig. 8.4).

Although vertical cliffs and deep mines provide one-stop exposures of long vertical sequences of strata, regional stratigraphic sequences can be discerned by detailed examination of smaller exposures in creek beds, road cuts and other limited spaces. This practice works because, even in relatively inactive continental interiors, strata are typically gently warped by stresses that were applied at continental margins during the formation of highly deformed mountain belts. Because the strata in places like Iowa or Ohio are slightly tilted, different layers intersect the land surface at different places, thus enabling stratigraphers to map the layers based on their exposures.

Close on the heels of this expansion of surface-based data has come an even greater expansion of our ability to penetrate and sample unseen strata below our feet. During the latter nineteenth and early twentieth centuries, techniques and equipment for drilling into the uppermost 30,000 feet of the continental crust were developed, spurred by the monetary incentives provided by the prospect of producing oil and gas. In North America, for example,

---

[5]Claude C. Albritton, *The Abyss of Time: Changing Conceptions of Earth's Antiquity After the Sixteenth Century* (Mineola, N.Y.: Dover Publications, 1980); and Martin J. S. Rudwick, *Bursting the Limits of Time: The Reconstruction of Geohistory in the Age of Revolution* (Chicago: University of Chicago Press, 2003).

[6]H. C. Sorby, "On the Microscopical Structure of Crystals, Indicating the Origin of Minerals and Rocks," *Quarterly Journal GSL* 14 (1858): 453-500.

hundreds of thousands of oil and gas wells have returned data from the subsurface. More than 500,000 wells are currently (2007) producing oil and gas in the United States![7] Data from these wells are often saved in the form of *drill cores*, thin cylindrical columns of rock that are bored by specially designed, hollow

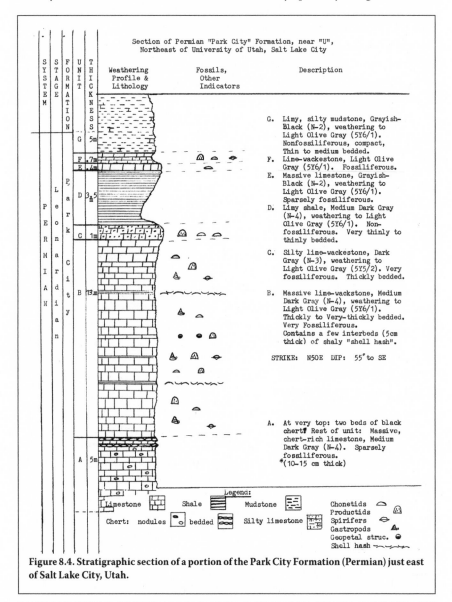

Figure 8.4. Stratigraphic section of a portion of the Park City Formation (Permian) just east of Salt Lake City, Utah.

[7]American Petroleum Institute, Web article, "About Oil and Natural Gas: Industry Sectors," (September 7, 2007), under <www.api.org>. Washington, D.C., American Petroleum Institute.

drill bits. Cores may be hundreds of feet in length and continuous. Cores are stored in *core libraries* that are maintained as in-house reference sources by oil and gas companies, or as public resources by state geologic agencies or universities. Most readers of this book have access to a major core library within a day's drive of their residence.

As drilling techniques were refined, sedimentary geologists learned to extract small fossil skeletons, *microfossils*, from portions of cores. The study of these fossils, termed *micropaleontology*, became of great industrial importance. Several different tribes of ancient creatures are useful, depending on the type of rock sampled and its relative position in the rock column. Major microfossil groups with industrial biostratigraphic application include diatoms, coccolithophorids, conodonts, radiolarians, ostracods and foraminifera.

Because coring adds expense to the drilling procedure, not all wells are cored. However, all wells return pulverized rock debris that during drilling rises to the surface where it is continuously sampled. These samples, called *well cuttings,* provide data on rock layers as they are penetrated. Finally, characteristics of underground rock layers can be determined by lowering devices called *sondes* into the borehole. Sondes are equipped with sensors to measure physical and chemical properties of rocks, such as their electrical properties, overall porosity, presence of fluids and natural radioactivity. Data obtained by sondes are often termed *wireline data.*[8]

As one example of a mineral constituent that is easily sensed remotely, consider the mineral sylvite, potassium chloride (KCl). Sylvite is unique among the many naturally precipitated rock salts. A portion of the potassium that sylvite contains is radioactive potassium 40 (^{40}K). Thus, a gamma ray detector lowered into the drill hole in a sequence of salty deposits will register an enormous spike in radiation as it passes through a layer that is predominantly sylvite. On the basis of gamma rays, sylvite layers can be discerned easily in the A-1 Evaporite unit of the Salina Group in the Michigan Basin, for example (see chap. 12).[9]

During the 1940s, drilling techniques were extended offshore. Many thousands of wells have been drilled in the continental shelf of North America alone. About twenty-five percent of U.S. oil and gas production comes from offshore wells. This drilling effort was taken into the deep ocean through the auspices of the Deep-Sea Drilling Project (DSDP) and its successor, the

---

[8]R. L. Brenner and T. R. McHargue, *Integrative Stratigraphy* (Englewood Cliffs, N.J.: Prentice-Hall, 1988).

[9]R. T. Lilienthal, *Stratigraphic Cross-Sections of the Michigan Basin,* Division Report of Investigations 19 (Lansing: Michigan Geological Survey, 1978); and R. C. Elowski, *Potassium Salts (Potash) of the Salina A-1 Evaporite in the Michigan Basin,* Division Report of Investigations 25 (Lansing: Michigan Geological Survey, 1980).

Ocean Drilling Project (ODP). The first major deep-water scientific drilling vessel, the *Glomar Challenger,* operated between 1968 and 1983 and drilled more than 500 holes in the seafloor beneath water as deep as 24,000 feet.[10] Its successor vessel, the *JOIDES Resolution* (figs. 8.5 and 8.6), began operation in 1985 and continues this work until the present day. Future deep-water drilling vessels are under design.[11]

Figure 8.5. Deep-sea scientific drilling vessel *JOIDES Resolution.* **Courtesy of International Ocean Drilling Program/Texas A & M University.**

Our three-dimensional picture of subsurface layering can now be augmented by techniques that do not require drilling. During the twentieth century, geophysicists learned how to create artificially generated *seismic waves* within the Earth and use these to sense the three-dimensional structure of layered rocks. Seismic waves, which behave in a fashion similar to sound waves, pass through layers of rock, but a portion of each wave's energy is reflected off the interface between two different layers. The rock discontinuity is termed a *seismic reflector.* For land-based surveys, seismic energy is put di-

---

[10]Roger Revelle, "The Past and Future of Ocean Drilling," in *The Deep-Sea Drilling Project: A Decade of Progress,* ed. J. E. Warme, R. G. Douglas and E. L. Winterer (Tulsa, Okla.: Society of Economic Paleontologists and Mineralogists, 1981).

[11]A. E. Maxwell, "An Abridged History of Deep Ocean Drilling," *Oceanus* 36 (1994): 8-12; and H. V. Thurman and A. P. Trujillo, *Introductory Oceanography* (Upper Saddle River, N.J.: Pearson Prentice Hall, 2004).

Figure 8.6. Shipboard core laboratory, *JOIDES Resolution.* **Courtesy of International Ocean Drilling Program/Texas A & M University.**

rectly into the ground by means of explosions or vibration from a mechanical vibrator. Distance to an individual reflector is a function determined by travel time through the rock layer. Travel times are recorded by means of series of detectors termed *geophones* strung along a calibrated line that may be a few miles in length. Depths to reflectors are calculated, and a series of images of *stacked reflectors* is produced. Seismic surveys can also be carried out at sea easily through energy released from explosive charges or from large air guns towed behind research vessels. Geophones are trailed behind the vessel, and repeated transits over a region provide a dense data set revealing region-wide stratigraphy. Collection and analysis of huge seismic data sets are standard and critical techniques in the oil and gas industry. With the advent of computer handling of large data sets, *three-dimensional seismic reflector volumes* can be produced that track reflectors over large spatial regions. Well data provide cross-checking of the interpretation of the reflector images.

Seismic techniques have proven extremely useful for understanding broad-scale patterns of sediment layering and thickness in regions where drilling costs are very high, particularly in offshore areas of deltas and continental shelves where hydrocarbon exploration is active and ongoing (fig. 8.7). Large volumes of seismic data have enabled Earth scientists to see that the continental shelves are draped with hundreds to tens of thousands of feet of sediment

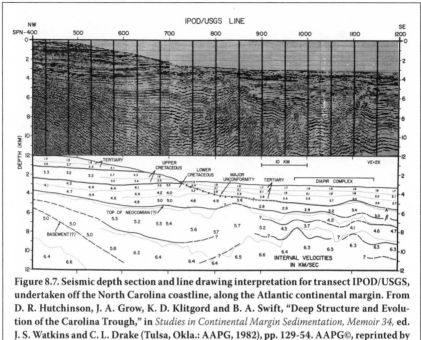

Figure 8.7. Seismic depth section and line drawing interpretation for transect IPOD/USGS, undertaken off the North Carolina coastline, along the Atlantic continental margin. From D. R. Hutchinson, J. A. Grow, K. D. Klitgord and B. A. Swift, "Deep Structure and Evolution of the Carolina Trough," in *Studies in Continental Margin Sedimentation, Memoir 34,* **ed. J. S. Watkins and C. L. Drake (Tulsa, Okla.: AAPG, 1982), pp. 129-54. AAPG©, reprinted by permission of the AAPG, whose permission is required for further use.**

and sedimentary rock. The Atlantic margin of North America, for example, is a wedge of sediments thickening seaward, ultimately achieving thicknesses of ten miles or more, blanketing the preexisting basement rocks.[12] Selective drilling into these blankets permits the stratigrapher to understand particulars of individual layers.

FLOOD GEOLOGY: CATASTROPHIC NOACHIAN SEDIMENTATION

Large, regionally coherent blankets of sedimentary rock cover the interiors of continents, extend out over the continental shelves, and are commonly folded or otherwise deformed in mountains. As noted in chapter five, however, a separate school of stratigraphy emerged during the twentieth century that vigorously asserted and continues to assert that mainstream stratigraphy is grossly off-base. The economic successes of modern stratigraphy were held to be mainly accidental: "The uniformitarian hypothesis and the evolutionary framework of geological ages have been shown to be largely irrelevant to the actual practice of petroleum exploration."[13] Beginning with George McCready

[12]J. Kennett, *Marine Geology* (Englewood Cliffs, N.J.: Prentice-Hall, 1982).
[13]John C. Whitcomb Jr. and Henry M. Morris, *The Genesis Flood: The Biblical Record and Its*

Price and Byron Nelson, this school of stratigraphy emphasized the work that Noah's Flood could have accomplished in the rock record.[14] This viewpoint is often labeled "Flood geology" and its advocates "Flood geologists." We follow this convention.

Advocates of Flood geology make several claims regarding the stratigraphic record, most of which are direct or indirect critiques of mainstream geological interpretations. First, most if not all sedimentary rock strata are held to have been deposited by a single planetary catastrophe. Price wrote in 1923:

> our second corollary is that this hypothesis of a world catastrophe *deals with the world as a whole*, that is, it deals with the world in its planetary aspects; and therefore this catastrophe must have been of an astronomical character, and must have an astronomical cause. In other words, to spoil this ideal world, and to do it suddenly, would require an *astronomical cause*, something that would disturb this delicate equilibrium existing between water and land, thus destroying the ideal climate, and incidentally destroying the plants and animals existing upon it. But the only astronomical cause which we can readily imagine as being competent to bring about such a result would be something in the nature of a *jar or a shock from the outside*, which would produce an abnormal tidal action, resulting in great tidal waves sweeping twice daily around the earth from east to west, this wave traveling 1000 miles an hour at the equator.[15]

A contemporary flood advocate, Larry Vardiman, wrote in 1997:

> On average, the continental crust of the earth is covered with about a mile of sediments which appear to have been formed by the Genesis Flood. . . . It appears that a slurry of mud and sand was formed in the oceans by the action of the Flood events, which was deposited onto the continents and ocean floors to form the sedimentary rock layers.[16]

In 1999, John Morris said that "The Flood would have totally restructured the surface of the globe," and as recently as 2005, William Hoesch and Stephen A. Austin termed the Flood "the greatest of tsunamis."[17]

Flood geologists have proposed a variety of causes of a great aqueous cat-

---

Scientific Implications (Philadelphia: Presbyterian & Reformed, 1961), p. 437.

[14]George McCready Price, *Q.E.D., or New Light on the Doctrine of Creation* (New York: Fleming H. Revell, 1917), and *The New Geology* (Mountain View, Calif.: Pacific Press, 1923); and Byron C. Nelson, *The Deluge Story in Stone: A History of the Flood Theory of Geology* (Minneapolis: Augsburg, 1931).

[15]Price, *The New Geology*, p. 682 (italics in original).

[16]Larry Vardiman, "Global Warming and the Flood," *Institute for Creation Research Impact Series* 294 (1997): i-iv.

[17]John D. Morris, "Where Was the Garden of Eden Located?" *Back to Genesis* 132 (1999); and W. A. Hoesch and Steven A. Austin, "Do Tsunamis Come in Super-Size?" *Institute for Creation Research Impact Series* 382 (2005): iii.

aclysm. These have been summarized by Austin and his colleagues and by Brown.[18] Austin and his colleagues, for example, proposed a "thermal runaway" model that accounts for vast amounts of rainfall plus rapid marine transgression in terms of catastrophic plate tectonics. In recent years, many Flood geologists have modified Price's position by postulating that some strata were deposited prior to the Flood and some after the Flood.[19]

Second, Flood geologists from Price to the present have consistently advocated the claim that the presence of great numbers of preserved animal and plant remains in the sedimentary record necessitates very rapid sedimentation and burial. This is especially thought to be evident for spectacularly preserved mass mortality layers or "fossil graveyards."

Third, Price and many Flood geologists following him have placed great weight on their claim that the ordering to the fossil record, which enabled nineteenth-century biostratigraphers to correlate strata from one region to the next, is fallacious or even a deceit.

> What Do the Formations Represent? But the question very naturally arises, What do they represent? The answer to this is just as obvious: they simply represent a taxonomic or classification series of the life of the ancient world, just as living samples might be made up from here and there all over our modern earth to represent the life of the world today, these samples being each a representative local fauna or flora from some particular locality here or there over the modern world. For it is simply these buried local faunas and floras with which we are dealing when we speak of the various geological formations; and we have been piecing these scattered formations into systems and into groups of strata, under the impression that they represent various time-values, when they can not represent anything of the kind. The whole geological series is just as purely constructive, just as wholly artificial, as would be a corresponding series of the living plants and animals of our modern world, which might be made up by carefully gathering and arranging many thousands of local faunas and floras from scattered localities all over the earth. There would be no essential differences between the two series, save that the geological one represents dead (and often extinct) forms, while the modern one would represent living ones. The

---

[18]S. A. Austin, J. R. Baumgardner, D. R. Humphreys, A. A. Snelling, L. Vardiman and K. R. Wise, "Catastrophic Plate Tectonics: A Global Flood Model of Earth History," in *Proceedings of the Third International Conference on Creationism, Technical Symposium Sessions*, ed. R. E. Walsh (Pittsburgh: Creation Science Fellowship, 1994), pp. 609-21; and Walter Brown, *In the Beginning: Compelling Evidence for Creation and the Flood*, 7th ed. (Phoenix: Center for Scientific Creation, 2001).

[19]Austin et al., "Catastrophic Plate Tectonics." See also Leonard Brand, *Faith, Reason, and Earth History* (Berrien Springs, Mich.: Andrews University Press, 1997); and Ariel V. Roth, *Origins: Linking Science and Scripture* (Hagerstown, Md.: Review and Herald Publishing Association, 1998).

one would essentially parallel the other, and would also just as clearly represent a "history of creation" as would the other.[20]

For Price and his successors, most of the geological strata were arranged during a single geological event (the Flood). Thus, the contained fossils merely represent samples of the destroyed biota prior to the event. The histories pieced together from this stacked set of samples are thus viewed as accidental and/or contrived.

What are some predicted features of a global catastrophic stratigraphic record? Since the late 1970s, Flood advocate Walter Brown has promoted the effects of *liquefaction* on formation and delivery of flood sediments. Liquefaction is well documented today in cases where earthquake shocks provide energy into liquefied sediment. Such sediment flows easily and provides an analog to hypothesized planetary water waves bearing immense slurries of sediment. Furthermore, as these slurries settle, packages of sediment often naturally segregate in a phenomenon that Brown terms "lensing." Brown believes that liquefaction and lensing furnish the best explanation for many interesting features of the sedimentary record as indicated by the following excerpts:

> One thick and extensive sedimentary layer has remarkable purity. The St. Peter sandstone, spanning about 500,000 square miles in the central United States, is composed of almost pure quartz, similar to sand on a white beach. It is hard to imagine how any geologic process, other than global liquefaction, could achieve this degree of purity over such a wide area. Almost all other processes involve mixing, which destroys purity.
>
> Sedimentary layers usually have boundaries that are sharply defined, parallel, and nearly horizontal. These layers are often stacked vertically for thousands of feet. If layers had been laid down thousands of years apart, erosion would have destroyed this parallelism. Liquefaction, especially liquefaction lenses, explain this common observation.
>
> Varves are extremely thin layers which evolutionists claim are laid down annually in lakes. By counting varves, they believe time can be measured. However, varves are too uniform, show no evidence of the slightest erosion, and are deposited over wider areas than where streams enter lakes (which is where most deposits occur in lakes). Lakes would not produce varves. Varves are better explained by liquefaction.[21]

In reply, we note that the St. Peter Sandstone is composed of pure quartz, and its sand grains are very well sorted by size. The grains are all extremely rounded. The formation contains internal sedimentary structures like low-angle tabular cross-stratification that forms today in beach and shallow

---

[20]Price, *The New Geology*, p. 614.
[21]Brown, *In the Beginning*, p. 143.

Figure 8.8. Low-angle, planar cross-stratification in modern coastal beach sands, Presque Isle, Erie, Pennsylvania. Photo by D. A. Young.

shoreface settings (figs. 8.8 and 8.9). Mainstream sedimentary geologists maintain that the purity, sorting and structural characteristics of the deposit demand *consistent conditions operating over time.* We find it difficult to image a violent short-term event that churned and dumped vast quantities of sediment capable of such discriminating selection, even granting Brown's lensing phenomenon. Such an event should mingle sedimentary particles of all shapes and compositions.

Second, as we have seen in our discussion of the Grand Canyon strata, sedimentary layers *are* often distorted, forming contacts with obvious angular unconformities (figs. 2.2 and 3.6). In many places distinct erosional surfaces are common and *easily* seen.

Third, thinly laminated muds and silts are typical deep-water sediments in lake basins. Again it is difficult to imagine a single, short-term, catastrophic depositional model that can account for thick rhythmic sequences of fine-grained laminae and near-contemporaneous deposition of thick beds of pure coarse-grained rocks and nearly pure marine limestone.

Other Flood geologists have outlined various predictions based on such liquefied, catastrophic planetary slurries, acting together with some sort of hydrodynamic sorting.[22] Predicted features of the sedimentary record include the ab-

---

[22]M. E. Clark and H. D. Voss, "Resonance and Sedimentary Layering in the Context of a Global

sence of growth structures such as fossil tree roots in sediment, the presence of mass "graveyards" of animals and plants killed and rapidly buried, the presence of "polystrate" structures, particularly large fossils, which extend vertically through many beds (demonstrating rapid sedimentation), and some sort of sorting of sedimentary materials by specific gravity. We address these claims in chapters nine, ten and twelve.

Flood geologists associated with the Institute for Creation Research (ICR) have invested much time and effort during the past thirty years in reinterpreting the Paleozoic strata visible in the Grand Canyon.[23] Their efforts are much more field based than are those of many Flood geology proponents and are much more

Figure 8.9. Low-angle, planar cross-stratification in St. Peter Sandstone (Ordovician), Graham Cave, Missouri. Photo by R. F. Stearley.

geologically informed. In the ICR Grand Canyon studies, physical descriptions of rock layers are accurate. Unconformities are documented but explained as resulting from extremely brief intervals of erosion. Models are presented for layering based on transgressions and regressions of the waters associated with a global flood event. Because mainstream stratigraphers interpret these units as constructed over long time periods under diverse ecological conditions, alternative Flood-depositional interpretations for their origin must be explained and defended. For example, the Coconino Formation (fig. 8.2) is interpreted by mainstream stratigraphers as the deposit of a subaerial sand dune complex, whereas Austin and Brand interpret the Coconino Formation as the result of

Flood," in *Proceedings of the Second International Conference on Creationism*, vol. 1: *Technical Symposium Sessions*, ed. R. E. Walsh and C. L. Brooks (Pittsburgh: Creation Science Fellowship, 1990), pp. 53-64.

[23]S. A. Austin, "Interpreting Strata of Grand Canyon" and "A Creationist View of Grand Canyon Strata," in *Grand Canyon: Monument to Catastrophe*, ed. S. A. Austin (Santee, Calif.: Institute for Creation Research, 1994), pp. 21-56 and pp. 57-82. See also W. R. Barnhart, M. L. Folsom and K. P. Wise, "Fossils of Grand Canyon," also in *Grand Canyon: Monument to Catastrophe*, pp. 133-52.

catastrophic wave-transported subaqueous sand bodies.[24] We return to this particular deposit in chapter ten.

FLOOD-GEOLOGICAL CRITIQUE OF "UNIFORMITARIAN" LOGIC

Not surprisingly, those who insist that the Earth is very young do not accept the approach to sedimentary rocks taken by mainstream geologists. There has been a prodigious effort by Flood geologists to discredit the evidences that indicate the antiquity of sedimentary rock sequences. In addition, there has been a great deal of young-Earth creationist literature that, while ignoring, discrediting or explaining away the evidences for antiquity, focuses attention on several features in the sedimentary rocks, such as "polystrate fossils," that supposedly can be accounted for only in terms of very rapid or catastrophic deposition. The evidences of catastrophic deposition are then used as one supposed proof of the young-Earth position.

The Flood-geology version of the creationist argument typically follows these lines: (1) Mainstream geologists believe that sedimentary rocks must be interpreted in light of modern sedimentary processes. (2) Mainstream geologists believe that most modern processes are all slow and that since catastrophes rarely occur today, they are only sporadically represented in the rock record. (3) Sedimentary rocks actually show a number of features that can be accounted for only by very rapid catastrophic deposition. (4) The evidences of catastrophic deposition contradict modern geological theory and practice. (5) Therefore, young-Earth creationists are right and mainstream geologists are wrong. (6) Many if not all highly fossiliferous beds imply very rapid burial under water. (7)The only aqueous catastrophe of worldwide proportions that could account for the worldwide distribution of evidences for catastrophic deposition is the Noachian Flood. (8) Scripture indicates that the Flood affected the Earth only for one year, that it occurred only a few thousand years ago, and that it occurred only a few thousand years after the creation in six twenty-four-hour days. (9) Therefore, the Earth is only a few thousand years old.

In very general terms we can say that the premises of the arguments are faulty, and, therefore, the conclusion is by no means certain. As regards premise 3, we agree that many sedimentary rocks around the world do contain features that were produced by very rapid or catastrophic deposition. But the premise is misleading because it tells only half of the truth. Many other sedimentary rocks around the world contain features that must result from extremely slow rates of deposition. Any valid theory of the origin of sedimentary rocks must not ignore *either* of these two groups of features.

---

[24]Austin, "Interpreting Strata of Grand Canyon," and Brand, *Faith, Reason, and Earth History.*

Regarding premise 2, it is not true that modern geologists entertain the idea that most modern-day geologic processes are slow (see chap. 16). Neither is it true that mainstream geologists reject the idea that catastrophes rarely occur today or that they rarely occurred in the past. Geologists certainly accept the idea that the average rates of sedimentary depositional processes over long stretches of time are quite slow, but it is also accepted that there may be brief, spasmodic episodes when rates of deposition may be extremely high, even catastrophic. We will look at some examples of modern catastrophic deposition later in this chapter. Since premises 2 and 3 are incorrect, conclusion 4, that evidences of catastrophic deposition contradict modern geologic theory and practice, does not follow. Thus, neither is conclusion 5—that creationists are right and modern geologists are wrong—established.

Fossil accumulations contain internal evidences as to circumstances of deposition. In some cases deposition is rapid, but in other cases the deposit may represent a "trap" that operated over some time. Thus premise 6 paints a simplistic and misleading view of the fossil record. We examine such cases in depth in chapter nine. We also note with regard to premise 7 that there exist other reasonable explanations besides a global Flood that can explain the worldwide distribution of evidences for catastrophic deposition. Because of the incorrect premise, conclusion 9 is not established.

In chapters nine and ten, we look in greater detail at some creationist arguments about sedimentary rocks and attempt to evaluate them. We now turn to the claim that mainstream geology rejects catastrophes.

ARE MODERN GEOLOGISTS ANTI-CATASTROPHISTS?

We first briefly address the claims put forth by Flood geologists concerning premise 2 and conclusion 4 above. Over the course of the twentieth century, sedimentologists have discovered abundant evidence for the effects of catastrophes in Earth history. Although nineteenth-century uniformitarians attempted to downplay the role of catastrophes, such dramatic "events" are now part of geologic orthodoxy (see chap. 16).

During the early twentieth century, evidence accumulated that sporadic *rapid submarine gravity-driven flows* occur naturally. A notable example occurred during the Grand Banks earthquake of 1929, when a large mass of sediment along the continental slope south of the Grand Banks of Newfoundland, extending for 50 miles and up to 1,000 feet thick, slid abruptly downslope, generating a rapidly moving slurry of water and mud. The slurry proceeded to snap several trans-Atlantic telegraph cables in succession. The velocity of the main flow was at least as high as 37 knots, or 19 meters per second.[25] Rapid subaqueous gravity-driven flows were linked to the formation of certain marine sandstones by Kuenen and Migliorini in a foundational 1950 paper.[26] This connection sparked a minor industry within the sedimentological community and led to thousands of scientific papers devoted to the topic of subaqueous gravity-driven deposits before the end of the twentieth century. The Ocean Drilling Project (ODP) has cored and analyzed numerous seafloor deposits that originated as subaqueous gravity flows such as the Amazon fan.[27]

British biostratigrapher Derek Ager drew attention to rare, violent events such as storms in the global geologic record in his highly influential volume, *The Nature of the Stratigraphic Record*.[28] Ager provided many examples of well-documented cases of catastrophic sedimentation. For example, he discussed giant "boulder beds" in the Tertiary deposits of Ecuador, in which "boulders" as much as two miles in diameter were positioned along a paleoscarp and were visible in outcrop for a distance of 200 miles! Ager gave chapters in the book titles like "Catastrophic Uniformitarianism" in an effort to prod fellow stratigraphers into considering the impact of the rare but violent event in Earth history. Ager was pleasantly surprised to find that he received a lot of support from other sedimentary geologists.

The 1980s and 1990s saw further interest in rare, violent events recorded in the stratigraphic record. In 1982, Robert H. Dott, in his presidential address to the Society of Economic Paleontologists and Mineralogists, emphasized the importance of unusual sedimentation events in the rock record, reflecting the influence of volcanic eruptions, violent storms, tsunamis and asteroid impacts. Speaking and writing in the climate of renewed claims for Flood geology, however, Dott attempted to divorce his terminology from potential catastrophist misuse, and he advocated thinking in terms of *episodic events* that, although rare, were to be expected over long time spans.[29] Since Dott's address, the field of episodic stratigraphy has mushroomed, with regular appearances of major texts like *Cycles and Events in Stratigraphy*, and papers like that of Shiki that

[25]R. G. Walker, "Mopping Up the Turbidite Mess," in *Evolving Concepts in Sedimentology*, ed. Robert N. Ginsburg and Francis J. Pettijohn (Baltimore, Md.: Johns Hopkins University Press, 1973), pp. 1-37.

[26]Philip H. Kuenen and C. I. Migliorini, "Turbidity Currents as a Cause of Graded Bedding," *Journal of Geology* 58 (1950): 91-127.

[27]R. N. Hiscott, C. Pirmez and R. D. Flood, "Amazon Submarine Fan Drilling: A Big Step Forward for Deep-Sea Fan Models," *Geosciences Canada* 24 (1997): 13-24.

[28]Derek V. Ager, *The Nature of the Stratigraphic Record* (New York: John Wiley, 1973), and *The Nature of the Stratigraphic Record*, 2nd ed. (New York: John Wiley, 1981).

[29]Robert H. Dott, "Episodic Sedimentation—How Normal Is Average? How Rare is Rare? Does It Matter?" *Journal of Sedimentary Petrology* 53 (1982): 5-23.

emphasized the role of tsunamis in generating deposits.[30]

Evidence of *catastrophic astronomical collisions* has also been recorded in the rock record. Globally, the stratigraphic boundary between rocks of the Cretaceous and Tertiary Systems records a large turnover in the world's biota, as determined from fossils entombed in rocks directly above and below the boundary. During the 1970s, study of the boundary section in Italy drew Walter Alvarez into a profound puzzle concerning the origin of a narrow band of clay positioned directly at the boundary. Study of trace metals in the clay suggested that it had originated as fallout from major impact of an extraterrestrial body with the Earth.[31] Similar clay layers have been located at the same stratigraphic position in many places around the world. The "impact hypothesis" was highly controversial but generated a large interest in the study of cratering processes and potential products, including blast-derived and mega-tsunami sediments.[32] Eventually, a probable crater was located at Chicxulub in the Yucatan Peninsula that may correspond to the end-Cretaceous impact.[33] Twenty-five years after the original proposal, most geologists are convinced that a large-scale impact did occur at the end of the Cretaceous Period but disagree on its significance for the history of life.

The tsunami generated by the presumed Chicxulub impact is calculated to have had a height between 300 and 600 feet as it struck the northern shore of the Gulf of Mexico and to have run inland perhaps as far as 200 miles beyond the Mississippi River embayment![34] Resulting tsunami deposits surrounding the Gulf of Mexico average a few meters in thickness. However, localized gravity-flow units now exposed in western Cuba, interpreted as catastrophic deep-sea tsunami beds, range up to 2,200 feet thick.[35]

[30]G. Einsele, W. Ricken and A. Seilacher, eds., *Cycles and Events in Stratigraphy* (Berlin: Springer-Verlag, 1991); and T. Shiki, "Reading the Trigger Records of Sedimentary Events—A Problem for Future Studies," *Sedimentary Geology* 104 (1996): 249-55.

[31]L. Alvarez, W. Alvarez, F. Asaro and H. V. Michel, "Extraterrestrial Cause for the Cretaceous-Tertiary Extinction," *Science* 208 (1980): 1095-108.

[32]A. R. Hildebrand and W. V. Boynton, "Proximal Cretaceous-Tertiary Impact Deposits in the Caribbean," *Science* 248 (1990): 843-47; and F. J.-M. R. Maurrasse and G. Sen, "Impacts, Tsunamis, and the Haitian Cretaceous-Tertiary Boundary Layer," *Science* 252 (1991): 1690-93.

[33]A. R. Hildebrand, G. T. Penfield, D. Kring, M. Pilkington, A. Carmargo, S. B. Jacobsen and W. Boynton, "Chicxulub Crater: A Possible Cretaceous-Tertiary Boundary Impact Crater on the Yucatan Peninsula, Mexico," *Geology* 19 (1991): 867-71; and J. Morgan, M. Warner and R. Grieve, "Geophysical Constraints on the Size and Structure of the Chicxulub Impact Crater," in *Catastrophic Events and Mass Extinctions: Impacts and Beyond*, GSA Special Paper 356, ed. C. Koeberl and K. G. MacLeod (Boulder, Colo.: GSA, 2002), pp. 39-46.

[34]T. Matsui, F. Imamura, E. Tajika, Y. Nakano and Y. Fujisawa, "Generation and Propagation of a Tsunami from the Cretaceous-Tertiary Impact Event," in *Catastrophic Events and Mass Extinctions: Impacts and Beyond*, GSA Special Paper 356, ed. C. Koeberl and K. G. MacLeod (Boulder, Colo.: GSA, 2002), pp. 69-77.

[35]R. Tada et al., "Complex Tsunami Waves Suggested by the Cretaceous-Tertiary Boundary Deposit at the Moncada Section, Western Cuba," in *Catastrophic Events and Mass Extinctions:*

The revelation of the ancient catastrophe that occurred at the end of the Cretaceous Period has stimulated closer examination of many other sections containing large-scale debris flows or boulder beds. For example, the Late Devonian stratigraphic record of Nevada and western Utah contains a large megabreccia that has been linked to multiple comet showers. In turn, these comet impacts have ramifications for Late Devonian climate and sea-level fluctuations.[36] Interestingly, since their recognition, megabreccia beds have been claimed by Flood geologists as evidence for a recent global Flood.[37]

Thus, evidence has accumulated over the course of the twentieth century for local and for large-scale catastrophes that left clues within the sedimentary rock record.[38] Far from suppressing or denying this accumulating evidence, the community of sedimentary geologists has incorporated these new, exciting data into more complete models of erosion and deposition. These additional interpretive models, however, have not negated well-understood mechanisms for the formation of many "ordinary" types of sedimentary rocks such as shallow marine limestones, fossil reefs and regional salt deposits. The vast majority of sedimentary geologists and stratigraphers see no need to explain all, or even most, strata on the basis of a rare catastrophic event—especially on the basis of a single planetary cataclysm.

FAUNAL SUCCESSION IN THE ROCK RECORD

Although any one locality on the face of the Earth preserves but a portion of the overall history of the planet, the order to the fossils, first discovered in the late 1700s and early 1800s, remains uniform. Thus, gaps will occur locally within the overall sequence, but the sequence has been found to be reliably duplicated over the Earth. Gaps do not present a problem to mainstream stratigraphers, for they do not expect all places on the Earth to be uniformly acquiring sediment all the time. At the present time, for example, Pike's Peak, Colorado, is not accumulating a fossil record—all animal and plant remains

Impacts and Beyond, GSA Special Paper 356, ed. C. Koeberl and K. G. MacLeod (Boulder, Colo.: GSA, 2002), pp. 109-23.

[36]J. Warme and C. A. Sandberg, "Alamo Megabreccia: Record of a Late Devonian Impact in Southern Nevada," *GSA Today* 6 (1996): 1-7. See also C. A. Sandberg, J. R. Morrow and W. Ziegler, "Late Devonian Sea-Level Changes, Catastrophic Events, and Mass Extinctions," in *Catastrophic Events and Mass Extinctions: Impacts and Beyond*, GSA Special Paper 356, ed. C. Koeberl and K. G. MacLeod (Boulder, Colo.: GSA, 2002), pp. 473-87.

[37]A. V. Chadwick, "Megabreccias: Evidence for Catastrophism," *Origins* 5 (1978): 39-46; Brand, *Faith, Reason, and Earth History*; and Hoesch and Austin, "Do Tsunamis Come in Super-Size?"

[38]C. C. Albritton, *Catastrophic Episodes in Earth History* (London: Chapman and Hall, 1989); and W. A. Berggren and J. A. Van Couvering, eds., *Catastrophes and Earth History: The New Uniformitarianism* (Princeton, N.J.: Princeton University Press, 1984).

are on their way downhill, if they survive at all!

The order to the fossil record has been repeatedly verified by thousands of professional paleontologists and stratigraphers during the past two hundred years and duplicated by many thousands of amateur fossil collectors. With the advent of micropaleontology, the order of the fossil record could be applied to observations of layered rocks obtained from drill cores. The ability to target particular levels within strata based on microfossils has been of inestimable value in the search for fossil hydrocarbons.

Unfortunately, the law of faunal (or biological, fossil) succession, which is put to practical use daily by geological detectives attempting to pin down the location of aquifers or oil- and gas-producing strata deep in the subsurface, has been treated as anathema by some twentieth-century Christian catastrophists.

FLOOD GEOLOGY AND THE BIOSTRATIGRAPHIC RECORD: ATTACK APPROACH

Twentieth-century young-Earth creationist authors often accuse geologists of piecing together the fossil record, crazy-quilt style, to fit a preconceived notion of organic evolution. In effect, these modern critics claim that the pattern of fossils depicted in geology textbooks is a fiction, produced for evolutionary propaganda purposes.

As previously discussed, George McCready Price vigorously insisted that fossil faunas were instantaneous "snapshots" of localized faunas preserved by Noah's Flood. As a corollary to this proposal, Price maintained that stratigraphic layers were sorted by geologists to fit preconceived notions of the history of life. He likened the work of the stratigraphic geologist to that of a librarian sorting file cards. In 1917, Price already elaborated this "card-sorting" activity:

> The geological series is merely an old-time classification series, a classification of the forms of life that used to live on the earth, and is of course just as artificial as any similar arrangement of the modern forms of life would be. We may illustrate the matter by comparing this series with a card index. The earlier students of geology arranged the outline of the order of fossils by a rather general comparison with the series of modern life forms, which happened to agree fairly well with the order in which they had found the fossils occurring in England and France. But only a block out of the middle of the complete card index could be made up from the rocks of England and France; the rest has had to made up from the rocks found elsewhere. Louis Agassiz did Herculean work in rearranging and trimming this fossil card index so as to make it conform better, not only to the companion card index of the modern forms of life, but also that of the embryonic series. From time to time even now readjustments are made in the details of all three indexes, the fossil, the modern, and the embryonic, the

method of rearrangement being charmingly simple: *just taking a card out of one place and putting it into another place* where it belongs. . . . In view of these facts, we need not be concerned as to the fate of the geological classification of the fossils. It is a purely artificial system.[39]

In Price's 1923 magnum opus, *The New Geology,* he repeated the "card catalog" critique and went further, to elaborate his "law of conformable stratigraphic sequence" by claiming that "any kind of fossiliferous beds whatever, 'young' or 'old' may be found occurring conformably on any other fossiliferous beds, 'older' or 'younger.'"[40]

Price based this law on two phenomena. First, the presence of gaps in the global fossil sequence within local stratigraphic successions, and second, instances where thrust faults had emplaced older strata, with their entombed biota, on top of younger strata. His emphasis on local gaps was misplaced for reasons elaborated above. His antagonism toward thrust faults was also misplaced, because thrust faults do exist and are empirically verifiable through careful field observations (fig. 8.10). Price was so convinced that geologists were practicing a deception that he could not make himself confront the massive evidence for the unity of the order to the fossil record.

Price's critique of the law of faunal (fossil) succession was enthusiastically adopted by many twentieth-century advocates of Flood geology. In 1931, Lutheran minister Byron Nelson quoted Price favorably in *The Deluge Story in Stone.* Nelson illustrated the seeming problem of reversal of order in deformed, thrust-faulted regions like the Northern Rockies and Switzerland through a series of remarkable photographs, matched against a standard geologic column, with lines connecting the strata of the column to various layers in the photographs. The result was an impressive tangle of criss-crossed lines resembling a telephone operator's console from the 1940s.

With *The Genesis Flood,* published in 1961, Whitcomb and Morris can be credited with the modern revival of the approach of Price and Nelson, arguing that strata are ordered by deceitful geologists, based on a preconceived notion of the evolution of life:

> The rock systems of geology and their corresponding geologic ages have for many years been worked up in the form of a geologic timetable. For a typical example, see figure 5. Such a presentation obviously indicates a gradual progression of life from the simple to the complex, from lower to higher, and therefore implies organic evolution. This is considered by geologists to be a tremendously important key to the interpretation of geological history.
>
> Of course, it is maintained by many stratigraphers that other factors, espe-

---

[39]Price, *Q.E.D.*, pp. 120-22 (italics in original).
[40]Price, *The New Geology,* pp. 294, 638.

Figure 8.10. Overthrust fault in Lower Paleozoic rocks, I-75, north of Knoxville, Tennessee. The fault runs diagonally from upper left to lower right of the road cut and can be detected where the more steeply dipping layers on the right rest on gently dipping layers on the left. The stack of layers on the right side of the roadcut were thrust upward relative to the stack of layers on the left. A very small fold caused by fault movement is visible immediately above the fault surface toward the upper left. Photo by D. A. Young.

cially that of superposition of the strata, are also important in geologic correlation and that, in general, these factors justify the usual assignment of ages to strata on the basis of their fossil contents. The usual situation, however, is that only a few formations are ever superposed in any one locality and that it is very difficult or impossible to correlate strata in different localities by this principle of superposition. The fossils must be resorted to, and the fossil sequence is assumed to accord with the principle of evolution.[41]

In 1974, Morris more explicitly argued in *Scientific Creationism* that this constitutes circular reasoning:

And now, finally, we begin to recognize the *real* message of the fossils. There is no truly objective time sequence to the fossil record, since the time connections are based on the evolutionary assumption, which is the very point in question. The relative positioning of the fossiliferous strata, therefore, must be strictly a function of the sedimentary and other processes which deposited them.[42]

---

[41]Whitcomb and Morris, *The Genesis Flood*, pp. 133-35.
[42]Henry M. Morris, *Scientific Creationism*, 1st ed. (San Diego: Creation-Life Publishers, 1974), p. 96.

Here is obviously a powerful system of circular reasoning. Fossils are used as the only key for placing rocks in chronological order. The criterion for assigning fossils to specific places in that chronology is the assumed evolutionary progression of life; the assumed evolutionary progression is based on the fossil record so constructed. The main evidence for evolution is the assumption of evolution![43]

During the 1970s, many young-Earth creationists followed Whitcomb and Morris by advocating that the ordering to the fossils and their containing strata was a fiction conceived by atheists bent on proving preconceived notions of evolution.[44] The claim was repeated by Wilder-Smith in 1981:

Certainly the index fossil method is the most important method for the dating of formations known to modern geology. It has served more than all other dating methods to establish evolutionary theory. It must be remembered that, as applied today, it always supports evolutionary theory! Of course, the reason for this is by now perfectly clear: the method assumes that evolutionary theory is experimentally correct so that a suitable family tree can be set up depending on evolutionary concepts. Then it confirms the veracity of the evolutionary theory on the basis of the evolutionary family tree . . . that is, the index fossil method is calibrated against the theory of evolution . . . then it proceeds to calibrate evolutionary theory against the index fossil method. Is it surprising that the theory of evolution confirms the index fossil method and vice versa? Neodarwinian theory has been thriving on this circular thinking between theory and practice and practice and theory for over 130 years.[45]

However, we saw in chapter four that the founders of biostratigraphy were actively sorting out the pattern of extinct life forms more than a generation prior to the publication of Darwin's *The Origin of Species,* and were in fact indifferent or hostile to the concept of organic evolution. The pattern of preserved organic remains in rocks is an empirical one. Any valid theory of the history of the Earth must take into account this empirical pattern.

FLOOD GEOLOGY AND THE BIOSTRATIGRAPHIC RECORD: THE "ECOLOGICAL ZONATION" VERSION

Even as Price and Nelson were vigorously promoting the thesis that the order to the fossil record is a fiction contrived by atheistic scientists, some of their contemporary Flood geologists were grappling with the fact of biotic succession in stratigraphy. Notably, Harold Clark propounded a theory in *The New*

---

[43]Ibid., p. 136.
[44]R. L. Wysong, *The Creation-Evolution Controversy* (Lansing, Mich.: Inquiry Press, 1976).
[45]A. E. Wilder-Smith, *The Natural Sciences Know Nothing of Evolution* (San Diego: Master Books, 1981), p. 108 (ellipses are original to Wilder-Smith's text).

Diluvialism (1946) that the organization of the fossil record basically conforms to an expectation of preservation order according to pre-Flood ecological zonation. Clark, a Seventh-day Adventist, began his career as an intellectual protégé of Price. However, a visit to the oilfields of Oklahoma and Texas during the summer of 1938 persuaded him that the order to the fossil record is empirical and provided practical benefits to those exploring the subsurface.[46] Clark was forced to part company with Price, resulting in animosity on Price's part. Clark continued to look up to his former mentor and furnished a one-hundred-page eulogy in 1966 for Price, *Crusader for Creation*.[47]

Clark's "ecological zonation theory" for Flood-buried animals and plants has gradually gained wider acceptance within the young-Earth creationist community.[48] Ariel Roth wrote:

> As we consider how the flood might have caused the sequence found in the fossil record, we should differentiate between the familiar, small local floods and an unfamiliar worldwide event as described in Genesis. Sometimes we think of a flood as washing sediment from a higher area into a lower one and mixing everything in a disorganized pattern. However, flood deposits are often well-sorted, forming widespread flat layers. On a larger scale, mixing is even more difficult. A sequence of fossils would result as slowly rising floodwaters sequentially destroyed the various preflood landscapes along with their unique organisms, redepositing them in order in large depositional basins of the continents. . . . the order of the fossils in such sedimentary basins would reflect the order of the eroded landscapes destroyed by the gradually rising waters.[49]

Adherents of the ecological zonation theory claim that most large-scale features of the fossil record can be harmonized within the restrictions of a single year-long Flood event. Thus, Lower Paleozoic strata, which lack any record of land plants or land animals, are thought to be the preserved remains of pre-Flood deeper-water marine communities. At the other end of the geological spectrum, Cenozoic strata with their rich record of fossil mammals are claimed to result from concentrations of mobile and intelligent animals which could seek higher ground during the Flood event and so were overwhelmed last. Dinosaurs and other Mesozoic reptiles were not so mobile and so appar-

---

[46]Ronald L. Numbers, *The Creationists: The Evolution of Scientific Creationism* (New York: Alfred A. Knopf, 1992).

[47]Harold W. Clark, *Crusader for Creation: The Life and Writings of George McCready Price* (Mountain View, Calif.: Pacific Press, 1966).

[48]H. M. Morris, ed., *Scientific Creationism*, 2nd ed. (El Cajon, Calif.: Creation-Life Publishers, 1982); Gary E. Parker, "Part II: The Life Sciences," in *What Is Creation Science?* ed. H. M. Morris and G. E. Parker (El Cajon, Calif.: Master Books, 1987), pp. 31-184; Brand, *Faith, Reason, and Earth History;* and Roth, *Origins.*

[49]Roth, *Origins*, p. 170.

ently could not find the highest refuges from the catastrophic Flood, and so were entombed in an intermediate stratigraphic position.

However, there are many blatant incongruities in this explanation. For example, pterosaurs (extinct flying reptiles) are limited in their stratigraphic distribution to the Mesozoic. Presumably they were as mobile as many birds and even more mobile than many mammals and so should be preserved well throughout the Cenozoic. As a second example, why are there absolutely no angiosperm fossils preserved in Carboniferous coal deposits? In today's world, angiosperms are by far the most numerous plant taxa, with more than 250,000 species documented, and occupying all sorts of habitats, including the coastal marine realm. Why did a catastrophic global Flood not mix a few angiosperms with standard Carboniferous plant communities? As a third incongruity, there are many examples in the rock record of marine fossiliferous successions overlying terrestrial fossil-bearing strata. In the western United States, for example, thick sequences of Cretaceous System rocks with abundant marine fossils overlie the terrestrial vertebrate (dinosaur) fossil-bearing strata of the Upper Jurassic.

To their credit, some Flood geologists recognize that the theory of ecological zonation does not magically remove all paradoxes of Flood geology at one blow. Ariel Roth, for example, acknowledged that the sedimentary record demonstrates abundant marine fossils at several stratigraphic levels. To resolve this discrepancy, Roth proposed that before the Flood, there were major seas, more extensive than the present-day Caspian Sea, for example, existing at various levels on the continents. The biota of these perched seas would be preserved at differing heights from the biota of the pre-Flood ocean floor.[50]

(Controversial) Nature of the Stratigraphic Record?

Much of the surface of the continents and the continental shelves is covered with thick stacks of lithified sedimentary rock deposited in layers. The layers are characterized in terms of bulk chemistry, mineralogy, color, texture, internal features like laminae and traces of burrowing, and body fossils. Beds are separated by discontinuities termed *bedding planes.*

Flood geologists feel compelled to explain most of these bedded rocks as resulting from the Flood of Noah. It appears that many recent seven-day creationists believe that because animal death stemmed from Adam's fall, the fossils entombed in rock must be post-Fall, and the most likely candidate for their formation in a limited time span is Noah's Flood.

If Noah's Flood was competent to deposit thousands of feet of sediment,

---

[50]Ibid.

it must have also been competent to erode vast volumes of sand and mud that comprise much of the sediment. It must also have been competent to smear this soft material over vast regions of the Earth. Thus, Noah's Flood is depicted as a planetary catastrophe with monstrous waves reinforcing and canceling each other, and racing around the planet at high speed.[51] On the other hand, there are scriptural and geological objections to the planetary Flood model. First, Scripture speaks of the general location of the Garden of Eden with reference to the Tigris and Euphrates Rivers.[52] One may surmise that the site of the Garden now lies buried below the Persian Gulf, which has been covered by seawater since the end of the Pleistocene Epoch. However, the Tigris and Euphrates lie on top of thousands of feet of fossiliferous oil-bearing strata, which Flood geologists claim were emplaced during Noah's Flood. The Bible itself appears to argue that Noah's Flood occurred subsequent to the deposition and lithification of these strata! This difficulty is acknowledged by Flood geologists, who seem to think that the modern Tigris and Euphrates Rivers were named by Noah's descendants after previous rivers from the antediluvial world.[53]

In the second place, geologists wonder how a powerful agent like the Flood described above can deposit sedimentary layers in which great consistency in particle size, color, texture and fossil content can be maintained. How can one layer be a distinctive tan limestone that bears fossil corals and other marine organisms, whereas its upstairs neighbor is a distinctive brown mudstone that contains terrestrial plants? How can layers that are pure accumulations of salt or gypsum, minerals that are produced by evaporating seawater, occur amid the other layers described? Why are not all these elements mingled into chaotic jumbles?

Flood geologists believe that fossil preservation requires rapid burial. This is true to some extent, but rapid burial need not be applied over the entire

[51]Price, *The New Geology;* Nelson, *The Deluge Story in Stone;* Harold Clark, *The New Diluvialism* (Angwin, Calif.: Science Publications, 1946); Whitcomb and Morris, *The Genesis Flood;* Henry M. Morris, "Sedimentation and the Fossil Record: A Study in Hydraulic Engineering," in *Why Not Creation?* ed. W. E. Lammerts (Grand Rapids: Baker, 1970), pp. 114-37; Clifford Burdick, "Streamlining Stratigraphy," in *Scientific Studies in Special Creation,* ed. W. E. Lammerts (Grand Rapids: Baker, 1971), pp. 125-35; Harold W. Clark, "The Mystery of the Red Beds," in *Scientific Studies in Special Creation,* pp. 156-64; Clark and Voss, "Resonance and Sedimentary Layering"; M. E. Clark and H. D. Voss, "Toward an Understanding of the Tidal Fluid Mechanics Associated with the Genesis Flood," in *Proceedings of the Third International Conference on Creation,* pp. 151-67; Vardiman, "Global Warming and the Flood"; Henry M. Morris, "Why Christians Should Believe in a Global Flood," *Back to Genesis* 116 (1998): a-c; Brown, *In the Beginning;* Hoesch and Austin, "Do Tsunamis Come in Super-Size?"
[52]Carol Hill, "The Garden of Eden: A Modern Landscape," *Perspectives on Science and Christian Faith* 52 (2000): 31-46.
[53]J. Morris, "Where Was the Garden of Eden Located?" *Back to Genesis,* no. 132 (1999).

THE BIBLE, ROCKS AND TIME

stratigraphic record in one geologic instant. The record may be a succession of local instances of burial. Many dense skeletal fossil accumulations, when closely examined, are explainable as natural built-in traps for organic remains or are only localized mass mortality settings.

Research by thousands of biostratigraphers during the past two hundred years has verified that a well-defined order to the fossils occurs through the succession of strata. Contrary to the repeated claims of many contemporary seven-day creationist authors, the ordering of the fossil record is not a fiction born of a desire to prove Darwinism. The record first came to view through the hard labor of many, including numerous Christians, a generation prior to Darwin. Other creationists attempt to explain the ordering of fossils as a result of the preservation of pre-Flood ecological zonation. Although this approach obviously has much more merit than that of the attempt to discredit biotic succession, it has to be continually "tweaked" in order to explain the particularities of the succession of fossils. How much tweaking can the record accommodate before the model fails?

During the nineteenth century, prior to the development of geochronometers such as radiometric dating, geologists became convinced that the record gave abundant proofs for the passage of time while the layered rocks were forming, far too much time to be accommodated during a single year-long Flood. We examine a few select cases exhibiting the passage of time in chapter ten. Thus we, along with all other mainstream geologists, insist that the overall stratigraphic record overwhelmingly testifies to the passage of long time intervals rather than testifying to the power of a single, short-term planetary catastrophe.

9

Fossil Graveyards

A Rumble in the Jumble?

IN THIS CHAPTER WE CONSIDER FOSSILS. We define *fossil* as any type of evidence of past life, including the preserved hard parts of organisms such as skeletons and shells, preserved soft body parts, impressions of body outlines, and traces of activity like trackways or burrows (figs. 9.1 to 9.7). Most of this chapter examines in depth several particular cases of extravagant fossil preservation that Flood geologists call "fossil graveyards." Flood geologists claim that these graveyards are explicable only by a global-catastrophic model of massive death and near-instantaneous deposition. However, a global-catastrophic model does not explain these cases.

FOSSIL PRESERVATION

Since the end of World War I, paleontologists increasingly have undertaken studies on processes leading to the preservation of organic remains in the rock record. This subdiscipline of paleontology is termed *taphonomy* (deriving from Greek for "laws of burial"), and some paleontologists consider themselves *taphonomists.* Taphonomy covers all events or factors that would lead to differential preservation in the fossil record, including original organismal constitution and ecological habitat, death and decay, postmortem transport, burial, and diagenesis, or alteration through chemical change and/or pressure after burial.

Figure 9.1. Upper Ordovician limestone slab with fossil brachiopods on bedding plane, near Cincinnati, Ohio. From Calvin College geology collections, Clayton Camozzi, collector. This sample illustrates the phenomenon of preservation of original organismal hard parts, in this case, external shells.

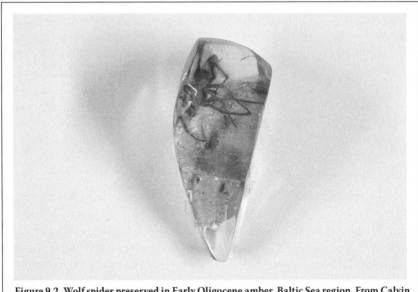

Figure 9.2. Wolf spider preserved in Early Oligocene amber, Baltic Sea region. From Calvin College geology collections, James A. Clark, collector. This example illustrates the phenomenon of preservation of original soft parts.

Figure 9.3. Silicified fossil branch from ash horizon, Brown's Creek Formation (Miocene) of southwestern Idaho. From Calvin College geology collections, R. F. Stearley, collector. This example illustrates a biological specimen possessing great initial porosity and permeability that has become infilled with precipitated microcrystalline quartz and thus permineralized.

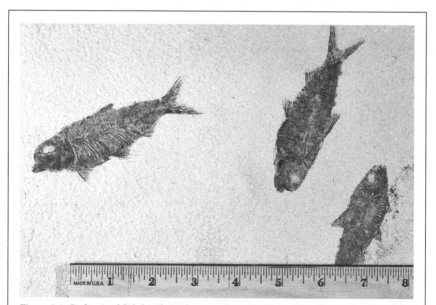

Figure 9.4. Carbonized fish fossils *Diplomystus,* from the Green River Formation (Eocene) of southwestern Wyoming. From Calvin College geology collections. This specimen and the specimen illustrated in figure 9.5 have been gently compressed in a low-oxygen environment; hard parts are preserved intact while original organic compounds have been altered into a tarry chemical mix.

Most current textbooks in paleontology routinely include discussions of taphonomic processes. Treatises on paleoecology devote many pages to discussion of the factors that alter an assemblage of organisms from a *biocoenosis* (life assemblage) to a *thanatocoenosis* (death assemblage).[1] Unfortunately, despite the large volume of both experimental and observational work regarding taphonomic processes, many catastrophic Flood advocates write and speak as though mainstream geologists are ignorant of the processes that affect or favor the preservation of the record of past life. Randy Wysong, a veterinarian, young-Earth creationist and Flood-geology advocate, for example, wrote:

> Creationists insist: "Fossilization is unnatural, abnormal, catastrophic, quick, unique, exceptional, cataclysmic. When we see fossilization world-wide, when we note that water is the agency that has presented the conditions for fossilization, then we must conclude that there was a world-wide water cataclysm in the past. The geological column is not a record of the coming of life, it is a record of its going, its departure, its demise." The scientific community is not naïve to this evidence. Some simply shelve it or ignore it in order to maintain the doctrine of uniformitarianism.[2]

Most organisms will not become fossils; this has always been true. We are blessed to have the record that we do have and should be thankful for the amazing beauty and intricacy of past organisms that it has provided to us. Despite the low overall frequency of preservation, however, there are many local situations today and in the past in which organisms or their traces are or were preserved. For example, we can point to proto-historic human remains frozen into glacial ice in the Austrian Alps and preserved by natural tanning in bogs in Denmark, as well as the molds in tephra surrounding the spaces occupied by corpses after the volcanic eruptions at Pompeii in A.D. 79. Like these

---

[1]Relevant works on paleoecology include J. Robert Dodd and R. J. Stanton, *Paleoecology: Concepts and Applications* (New York: John Wiley & Sons, 1981); C. Newton and L. Laporte, *Ancient Environments*, 3rd ed. (Englewood Cliffs, N.J.: Prentice-Hall, 1989); P. A. Selden and J. R. Nudds, *Evolution of Fossil Ecosystems* (Chicago: University of Chicago Press, 2004). Reviews of fossil preservation and taphonomy include J. Weigelt, *Recent Vertebrate Carcasses and Their Paleobiological Implications*, trans. Judith Schaefer (Chicago: University of Chicago Press, 1989); A. K. Behrensmeyer and A. Hill, eds., *Fossils in the Making* (Chicago: University of Chicago Press, 1980); P. Shipman, *Life History of a Fossil* (Cambridge, Mass.: Harvard University Press, 1981); S. K. Donovan, ed., *The Processes of Fossilization* (New York: Columbia University Press, 1991). Superior studies of death and decay in the marine realm are summarized in M. Brongersma-Sanders, "Mass Mortality in the Sea," in *Treatise on Marine Ecology and Paleoecology*, vol. 1: *Ecology*, ed. J. W. Hedgpeth (Baltimore, Md.: GSA, 1957), pp. 941-1010); Wilhelm Schaefer, *Ecology and Paleoecology of Marine Environments*, ed. G. Y. Craig, trans. Irmgaard Oertel (Chicago: University of Chicago Press, 1972). A recent review of settings of exceptional preservation was edited by D. J. Bottjer, W. Etter, J. W. Hagadorn and C. M. Tang, *Exceptional Fossil Preservation: A Unique View on the Evolution of Marine Life* (New York: Columbia University Press, 2002).

[2]R. L. Wysong, *The Creation-Evolution Controversy* (Lansing, Mich.: Inquiry Press, 1976), p. 361.

instances, fossil deposits contain clues that indicate the ordinary or extraordinary conditions under which a deposit came into being. For example, insects or spiders preserved intact in amber testify to an event in which the arthropod was trapped against a tree oozing resin (fig. 9.2). Broken and gnawed bones, often of selected body parts like limbs, accumulated in a cave testify to transport by predators or scavengers into the cave. However, advocates of Flood geology often ignore such clues or reinterpret them to speak uniformly to preservation under conditions of a global catastrophic Flood. For example, Barnhart, Folsom and Wise discussed the *entirety* of the fossils entombed in the mile-thick sequence of Paleozoic strata exposed in the Grand Canyon:

> Fossils could not have formed at all, unless some rather unusual events occurred. Decomposition and erosion are extremely efficient at destroying evidence of past life. Of the billions of human footprints made on beaches, both out of and within the water, none are known as trace fossils. To resist the effects of erosion, the sediment containing prints and trails must rapidly become somewhat solid, or trace fossils could never form. Of the millions of American bison killed in the last century, no fossils have been found. To avoid being eaten by carnivores and scavengers, and broken down by fungi and bacteria, a body must be placed in an environment where these decomposers cannot thrive. Yet, nearly everywhere you find oxygen, you'll find a carnivore, a scavenger, or, at the very least, decomposing fungi. Decomposing bacteria are found in the top few feet of soil, sand, and mud nearly everywhere on earth. Virtually, the only way for a body to avoid decomposition is to be buried quickly, completely, and deeply, thus sealing it from the decomposers. The commonness of body fossils throughout Grand Canyon argues that many, if not all, Grand Canyon rocks were deposited very quickly. The commonness of trace fossils throughout Grand Canyon means that many, if not all, sediments firmed up and, perhaps, turned into rock (i.e., lithified) rather rapidly. Thus, the fossils of Grand Canyon and the rocks in which they are found, were probably deposited and lithified very rapidly.[3]

This description fails to note that the "body fossils" in the various Grand Canyon rocks are almost all marine shells that are typically disarticulated and sorted by currents. The soft tissues have long ago disappeared. The use of the word *rapidly* is ambiguous here—there is no need to postulate near-instantaneous massive burial to preserve these durable hard parts. Similarly, the preservation of traces requires a consistent environment during the formation and lithification of the sediment (see discussion on *Diplocraterion* in chap. 10) but not catastrophic sedimentation.

---

[3]W. R. Barnhart, M. L. Folsom and K. P. Wise, "Fossils of Grand Canyon," in *Grand Canyon: Monument to Catastrophe*, ed. S. A. Austin (Santee, Calif.: Institute for Creation Research, 1994), p. 134.

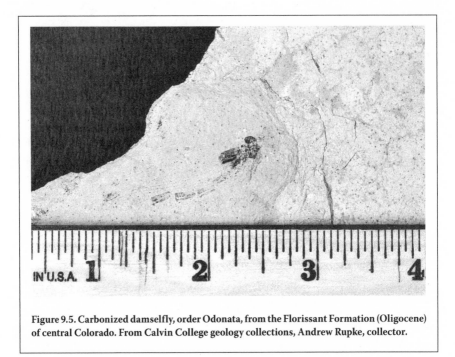

Figure 9.5. Carbonized damselfly, order Odonata, from the Florissant Formation (Oligocene) of central Colorado. From Calvin College geology collections, Andrew Rupke, collector.

Figure 9.6. Fossil U-shaped worm burrow *Diplocraterion,* from the Fairview Formation (Upper Ordovician) of northern Kentucky. From Calvin College geology collections, R. F. Stearley, collector. This structure preserves a record of past activity only; it is termed a *trace fossil.*

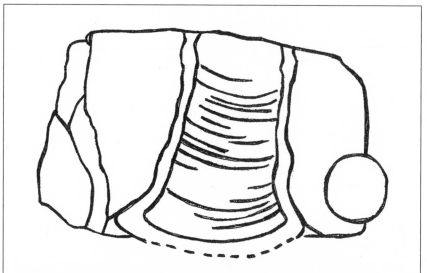

Figure 9.7. Sketch of *Diplocraterion* specimen in figure 9.6, emphasizing the external dwelling tube and the preserved imprints of burrowing activity (spreiten structures).

Figure 9.8. Mammoth *(Mammuthus columbi)* skeletons exposed by excavation at Hot Springs Mammoth Site, western South Dakota. Photo courtesy of Larry Agenbroad.

Advocates of Flood geology not only believe that fossil preservation in general requires extremely rapid burial, but that particularly dense accumulations of ancient remains must argue for highly unusual depositional settings. In their interpretation, "highly unusual" corresponds to "global catastrophic."

FOSSIL GRAVEYARDS?

Accordingly, proponents of Flood geology have made much over the existence of "fossil graveyards." Fossil graveyards are complex accumulations of large numbers of fossils in a relatively restricted area of sedimentary rock. Fossil graveyards or mass mortality layers are typically presented by Flood geologists as inexplicable under present-day, "standard" conditions:

> Never does one find, in the present era, great "graveyards" of organisms buried together and waiting fossilization. But this is exactly the sort of thing that is encountered in fossil deposits in many, many places around the world.[4]

> In the Earth's sedimentary rocks, there are buried vast numbers of plants and animals of all kinds, often in great fossil "cemeteries," where thousands, even millions, of organisms may be found crushed together and buried by sediments. Even after centuries of collecting, new "graveyards" continue to be found. It is a matter of the most elementary scientific logic to recognize that phenomena such as these must be attributed to very rapid burial, or other wise they could never have been preserved. And since most such fossil graveyards have been buried in water-laid sediments, they clearly give witness to the fact of aqueous catastrophism.[5]

> Where is one example of fossilization occurring on a scale today, equivalent to the magnitude seen in the plethora of fossil graveyards around the world? Conditions might fortuitously happen today that might result in the fossilization of a few animals (a cliff dropping in the water and burying a few fish, clams, insects, etc.), but never have we observed any cataclysm on a scale like that which would have been necessary in the past to account for the thousands of square miles of solid heaps of fossils.[6]

For example, Henry Morris and other young-Earth creationists, such as Wysong and Gish, have cited an extensive and impressive deposit of fossil herring-like fish, *Xyne grex*, from the Miocene Monterey diatomite of Southern California.[7] David Starr Jordan, a prominent ichthyologist of the early twen-

---

[4]J. C. Whitcomb Jr. and H. M. Morris, *The Genesis Flood: The Biblical Record and Its Scientific Implications* (Philadelphia: Presbyterian & Reformed, 1961), p. 156.

[5]Henry M. Morris, "Sedimentation and the Fossil Record: A Study in Hydraulic Engineering," in *Why Not Creation?* ed. W. E. Lammerts (Grand Rapids: Baker, 1970), pp. 114-37.

[6]Wysong, *The Creation-Evolution Controversy*, p. 361.

[7]Henry M. Morris, *Scientific Creationism*, 1st ed. (San Diego: Creation-Life Publishers, 1974), pp.

tieth century, president of Stanford University and antagonist of George Mc-Cready Price, documented a single layer of the Monterey diatomite at Lompoc, California, containing millions, perhaps a billion, individual *Xyne* spread over an area of approximately four square miles.[8] In 1974, Morris acknowledged that mass mortality of herring or other fish taxa is possible under natural conditions, such as those of a toxic "red tide," but he dismissed this possibility as a cause of the fossil deposit:

> The author failed to note, however, that while a "red tide" may produce vast numbers of dead fish, it does not produce *fossil* fish! The fish decay on the shore, or are eaten by scavengers, but they don't become fossils.[9]

We further examine this fossil setting, and Morris's claim, later in the chapter.

Creationist biologist Gary Parker provided further examples of this sort of claim:

> But nowhere on Earth today do we have fossils forming on the scale that we see in geologic deposits. The Karroo Beds in Africa, for example contain the remains of perhaps 800 billion vertebrates! A million fish can be killed in red tides in the Gulf of Mexico today, but they simply decay away and do not become fossils. Similarly, debris from vegetation mats doesn't become coal unless it is buried under a heavy load of sediment.[10]

Prominent Creationist debater and biochemist Duane Gish repeated these cases in *Evolution: The Fossils Still Say No!*

> While present geological processes may have operated at present rates for long periods of time, the advocates of this model contend that it is impossible to account for most of the important geological formations according to uniformitarian principles. These formations include the vast Tibetan Plateau, 750,000 square miles of sedimentary deposits many thousands of feet in thickness and now at an elevation of three miles; the Karoo Supergroup of Africa, which has been estimated by Robert Broom to contain the fossils of 800 billion vertebrate

---

97-98; Wysong, *The Creation-Evolution Controversy;* Henry M. Morris, *Scientific Creationism,* 2nd ed. (El Cajon, Calif.: Creation-Life Publishers, 1982); Duane T. Gish, *Evolution: The Challenge of the Fossil Record* (El Cajon, Calif.: Creation-Life Publishers, 1985), and *Evolution: The Fossils Still Say No!* (El Cajon, Calif.: Institute for Creation Research, 1995).

[8]D. S. Jordan, "A Miocene Catastrophe," *Natural History* 20 (1920): 18-22; D. S. Jordan and J. Z. Gilbert, "Fossil Fishes of Diatom Beds of Lompoc, California," *Stanford University Publications* (1920): 1-45; L. R. David, *Miocene Fishes of Southern California,* GSA Special Paper 43 (Washington, D.C.: GSA, 1943); L. R. David, "Fishes (Other Than Agnatha)," in *Treatise on Marine Ecology and Paleoecology,* vol. 2: *Paleoecology,* GSA Memoir 67, ed. H. S. Ladd (Washington, D.C.: GSA, 1957), pp. 999-1010.

[9]Morris, *Scientific Creationism,* 1st ed., p. 98.

[10]Gary E. Parker, "Part II: The Life Sciences," in *What Is Creation Science?* ed. H. M. Morris and G. E. Parker (El Cajon, Calif.: Master Books, 1987), p. 172.

animals; the herring fossil bed in the Miocene shales of California, containing evidence that a billion fish died within a four-square mile area; and the Cumberland Bone Cave of Maryland, containing fossilized remains of dozens of species of mammals, from bats to mastodons, along with the fossils of some reptiles and birds—including animals which now have accommodated to different climates and habitats from the Arctic region to tropical zones.[11]

Such "fossil graveyards" fall into three primary categories: small-scale natural traps which sample local fauna and flora that locomote or are transported into the trap site; familiar small- or large-scale sedimentary environments that promote accumulation of biotic remains, such as anoxic lake or pond bottoms, reefs and riverine sand bars; and extraordinary biotic accumulations resulting from local catastrophes. We term this last circumstance "localized mass mortality layers."

There are indeed large-scale extinction events that can be detected in the stratigraphic record.[12] For such an explanation to hold up, however, tight chronologic control must demonstrate the synchroneity of the mortalities. Such efforts have been exerted, for example, to demonstrate the suddenness and contemporaneity of the extinction of dinosaurs, pterosaurs, ammonites, rudist bivalves, many planktonic marine organisms and other organisms at the end of the Cretaceous Period.

LOCALIZED NATURAL TRAPS

Instances abound in which "fossil graveyards" represent *localized natural traps*—local environments of selective superior accumulation or preservation. A first example of a localized natural trap setting that we consider would be a low, moist area in a region undergoing severe drought. Today, animals congregate in such areas and may eventually suffer massive mortality. For example, Gary Haynes has studied modern African elephant mortality in Zimbabwe in such local concentrations and, in fact, documented burial of many skeletal remains in time spans of a few years to decades.[13] In many instances, modern elephants dig "wells" to obtain water. Individuals may in turn become entrapped within these wells and die in place. In some instances, entrapped elephants are

---

[11]Gish, *Evolution: The Fossils Still Say No!* pp. 48-49.

[12]G. Ryder, G., D. Fastovsky and S. Gartner. *The Cretaceous-Tertiary Boundary Event, and Other Catastrophes in Earth History,* GSA Special Paper 307 (Boulder, Colo.: GSA, 1996), p. 1; A. Hallam and P. B. Wignall, *Mass Extinctions and Their Aftermath* (Oxford: Oxford University Press, 1997); M. J. Benton, *When Life Nearly Died: The Greatest Mass Extinction of All Time* (London: Thames and Hudson, 2003); D. H. Erwin, *Extinction: How Life on Earth Nearly Ended 250 Million Years Ago* (Princeton, N.J.: Princeton University Press, 2006).

[13]G. Haynes, *Mammoths, Mastodons, and Elephants* (Cambridge: Cambridge University Press, 1991).

mired upright, unable to extract their legs from the wells. Studies of cases like these Zimbabwe waterhole mortality events are often helpful in interpreting the context of concentrations of bones in extremely localized pockets such as the Cleveland-Lloyd dinosaur quarry or the Nebraska Ashfall State Park mammalian deposit discussed below.

As an example of a well-documented natural trap that operated in the past, consider the Hot Springs Mammoth Site in western South Dakota.[14] Here, a small, well-defined and mapped freshwater sinkhole spring with steep walls entrapped more than 40 mammoths *(Mammuthus columbi)* plus smaller mammals during the terminal Pleistocene (fig. 9.8).[15] Associated molluscan remains are those of freshwater snails, including abundant *Physa,* and fingernail clams *(Pisidium)* plus a few terrestrial snails.[16] Pollen grains are mostly those of standard prairie grasses plus freshwater plants. The composite ecological assemblage is a consistent one representing a local biota with many similarities to that of today's regional grassland, with the addition of extinct Pleistocene megafauna. The bone deposit is *not* catastrophic in any physical sense, although the entrapment of each individual mammoth was certainly a catastrophe for the creature itself.

Another distinct class of trap setting is that of caves and fissures. Open cave systems are often occupied by denning or roosting animals, including such extinct creatures as giant ground sloths. Furthermore, many occupants are carnivores like foxes or owls, which bring back to the cave a local sample of the nearby biota. Small mammals such as packrats contribute to the return of the local samples.[17] Fissures may accumulate animals that fall into them or the carcasses of prey dropped by roosting carnivores.[18]

---

[14]L. D. Agenbroad, J. I. Mead and L. W. Nelson, eds., *Megafauna and Man: Discovery of America's Heartland,* Scientific Papers, Mammoth Site of Hot Springs 1 (Hot Springs, S.D.: Mammoth Site of Hot Springs, 1990); L. D. Agenbroad and J. I. Mead, eds., *The Hot Springs Mammoth Site: A Decade of Field and Laboratory Research in Paleontology, Geology, and Paleoecology* (Hot Springs, S.D.: Mammoth Site of Hot Springs, 1994).

[15]L. D. Agenbroad, "The Mammoth Population of the Hot Springs Site and Associated Fauna," in Agenbroad, Mead and Nelson, *Megafauna and Man,* pp. 32-39.

[16]J. Mead, J., R. Hevly and L. D. Agenbroad, "Late Pleistocene Invertebrate and Plant Remains, Mammoth Site, Black Hills, South Dakota," in Agenbroad and Mead, *The Hot Springs Mammoth Site,* pp. 117-35.

[17]J. L. Betancourt, T. R. Van Devender and P. S. Martin, eds., *Packrat Middens: The Last 40,000 Years of Biotic Change* (Tucson: University of Arizona Press, 1990); B. W. Schubert, J. I. Mead and R. W. Graham, *Ice Age Cave Faunas of North America* (Bloomington: University of Indiana Press, 2003); P. S. Martin, *Twilight of the Mammoths: Ice Age Extinctions and the Rewilding of North America* (Berkeley: University of California Press, 2005).

[18]A. J. Sutcliffe, *On the Track of Ice Age Mammals* (Cambridge, Mass.: Harvard University Press, 1985). Sutcliffe provided many examples of caves or fissures that accumulated samples of the surrounding biota during the Pleistocene "Ice Age."

Cumberland Cave, Maryland, belongs to the fissure-trap category. It contains a rich Pleistocene bone accumulation. When discovered in the early 1900s, the fissure was open at the crest of a ridge and filled with unstratified clays containing a diverse mammalian fauna.[19] Many of the bones in this deposit are broken, but none are water-worn. The fauna from Cumberland Cave includes peccaries, tapirs, wolverine, other mustelids, bears, aquatic mammals such as beaver, and mammals with preferences for open prairies like badgers, hares, antelope and coyotes.

Catastrophic Flood proponents Whitcomb and Morris described the diversity of mammalian remains at Cumberland Cave and commented that "this kind of thing does not lend itself well to uniformitarian interpretation but strongly suggests some sort of very unusual catastrophe(s)."[20] The mix of habitat associations represented by the fauna suggested to these authors as well as Gish the necessity of a catastrophic accumulation process.[21] However, such unfamiliar and seemingly nonintuitive assemblages represent typical Pleistocene Ice Age faunas that testify to somewhat different ecosystems than those of the present.[22] The Cumberland Cave assemblage is typical of Late Pleistocene assemblages from eastern North America. For example, tapirs are included within the Cumberland Cave fauna. Today tapirs do not range north of Mexico, but they are known from Pleistocene cave fills in Pennsylvania, West Virginia, Missouri, Kansas, Arizona and Oregon.[23]

Another sort of localized natural trap that Flood geology proponents have claimed to represent a catastrophic "fossil graveyard" is the substantial accumulation of Pleistocene mammalian and avian remains at the Rancho La Brea tar seeps in the Los Angeles basin.[24] Radiocarbon dates for this fascinating

---

[19]B. Kurtén and E. Anderson, *Pleistocene Mammals of North America* (New York: Columbia University Press, 1980).

[20]Whitcomb and Morris, *The Genesis Flood,* p. 158.

[21]D. T. Gish, *Evolution: The Challenge of the Fossil Record* (El Cajon, Calif.: Creation-Life Publishers, 1985); and Gish, *Evolution: The Fossils Still Say No!* See quote in text on p. 252.

[22]Sutcliffe, *On the Track of Ice Age Mammals;* Martin, *Twilight of the Mammoths;* R. W. Graham, "Response of Mammalian Communities to Environmental Changes During the Late Quaternary," in *Community Ecology,* ed. J. Diamond and T. J. Case (New York: Harper & Row, 1986), pp. 300-313; R. W. Graham, "Evolution of New Ecosystems at the End of the Pleistocene," in Agenbroad, Mead and Nelson, *Megafauna and Man,* pp. 54-60.

[23]G. Jefferson, "Late Cenozoic Tapirs (Mammalia: Perissodactyla) of Western North America," *Natural History Museum of Los Angeles County Contributions in Science* 406 (1989): 1-21; R. W. Graham, "Pleistocene Tapirs from Hill Top Cave, Trigg County, Kentucky, and a Review of Plio-Pleistocene Tapirs of North America and Their Paleoecology," in Schubert, Mead and Graham, *Ice Age Cave Faunas,* pp. 87-118.

[24]Kurtén and Andersen, *Pleistocene Mammals of North America;* G. Jefferson, *A Catalog of Late Quaternary Vertebrates from California,* Part Two: *Mammals,* Technical Reports 7 (Los Angeles: Natural History Museum of Los Angeles County, 1991); C. Stock and J. M. Harris, "Rancho La Brea: A Record of Pleistocene Life in California," *Natural History Museum of Los Angeles*

deposit range between roughly 10,000 to about 50,000 years ago.

At La Brea, animals were entrapped in asphaltic quicksand near ephemeral ponds and extraordinarily preserved by the seeping asphalt. Some remains are abraded and probably represent carcasses washed downstream into the deposit. Abundant fragments of insect carapaces, particularly those of carrion beetles and flies, testify to prolonged exposure to air and rotting. Freshwater mollusks, frogs and toads, terrestrial snakes, and terrestrial plant remains are included in the preserved biota. Mammalian remains are dominated by hundreds of individual dire wolves, *Canis dirus*, but numerous other species are common in the deposit, including several other extinct Pleistocene megafauna. Birds are well represented, with more than 130 species present, including extinct condors. The deposit represents a local ecological assemblage accumulated over some millennia; however, carnivores and scavengers, both mammalian and avian, are vastly over-represented.[25] Although it is difficult to imagine a global flood producing a carnivore-dominated deposit, the common representation of entrapped animals and/or carcasses in turn attracting predators which became trapped as well functions admirably as a model for the sorting mechanism for the overall fauna.

Whitcomb and Morris commented briefly on this deposit:

> One might, for example, discuss at length such marvels as the La Brea Pits in Los Angeles, which have yielded tens of thousands of specimens of all kinds of living and extinct animals (each of which, by the unbelievable uniformitarian explanation, fell into this sticky graveyard by accident—one at a time!)[26]

However, contra Whitcomb and Morris, anyone visiting the site can observe that the deposit does *not* contain "all kinds of living and extinct animals." There are no dinosaurs or pterosaurs, for example, nor are there any marine creatures whatsoever. The deposit represents an intelligible sample, largely mammals and birds, of a local ecological assemblage.

One young-Earth creationist, Jeremy Auldaney, has attempted to analyze the La Brea deposits. He has competently marshaled evidence that many of the faunal elements were washed downslope, came to rest in isolated pockets, and were later impregnated with the asphalt and so preserved.[27] Working within a young-Earth perspective, he has judiciously claimed that La Brea represents

---

County Science Series 37 (1992): 1-113; Selden and Nudds, *Evolution of Fossil Ecosystems.*

[25]H. Howard, "A Comparison of Avian Assemblages from Individual Pits at Rancho La Brea, California," *Natural History Museum of Los Angeles County Contributions in Science* 58 (1962): 1-24; Stock and Harris, "Rancho La Brea."

[26]Whitcomb and Morris, *The Genesis Flood*, pp. 160-61.

[27]J. Auldaney, "Catastrophic Fluvial Deposition at the Asphalt Seeps of Rancho La Brea, California," in *Proceedings of the Third International Conference on Creationism, Technical Symposium Sessions*, ed. R. E. Walsh (Pittsburgh: Creation Science Fellowship, 1994), pp. 25-36.

a post-Flood deposit. However, he further concluded that these post-Flood remains must represent repeated instances of "catastrophic" sedimentation. Because Auldaney rejected the evidence of radiocarbon dating, he felt comfortable placing the deposit within the past 5,000 years. However, the testimony of insects as well as scavenging raptors in the deposit argues that many of the carcasses were entrapped and only slowly were incorporated into the tarry sand. Of course, a critical issue is the acceptance or nonacceptance of the radiometric dates for the deposit (see chaps. 14-15).

FOSSIL ACCUMULATIONS IN LONG-LIVED, REGIONAL-SCALE DEPOSITIONAL ENVIRONMENTS

In this section, we look at some examples of larger-scale environments that have accumulated large samples of regional biota. Such environments represent long-lived basins that received large quantities of sediment and entombed large numbers of the remains of organisms. We begin with reference to the elegantly preserved organisms of the Solnhofen Limestone of southern Germany, which serves as a springboard to the remaining discussion. We then examine in detail five regional-scale settings that have been claimed by proponents of Flood geology to represent catastrophic accumulations.

The famous fossil deposits of Solnhofen, Germany, cover an area of 45 by 20 miles in the southern Franconian Alb and are 100 to 300 feet thick (figs. 9.9 and 9.10).[28] These deposits consist of extremely fine-grained, laminated calcareous sediments (Plattenkalk) resulting from the slow accumulation of suspended, chemically precipitated lime mud in a region of restricted water circulation. The sediments lack traces of disturbance by organisms (bioturbation); organisms are commonly in perfect articulation with no trace of scavenging. The outlines of soft tissues, such as the wing membranes of pterosaurs, the ink sacs of squids or the imprints of bird feathers, are often preserved in perfect detail. The entombed biota include a marginal-marine community of brown algae (often with encrusting invertebrates), sponges, jellyfish, corals, brachiopods, crabs, gastropods, squids, sea urchins, sharks and bony fishes, as well as organisms that flew overhead: insects, pterosaurs and antique toothed birds. Preserved leaves indicate that the surrounding land vegetation was adapted to arid conditions.[29]

---

[28]K. W. Barthel, N. H. M. Swinburne and S. C. Morris, *Solnhofen: A Study in Mesozoic Paleontology* (Cambridge: Cambridge University Press, 1990); W. Etter, "Solnhofen: Plattenkalk Preservation with *Archaeopteryx*," in *Exceptional Fossil Preservation: A Unique View on the Evolution of Marine Life*, ed. D. J. Bottjer, W. Etter, J. W. Hagadorn and C. M. Tang (New York: Columbia University Press, 2002), pp. 327-52; Selden and Nudds, *The Evolution of Fossil Ecosystems.*

[29]G. Viohl, "Geology of the Solnhofen Lithographic Limestone and the Habitat of *Archaeopteryx*," in *The Beginnings of Birds*, ed. M. K. Hecht, J. H. Ostrom, G. Viohl and P. Wellnhofer (Eichstätt, Germany: Freunde des Jura-Museums Eichstätt, 1985), pp. 31-43.

Figure 9.9. One of many quarries of "lithographic" limestone in the vicinity of Solnhofen, Bavaria. Note the thin, slabby nature of loosened rock fragments. Photo by S. O. Moshier. Used by permission.

In this sedimentary rock formation, the uniform, extremely fine grain size of the particles (1 to 3 microns), the preservation of extremely fine laminations, and the chemical composition of the sediments provide limits to the possibilities of inferred depositional setting. In particular, it indicates a restrictive, warm, quiet-water setting that would permit the precipitation and settling of fine lime particles in the absence of coarse-grained sediment input.[30] Preserved trackways in these sediments often lead to the body fossil of a creature such as a horseshoe crab. Such tracks are irregular in their course and are thought to be the final march of an organism in the process of dying. Occasionally, schools of the fish *Leptolepides sprattiformis* numbering in the tens to hundreds are located on a single layer—a small-scale version of the Lompoc *Xyne* fish kill. The absence of burrowing creatures, the lack of scavenging of remains, and the evidence of dying, in conjunction with the sedimentological character of the rock, have led paleontologists to interpret the assemblage as forming in a quiet lagoon wherein the bottom waters were toxic. Coccolithophorid algal remains suggest the possibility that this toxicity was

---

[30]Ibid.; Barthel, Swinburne and Morris, *Solnhofen.*

due to algal blooms, described and explained later in this chapter. Some layers, such as the *Leptolepides* mortality layers, require "rapid" burial of fossils, but in pulses of settling of lime particles into layers measured in centimeters, not thousands of feet![31]

The detailed preservation of delicate elements such as the ink sacs of squid or the wing membranes of pterosaurs could be claimed by Flood geologists to demand catastrophic rapid burial. In fact, a rapidly moving slurry of particles formed by a fast-paced, universal Flood would tend to dismember or disaggregate organisms while entombing them.

We now turn our attention to five cases of large-scale deposits that contain abundant fossil remains and are often cited in the Flood literature as "fossil graveyards" inexplicable by standard processes. These are the Eocene Green River Formation of Wyoming and surrounding areas; the Late Eocene Floris-

Figure 9.10. Exposure of thinly bedded Solnhofen Limestone (Jurassic) of southwestern Germany famed for its well-preserved fossils, including *Archaeopteryx*. Photo by S. O. Moshier. Used by permission.

[31] P. H. Busonje, "Climatological Conditions During Deposition of the Solnhofen Limestones," in *The Beginnings of Birds*, ed. M. K. Hecht, J. H. Ostrom, G. Viohl and P. Wellnhofer (Eichstätt, Germany: Freunde des Jura-Museums Eichstätt, 1985), pp. 45-65; Viohl, "Geology of the Solnhofen Lithographic Limestone."

sant Formation of central Colorado; the Jurassic Morrison Formation of the Rocky Mountain region of the United States; the Permo-Triassic Karoo beds of South Africa; and the vast stretch of permafrost with entombed Pleistocene megafauna in Siberia.

THE GREEN RIVER FORMATION

The Green River Basin in southwestern Wyoming is a broad regional depositional basin that has accumulated the remains of millions of aquatic organisms. The Green River Basin contains sediments, collectively termed the Green River Formation, that were deposited in an intermontane basin about 160 miles long by 60 miles wide (figs. 9.11 and 9.12).[32] Within this basin, the maximum thickness of the Green River Formation is around 2,500 feet. Similar intermontane basins in northwestern Colorado and northeastern Utah were filled with equivalent sediments and biota during the same time interval.[33]

Fossil fishes were collected from the Green River Formation during the 1850s and returned to Philadelphia for study by paleontologist Joseph Leidy.[34] Ferdinand Vandiveer Hayden's Geological and Geographic Survey of the Territories, funded by Congress between 1867 and 1878, surveyed and named the Green River Formation. In Hayden's report of 1869, he stated:

> A little east of Rock Spring Station a new group commences composed of thinly laminated chalky shales, which I have called the Green River shales, because they are best displayed along Green River. They are evidently of purely freshwater origin and of middle Tertiary age. The layers are nearly horizontal and, as shown in the valley of the Green River, present a peculiarly banded appearance. When carefully studied these shales will form one of the most interesting groups in the west.[35]

Hayden was an able and tireless field geologist. Note his confident identification of the Green River Formation sediments as those of a freshwater lake.

The geology of these deposits was studied extensively by W. H. Bradley during the early twentieth century and by many others since.[36] Since the time of

---

[32]H. W. Roehler, *Introduction to Greater Green River Basin Geology, Physiography, and History of Investigations*, Professional Paper 1506-A (Washington, D.C.: USGS, 1992).

[33]K. R. Newman, "Geology of Oil Shale in Piceance Creek Basin, Colorado," in *Colorado Geology*, ed. H. C. Kent and K. W. Porter (Denver: Rocky Mountain Association of Geologists, 1980), pp. 199-203; W. L. Stokes, *Geology of Utah*, Occasional Paper 6 (Salt Lake City: Utah Museum of Natural History, 1986).

[34]L. Grande, *Paleontology of the Green River Formation, with a Review of the Fish Fauna*, Bulletin 63 (Laramie: Geological Survey of Wyoming, 1984).

[35]F. V. Hayden, *U.S. Geological and Geographical Survey of the Territories, Third Annual Report* (1869), p. 190.

[36]W. H. Bradley, *The Varves and Climate of the Green River Epoch*, Professional Paper 158-E

Figure 9.11. View to north from city of Green River, Wyoming. The pale cliffs in the distance consist of Green River Formation lacustrine sediments (Eocene). Photo by R. F. Stearley. See also figure 9.4.

Hayden, mainstream sedimentary geologists and paleontologists have consistently maintained that these sediments accumulated about 50 million years ago in an ancient system of lakes that were active for several millions of years. The lakes varied from freshwater to alkaline in composition during their history and from time to time became very dry and salty.[37] The deep-water portions of the lakes received hundreds to thousands of feet of fine-grained calcareous sediments (marls) that are often laminated at a fine scale. Such lamination is common in lake beds and represents cyclic fluctuations in particle size and/or composition of sediment reaching the lake bottom. Such fragile laminations require quiet-water conditions for the settling and preservation of fine particles. Fish and other remains that occur in the laminated sediments are often

(Washington, D.C.: USGS, 1929), pp. 87-110; W. H. Bradley, *Origin and Microfossils of the Oil Shale of the Green River Formation of Colorado and Utah,* Professional Paper 168 (Washington, D.C.: USGS, 1931).

[37]R. C. Surdam and C. A. Wolfbauer, "Green River Formation, Wyoming: A Playa-Lake Complex," *Bulletin GSA* 86 (1975): 335-45; Newman, "Geology of Oil Shale in Piceance Creek Basin"; R. Sullivan, "Origin of Lacustrine Rocks of Wilkins Peak Member, Wyoming," *AAPG Bulletin* 69 (1985): 913-22; H. P. Buchheim, "Eocene Fossil Lake, Green River Formation, Wyoming: A History of Fluctuating Salinity," in *Sedimentology and Geochemistry of Modern and Ancient Saline Lakes,* SEPM Special Publication No. 50, ed. R. W. Renaut and W. M. Last (Tulsa, Okla.: Society for Sedimentary Geology, 1994), pp. 239-47.

well preserved with scales in place and are stained brown from organic residue of the tissues (fig. 9.4).

The Green River beds are famous among rockhounds and scientists for their elegantly-preserved fauna and flora, including a well-defined suite of freshwater fishes; lake-margin insects like flies and mosquitoes; freshwater aquatic vegetation like cattails and horsetails; freshwater clams and snails, crayfish, snakes, wading birds, turtles and bones of a few mammals that were washed into the lake where rivers entered.[38] The assemblage contains no anomalous remains but represents an intact ecological assemblage that can be further subdivided into deep-basinal, lake margin and fluvial associations.

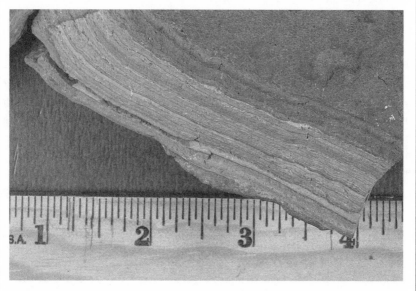

Figure 9.12. Green River Formation hand specimen exhibiting varves. Photo by R. F. Stearley.

Catastrophic Flood proponents have devoted much attention to these sediments during the past fifty years, with some disagreement on their interpretation. Early young-Earth creationists, such as Whitcomb and Morris and Walt Brown, believed these to be produced by the global Flood of Noah, while some young-Earth creationists of the next generation, such as Austin and Brand, viewed these as related to a post-Flood lake. We first examine the attempt by

---

[38]Grande, *Paleontology of the Green River Formation.*

early creationists to explain these deposits as those relating to a worldwide catastrophic Flood.

Whitcomb and Morris briefly described the collective Green River biota from Lincoln County, Wyoming. We treat their discussion in some detail because we believe it to be illustrative of their overall approach to the fossil record. They began their discussion with a secondhand description of the "deposits found in Lincoln County, Wyoming."[39] Their description was not taken from professional literature or personal inspection of sites, but rather a three-page popular article from *Compressed Air Magazine* (1958), titled "Fishing for Fossils."[40] The article included errors, such as a report that the deposit contained "deep sea bass," and it also lacked important details. For example, it reported that "mollusca, crustaceans, birds, turtles, mammals and many varieties of insects" were members of the fauna without further elaboration. The mollusks and crustaceans are all *freshwater* forms and the birds are typically those associated with aquatic habitats.

The excerpt from the article includes the sentence, "The occurrence of these confirms the geological theory that the climate was tropical and quite unlike the blizzard-ridden mountains of Wyoming today." On this point the original article is accurate: the assemblage uniformly reflects a warmer climate than that of today. However, there is no mixture in the Green River Formation with cold-loving forms such as might be expected in a catastrophic, planetary-Flood death assemblage.

After quoting from "Fishing for Fossils," Whitcomb and Morris commented:

> It is not easy to imagine any kind of "uniform" process by which this conglomeration of modern and extinct fishes, birds, reptiles, mammals, insects and plants could have been piled together and preserved for posterity. Fish, no less than other creatures, do not naturally become entombed like this but are usually quickly devoured by other fish after dying.[41]

Unfortunately, as previously outlined, the assemblage is *not* a "conglomeration" that is "piled together," but rather a diverse natural assemblage related to the ecology of a large, alkaline, long-lived lake. Modern lakes that exhibit similarity to this ancient lake include some of the alkaline East African rift lakes, such as Lake Turkana and Lake Natron.[42] The interest should rather lie in the question of why, if a planetary Flood were truly responsible for the "conglomeration" of fossils, there are no trilobites, corals, marine

---

[39]Whitcomb and Morris, *The Genesis Flood*, pp. 156-57.
[40]Anonymous, "Fishing for Fossils," *Compressed Air Magazine* (1958): 23-25.
[41]Whitcomb and Morris, *The Genesis Flood*, p. 157.
[42]Buchheim, "Eocene Fossil Lake."

mollusks, crabs, lobsters, other marine crustacea, dinosaurs, exotic marine Mesozoic reptiles, pterosaurs, mammoths, mastodons, ground sloths, other large Pleistocene megafauna or human remains entombed with the regional lacustrine biota!

The circumstances of good preservation of these animals and plants are indeed unusual but are decipherable. Most of the well-preserved remains are entombed in fine-grained, laminated marlstones, evidently deposited in quiet waters at the center of a deep-water basin. Several lines of evidence point to the conclusion that these deeper waters were devoid of oxygen and possibly toxic, thus completely excluding scavengers.[43] Remains entombed at the lake margin, clearly marked by abundant lake crustaceans, snails, wading birds and alligators, are also well-preserved in some horizons. Grande and Buchheim suggested that fluctuations in dissolved oxygen plus high alkalinity may have reduced scavenging. Catastrophic explanations for these preservational circumstances are actually *contra-indicated.*[44]

During the past two decades, several Flood geologists, including Austin and Brand, have revised the "conglomeration" depiction by Morris for the Green River sedimentary rocks and those of other Cenozoic intermontane lakes. Brand, for example, stated:

> Thus many indications show that it was a lake with the animals living in normal ecological relationships. That is how most geologists would interpret it. But since this deposit is in the Eocene, the upper part of the geologic column, a catastrophic geologist is likely to conclude that this was indeed a lake that existed in the time after the worldwide catastrophe. He or she would predict that when we understand all of the evidence, it would indicate that the lake was filled in to form the Green River Formation in a much shorter time frame than usually is believed.[45]

However, as the Earth chronology of Austin and Brand permits only a few thousands of years of Earth history following the Flood, they are committed to the notion that all of the Green River lacustrine sediments must have been deposited within a few thousand years. This, in turn, implies an average sedimentation for these fine-grained deposits greater than one foot per year, given the known thickness of the Green River Formation! We will further examine the Green River sediments in chapter ten.

---

[43]R. L. Elder and G. R. Smith, "Fish Taphonomy and Environmental Inference in Paleolimnology," *Palaeogeography, Palaeoclimatology, Palaeoecology* 62 (1988): 577-92.

[44]L. Grande and H. P. Buchheim, "Paleontological and Sedimentological Variation in Early Eocene Fossil Lake," *University of Wyoming Contributions to Geology* 30 (1994): 33-56.

[45]Brand, *Faith, Reason, and Earth History,* p. 222.

THE FLORISSANT FORMATION

According to mainstream geology, the Florissant Formation was also deposited in an ancient intermontane lake basin.[46] The basin, located in south-central Colorado, was much smaller than that of the Green River Formation and is approximately 1 mile wide and 15 miles long. Its depositional context was that of an ancient valley, dammed by volcanic mudflows and impounding still waters behind these dams. Radiometric dates on volcanic rocks bracketing the formation fix the age of this lake at around 35 million years.

Whitcomb and Morris believed that this deposit cannot be explained under standard geologic models. They discussed the remarkable fossil beds of Florissant, where well-preserved insects, mollusks, fish, birds and plants, including nuts and blossoms, are found in profusion. They began by extensively quoting a scientific article by R. D. Manwell. Whitcomb and Morris then duplicated the strategy they used with the Green River beds by stating:

> Again, one must realize the difficulty of trying to account for such phenomena on the basis of continuity with present processes. The general sort of explanation postulated for the Florissant deposits has to do with volcanic dust showers over a body of water, but no one can point to similar phenomena creating similar deposits today.[47]

The Florissant Formation consists of fine-grained and finely laminated shales interbedded with thickly-bedded pumice-bearing conglomerates. The shales accumulated in a water-filled basin that formed within a volcanic setting. The waters were impounded behind a natural dam formed by the Guffey volcano lahar breccias of the lower Thirtynine Mile Andesite. From time to time, lahars flowed into the basin, leaving their record of tephra-bearing conglomerates and breccias. These sediments eventually overtopped the entire complex and ended the life of the basin.[48]

Three major vertical sections of "paper shale" comprise the beds interpreted as lake sediments. These lake beds, consisting of many hundreds of alternating thin (1 mm or less) laminations of freshwater diatoms and mud formed from water-altered volcanic ash, are termed the Florissant Formation. Alternation of diatom-rich versus mud-rich laminae is clearly cyclic at this site, but there is yet no agreement on the timetable of these cycles. The extremely well-preserved fossil leaves, mollusks and arthropods are coated with

---

[46]H. W. Meyer, *The Fossils of Florissant* (Washington, D.C.: Smithsonian Books, 2003).

[47]Whitcomb and Morris, *The Genesis Flood*, p. 158.

[48]R. C. Epis, G. R. Scott, R. B. Taylor and C. E. Chapin, "Summary of Cenozoic Geomorphic, Volcanic and Tectonic Features of Central Colorado and Adjoining Areas," in *Colorado Geology*, ed. H. C. Kent and K. W. Porter (Denver: Rocky Mountain Association of Geologists, 1980), pp. 135-56; Meyer, *The Fossils of Florissant*.

diatoms. It is now thought that diatom mats were the major agent responsible for the exquisite preservation of very delicate features. On the other hand, the obvious lahar beds contain few fossils and those present are very poorly preserved. A lot of the early *popular* literature portrayed the shales as forming under blasts of volcanic ash, thereby misrepresenting the actual rocks and the conditions under which they formed.[49] Whitcomb and Morris were unfortunately misled by these early mistaken descriptions.

The "paper shales" contain many thousands of fossils, flattened along bedding planes. The Florissant biota includes more than 150 plant taxa represented by thousands of leaves and includes prominent shoreline freshwater aquatic vegetation such as cattails, horsetails, mosses and ferns as well as many species of conifers and broadleaved trees like oaks and elms. There are more than 1,500 described taxa of insects and spiders (fig. 9.5), making this deposit one of the most remarkable in the world for its testimony to the history of arthropod communities. Freshwater clams and snails as well as a sparse fish fauna of bowfins, suckers, catfish and pirate perch and rare avian and mammalian remains complete a picture of a small intermontane lacustrine fauna.[50] As in the case of the Green River Formation, the preserved biota provide a view into a more-or-less intact ecosystem rather than a catastrophic mixture of unconnected forms.

THE MORRISON FORMATION

The Morrison Formation is a widespread stratum of colorful, soft mudstones and siltstones, with some sandy layers, draped over much of the intermountain region of western North America from northern New Mexico and Arizona northward to southernmost Canada.[51] Stratigraphically, the Morrison Formation underlies the Green River Formation with other for-

---

[49]See Meyer, *The Fossils of Florissant*, in which errors in popular literature are pointed out.

[50]M. V. H. Wilson, "Paleogene Insect Faunas of Western North America," *Quaestiones Entomologicae* 14 (1978): 13-34; W. D. Tidwell, *Common Fossil Plants of Western North America* (Washington, D.C.: Smithsonian Institution Press, 1998); Meyer, *The Fossils of Florissant*.

[51]P. Dodson, R. T. Bakker, A. K. Behrensmeyer and J. S. McIntosh, "Taphonomy and Paleoecology of the Dinosaur Beds of the Jurassic Morrison Formation," *Paleobiology* 6 (1980): 208-32.; A. E. Berman, D. Poleschook Jr. and T. E. Dimelow, "Jurassic and Cretaceous Systems of Colorado," in *Colorado Geology*, ed. H. C. Kent and K. W. Porter (Denver: Rocky Mountain Association of Geologists, 1980), pp. 111-28; Stokes, *Geology of Utah*; L. Hintze, *Geologic History of Utah*, Geology Studies Special Publication 7 (Provo, Utah: Brigham Young University, 1988); F. Peterson, "A Synthesis of the Jurassic System in the Southern Rocky Mountains," in *Sedimentary Cover—North American Craton, U.S., The Geology of North America*, vol. D-2, ed. L. L. Sloss (Boulder, Colo.: GSA, 1988), pp. 65-76; D. L. Baars, *The Colorado Plateau: A Geologic History* (Albuquerque: University of New Mexico Press, 2000); and Selden and Nudds, *Evolution of Fossil Ecosystems*.

mations between the two. The Morrison varies in thickness from 100 to 800 feet and contains numerous remains of dinosaurs and other terrestrial vertebrates, often found by collectors in localized concentrations ("bone beds") such as those at Morrison, Colorado, which gave the formation its name; Como Bluff in Wyoming; the Cleveland-Lloyd quarry of east-central Utah; and Dinosaur National Monument on the northeastern Utah/ northwestern Colorado state line (fig. 9.13). The Morrison Formation has been prospected for fossils since the late 1870s and has been the location for a great deal of colorful history within the paleontological community, including the infamous "bone war" between O. C. Marsh and E. D. Cope during the late nineteenth century.[52]

Whitcomb and Morris drew attention to amassed bone assemblages within the Morrison Formation.[53] In particular, they reported on the truly remarkable accumulation of large bones, mostly sauropod, exposed at Dinosaur National Monument (DNM) (fig. 9.14). After producing numerous nearly complete skeletons that were shipped to many institutions, the quarry here has been converted into a living museum, where visitors can see more than 2,000 bones exposed on a large bed of tilted sandstone. Whitcomb and Morris considered this bone bed, obviously representing an assemblage of carcasses transported into this site at some time in the past, as blatant evidence for a catastrophic depositional event. To reaffirm their claim of the DNM deposit as a "dinosaur graveyard," they quoted DNM dinosaur paleontologists J. M. Good, T. E. White and G. F. Stucker:

> The quarry area is a dinosaur graveyard, not a place where they died. A majority of the remains probably floated down an eastward-flowing river until they were stranded on a shallow sandbar. Some of them, such as the stegosaurs, may have come from far-away dry-land areas to the west. Perhaps they drowned trying to ford a tributary stream or were washed away during floods. Some of the swamp dwellers may have mired down on the very sandbar that became their grave while others may have floated for miles before being stranded.[54]

Whitcomb and Morris next commented that "one could hardly ask for a better description of the way in which these great reptiles were overwhelmed, drowned and buried by the Deluge waters."[55]

---

[52]J. N. Wilford, The Riddle of the Dinosaur (New York: Alfred Knopf, 1985); M. Jaffe, The Gilded Dinosaur: The Fossil War Between E. D. Cope and O. C. Marsh and the Rise of American Science (New York: Crown Publishers, 2000); L. West and D. Chure, Dinosaur: The Dinosaur National Monument Quarry (Jensen, Utah: Dinosaur Nature Association, 1984).

[53]Whitcomb and Morris, The Genesis Flood, pp. 280-82.

[54]Ibid., p. 280.

[55]Ibid., p. 280.

Figure 9.13. Dinosaur National Monument, Utah/Colorado, external view of site. Photo courtesy of Dinosaur National Monument.

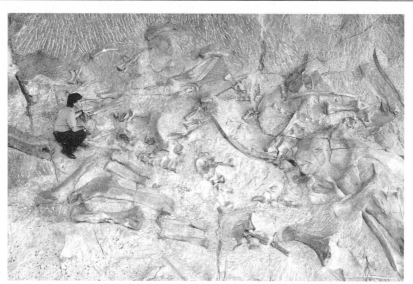

Figure 9.14. Internal view of working site at Dinosaur National Monument, Utah/Colorado. Photo courtesy of Dinosaur National Monument.

Enlarging on this particular case, Morris collected several quotes from the literature of dinosaur paleontology that emphasized catastrophic bone accumulations.[56] One of this compendium of sites, the famous Bone Cabin Quarry site in southeastern Wyoming, like Dinosaur National Monument, represents a fauna from the Morrison Formation. Morris quoted Edwin Colbert's excellent history of dinosaur discoveries, *Men and Dinosaurs* (1968, p. 151), regarding the famous Bone Cabin site in southeastern Wyoming, to this effect:

> At this spot the fossil hunters found a hillside literally covered with large fragments of dinosaur bones. . . . In short, it was a veritable mine of dinosaur bones. . . . The concentration of the fossils was remarkable; they were piled in like logs in a jam.[57]

We acknowledge that this site is a remarkable one but the second half of this quote actually falls on page 173 of Colbert's text and *applies to a different quarry in Wyoming, the Howe Quarry.*

Young-Earth creationists continue to reference the Morrison Formation, and especially the collective deposit at Dinosaur National Monument, as evidence of a worldwide catastrophe, and not some putative high-water event in a sandy stream. Parker stated, for example:

> Some geologic formations are spread out over vast areas of a whole continent. For example, there's the Morrison Formation, famous for its dinosaur remains, that covers much of the mountainous West . . . but slow sediment build up could not possibly produce such widespread deposits, such broadly consistent sedimentary and paleontological features, as we see in the Morrison and St. Peter's Formations. In this case, knowledge of the present tells us something happened on a much larger scale in the past than we see it happening anywhere today.[58]

Amplifying this chain of argument, William Hoesch and Stephen Austin claimed that the large bone concentration remaining intact within the tilted beds of the Morrison Formation exposed at Dinosaur National Monument gives eloquent witness to the power of Noah's Flood:

> This tangled knot of dinosaur bones represents a classic "mass burial" deposit, a trademark of what geologists call the Morrison Formation. Extending from New Mexico to Canada, the Morrison Formation covers about 700 thousand square miles and has been assigned to the Jurassic System. How did such a burial take place? We seek to find the real significance of the deposit at Dinosaur National Monument (DNM) and to dispel myths that our culture has delivered to us.[59]

[56]Morris, *Scientific Creationism,* 1st ed. and 2nd ed.
[57]Morris, *Scientific Creationism,* 2nd ed., p. 98; quoting Colbert, *Men and Dinosaurs,* p. 151.
[58]Parker, "Part II: The Life Sciences," pp. 172-73.
[59]W. A. Hoesch and S. A. Austin, "Dinosaur National Monument: Jurassic Park or Jurassic Jum-

The specific "myths" that Hoesch and Austin disputed include those surrounding the depiction of Jurassic dinosaur communities that were published during the middle twentieth century (which they term "Jurassic Park"). Such depictions are epitomized by a famous mural by Rudolf Zallinger on display at Yale's Peabody Museum in which *Brontosaurus* stands in the waters of a lake, supposedly in need of watery support for its massive bulk. Mid-twentieth-century depictions of large sauropods often placed these in bodies of water, a misconception based on a misinterpretation of sauropod biology and trackways.[60] Most paleontologists acknowledge that dinosaurs and their communities have been publicly depicted in cliché style.[61]

Hoesch and Austin also wished to reassess the circumstances surrounding the entombment of the fossils exposed at Dinosaur National Monument. They did so in a series of six "facts." They first noted (Fact 1) that fossil shells of the freshwater clam *Unio* are much more common in these sites than are dinosaur bones. Many shells are disarticulated and show clear evidence of transport; others are articulated and were obviously very rapidly buried: "The numbers of these clams, and their manner of burial, remind us that the real story at DNM is first and foremost one of death, transport, and rapid burial."[62]

They next alerted the reader to the fact (Fact 2) that the mud transported within much of the Morrison Formation is actually altered volcanic ash, perhaps due to catastrophic volcanism. Their third point (Fact 3) is that the DNM deposit "represents a water-transported and processed assemblage, not an in situ ecosystem." Many items show abrasion or evidence of transport. Their fourth major point (Fact 4) is that, at DNM, the dinosaur bones are concentrated in three horizons, each distinct in its geometry and sediment style, with underlying scour contacts into the lower sediments. Sand grains within these horizons are composed of altered tuff and chert. Hoesch and Austin believed the sediment source to be catastrophic, a circumstance modeled by the volcanic mudflows (lahars) of the Mount St. Helens eruption. "The deposit itself gives us an impression of a very catastrophic water burial event."[63] Fact 5 is that "food requirements for the giant herbivores imply abundant vegetation, yet fossil evidence for localized swamps or for in situ flourishing of plants is scant to nonexistent."[64] This point becomes a critique of the swampy "Jurassic Park" scenario. Their sixth and final major fact is that this mass bone accumu-

---

ble?" *Institute for Creation Research Impact Series* 370 (2004): i.

[60]Dodson, Bakker, Behrensmeyer and McIntosh, "Taphonomy and Paleoecology of the Dinosaur Beds"; S. Lucas, *Dinosaurs: The Textbook* (Dubuque, Iowa: William C. Brown, 1994).

[61]M. J. S. Rudwick, *Scenes from Deep Time* (Chicago: University of Chicago Press, 1992).

[62]Hoesch and Austin, "Dinosaur National Monument," p. iii.

[63]Ibid., p. vi.

[64]Ibid.

lation "represents a mystery that lacks a satisfactory explanation." They note that about twenty such extraordinary bone quarries exist, each separated from one another by vast reaches of Morrison Formation generally lacking fossils. They concluded,

> Why does the public not receive frequent reminders of the facts so obvious with "The Wall" at DNM? Why does a coherent dinosaur "environment" seem so elusive? "Jurassic Park" is too peaceful a picture here. Clams, snails, and dismembered dinosaurs within the same deposit demonstrate a watery catastrophe. "Jurassic Jumble" is more appropriate.[65]

In reply, we note that beside freshwater clams, a whole host of other strictly freshwater invertebrates are documented from the Morrison Formation. These include abundant freshwater snails, ostracods, crayfish and caddisfly larval cases. The formation as a whole includes frogs, turtles, true lizards, crocodiles, some pterosaurs, some archaic mammals and a variety of fishes, including lungfish. And, despite a general paucity of plant megafossils, well-preserved flora now include horsetails, ferns, cycads, ginkgos and conifers—a typical Mesozoic biostratigraphic assemblage.[66] The sample preserved at DNM, while certainly biased and not representing a simple ecosystem, is just as certainly an intelligible sample of a larger ecosystem.

The significant volcanogenic component of the fine clastic sediments of the Morrison Formation has been appreciated for some decades by mainstream geologists, who have related it to regional volcanism during its deposition.[67] The abundance of chemically altered volcanic ash within the fine-grained portion does not in any respect require a global catastrophe for its production! On the other hand, the fine sediments contain broad color banding indicative of fossil soil horizons, including well-preserved calcretes, primary carbonate soil layers that form in arid soils. The Morrison sediments also include thin localized limestone interbeds that contain evidence of their origin as shallow ponds or lakes within an arid floodplain setting: fossil charophytes (freshwater algal remains) plus a suite of freshwater aquatic invertebrates and vertebrates noted previously.

Many of the skeletons of dinosaurs exposed at DNM demonstrate evidences of subaerial exposure prior to burial, including the rigid positions of the dinosaurs and weathering of upper surfaces of bones but not lower. Other skeletons

---

[65]Ibid., p. vii.

[66]Tidwell, *Common Fossil Plants of Western North America;* Selden and Nudds, *Evolution of Fossil Ecosystems.*

[67]Berman, Poleschook and Dimelow, "Jurassic and Cretaceous Systems of Colorado"; Peterson, "A Synthesis of the Jurassic System in the Southern Rocky Mountains."

are unweathered and indicate that a fresh carcass was traveling.[68] Sandy sediment layers with scoured bases traditionally interpreted as sandstone channels exhibit a lack of extremely delicate bones and structures such as mud fragments, silt drapes and root casts, all of which are sedimentological features of braided streams in arid terrains experiencing periodic high floods followed by dry periods.[69] Thus, the early view of the DNM bone-bearing beds as former sand bars within a river floodplain is really not that far off the mark.

Similar conditions to those of DNM are seen at the dinosaur quarries in the Morrison Formation located at Cañon City, Colorado. Dinosaur bone is concentrated in sandstone beds with lenticular cross-sections, interpreted as river channels, in an overall section dominated by mudstone. Dinosaur skeletons here are in all stages of disarticulation, demonstrating that the timing of death of individuals was not the same.[70] Some bones are pristine with little wear; others are heavily worn. There are even rounded bone pebbles, indicating significant transport and erosion.

Not all Morrison Formation dinosaur bone concentrations are mimics of the DNM situation. The large accumulation of dinosaur bones at the Cleveland-Lloyd quarry in eastern Utah occurs in soft mudstone, associated with charophytes and turtles.[71] All dinosaur remains here are disarticulated, but delicate bones are preserved, in both these respects in contrast to the fossils at DNM. The environment of deposition is thus very different from that of DNM and is interpreted by mainstream geologists as that of a freshwater bog or shallow pond within the generally arid Upper Jurassic regional floodplain. The deposit may well represent a "water hole" trap active during an intense drought, analogous to the Zimbabwe water hole traps studied by Haynes.[72] Interestingly, whereas the DNM dinosaurs are primarily sauropods or stegosaurs, the deposit at Cleveland-Lloyd is dominated by carnivorous dinosaurs. Thus, neither represents an intact, "instantaneous" snapshot of the terrestrial community of its day, but both represent biased samples.[73]

The Morrison Formation also hosts more than thirty small dinosaur

---

[68]R. Lawton, "Taphonomy of the Dinosaur Quarry, Dinosaur National Monument," *University of Wyoming Contributions to Geology* 15 (1977): 119-26; Dodson, Bakker, Behrensmeyer and McIntosh, "Taphonomy and Paleoecology of the Dinosaur Beds."

[69]West and Chure, *Dinosaur*; Stokes, *Geology of Utah*; Lawton, "Taphonomy of the Dinosaur Quarry."

[70]E. Evanoff and K. Carpenter, "History, Sedimentology, and Taphonomy of Felch Quarry 1 and Associated Sandbodies, Morrison Formation, Garden Park, Colorado," *Modern Geology* 22 (2001): 145-69.

[71]Stokes, *Geology of Utah*.

[72]Haynes, *Mammoths, Mastodonts, and Elephants*.

[73]Dodson, Bakker, Behrensmeyer and McIntosh, "Taphonomy and Paleoecology of the Dinosaur Beds."

trackways and a few large ones.[74] The largest footprint site is the Purgatoire Valley site of southeastern Colorado. At this site, more than 1,300 footprints are visible to the casual visitor. The trackway site extends more than 1,000 feet in its longest direction. Significantly, the trackway site at Purgatoire Valley resembles the Cleveland-Lloyd site in terms of sedimentology and fossil association. The Purgatoire site has abundant fossils of charophyte algae, freshwater plants, freshwater snails and clams, freshwater crustaceans, and fishes. The local environment is pictured by mainstream paleontologists and geologists as that of a shallow freshwater-to-alkaline lake. Significantly, some dinosaur footprints at this site contain crushed clams and/or flattened vegetation.

Much of the bias of the dinosaur fauna exposed at DNM is due to transport, as Hoesch and Austin and mainstream geologists all agree. The disagreement surrounds the magnitude of the event or events causing the transport of these bones. If one insists that the Flood of Noah deposited the entire Morrison Formation as a blanket deposit, then one has to explain how *one process* can transport and concentrate assemblages of large bones "like logs in a jam" and simultaneously deposit localized fine, alkaline sediment packages with assemblages of freshwater fossils and long trackways of dinosaur footprints with crushed clams and plants. The detectable imprint of multiple processes acting over a landscape, sampling various components of an active ecosystem, makes much more sense.

The Morrison Formation, moreover, occurs within a larger stratigraphic context. At Dinosaur National Monument, the Morrison Formation is underlain by a mile of Cambrian through Lower Jurassic strata.[75] The Morrison, in turn, is overlain by a mile of Cretaceous strata, mostly the Mancos Shale with its entombed fauna of marine invertebrates and vertebrates. This two-mile-thick sedimentary package overlies 24,000 feet of Precambrian sedimentary rocks of the Uinta Mountain Group. Thus, even if a Flood geologist sets aside the Uinta Mountain Group as "pre-Flood," he or she must account for the coherence of the Morrison Formation in its entirety within a two-mile-thick sequence of pure sandstones, large beds of limestone with open-marine faunas, and marine shales with distinct faunas. A process so violent as to deposit the whole two-mile stack of sediments should commingle freshwater and marine elements. Given the doubtful assumption that trackways could be preserved by such a violent process, why should the tracks cohere with the rest of the Morrison biota?

---

[74]M. Lockley and A. P. Hunt, *Dinosaur Tracks and Other Fossil Footprints of the Western United States* (New York: Columbia University Press, 1995).
[75]Hintze, *Geologic History of Utah.*

THE KAROO BASIN

The Karoo Basin strata of South Africa are consistently referred to by Flood geology proponents as evidence of some sort of catastrophe.[76] The basin covers a vast area of more than 200,000 square miles. The Beaufort Beds, which contain an excellent record of Permian and Lower Triassic vertebrate life, are approximately two miles thick.[77] The deposits are interpreted by mainstream geologists as having formed in a vast set of interlaced floodplain environments with meandering streams, much like that of the Morrison Formation. As in the case of the Morrison Formation, well-preserved hard parts of organisms are concentrated in restricted zones. And like the Morrison Formation, the Beaufort beds yield the remains of terrestrial vertebrates plus freshwater aquatic forms. Aquatic vertebrates include many primitive amphibians as well as a distinctive freshwater fish fauna. In the case of the Beaufort Beds, however, the vertebrates are overwhelmingly those of the "mammal-like reptiles," otherwise known as *therapsids*. Thousands of skulls and skeletons of these bizarre tetrapods, representing more than 100 taxa, have been collected. A census of Karoo vertebrates currently underway at the Bernard Price Institute of the University of the Witwatersrand, directed by Bruce Rubidge, estimates that the total number of specimens located to date is around 27,000, including all sorts of fragments.[78] Unfortunately, in 1959, in an overview of the fossil record, Norman Newell repeated an undocumented speculation by Robert Broom:

> Robert Broom, the South African paleontologist, estimated that there are 800,000 million skeletons of vertebrate animals in the Karoo Formation. While such estimates, are, of course, not highly accurate, they stress the vast difference between the known paleontological sample and the astronomic numbers of fossils remaining in the rocks.[79]

The first sentence (only) of Newell's quote above has been repeated without caution throughout the Flood geology literature. The survey underway gives no support for Broom's extravagant estimate.

Supposing Broom's speculation were correct, what would this really mean for the advocates of Flood geology? How could 800 billion individual large ver-

---

[76]Whitcomb and Morris, *The Genesis Flood;* Wysong, *The Creation-Evolution Controversy;* H. M. Morris and G. E. Parker, *What Is Creation Science?* (El Cajon, Calif.: Master Books, 1987); Gish, *Evolution: The Challenge of the Fossil Record;* Gish, *Evolution: The Fossils Still Say No!*

[77]Benton, *When Life Nearly Died;* B. Rubidge, "Re-uniting Lost Continents: Fossil Reptiles from the Ancient Karoo and Their Wanderlust," 27th Du Toit Memorial Lecture, *South African Journal of Geology* 108 (2005): 135-72; Erwin, *Extinction.*

[78]B. Rubidge, personal communication, 2006.

[79]N. O. Newell, "Adequacy of the Fossil Record," *Journal of Paleontology* 33 (1959): 488-99. The quotation is from p. 492.

tebrates live in the limited ecosystem? What sort of population density would this imply? And how could 800 billion skeletons of terrestrial vertebrates be concentrated by a violent "water cataclysm" without any admixture of marine forms? As in the case of the Morrison Formation, this concentration of fossil remains is more realistically interpreted as the record of an ancient terrestrial ecosystem that existed over a long period of time.

ICE AGE FAUNA OF THE ARCTIC

The northern Siberian plain today consists of permafrost, soil frozen for much of the year. Permafrost provides a superior tomb for the remains of many species of Pleistocene megafauna, including such animals as moose, reindeer, horse, bison, musk ox, and wooly rhinoceros.[80] This fauna is especially rich in fossil mammoths, *Mammuthus* sp.[81] These dramatic fossils have been a major curiosity for centuries. During the eighteenth and nineteenth centuries, recovery of mammoth bones and tusks blossomed into an enormous commercial enterprise. The tusks were harvested as a source of ivory. According to the best estimate by Tolmachoff, the tusks of more than 46,000 individual mammoths had been harvested by 1913![82] Many thousands of fossil mammoths and other Pleistocene mammals certainly remain to be collected over the length of this broad region. Estimates range as high as hundreds of thousands of mammoths.

Central and northern Alaska likewise are covered with large regions of permafrost with entombed Pleistocene megafauna.[83] The sediments with their contained record of pollen and bones testify to a common Arctic borderland ecosystem during the end of the Ice Age, a few tens of thousands of years ago. During this interval, sea level would be lowered by a few hundred feet due to the storage of water as vast ice sheets. The continental shelves of northeastern Siberia and Alaska would merge and the northern ecological belt would extend uninterrupted between the two now-separated landmasses. Guthrie has termed this common ecosystem the "mammoth steppe," and we will refer to it as such in our discussion.

Although by far most permafrost remains are skeletons or isolated bones, there are dozens of examples of frozen carcasses of wooly mammoth, wooly

---

[80]T. Nilsson, *The Pleistocene: Geology and Life in the Quaternary Ice Age* (Dordrecht: D. Riedel, 1982).

[81]Sutcliffe, *On the Track of Ice Age Mammals;* Haynes, *Mammoths, Mastodonts, and Elephants.*

[82]I. P. Tolmachoff, "The Carcasses of the Mammoth and Rhinoceros Found in the Frozen Ground of Siberia," *Transactions of the American Philosophical Society,* new series 23 (1929): 1-74.

[83]Sutcliffe, *On the Track of Ice Age Mammals;* R. D. Guthrie, *Frozen Fauna of the Mammoth Steppe* (Chicago: University of Chicago Press, 1990).

rhinoceros, horse and bison that commonly include hide and fur.[84]

Proponents of Flood geology have long seized on this phenomenon as potential evidence of a global catastrophe.[85] The Victorian catastrophist Henry Howorth wrote *The Mammoth and the Flood* in 1887, arguing that frozen mammoths in Siberia testify to an extremely rapid onset of frigid glacial conditions, trapping the unwitting mammoths.[86] Howorth helped to propagate the misunderstanding that the frozen mammoth flesh was often so fresh as to be edible, and that the vegetation found in mammoth stomach contents differed radically from that of the region today. Howorth's evidences were subjected to severe critique by Basset Digby and I. P. Tolmachoff, who pointed out that frozen carcasses were quite rare; that individuals demonstrated clear evidence of miring in thawed permafrost or burial under cliff collapse; and that the mammoth diet corresponded to cold steppe vegetation not unlike that of Siberia today.[87]

Unfortunately, the resuscitation of catastrophist geology that began with the publication of *The Genesis Flood* included Howorth's arguments, uncritically adapted. Whitcomb and Morris repeated an estimate of perhaps 5,000,000 interred mammoths in Siberia and Alaska.[88] And Wysong stated that "about 1/7 of the Earth's surface, from Siberia into Alaska, is a frozen muck containing the remains of millions of mammoths. Some were frozen so rapidly that their flesh is still edible today."[89]

The Seventh-day Adventist Richard Ritland took these efforts to task in his 1970 publication, *A Search for Meaning in Nature*.[90] As an undergraduate at Walla Walla College, Ritland studied geology under Harold Clark, the associate of George McCready Price and advocate of the ecological zonation theory of Flood deposits.[91] He eventually completed a Ph.D. in vertebrate paleontology at Harvard University under Alfred S. Romer and became the director of the Geo-

---

[84]Tolmachoff, "The Carcasses of the Mammoth and Rhinoceros"; Sutcliffe, *On the Track of Ice Age Mammals*; Guthrie, *Frozen Fauna of the Mammoth Steppe*; Haynes, *Mammoths, Mastodonts, and Elephants*.

[85]Davis A. Young, *The Biblical Flood: A Case Study of the Church's Response to Extrabiblical Evidence* (Grand Rapids: Eerdmans, 1995).

[86]H. Howorth, *The Mammoth and the Flood: An Attempt to Confront the Theory of Uniformity with the Facts of Recent Geology* (London: Sampson Low, Marston, Searle and Rivington, 1887).

[87]B. Digby, *The Mammoth and Mammoth-Hunting in Northeast Siberia* (New York: D. Appleton, 1926); Tolmachoff, "The Carcasses of the Mammoth and Rhinoceros"; A. Lister and P. Bahn, *Mammoths* (New York: Macmillan, 1994).

[88]Whitcomb and Morris, *The Genesis Flood*.

[89]Wysong, *The Creation-Evolution Controversy*, p. 356.

[90]R. M. Ritland, *A Search for Meaning in Nature: A New Look at Creation and Evolution* (Mountain View, Calif.: Pacific Press Publishing, 1970).

[91]Ronald L. Numbers, *The Creationists: The Evolution of Scientific Creationism* (New York: Alfred A. Knopf, 1992).

science Research Institute at Andrews University, a major Seventh-day Adventist institution. Ritland remained on friendly terms with many of the Flood geologists of the latter half of the twentieth century, but he began to critique many of their arguments for a global Flood. Ritland was acquainted firsthand with Alaskan Pleistocene mammalian remains and their enclosing deposits through field mapping of permafrost sites in central Alaska. He devoted a chapter of *A Search for Meaning in Nature* to a thorough critique of the notion that the mammoth carcasses testified to a violent quick-freeze following Noah's Flood.

Ritland's efforts notwithstanding, some advocates of Flood geology have continued to place the frozen mammoths into a framework that is based on a global catastrophe. Walt Brown wrote that the mammoth steppe fauna were buried and quickly frozen during a global Flood by violent outpouring of water vapor that froze in the upper atmosphere and returned as muddy hail.[92] Brown has developed a variant of Flood geology that he terms the "Hydroplate Theory," in many respects a resurrection of Burnet's ideas discussed in chapter two. Brown, like Burnet, postulated a primitive shell of water below Earth's primordial crust. During a global cataclysm, Earth's crust was shattered, initiating plate tectonics as we know it and venting this water under great pressure:

> All along this globe-circling rupture, whose path corresponds to today's Mid-Ocean Ridge, a fountain of water jetted supersonically into and far above the atmosphere. Much of this water fragmented into an "ocean" of droplets that fell as rain great distances away. This produced torrential rains such as the Earth has never experienced—before or after. Some jetting water rose above Earth's atmosphere where it froze and then fell on various regions of the Earth as huge, cold masses of muddy "hail". That hail buried, suffocated, and froze many animals, including some mammoths.[93]

In support of his scenario, Brown claimed that wooly mammoths were tropical and not cold-adapted. He argued that the entire proboscidean tribe is and was tropical, and that the hair of the wooly mammoth was not true fur and could not provide adequate insulation. However, contrary to his claims, the mammoth coat includes a dense underfur two to eight inches thick from which the long hairs emerged.[94] Long hair below the belly formed a skirt much like that of the present-day muskox. There is a subcutaneous fat layer as well, which Brown dismissed, that would nonetheless provide insulation. Low surface-to-volume ratios for elephants, as opposed to caribou or moose, actu-

[92]Walter Brown, *In the Beginning: Compelling Evidence for Creation and the Flood*, 7th ed. (Phoenix: Center for Scientific Creation, 2001).

[93]Ibid., p. 101.

[94]Haynes, *Mammoths, Mastodonts, and Elephants*; Lister and Bahn, *Mammoths*.

ally favor heat conservation. Contemporary African elephants, in fact, have a problem with overheating; their large ears function as heat radiators. Wooly mammoths possessed small ears to minimize heat loss. All these considerations combine to demonstrate that Brown's claim that wooly mammoths were not adapted for a cold climate is totally unjustified.

Brown claimed that mammoths in an arctic steppe could not locate sufficient salt to satisfy their salt requirement, nor could they obtain adequate water. Yet large contemporary caribou herds manage to locate these resources, as do musk oxen, which occupy arctic lands frozen for most of the year.

The sediments that enclose wooly mammoth remains are a major sticking point for Brown:

> Muck is a major geological mystery. It covers one-seventh of the Earth's land surface—all surrounding the Arctic ocean. Muck occupies treeless, generally flat terrain, with no surrounding mountains from which the muck could have eroded. Russian geologists have in some places drilled through 4000 feet of muck without hitting solid rock. Where did so much eroded material come from? What eroded it?[95]

The unidimensional label "muck" is deficient in describing the diversity of sediments that are the site of the modern permafrost and their entombed fauna. Much of the muck consists of glacially ground silt that has been reworked by wind and/or solifluction in the permafrost. Solifluction refers to ice thawing and then refreezing, creating a gelatinous soil that can flow. In river valleys and steep hillsides, solifluction can lead to slumping. Slumping is in fact claimed to be responsible for the entrapment of several mammoths, including one of the most famous of the frozen mammoths, the Beresovka mammoth. This individual, a male in its late teens, was in sitting posture in the frozen sediments. Its forelegs were outstretched and are interpreted as a struggling posture. A femur, the pelvis, and some ribs were fractured, suggesting a fall as the slope collapsed. There is some evidence that this individual suffocated while struggling in the collapsing sediments.

Other sediments of the permafrost have been reworked by running water. Many mammoth bones have actually come from stream gravels within the permafrost plain. For example, excavations by Russian scientists in the bed of the Berelekh river during the 1970s recovered more than 8,000 mammoth bones, representing about 140 individuals. Small amounts of frozen soft tissues were recovered with the bones. Other mammoth remains have been discovered by gold miners who were disaggregating old stream gravels for the interred gold.

---

[95]Brown, *In the Beginning*, p. 163.

Some of the tundra sediments are wind-sorted silts, termed *loess*. Brown provided a nice description of loess deposits and illustrated examples from the arctic periphery. He thought that such wind-derived sediments meshed with his proposal of a violent muddy hailstorm as the source of the Pleistocene sediments. Loess and muddy deposits, however, are radically distinct.

Brown believed that he undermined the concept of the mammoth steppe ecosystem by ascertaining that wooly mammoths were adapted to warm climates. He also felt that his model of sediment and ice delivery to what is now the arctic borderland is superior to conventional models of sediment formation, but Brown has not made a convincing case for either claim.

Atmospheric scientist and Flood advocate Michael Oard has developed a comprehensive model for a single-phase, short "ice age," claiming that "there was only one Ice Age, brought on by the unique conditions that followed the global Flood."[96] Calculations based on hypothesized post-Flood climatic and ocean-temperature conditions lead to Oard's bold claim that the peak of ice-sheet cover in the northern hemisphere could have been reached in 500 years, with catastrophic deglaciation occurring during the following 200 years, for a total "Ice Age" of 700 years! He claimed that the mammoth steppe fauna entombed within the arctic permafrost flourished during this time interval and beyond, centuries after the great Flood of Noah. According to Oard, they were then wiped out by a series of intense dust storms, engendered by radical climate change.

Oard has marshaled many lines of evidence to support the notion that the northern Siberian and Alaskan mammoths did not all die at once. These include the general lack of carcasses as opposed to cleaned skeletal remains; symptoms of scavenging on both carcasses and bones; the presence of abundant fly pupae in bones and carcasses indicating rotting; and evidences such as stomach contents or pelt condition suggesting that animals died during different seasons. Oard's arguments, which summarize many painstaking studies by mainstream paleontologists, are cogent and irrefutable.

Although Oard has made a strong case for the existence of the mammoth steppe ecosystem, and against an instantaneous "quick-freeze" scenario for its demise, his arguments for extinction at the hands of several catastrophic dust storms are very weak. In fact, he has conceded that numerous mammoth fossil finds occur in old stream beds or in collapsed soil layers. The presence of loess, that is, glacially ground, windblown silt, over much of the arctic steppe is not in and of itself evidence for catastrophic dust storms as the agent of extinction. Oard was certainly correct in saying that climatic change affected the

---

[96]Michael Oard, *Frozen in Time* (Green Forest, Ark.: Master Books, 2004).

Pleistocene megafauna, but he has not come close to demonstrating that vast dust-storms were the immediate agent of death and preservation.

We summarize our resume of these five major cases. The Green River Formation and the Florissant Formation represent sediments and entombed fossils deposited in ancient lake systems. They possess diagnostic freshwater biotas that, in composite, present wonderful pictures of past ecosystems. The Morrison Formation and the Beaufort Beds represent ancient alluvial floodplains, predominantly siltstones and mudstones with colorful soil horizons and localized sand layers or string deposits representing ancient stream channels in which bones of terrestrial animals are concentrated. The frozen Siberian mammoths are admittedly intriguing, but the evidence for rotting of carcasses in place prior to burial, as well as the nature of the enclosing sediments, argues for some sort of regional prevalence for miring in solifluction zones. Although all of these examples are fascinating, they do not demand catastrophic deposition of thousands of feet of sediment over the span of one year.

LOCALIZED MASS MORTALITY AND THE ROCK RECORD

We now turn our attention to real examples of catastrophic mortality in the rock record. We acknowledge that, in some instances, the formation of fossils in general, and fossil graveyards in particular, requires some kind of "rapid" burial process. But, contrary to the impression given by creationists, geologists have long recognized that this is often the case! The fact that a fossil requires rapid burial in order to have a chance of forming, however, by no means implies a global catastrophe. We simply may be talking of *localized,* brief catastrophes, and it must be stressed that such events *are* occurring at the present time. These catastrophes are observable processes. Storms, earthquakes, tsunamis, volcanic eruptions, floods, mudslides and the like are all examples of brief, local catastrophes that may be responsible for much fossil preservation.

Let us consider some examples of recurring phenomena that result in mass killings of animals and plants and thus are fully capable of producing more "fossil graveyards."[97] One very important cause of mass mortality in the world ocean today is the phenomenon of algal blooms. In these, the sea is typically discolored brown or red by dinoflagellates or diatoms that have been stimulated through a combination of nutrient abundance and appropriate water temperatures. These blooms may include forms that are toxic to other sea creatures or terrestrial creatures that eat them. Such toxic blooms include

---

[97]M. Brongersma-Sanders, "Mass Mortality in the Sea"; Schaefer, *Ecology and Paleoecology of Marine Environments;* Elder and Smith, "Fish Taphonomy and Environmental Inference."

the infamous "red tides" of various coastal areas. Furthermore, as the abundant algae die, decomposing bacteria attack their remains, using up dissolved oxygen and rendering the waters disaerobic. Thus mass mortality can occur through lack of oxygen as well as by toxicity. For example, the bottom waters of the Gulf of Mexico offshore of New Orleans suffer from anoxia every summer due to algal blooms promoted by agricultural runoff, leading to the well-publicized "dead zone" that, in some years, has been as large as 7,000 square miles![98] The Gulf of Luanda in west Africa suffered a mass mortality incident in 1951 in which fishes in the upper water column died due to poisoning from algal blooms, whereas fishes in the lower water column perished due to suffocation. In turn, anoxic bottom waters prevent scavenging of remains and promote good preservation. In waters that are sufficiently deep, gas bubbles will not form and the dead fish will remain on the sea floor to be buried.[99]

In extreme instances, fish may die by the millions, and creatures such as shrimp, oysters, seals, penguins, turtles, crabs and barnacles accumulate dead on the seafloor. Birds may be poisoned through eating poisoned fish. In such instances, potential scavengers are often killed as well.

There are also many other causes of mass mortality. Great varieties of plants and animals have been entombed in falling volcanic ash, in fluidized ash (volcanic mudslides) or in flowing lava. Others have been killed by heat as lava entered the sea. Great numbers of fish have suffocated in ash-laden streams. One may recall the plight of salmon during the 1980 eruptions of Mount St. Helens in Washington. Whitcomb and Morris suggested that fossilization by volcanic ash-falls over bodies of water is not observable today and is therefore inconsistent with modern uniformitarian geology. This simply is not true. The 1912 Katmai, Alaska, eruption created great quantities of ash that fell over streams and land. A diverse group of plants and animals, including insects, have been well preserved in the ash.[100] Ash from Mount Vesuvius has destroyed animals in the Gulf of Naples.

There are parallel instances in the fossil record of volcanic catastrophes resulting in localized mass mortality. For example, Ashfall Beds State Historical Park in Nebraska contains a Miocene fauna that was entrapped in a small pond and buried by a six-foot thick bed of pure volcanic ash. More than 200 complete skeletons of birds, land tortoises, alligators and mammals, especially the rhinoceros *Teleoceras,* and several species of horse are exceptionally

---

[98]Boyce Thorne-Miller, *The Living Ocean: Understanding and Protecting Marine Biodiversity,* 2nd ed. (Washington, D.C.: Island Press, 1999).

[99]Schaefer, *Ecology and Paleoecology of Marine Environments;* Elder and Smith, "Fish Taphonomy and Environmental Inference."

[100]V. A. Krasilov, *Paleoecology of Terrestrial Plants* (New York: Halstead, 1975), pp. 61-62.

Figure 9.15. Intact, fully articulated barrel-bodied rhinoceros skeletons *(Teleoceras major)* buried in pure volcanic ash at the Ashfall Fossil Beds State Historical Park in northeast Nebraska. The site is mid-late Miocene in age, approximately 11.8 million years old. The fossil skeletons are part of a permanent, in situ display at the Ashfall Park. Photo courtesy of Ashfall Fossil Beds State Historic Park, and the University of Nebraska State Museum.

preserved.[101] Volcanic activity, like algal blooms, is clearly responsible for many instances of mass mortality today and in the past. Global-scale catastrophes need not be presumed in order to account for local volcanic-related phenomena such as that of Ashfall Beds State Park (fig. 9.15).

Earthquakes, particularly those located under the sea, have killed life in catastrophic proportions. Shock waves from the earthquakes have killed tremendous numbers of organisms, especially fish. Local rapid uplift of the seabed above sea level has killed all bottom-dwelling organisms such as barnacles, mussels, anemones and others. Earthquakes can promote temporarily rapid sedimentation. For example, sea-floor sediment can be briefly stirred up by the shock of earthquakes and rapidly bury dead or dying organisms killed by the shock. Several examples of mass kills triggered by earthquakes have been recorded from such earthquake-prone areas as Alaska, Japan, Chile, Mexico, Peru and the Mediterranean area.

Severe temperature changes have also been responsible for great catastrophic mortalities. Other mortalities have been caused by lack of oxygen presence or noxious gases. Severe storms also are responsible for catastrophic kills.

There is absolutely no question that catastrophes are constantly occurring and that in many of these catastrophes enormous numbers of a wide variety of organisms are not only killed but also buried by rapidly deposited sediment. In short, fossils and fossil graveyards are being formed today. Young-Earth creationists are correct in saying that mass accumulations of dead organisms often occur as a result of a catastrophe, but wrong in assuming that we have no evidence supporting local catastrophes in the rock record and that, therefore, only the Genesis Flood could account for these deposits.

We next turn our attention to two well-defined cases of mass mortality in the fossil record that have been claimed to be the result of a global catastrophe by Flood geology proponents. These are the fossil insect beds of the Belmont Chert in Australia and the catastrophic mass mortality layer of fossil herring-like fishes, *Xyne grex*, discovered in the Lompoc diatomite. In both instances, we accept the notion that the deposits record an episode of mass mortality; in both instances, the context dictates that the catastrophic event was local, not global.

AUSTRALIAN INSECT MASS MORTALITY LAYER

Andrew Snelling, an Australian geologist working for the Institute for Creation Research, has drawn attention to an impressive fossil insect bed near

---

[101] M. Voorhies, "Dwarfing the St. Helens Eruption: Ancient Ashfall Creates a Pompeii of Prehistoric Animals," *National Geographic* 159 (1981): 66-75; B. MacFadden, *Fossil Horses: Systematics, Paleobiology, and Evolution of the Family Equidae* (Cambridge: Cambridge University Press, 1994).

Newcastle, Australia, approximately 90 miles north of Sydney. All insect remains come from a unit informally named the Belmont Chert, a layer 2.5 feet thick, extending over an area of at least one by six miles.[102] The Belmont Chert was extensively sampled during the first half of the twentieth century by local collectors, who donated many of their specimens to the Australian Museum. In turn, the insects were described by R. J. Tillyard, C. Davis and J. W. Edwards in a series of significant papers.[103] The insect fauna was later summarized by Riek, who listed 145 species assignable to 97 genera.[104] By far most insect remains entombed here are those of isolated wings. On the basis of an average number of insect wing impressions in the Belmont Chert, approximately ten to twenty wings per cubic foot of rock, Knight estimated that hundreds of millions of wings were present per square mile.[105]

Associated fossils include fish scales, microcrustaceans and plant debris. The deposit is interpreted to have resulted from an extensive fall of volcanic ash into a freshwater lake, not unlike the Florissant setting. Snelling gave the following interpretation:

> These swarms of insects, whose original ancestors had been created and then diversified as they had reproduced after their "kinds", were catastrophically destroyed and entombed by a volcanic blast during a water cataclysm. This Australian fossil insect bed, therefore, bears eloquent testimony to the devastation during the Genesis Flood.[106]

Snelling cited as further evidence of cataclysms a conglomerate bed below the Belmont Chert and coal seams stratigraphically above the Belmont Chert. The Belmont Chert itself is viewed as a "fleeting stage in a far greater water cataclysm."

A broader view of the section containing the Belmont Chert is instructive. The Belmont Chert is a tiny member within the overall Newcastle Coal Measures.[107] The Newcastle Coal Measures contain about 1,300 feet of strata and include thirteen prominent coal seams plus sandstone layers, conglomerate layers and tuffs like the Belmont Chert. Many of these coal seams, including the Victoria Tunnel Coal and the Lower Pilot Coal, preserve intact stumps

---

[102]A. A. Snelling, "An Australian Fossil Insect Bed Resulting from Cataclysmic Destruction," *Institute for Creation Research Impact Series* 329 (2000): i-iv.

[103]O. LeM. Knight, "Fossil Insect Beds of Belmont, N.S.W.," *Records of the Australian Museum* 22 (1949): 251-53.

[104]E. F. Riek, "Undescribed Fossil Insects from the Upper Permian of Belmont, New South Wales," *Records of the Australian Museum* 27 (1968): 303-10.

[105]Knight, "Fossil Insect Beds of Belmont, N.S.W."

[106]Snelling, "An Australian Fossil Insect Bed," p. iii.

[107]I. G. Percival, *The Geological Heritage of New South Wales*, vol. 1 (Sydney: New South Wales National Parks and Wildlife Service, 1985).

with lateral root systems in place. In turn, the Newcastle Coal Measures are but a portion of the overall regional stratigraphic sequence.

We can only agree that the circumstances surrounding the emplacement of the tuffaceous chert with its entombed insects were catastrophic. However, we cannot reconcile the larger context with that of a global catastrophe. A watery catastrophe that exerted enough force to emplace thick layers of sand and mats of plant debris that would later become individual coal seams, would rework and homogenize a thin, fragile flow of volcanic ash. The ash fall would require time to convert to a hard, siliceous tuff prior to continued deposition! Again, why does the fauna consist only of insects, fish scales, lacustrine microcrustaceans and plant debris? We are once again provided with a biological sample, a "snapshot," representing the life connected with or in a small lake. The connection of the Belmont Chert and its wonderful preserved biota to a "greater water cataclysm" is undemonstrated.

RETURN TO THE LOMPOC DIATOMITE MASS MORTALITY LAYER

The mass mortality layer at Lompoc, California, containing the herring-like fish *Xyne grex* documented by Jordan and David and cited by Morris as an example of an unexplainable catastrophe, occurs within a larger and well-studied geologic context.[108] This context is that of the highly fossiliferous siliceous shales of the Monterey Formation. The Monterey Formation is a widespread sedimentary unit of considerable thickness in the coastal borderland basin province of southern California. In many localities it is very oily and has been mapped, sampled, drilled and analyzed for many decades.[109] In some portions its siliceous shales are so diatom-rich as to deserve the name *diatomite.* Because the formation was draped over a series of small widening basins, it varies in thickness but in many places exceeds a mile in total thickness. At Lompoc, the Monterey Formation is approximately 5,000 feet, and the extensive nearly pure diatomite section is nearly 1,000 feet in thickness.[110]

The Monterey Formation is well known among professional paleontologists for its excellent fossil preservation. It hosts a rich fauna of marine invertebrates, vertebrates, macroalgae, diatoms and foraminiferans. The skeletal remains of marine mammals and marine birds such as albatrosses are not un-

---

[108]Jordan, "A Miocene Catastrophe"; David, *Miocene Fishes of Southern California;* David, "Fishes (other than Agnatha)"; Morris, *Scientific Creationism.*

[109]R. M. Norris and R. W. Webb, *Geology of California,* 2nd ed. (New York: John Wiley, 1990); C. Isaacs, "Depositional Framework of the Monterey Formation, California," in *The Monterey Formation: From Rocks to Molecules,* ed. C. M. Isaacs and J. Rullkoetter (New York: Columbia University Press, 2001), pp. 1-30.

[110]David, *Miocene Fishes of Southern California.*

common in this deposit.[111] Faunal composition, overall abundance of diatoms and chemical composition of the sediments suggests an oceanic upwelling system somewhat like that of coastal California today, but under the influence of a warmer climate.[112] Within this upwelling context, ocean productivity was especially high at Lompoc, leading to the deposition of this remarkable diatom deposit.[113]

The Lompoc locality within the Monterey Formation contains many fossil fish taxa, belonging to two families of sharks and at least 28 families of bony fishes (osteichthyans).[114] All Lompoc taxa are marine and form a coherent ecological assemblage representing an offshore mid-water fauna with admixture of deeper-water and open-ocean pelagic forms. For example, the assemblage includes fossil tarpon, a myctophid, cod, drum, jack, tuna, rockfish and flounder, among others. The larger Monterey Formation contains other marine fish taxa that indicate that some localities represent deeper-water assemblages. For example, localities in the Santa Monica Mountains have provided several specimens of viperfish, family *Chauliodontidae*.[115] Collectively, the fish fauna indicates somewhat warmer conditions than that of today, but the fauna is plainly a northeastern Pacific one with many similarities to that of the present Pacific coast of North and Central America.[116]

Globally, mass mortalities of mid-water or pelagic schooling fishes are not

[111]L. Barnes, "Outline of Eastern North Pacific Fossil Cetacean Assemblages," *Systematic Zoology* 23 (1976): 321-43; L. Barnes, R. E. Raschke and S. A. McLeod, "A Late Miocene Marine Vertebrate Assemblage from Southern California," *National Geographic Research Reports* 21 (1985): 13-20; H. Howard, "Late Miocene Marine Birds from Orange County, California," *Natural History Museum of Los Angeles County Contributions in Science* 290 (1978): 1-26.

[112]S. A. Graham and L. A. Williams, "Tectonic, Depositional, and Diagenetic History of Monterey Formation (Miocene), Central San Joaquin Basin, California," *AAPG Bulletin* 69 (1985): 385-411; M. B. Lagoe, "Depositional Environments in the Monterey Formation, Cuyama Basin, California," *AAPG Bulletin* 96 (1985): 1296-312; D. Z. Piper and C. M. Isaacs, *Geochemistry of Minor Elements in the Monterey Formation, California: Seawater Chemistry of Deposition*, Professional Paper 1566 (Washington, D.C.: USGS, 1995); Isaacs, "Depositional Framework of the Monterey Formation."

[113]F. M. Govean and R. E. Garrison, "Significance of Laminated and Massive Diatomites in the Upper Part of the Monterey Formation, California," in *The Monterey Formation and Related Siliceous Rocks of California*, ed. R. E. Garrison and R. G. Douglas (Los Angeles: Society of Economic Paleontologists and Mineralogists, 1981), pp. 181-98; Isaacs, "Depositional Framework of the Monterey Formation."

[114]Jordan and Gilbert, "Fossil Fishes of Diatom Beds of Lompoc, California"; David, *Miocene Fishes of Southern California*.

[115]David, *Miocene Fishes of Southern California*; J. M. Crane, "Late Tertiary Radiation of Viperfishes (Chauliodontidae) Based on a Comparison of Recent and Miocene Species," *Natural History Museum of Los Angeles County Contributions in Science* 115 (1966): 1-29.

[116]David, *Miocene Fishes of Southern California*; L. G. Allen, D. J. Pondella II and M. H. Horn, *The Ecology of Marine Fishes: California and Adjacent Waters* (Berkeley: University of California Press, 2006).

uncommon at present and are often due to toxic dinoflagellate blooms ("red tides") or other algal over-productivity, as discussed previously. Excess algal productivity may not result in direct toxicity but can render the bottom waters of a localized embayment or broader region anoxic through oxidation of algal organic material. Brongersma-Sanders discussed the Monterey shales and their fish remains and made direct comparisons to contemporary mass mortalities in various regions. She noted the massive die-off of sardines entering Walvis Bay during December, 1924, and compared the layer of fish skeletal remains there to that of the Monterey Formation: "In the *Xyne grex* layer of the Monterey Shales (Jordan, 1920), millions of a herringlike fish lie buried; this layer is an exact equivalent of the layer that was deposited in Walvis Bay in December 1924."[117]

The Monterey Formation and its entombed fossils form a coherent sedimentary and paleoecologic package. Its vertebrate biota and paleoecology can be interpreted by comparison to modern offshore conditions along the southern California coast, where seasonal marine upwelling promotes algal blooms and a strong pelagic fishery.[118] In this ecologic setting, the overall fish fauna of the Monterey Formation forms a consistent picture of a former ecosystem. The mass mortality layer of *Xyne* represents a rare but not unusual event. In contrast to the repeated claims by many Flood proponents, not only is this fossil fish layer explicable by reference to contemporary circumstances, but in fact the nature of the entire Miocene package at this location demands a long-term, stable system with parameters similar to those of the local community today. It is not consistent with deposition under conditions of a violent global catastrophe.

Conclusion

In this chapter we have reviewed, in some detail, several different types of fossil deposit, formed under differing circumstances. Some deposits reviewed are considered *localized natural traps*. These include the La Brea "Tar Pits"; caves and fissures such as Cumberland Cave, Maryland; and ancient drying waterholes such as the Cleveland-Lloyd Dinosaur Quarry in Utah. Other fossil deposits represent more *widespread, regional accumulations* of animal and/or plant remains that were preserved in a recognizable *ancient, bounded ecosystem.* Such larger accumulations include those of well-understood ancient lakes, such as the Green River lake beds of southwestern Wyoming or the Florissant Formation of central Colorado; restricted coastal lagoons such as the

---

[117]Brongersma-Sanders, "Mass Mortality in the Sea," p. 970.
[118]Allen, Pondella and Horn, *The Ecology of Marine Fishes.*

fine-grained limestones of the Solnhofen Formation and their elegant biota; ancient floodplains with river channels such as those of the Morrison Formation in the western United States or the Karoo Formation of South Africa; and the cold steppe environment of the circum-arctic, with its large numbers of well-preserved bony remains as well as frozen mummies representing an Ice-Age fauna. A third major category of fossil deposit is that of a *truly catastrophic but spatially-restricted death assemblage,* representing an abrupt environmental impact on a small-to-moderate-sized region. Examples of this third category include the fossil insect-bearing Belmont Chert, near Sydney, Australia; the Miocene burial of animals under volcanic ash preserved at Ashfall Beds State Historical Park in Nebraska; and the mass mortality layer of the fish *Xyne* from the Lompoc Diatomite in California. In the Flood-geology literature, all of these distinct preservational environments typically are lumped together as "fossil graveyards."

We also have seen that in many instances, such "fossil graveyards" have been further misrepresented by Flood geology advocates as "conglomerations," "tangled knots" or "jumbles," "piled up," of "all kinds of living and extinct animals." Although many of these assemblages clearly demonstrate transport of the various elements, others demonstrate a notable lack of transport. All contain clues or signals to the circumstances surrounding their composition and formation, clues that do not testify to a global catastrophe. We liken these signals to a low rumble audible to those willing to listen. There is a rumble in the jumble.

In chapter eight, we noted that one group of catastrophic Flood geologists, including Harold Clark, Ariel Roth and Leonard Brand, interpret many deposits like the Morrison Formation or the Green River Formation as truly the preserved remains of ancient ecosystems. In so doing, they often accept many of the conclusions of mainstream geologists and paleontologists. However, they perceive these fossil ecosystems as either near-instantaneously overwhelmed ecosystems of an antediluvian world (e.g., Morrison Formation), or preserved ecosystems that flourished during the few thousands of years since the great Flood (e.g., Siberian mammoth steppe, Green River Formation lakes). Do these sedimentary units contain internal or external evidence as to the time required for their formation?

We maintain that features internal and external to sedimentary deposits provide a context that is in general noncatastrophic. In the next chapter, we look at many such features that give clues as to time involved in the formation of these deposits.

10

SANDS THROUGH
THE HOURGLASS

Sedimentation, Ancient Environments and Time

SUPPOSE ONE WERE TO STROLL ALONG THE BEACH and come across an hourglass (a genuine hourglass, that keeps time for one hour) in which sand is in the process of draining from the upper chamber into the lower. The base of the hourglass is embossed with the name of the manufacturer: "Bascom & Son." Approximately half of the sand has drained from the upper chamber into the lower. What could one conclude about the past history of the hourglass?

First, one could guess that the hourglass came into existence a split second prior to our observation, with the sand evenly distributed between the upper and lower chambers. If this were the case, then a casual examination of the hourglass could lead the observer to the very erroneous conclusion that a half-hour had elapsed while the hourglass sat on the beach.

One could check this hypothesis. If "Bascom & Son" were determined to be a legitimate manufacturer and the company acknowledged that the hourglass was genuine, then a simpler hypothesis would be that the hourglass was indeed a true artifact and not a magical device. As a true artifact, it would have been set to go approximately a half-hour before our observer encountered it. Of course, in this case the observer is relying on the constancy of gravitational attraction, what he or she knows about the diameter of sand grains in the hourglass and so on.

However, suppose that a respected authority on trumpet playing arrives and insists that the hourglass had been keeping time for only a few minutes. Our trumpeter is a noted skeptic and believes that the sand was forced from the upper chamber at a phenomenally high rate through the narrow neck of the glass, and that the original assessment of a half-hour of time elapsed was biased by the "uniformitarian" assumptions of the observer. How would our observer reply?

Taking the challenge of the trumpeter seriously, an observer could examine the hourglass in detail. By going to the laboratory, one could model the flow of sand under high pressure and realize that such a forced flow would severely scratch glass. Does our hourglass exhibit the scratches indicative of high flow? One could also examine the compaction and surface characteristics of sand accumulated under high flow. Does the sand in the lower chamber of the hourglass exhibit the expected characteristics of sand forced together under great pressure?

A mainstream sedimentary geologist is in many cases a lot like our observer. He or she sees a fairly straightforward accumulation of sediment (sand) with some sort of criteria for assessing elapsed time. Challenges come in two forms from young-Earth creationists: some natural phenomena are claimed to simply have occurred more or less instantaneously, regardless of any physical evidence to the contrary, because it is within God's omnipotent power to accomplish such things instantaneously, and/or other phenomena are held to have occurred at incredibly fast rates, much faster than our "naive" observer is capable of imagining. In fact, in many real cases the mainstream sedimentary geologist has laid out basic lines of empirical evidence (analogous to our determination that the hourglass was of human manufacture and was not severely scratched by sand forced through the neck at an impossible rate) long before the objections were raised by the skeptic Flood geologist. In other cases the skeptical objections have had a salutary effect of forcing the mainstream geologist to lay out evidence plainly. In this chapter we give several examples of different lines of evidence that exhibit elapsed time in the formation of a given sedimentary deposit.

SMALL-SCALE INTRINSIC TIME KEEPERS

We first turn our attention to features such as fossils, rock structures and horizons that provide a relatively easy evaluation of the passage of time during their formation. Such features are small-scale and can be examined efficiently at one or a few outcrops. The first such feature discussed below clearly indicates an interval of very rapid deposition, in which several feet or tens of feet of sediment were deposited over a relatively short span of time. The next four diverse small-scale features, however, attest to slow or sporadic episodes of deposition. All the features discussed here are not rare and can be observed in outcrop in various locations.

Features like the ones illustrated below are often termed "intrinsic time-keepers" in textbooks of stratigraphy. They provide prima facie evidence for the passage of time during the deposition of beds and are crudely analogous to the hourglass example. There are many more categories of easily observed intrinsic timekeepers than those discussed here. While Flood geologists have readily promoted examples such as the polystrate trees referred to below as evidence for rapid deposition, they routinely ignore or attempt to explain away the multiple kinds of small-scale features that attest to longer time periods involved during sediment deposition.

Polystrate Fossils

Young-Earth creationists have pointed to the existence of fossil trees embedded upright through several feet of massively bedded sandstone as indicative of an extremely rapid rate of burial. A classic case in point is the locality at Joggins, Nova Scotia, on the coast of the northern arm of the upper Bay of Fundy.[1] The tree trunks here are typical Carboniferous lycopods and associated flora. Some of the completely enclosed, upright trunks are more than thirty feet tall. Flood geologists have argued that this phenomenon signifies extremely rapid sedimentation.[2] Possibly many of these fossil trees were rafted into place and buried in an upright position. Flood geologist Harold Coffin has done some credible work relating these upright tree trunks to upright floating tree trunks in Spirit Lake, Washington, following the catastrophic eruption of Mount St. Helens in 1980.[3] Flood geologists believe that this occurrence testifies to a global catastrophic Flood.

---

[1]On the Joggins occurrence, see Marshall Kay and Edwin H. Colbert, *Stratigraphy and Life History* (New York: John Wiley & Sons, 1965).

[2]N. A. Rupke, "Prolegomena to a Study of Cataclysmical Sedimentation," in *Why Not Creation?* ed. W. E. Lammerts (Philadelphia: Presbyterian & Reformed, 1970), pp. 141-79; R. L. Wysong, *The Creation-Evolution Controversy* (Lansing, Mich.: Inquiry Press, 1976); H. M. Morris and G. E. Parker, *What Is Creation Science?* (El Cajon, Calif.: Master Books, 1987); M. E. Clark and H. D. Voss, "Resonance and Sedimentary Layering in the Context of a Global Flood," in *Proceedings of the Second International Conference on Creationism*, vol. 2: *Technical Symposium Sessions*, ed. R. E. Walsh and C. L. Brooks (Pittsburgh, Penn.: Creation Science Fellowship, 1990), pp. 53-64; and Ariel V. Roth, *Origins: Linking Science and Scripture* (Hagerstown, Md.: Review and Herald Publishing Association, 1998).

[3]H. G. Coffin, "Erect Floating Stumps in Spirit Lake, Washington," *Geology* 11 (1983): 298-99, and "Mt. St. Helens and Spirit Lake," *Origins* 19 (1983): 9-17.

The sand surrounding the trees at Joggins certainly accumulated rapidly. Although we concur with the assessment of Flood geologists that these fossil trees were rapidly buried, we, along with other mainstream geologists, doubt that a global Flood provides the best model for the formation of these layered rocks. The polystrate trees at Joggins, for example, have been illustrated in mainstream geology texts as evidence for instances of rapid deposition in a deltaic complex without implication of a global Flood. Even Charles Lyell visited Joggins on his travels to North America during the nineteenth century and did not return to England as a catastrophist.

Ancient Root Systems

In contrast to the situation at Joggins, other buried fossil plants clearly testify to protracted spans of time involved in their original growth and emplacement. One type of evidence lies in the preservation of plant root zones under coal beds. For example, in central Michigan the sedimentary rocks of the Pennsylvanian Saginaw Formation (fig. 10.1) host root zones that are plain to the observer at various locales (figs. 10.2 and 10.3). The intact growth positioning of many roots and rootlets clearly requires a period of some time, and a time of relative calm. Does such a horizon require a few weeks, a few months or a few years to develop?

Despite the fact that such horizons are moderately common and have been examined at many localities by thousands of geological students and professionals over many decades, some Flood geologists claim that these do not exist![4]

Fossil Marine Worm Dwelling Burrows

Today, many types of marine creatures dwelling in nearshore surf-zone environments dig vertically-oriented burrows for protection and anchorage. The common marine lugworm, *Arenicola*, for example, digs a U-shaped burrow with two exits. The worm remains within the burrow, extracting nourishment from detritus brought through the water. Comparable U-shaped burrows are fairly common in sedimentary rock horizons that are interpreted as having been deposited in ancient shallow-water settings. The trace fossil *Diplocraterion* is a well-documented U-shaped burrow that retains relic arcuate excavation marks termed *spreiten* structures (figs. 9.6 and 9.7). The excavation marks on the illustrated example indicate that the burrow was lengthened in a downward direction. Mainstream sedimentary geologists interpret these in a straightforward manner: the vertical tubes are parts of a dwelling that, at

---

[4]Clark and Voss, "Resonance and Sedimentary Layering."

Figure 10.1. Outcrop of a portion of the Saginaw Formation (Pennsylvanian), near Grand Ledge, central Michigan. In vertical ascending order, the rock types visible include a gray shale, a tan fine-grained sandstone, a thin coal seam and an upper sandstone. These are considered part of a cyclic sedimentary succession, a "cyclothem." Photo by R. F. Stearley.

Figure 10.2. Carbonized fossil roots preserved in place on bedding plane in the tan sandstone demonstrated in figure 10.1. Photo by R. F. Stearley.

Figure 10.3. Carbonized fossil roots preserved in vertical position in the tan sandstone demonstrated in figure 10.1. Photo by R. F. Stearley.

some time in the past, connected to a stable sea floor. The indwelling worm probably took days or weeks to dig the dwelling deeper in response to growth or other biological factors. Thus, the illustrated *Diplocraterion,* obtained from an Upper Ordovician sedimentary horizon outside Covington, Kentucky, demonstrates that during the deposition of this sequence of limey shales the sea floor was stable and deposition was minimal for some span of time.

Some contemporary Flood geologists are aware that such burrows exist, but attempt to find an alternative explanation for their formation. Steven Austin suggested that *Diplocraterion* located in strata exposed in the Grand Canyon of the Colorado River were actually "escape structures," reflecting the frantic attempts of a worm to dig its way out.[5] However, this explanation glosses over the fact that the relic digging marks lie *above* the dwelling tube, indicating downward extension, not upward struggle!

Fossil Coral Growth in Place

Ancient reefs, formed from the skeletal remains of corals, algae, calcified sponges, mollusks and other organisms, have been documented from several stratigraphic intervals in many sedimentary sequences from around the world. The Great Lakes region of North America contains a rich record of such fossil reefs. Many have been extensively quarried for limestone for cement manufacture or for building stone. Numerous small "patch reefs," components of Devonian limestone strata, are exposed in large quarries along the northwestern shore of Lake Huron in Michigan (fig. 10.4).

The colony in figure 10.5 demonstrates typical coral growth. Coral colonies branch outward and upward as individual polyps reproduce by budding. At times, the colony dies back but upward growth resumes from a few surviving polyps or by new settling of coral larvae. This branching growth takes weeks, months or years. Furthermore, modern corals grow in clear waters. Coral polyps are fouled by turbidity and cannot flourish in muddy waters. How could these coral colonies seemingly form during the middle of a catastrophic deposition of a slurry of sediment? How can one reconcile the apparent evidence for extended time of formation in clear water with a claim of rapid formation in turbid mud?

From the standpoint of the Flood geologist, an alternative interpretation might be something to the effect that coral colonies like this one did not grow in place but were transported from elsewhere during the great Flood. If this be the case, then how is it that such colonies were transported with intact growth

---

[5]Stephen A. Austin, "Interpreting Strata of Grand Canyon," in *Grand Canyon: Monument to Catastrophe,* ed. S. A. Austin (Santee, Calif.: Institute for Creation Research, 1994), pp. 21-56.

Figure 10.4. LaFarge limestone quarry, Alpena, Michigan. Limestone visible in this picture belongs to the Upper Devonian Traverse Group. Photo by R. F. Stearley.

Figure 10.5. Coral growth framework exposed in quarry wall, LaFarge quarry. Photo by R. F. Stearley.

framework and with the broad and heavier side upward rather than downward? We return to a larger-scale examination of entire fossil reefs below.

Paleokarst

Karst refers to two interrelated phenomena: a set of surface features formed by the dissolution of solid rock, and the processes of solution responsible for these dissolution cavities. Karst features include landforms like sinkholes and caverns and are generally, but not solely, formed on terrains dominated by limestone. *Paleokarst* features are karst features identified in the rock record.

Karst features result from the exposure of *solid* limestone to subaerial chemical weathering and above-ground and underground water flow over prolonged time. Thus, for the Flood geologist, karst features should be restricted to solid rocks formed prior to the great Flood, and to the uppermost portions of the thick rock column that solidified after the Flood. Karst features such as erosional surfaces, sinkholes, caverns and collapse structures would not have time to form on rock units in the *middle* of the Flood sequence. Yet such features are quite common in the rock record and are not restricted to specific stratigraphic horizons.[6] These features testify to an episode of subaerial exposure and dissolution requiring years prior to the emplacement of overlying sediment.

Paleokarst features were explained and discussed by Daniel Wonderly in two volumes that critiqued Flood geology.[7] Wonderly, citing work by McKee and Gutschick, drew attention to unmistakable paleokarst features in the Redwall Formation, a thick limestone unit that lies in the middle of the Paleozoic sequence of the Grand Canyon.[8] Flood geologist Stephen Austin directly addressed this critique in his chapter on "Interpreting Strata of the Grand Canyon," admitting that "this surface of erosion is of considerable interest to both

---

[6]Noel P. James and Phillip W. Choquette, eds., *Paleokarsts* (New York: Springer-Verlag, 1988); Ajit Bhattacharayya and Chandan Chakraborty, *Analysis of Sedimentary Successions: A Field Manual* (Rotterdam: A. A. Balkema, 2000); E. Kosa and D. W. Hunt, "Heterogeneity in Fill and Properties of Karst-Modified Syndepositional Faults and Fractures: Upper Permian Capitan Platform, New Mexico, U.S.A.," *Journal of Sedimentary Research* 76 (2006): 131-51.

[7]Dan Wonderly, *God's Time-Records in Ancient Sediments* (Flint, Mich.: Crystal Press, 1977); Dan Wonderly, *Neglect of Geologic Data: Sedimentary Strata Compared with Young-Earth Creationist Writings* (Hatfield, Penn.: Interdisciplinary Biblical Research Institute, 1987); Davis A. Young, "The Discovery of Terrestrial History," in H. J. Van Till, R. E. Snow, J. H. Stek and D. A. Young, *Portraits of Creation* (Grand Rapids: Eerdmans, 1990), pp. 26-81.

[8]E. D. McKee and R. C. Gutschick, *History of the Redwall Limestone of Northern Arizona*, GSA Memoir 114 (Boulder, Colo.: GSA, 1969). See also S. Beus, "Redwall Limestone and Surprise Canyon Formation," in *Grand Canyon Geology*, ed. S. S. Beus and M. Morales, 2nd ed. (Oxford: Oxford University Press, 2003), pp. 115-35.

creationists and evolutionists."[9] Austin claimed that McKee and Gutschick *assumed* that the dissolution occurred between the times when the Redwall and overlying formations were deposited. Austin cited dissolution structures in rock units above the Redwall as potential portions of a larger network of solution-work extending from above. However, he neglected to inform his readers that the Surprise Canyon Formation, which overlies the Redwall, completely fills in the elaborate network of river channels, karst sinkholes and collapse features on the upper portion of the Redwall Formation. In places the solid limestone was incised downward up to 400 feet prior to being filled in. From such features McKee and Gutschick and, following them, Beus, observed and correctly inferred that a long interval of erosion occurred on the surface of the Redwall Formation, after which the surface was completely filled in by the overlying sediments.

The short-term catastrophic planetary washing of great slurries of sediment suggested by Flood geologists (see chap. 8) is incompatible with a rock record that contains evidences of episodic deposition interrupted by intervals of exposure, erosion, soil formation and dissolution (cave formation), all requiring vast amounts of time. Some Flood stratigraphers admit the existence of erosional surfaces that they presume developed during the great Flood.[10] However, karst surfaces and features form via *dissolution* and not by pure mechanical abrasion. While a global Flood model might permit episodes of abrasion, it does *not* accord well with multiple episodes of subaerial exposure and chemical dissolution.

LARGER-SCALE SYSTEMS THAT IMPLY PASSAGE OF CONSIDERABLE TIME

We next turn to some types of larger-scale sedimentary deposits that attest to the passage of considerable time during their formation. The examples below require scrutiny over a breadth of scales, from the microscopic to the individual outcrop or drill core, to broad regional integration of features. In textbooks, broad-scale sedimentary regimes that plainly can be related to ecological, chemical or physical conditions operating on Earth's surface are termed *environments of deposition.* Mainstream sedimentary geologists have identified a few dozen broad categories of such environments, with a host of minor variants on each theme.[11] We have selected four categories of such large-scale

---

[9]Austin, "Interpreting Strata of Grand Canyon."

[10]E. g., Morris and Parker, *What Is Creation Science?*; Austin, "Interpreting Strata of Grand Canyon"; and Roth, *Origins: Linking Science and Scripture.*

[11]M. E. Tucker, *Sedimentary Petrology*, 3rd ed. (Oxford: Blackwell Science, 2001); H. Blatt, R. J. Tracy and B. E. Owens, *Petrology: Igneous, Sedimentary and Metamorphic* (New York: W. H. Freeman, 2006); S. Boggs, *Principles of Sedimentology and Stratigraphy*, 4th ed. (Upper Saddle River, N.J.: Prentice-Hall, 2006); Donald R. Prothero and Fred Schwab, *Sedimentary Geology:*

features for explication. In each category, we examine a classic example or examples.

Carbonate Reefs and Other Organic Buildups

Large sedimentary buildups, composed of the preserved skeletons of marine corals, calcified sponges, encrusting algae, mollusks and many other marine taxa, are scattered through many stratigraphic horizons on all continents.[12] The individual components vary from buildup to buildup. Classically, sedimentologists have termed such buildups *reefs* if a clear resistance to wave and surf action is indicated; the term *bioherm* is often used for marine organic buildups that would provide minimal wave resistance.

We begin with a classic example of an ancient reef that can be visited easily, the Thornton Reef southeast of Chicago.[13] Like many Chicago-area fossil reefs, the Thornton reef is sited in the Silurian Racine Formation, which is composed mostly of dolomitic limestone. Many small and large quarries have been opened in the Racine Formation during the past two hundred years for construction stone, and the Thornton quarry is one of the largest. Small quarries were opened on this site during the early 1800s, and industrial-scale quarrying began during the late 1800s. Material Service Corporation has operated the Thornton quarry continuously since 1938. In the process of extensive excavation, the three-dimensional rock fabric of an ancient carbonate reef was peeled away. Generations of paleontologists and sedimentary geologists have studied the pattern of this reef rock. As a result, the reef can be mapped and visualized in three dimensions and many details of its original ecology pieced together.

An Introduction to Sedimentary Rocks and Stratigraphy (New York: W. H. Freeman, 1996); and Gerald M. Friedman, J. E. Sanders and D. C. Kopaska-Merkel, *Principles of Sedimentary Deposits* (New York: Macmillan, 1992).

[12]See P. H. Heckel, "Carbonate Build-ups in the Geologic Record: A Review," in *Reefs in Time and Space*, SEPM Special Publication 18, ed. L. F. Laporte (Tulsa, Okla.: Society of Economic Paleontologists and Mineralogists, 1974), pp. 90-154; James Lee Wilson, *Carbonate Facies in Geologic History* (New York: Springer-Verlag, 1975); J. A. Fagerstrom, *The Evolution of Reef Communities* (New York: John Wiley, 1987); P. Hallock, "Reefs and Reef Limestones in Earth History," in *Life and Death of Coral Reefs*, ed. C. Birkeland (New York: Chapman and Hall, 1997), pp. 13-42; Rachel Wood, *Reef Evolution* (Oxford: Oxford University Press, 1999).

[13]H. A. Lowenstam, "Niagaran Reefs of the Great Lakes Area," *Journal of Geology* 58 (1950): 430-87; J. J. C. Ingels, "Geometry, Paleontology, and Petrography of Thornton Reef Complex, Silurian of Northeastern Illinois," *AAPG Bulletin* 47 (1963): 405-40; H. B. Willman and E. Atherton, "Silurian System," in *Handbook of Illinois Stratigraphy*, Bulletin 95, ed. H. B. Willman et al. (Urbana: Illinois State Geological Survey, 1975), pp. 87-104; R. H. Shaver, "A History of Study of Silurian Reefs in the Michigan Basin Environs," in *Early Sedimentary Evolution of the Michigan Basin*, GSA Special Paper 256, ed. P. A. Catacosinos and P. A. Daniels (Boulder, Colo.: GSA, 1991), pp. 101-38; D. G. Mikulic and J. Kluessendorf, *Classic Silurian Reefs of the Chicago Area*, Michigan Basin Geological Society Field Excursion, June 27, 1998.

At Thornton, prominent vertical and horizontal zonation of marine fossils is encountered. In places, there are thick accumulations of specific biological types, like pentamerid brachiopods or pelmatozoan echinoderms. The southwestern edge of the lower portion of the reef is an elongate ridge of coral, approximately 1,400 feet long and trending northwest to southeast. The primary corals are colonial rugose corals informally termed *Pycnostylus* and tabulate corals such as *Halysites*. The corals are interbedded and in growth position. The coral colonies grew preferentially toward the southwest, yielding an asymmetrical shape to the ridge. This thick coral ridge is truncated and overlain by a grainy limestone consisting chiefly of the remains of stalked echinoderms. In turn, this limestone is topped by a complex ecological association of rugose and tabulate corals, massive calcareous stromatoporoid sponges and pentamerid brachiopods.

As in modern framework reefs, the above developmental stages exhibit a complex three-dimensional architecture with lots of natural cavities. A host of minor marine taxa fill in many of the cavities. Lenses of nearly pure accumulations of brachiopod shells, gastropods or other specific marine taxa occur. There are, moreover, no anomalous taxa such as fossil plants or mammals. The overall fabric is held together by the encrusting platy corals and calcareous sponges. Calcite cements within void spaces are structurally similar to those that form in modern shallow, well-circulating water typical of reef environments.

Modern reefs are subject to strong wave action and periodic violent storms. Back-reef areas and reef flanks grow by the spread of debris from the reef core. Such flanking beds are well developed on the Thornton Reef. Interstate 80 runs directly above the Thornton Quarry, and passengers (not the driver!) of cars can, with careful observation, see that these flank beds dip away from the axis of the quarry.

The overall upward transformation in the character of the buildup and the strong lateral variations in organic composition and degree of buildup closely mimic modern ecological zonation in well-studied reefs.

The Thornton carbonate buildup is part of a complex including many well-studied smaller carbonate buildups in the Racine Formation around Chicago. In total areal aspect, these form part of a large reef chain. The ecologic relationships demonstrable at Thornton are part of a coherent shallow-marine carbonate platform, similar to that of the modern Bahama Bank, but with differing ecological actors.

The Silurian beds of the Chicago region dip toward the center of Michigan at angles of less than one degree, as discussed in chapter twelve. Numerous similar reef structures, in strata that correlate to the Racine Formation, are present in northern Indiana and parts of Michigan. In many of the latter cases, the Racine Formation or its equivalents are overlain by hundreds to thousands of feet of additional sedimentary rock. The normal interpretation of this regional phenomenon is that the Midwest lay under a shallow sea during the Silurian Period. Sea level relative to the central continent was higher because there were no glacial ice caps at the time and because the North American continent was topographically lower than it is today. In this shallow sea a great chain of reefs grew. At later times other sedimentary regimes persisted, many of which were also shallow seas. These Silurian reefs were buried under later sediments; subtle warpings of the sedimentary layers positioned some, like those of Chicago, nearer the present-day surface than others. A long period of erosion, including erosion due to Pleistocene glacial advance, has exposed some of the reefs or brought them close enough to the surface to be attacked by quarry operators and geologists.

Reefs are naturally rich in cavities. Ancient reef rocks often retain these cavities and thus become superior oil reservoir rocks. At Thornton, many small cavities in the dolomite reef rock are filled with sludgy oil. Other Silurian reefs in the Michigan Basin host significant amounts of recoverable petroleum (chap. 12). A considerable amount of the world's oil production, from both the Arabian/Persian Gulf and from the Gulf Coast of Mexico, is obtained from highly porous mollusk-dominated reef rocks of Cretaceous age. The association with oil has of course spurred research on fossil reefs, and the scientific literature on these reefs is extensive.

The testimony to elapsed time during the formation of ancient reefs has been repeatedly pointed out. Wonderly, for example, devoted an extensive discussion to the large regional Devonian reef tract in Alberta, Canada.[14] These reefs are very oil-rich and have been intensively studied by oil geologists, stratigraphers and sedimentologists. Individual reefs may be several hundreds of feet in vertical (growth) extent; the reef tract in turn lies beneath at least a mile of strata.

Regionally integrated accumulations of bound marine framework organisms, often buried under hundreds to thousands of feet of overlying strata, are major "monkey wrenches" for any universal catastrophic scheme. Flood geologists have recognized that these structures must somehow be addressed. Since publication of *The Genesis Flood* in 1961, Flood geologists have employed two critical arguments against the existence of ancient reefs: they argue that such structures are misrepresented by mainstream geologists, and actually are collections of loose debris, or they argue that the frameworks are real but have

---

[14]Wonderly, *God's Time-Records in Ancient Sediments.*

been washed into place.[15] Flood geologist David D'Armond, for example, has argued that the Thornton reefs were washed en masse into their current position from origination points hundreds of miles distant.

There are, in fact, many documented accumulations of carbonate skeletal debris in the stratigraphic record that are certainly not true reefs. In some cases, these are understood, by analogy to modern carbonate banks, as local piles of skeletal particles that accumulated as a result of wave or current conditions; in other instances these are recognized as loosely bound accumulations that formed in situ as a result of high productivity. Carbonate sedimentologists designate the latter cases as "carbonate mud mounds," "bioherms" or by other labels, carefully distinguishing these from true wave-resistant framework structures.

As noted above, framework reefs normally accumulate skeletal debris around their flanks. Cursory discussions of such flank debris by Flood geologists do not negate the certain existence of core framework structure to these reefs. The argument noted above, that fossil reefs are mere accumulations of loose debris, fails under the weight of primary observation of tightly-bound growth structure in reef organisms, and the argument that entire reefs were washed into place boggles the imagination. In many cases the framework reefs are hundreds of feet thick and miles in lateral extent. These reefs are typically parts of long tracts of reefs, resembling the modern Bahama Bank—in some cases extensive enough to be comparable to the present-day Great Barrier Reef of Australia. Thus one must imagine entire chains of reefs to be washed into place, and with upward growth framework preserved intact.

Walter Brown claims that all preserved marine limestone and dolostone beds (presumably, including all carbonate reef rocks) are artifacts of the Great Flood. He explained that "the standard geological explanation is that those regions were covered by incredibly limy (alkaline) water for millions of years—a toxic condition not found anywhere on Earth today. Liquefaction would have quickly sorted limestone particles into vast sheets."[16] Unfortunately, the claim of toxicity is false. Marine limestones of various textures are forming today

---

[15]John C. Whitcomb Jr. and Henry M. Morris, *The Genesis Flood: The Biblical Record and Its Scientific Implications* (Philadelphia: Presbyterian & Reformed, 1961), pp. 407-8; S. E. Nevins, "Is the Capitan Limestone a Fossil Reef?" in *Speak to the Earth*, ed. G. E. Howe (Philadelphia: Presbyterian & Reformed, 1975), pp. 16-59; D. B. D'Armond, "Thornton Quarry Deposits: A Fossil Coral Reef or a Catastrophic Flood Deposit? A Preliminary Study," *Creation Research Society Quarterly* 17 (1980): 88-105; H. M. Morris and J. Morris, *Science, Scripture and the Young Earth*, 2nd ed. (El Cajon, Calif.: Institute for Creation Research, 1989).

[16]Walter Brown, *In the Beginning: Compelling Evidence for Creation and the Flood*, 8th ed. (Phoenix: Center for Scientific Creation, 2001), p. 144. Brown notes, inconsistently, that lime is being precipitated along the coasts of some Caribbean islands.

in warm carbonate banks, like the Great Bahama Bank, under limey but distinctly nontoxic conditions. The precipitation of calcium carbonate can be observed directly from above as white streaks in the water. The natural processes of marine limestone formation, with inclusion of marine fossils, is well understood from many studies conducted in warm waters around the globe.[17]

Bedded Precipitated Chemical Sediments: Evaporites

In southeastern Utah and neighboring portions of Colorado, a deep structure called the Paradox Basin is documented from numerous drilled wells. The Paradox Basin is a complex area of subsidence resulting from movement along multiple fault zones, but in its large-scale aspect forms an elongate trough or syncline trending northwest to southeast. The Paradox Basin began accumulating sediments during the Pennsylvanian Period and continued to accumulate sediment through the Permian Period. The basin covers an area of approximately 17,000 square miles, and the center of the basin accumulated well over a mile of sediment before a major break occurred in sedimentation, marked by a large unconformity. During its history the basin was restricted in lateral extent. Seawater covered the basin and deposited a thick sequence of halite (NaCl) and anhydrite ($CaSO_4$). At the center of the basin, these bedded salts are approximately one mile thick. There are minor beds of dark shale that appear cyclically between the much thicker pure salt sequences.[18]

In the same stratigraphic horizon, but laterally to the southwest, the deep-basin salts vanish and are replaced by coeval shallow-water carbonate beds with prominent algal mounds—carbonate bioherms. This belt is interpreted as the seaward margin of the restricted basin. Salinity was lower at this location, and algal bioherms were able to grow. The bioherms are very porous and host large volumes of oil and natural gas. This oil has led to intense drilling and study of the Paradox Basin. As of 2003, nearly 100 oil fields had produced over 450 million barrels of oil from this area.

---

[17]Cf. R. G. C. Bathurst, *Carbonate Sediments and Their Diagenesis* (Amsterdam: Elsevier Scientific Publishing, 1975); P. A. Scholle, D. G. Bebout and C. H. Moore, eds., *Carbonate Depositional Environments* (Tulsa, Okla.: AAPG, 1983); M. E. Tucker and V. P. Wright, *Carbonate Sedimentology* (Oxford: Blackwell Science, 1990); E. Insalaco, P. W. Skelton and T. J. Palmer, eds., *Carbonate Platform Systems: Components and Interactions*, GSL Special Publication 178 (Bath, England: The Geological Society, 2000).

[18]G. M. Stevenson and D. L. Baars, "The Paradox: A Pull-apart Basin of Pennsylvanian Age," in *Paleotectonics and Sedimentation in the Rocky Mountain Region, United States*, AAPG Memoir 41, ed. James A. Peterson (Tulsa, Okla.: AAPG, 1986), pp. 513-39; Lehi Hintze, *Geologic History of Utah*, Geology Studies Special Publication 7 (Provo, Utah: Brigham Young University, 1988); D. L. Baars, *The Colorado Plateau: A Geologic History* (Albuquerque: University of New Mexico Press, 2000); T. C. Chidsey, "An Up Close and Personal View of Cherokee Oil Field, San Juan County, Utah," *Utah Geological Survey Notes* 35 (2003): 1-3.

Mainstream geology relates such thick sequences of pure chemical salts to long episodes of evaporation of seawater in very restricted basins. Thus, deposits like those of the Paradox Formation are termed *evaporites*.[19] Sedimentary basins hosting thick evaporite sequences occur widely around the world during various geologic time periods. Evaporite units have been extensively studied in the subsurface because they deform easily, tending to seal faults or other cracks and thus form excellent hydrocarbon seals, as is the case in the Paradox Basin. Such sequences of nearly pure, marine chemical sediments such as halite, anhydrite, sylvite and others remain a major problem that Flood geology cannot explain.

Flood geologists complain that thick, relatively pure deposits of chemical salts do not form today, and that mainstream geological explanations for these deposits fail. Both Morris and Whitcomb have suggested that thick pure salt formations are more likely the result of chemical precipitation from very hot, supersaturated briny solutions arising originally from deep within the Earth. Austin and his colleagues have built this hydrothermal mechanism into their model of rapid-fire, catastrophic global tectonism during the great Flood.[20]

However, bedded chemically precipitated sediments in many, if not most cases, are sited in shallow-marine sequences that formed over the continental crust. As noted already, the Paradox evaporites formed in close proximity and eventually overlaid algal carbonate buildups that must have formed within the well-lit zone of a warm shallow sea. Similarly, bedded marine evaporites accompany the Silurian reefs of the Great Lakes region (see chap. 12) and the Devonian reefs of Alberta. In these situations there is absolutely no evidence for a spatial connection to a magmatic source for a hydrothermal fluid. Other evidence in the case of the Silurian salt of the Great Lakes region is provided by the chemical composition of fluid inclusions in the salt. Chemical analyses of the contents of tiny bubbles of brine permit determination of the composition of the waters at the time of salt precipitation. In turn, compositions are functions of the temperatures of formation. Satterfield and colleagues analyzed fluid inclusions from the Silurian F-1 evaporite near Detroit and deter-

---

[19]For further discussion of evaporites, see R. A. Berner, *Principles of Chemical Sedimentology* (New York: McGraw-Hill, 1971); B. C. Schreiber, "Arid Shorelines and Evaporites," in *Sedimentary Environments and Facies*, ed. H. G. Reading, 2nd ed. (Oxford: Blackwell Science, 1986); and J. Warren, *Evaporites: Their Evolution and Economics* (Oxford: Blackwell Science, 1999).

[20]Whitcomb and Morris, *The Genesis Flood*, pp. 412-17; Morris, *Scientific Creationism*, 2nd ed.; Morris and Morris, *Science, Scripture and the Young Earth*; S. A. Austin, J. R. Baumgardner, D. R. Humphreys, A. A. Snelling, L. Vardiman and K. R. Wise, "Catastrophic Plate Tectonics: A Global Flood Model of Earth History," in *Proceedings of the Third International Conference on Creationism, Technical Symposium Sessions*, ed. R. E. Walsh (Pittsburgh, Penn.: Creation Science Fellowship, 1994), pp. 609-21; S. E. Nevins, "Stratigraphic Evidence of the Flood," in *A Symposium on Creation III*, ed. D. W. Patten (Grand Rapids: Baker, 1971), pp. 33-65.

mined that the temperatures of formation ranged between 2° and 25° C, much too low to be derived from a magmataic source.[21]

Although Flood geologists feel that the purity and thickness of these precipitated salt deposits argue against their having formed slowly over time, the catastrophic global Flood model is incompatible with such pure deposits. Flood geologists envision a global catastrophe during which large volumes of slurried sediment are sloshed over wide areas. Stratigraphic sequences of several miles of thickness would be draped over eroded antediluvian terrain during this year-long event. In this type of violent episode, how can such deposits of salt remain pure? Surely the rapid, turbulent motions required to emplace all this sediment would have mixed all the components.

DESERT DEPOSITS—VAST SAND DUNE COMPLEXES

Another example of a sedimentary environment that demonstrates the passage of extended time during deposition is that of an ancient eolian (= aeolian) desert environment. We focus our discussion on large sand-dune fields, major components of such desert environments. The Arabic term *erg* has been coopted into the geological jargon for such large regional dune fields. Thick deposits of quartz sand that are interpreted as having formed in erg environments are known from around the world. They are particularly well exposed in the Colorado Plateau region of northern Arizona and southern Utah. For example, the Navajo Formation of Jurassic age is the prominent cliff-forming rock unit in Zion National Park, Capitol Reef National Park, Rainbow Bridge National Monument and Lake Powell National Recreation Area. The Coconino Formation of Permian age forms a massive, pale white sandstone cliff near the top of the Grand Canyon of the Colorado River (figs. 2.2 and 8.2). The Colorado Plateau region is home to several such sandstone formations that form scenic cliffs. These units were deposited from the Pennsylvanian through the Jurassic periods.[22]

---

[21]C. L. Satterfield, T. K. Lowenstein, R. H. Vreeland and W. D. Rosenzweig, "Paleobrine Temperatures, Chemistries and Paleoenvironments of Silurian Salina Formation F-1 Salt, Michigan Basin, U.S.A., from Petrography and Fluid Inclusions in Halite," *Journal of Sedimentary Research* 75 (2005): 534-46.

[22]H. E. Gregory, *Geology and Geography of the Zion Park Region of Utah and Arizona,* Professional Paper 220 (Washington, D.C.: USGS, 1950); W. L. Stokes, *Geology of Utah,* Occasional Paper 6 (Salt Lake City: Utah Museum of Natural History, 1986); Hintze, *Geologic History of Utah;* L. B. Clemmensen, H. Olsen and R. C. Blakey, "Erg-margin Deposits in the Lower Jurassic Moenave Formation and Wingate Sandstone, Southern Utah," *Bulletin GSA* 101 (1989): 759-73; M. A. Chan, "Erg Margin of the Permian White Rim Sandstone, SE Utah," *Sedimentology* 36 (1989): 235-51; R. C. Blakey, "Stratigraphy and Geologic History of Pennsylvanian and Permian Rocks, Mogollon Rim Region, Central Arizona and Vicinity," *Bulletin GSA* 102 (1990): 1189-217; L. T. Middleton, D. K. Elliot and M. Morales, "Coconino Sandstone," in *Grand Canyon Geology,* pp.

Mainstream sedimentologists feel that the eolian, that is, wind-blown, nature of such sand accumulations is well founded. The very fine sand of these formations has a uniform grain size that is characteristic of wind-blown sand in general. The grains consist of resistant quartz. Less resistant mica grains and ultra-fine clay particles have been abraded to oblivion and/or wafted off-site by wind. The surfaces of individual grains are well rounded. They exhibit a distinctive texture and are termed "frosted." Such "frosting" is actually a coating of microscopic silica that forms in arid environments in response to periodic, rare wetting events and subsequent drying. When taken together, all these aspects argue for a preliminary diagnosis of an eolian setting.[23]

In addition, the sands are stratified in bundles that exhibit steeply dipping cross-bedding. *Cross-bedding* refers to primary lamination formed by the resting of grains along planes that are inclined with respect to the bounding planes of the larger bundle (fig. 10.6 and 10.7). Many aquatic settings produce low-profile duneforms that, when excavated, demonstrate sand laminated in gently dipping cross-beds (figs. 8.8 and 8.9). However, steeply dipping cross-beds (> 20°) are typical of subaerial eolian dunes. The physical reason for this is easily grasped: sand accumulating under water is buoyed by the water and will not be subject to the friction required to maintain a steeply dipping, subaerial dune slope.

Flood geologists have attempted to address the rock composition and structures that are claimed as evidence for eolian deposition of rocks exposed in the scenic cliffs of the Colorado Plateau.[24] They are compelled to reexplain these regional-scale stratigraphic units in terms of water deposition. They have questioned the source of the sand and the significance of the high-angle cross-stratification. Were these perhaps eolian sands reworked by the great Flood? The catastrophist biologist/geologist Leonard Brand has undertaken some fine studies of the formation of tetrapod trackways under shallow aquatic conditions and concluded that the vertebrate trackways of the Coconino Formation could have resulted under conditions of a global Flood.[25]

163-79; M. Morales, "Mesozoic and Cenozoic Strata of the Colorado Plateau Near the Grand Canyon," in *Grand Canyon Geology*, pp. 212-21; Baars, *The Colorado Plateau*.

[23]G. M. Friedman, "Distinction between Dune, Beach and River Sands from their Textural Characteristics," *Journal of Sedimentary Petrology* 31 (1961): 514-29; M. E. Brookfield and T. S. Ahlbrandt, *Eolian Sediments and Processes* (Amsterdam: Elsevier Publishing, 1983); K. Pye and N. Lancaster, eds., *Aeolian Sediments: Ancient and Modern* (Oxford: Blackwell Scientific, 1993); A. S. Goudie, I. Livingstone and S. Stokes, eds., *Aeolian Environments, Sediments, and Landforms* (Chichester, England: John Wiley & Sons, 1999).

[24]H. W. Clark, "The Mystery of the Red Beds," in *Scientific Studies in Special Creation*, ed. W. E. Lammerts (Philadelphia: Presbyterian & Reformed, 1971), pp. 156-64; Austin, "Interpreting Strata of Grand Canyon"; Brown, *In the Beginning*.

[25]L. Brand, "Field and Laboratory Studies on the Coconino Sandstone (Permian) Vertebrate Footprints and their Paleoecological Implications," *Palaeogeography, Palaeoclimatology, Palaeo-*

In response to the Flood geology hypothesis, the general interpretation of an arid dryland setting for the deposition of the proposed eolian sands should be firmly established. In fact, mainstream sedimentology texts caution against jumping too quickly to the conclusion that any given sandstone body represents the deposit of an ancient dune field. There are, however, numerous ancillary features that support an arid, subaerial setting for the deposition of these sediments. Such features include minor intercalated alkali pond deposits; beds of caliche, a dryland concretionary carbonate soil; iron-stained rootlet traces; flood surfaces, marking localized flash-flood events; and beds of loess (loessite) among others.[26] Such features in combination lead sedimentary geologists to reject a hypothesis of aqueous deposition.

Furthermore, the paleontology of these rocks is consistent with an arid setting. In the Colorado Plateau region, most rock units between the Lower Permian and Upper Jurassic do not contain marine fossils. There are a few exceptions, such as the Upper Permian Kaibab Formation and the Jurassic Carmel Formation. Most of the formations from these time periods, however, contain terrestrial fossils. Triassic beds include, for example, the Chinle Formation, which hosts the "Petrified Forest" of Arizona, with its stout forest flora and accompanying terrestrial insect traces and small dinosaurs. The Coconino Formation and the Navajo Formation, interpreted as deposits of ancient coastal ergs, contain very few fossils at all. Most of these are tracks of terrestrial vertebrates such as lizards. These units host *no* marine fossils.

General climatic symptoms of the uppermost Paleozoic and Mesozoic rocks of the Colorado Plateau argue strongly for a warm and arid climate in this region during the deposition of these rocks. Most of the strata deposited during these time periods are sandstones, siltstones or mudstones with *zero* marine fauna and containing abundant signs of subaerial exposure such as soil horizons. Several prominent sandstone layers are located within this thick

---

ecology 28 (1979): 25-38; L. Brand and T. Tang, "Fossil Vertebrate Footprints in the Coconino Sandstone (Permian) of Northern Arizona: Evidence for Underwater Origin," *Geology* 19 (1991): 1201-4; L. Brand, "Variations in Salamander Trackways Resulting from Substrate Differences," *Journal of Paleontology* 70 (1996): 1004-10; Brand, *Faith, Reason and Earth History.*
[26]G. Kocurek, "Significance of Interdune Deposits and Bounding Surfaces in Aeolian Dune Sands," *Sedimentology* 28 (1981): 753-80; R. P. Langford, "Fluvial-Aeolian Interactions: Part I, Modern Systems," *Sedimentology* 36 (1989): 1023-35; R. P. Langford and M. A. Chan, "Fluvial-Aeolian Interactions: Part II, Ancient Systems," *Sedimentology* 36 (1989): 1037-51; Chan, "Erg Margin of the Permian White Rim Sandstone, SE Utah"; M. A. Chan, "Triassic Loessite of North-Central Utah: Stratigraphy, Petrophysical Character, and Paleoclimate Indications," *Journal of Sedimentary Research* 69 (1999): 477-85; J. L. P. Kessler, G. S. Soreghan and H. J. Wacker, "Equatorial Aridity in Western Pangea: Lower Permian Loessite and Dolomitic Paleosols in Northeastern New Mexico, U.S.A.," *Journal of Sedimentary Research* 71 (2001): 817-32; A. M. Alonzo-Zarza and L. H. Tanner, eds., *Paleoenvironmental Record and Applications of Calcretes and Palustrine Carbonates,* Special Paper 416 (Boulder, Colo.: GSA, 2006).

Figure 10.6. Steeply dipping planar cross-stratification in Navajo Formation (Jurassic) exposed at Checkerboard Mesa, Zion National Park, Utah. Photo by R. F. Stearley.

Figure 10.7. Weathered exposures of Navajo Formation near east entrance, Zion National Park, Utah. Photo by R. F. Stearley. Contrast the angles of dip in the cross-stratified beds with those demonstrated in figures 8.8 and 8.9.

stack of strata. Abundant evidences support the mainstream view that these sandstone layers were deposited as part of large ergs adjacent to the seacoast. At intervals the sea covered the arid coast and marine rocks containing abundant marine fossils were deposited. The overall interpretation of several of these large regional sandstone units as the deposits of a widespread coastal erg is well justified.

Lake Beds/Varves

The Green River Formation and its fossils were introduced and reviewed in chapter nine. These sedimentary rocks, covering extensive tracts of southwestern Wyoming, northwestern Colorado and northeastern Utah, have been recognized by mainstream geologists since the late 1800s to be the deposits of ancient, long-lived lakes.

As previously discussed, the flora and fauna of these deposits are major clues to the interpretation of an ancient lake setting.[27] The extensive clam and snail fauna is strictly freshwater. The large arthropod fauna includes spiders, freshwater crayfish and plenteous insects associated with freshwater habitats such as mayflies, water striders and damselflies, in addition to creatures such as crickets, cockroaches, many kinds of beetles, butterflies and moths, and wasps. Plant remains include many well-preserved leaves and stems of such freshwater forms as cattails (*Typha* sp.), horsetails *(Equisetum)* and lilypads, as well as upland plants such as pines, maples, poplars and sycamores. Aquatic birds such as frigatebirds, grebes and long-legged wading birds are well preserved in these sediments. Many fossil turtles are found in these sediments, in addition to preserved skeletons of frogs and crocodiles. The fossil fishes from these sediments are known from rock-and-gem shops worldwide. Whether examined at the outcrop scale or viewed as a regional pattern, the consistent and overwhelming impression is that of a diverse fauna from a large system of ancient lakes.

Much of the Green River Formation consists of varved marls, often dolomitic and often containing much kerogen. Other beds are predominantly evaporitic. The sodium carbonate mineral trona, for example, is mined from these beds for glass-making and other uses. Mainstream sedimentologists interpret the evaporite beds as markers of intervals of dryness and salinity in these ancient lakes.[28]

---

[27]The best summary of the entire fauna is in L. Grande, *Paleontology of the Green River Formation, with a Review of the Fish Fauna,* Bulletin 63 (Laramie: Geological Survey of Wyoming, 1984). The bibliography of this work contains dozens of references to particular sectors of the Green River biota.

[28]H. P. Eugster and R. C. Surdam, "Depositional Environments of the Green River Formation of

The varved beds have been interpreted as bedded lake-bottom calcareous muds since their description in Hayden's report of 1869 (see chap. 9). *Varves* are extremely thin beds of fine sediment, typically forming as couplets with contrasting color and composition. Beginning with Whitcomb and Morris in 1961, many Flood geologists have opined that varves are definitely *not* deposits formed from settling in lakes, and in fact could only have resulted from catastrophic sedimentation.[29] Walter Brown, for example, claimed that liquefaction would sort sediments into "thousands of thin layers" and that "lakes would not produce varves. Varves are better explained by liquefaction."[30]

The term *varve* is a Swedish word originally applied to couplets of coarse and fine material deposited over an annual cycle in seasonal-meltwater environments at the margins of melting glaciers.[31] The term has subsequently been applied to several kinds of bimodally deposited thin layers. In hardwater lakes, the two types of layers typically encountered include a pale marl layer, denoting a season of high precipitation of carbonate, and a darker carbon-rich mud layer, denoting a season of die-back in organic production. While a couplet may represent the products of an annual cycle in lakes, there may in fact be more than one couplet representing a year. Over the past several decades, sampling of lake-bottom sediments has increased in sophistication. Studies conducted on modern lakes in North America, South America, Europe and East Africa have confirmed that varves are typical deposits of the deeper portions of lakes where the waters are quiet and sediments remain undisturbed.[32] In many cases the varve record can be tied into dated human impacts on lakes, such as the commencement of industrialization or eutrophication, and hence the annual nature of the varve cycle can be confirmed.

Wyoming: A Preliminary Report," *Bulletin GSA* 84 (1973): 1115-120; R. Sullivan, "Origin of Lacustrine Rocks of Wilkins Peak Member, Wyoming," *AAPG Bulletin* 69 (1985): 913-22; H. P. Buchheim, "Eocene Fossil Lake, Green River Formation, Wyoming: A History of Fluctuating Salinity," in *Sedimentology and Geochemistry of Modern and Ancient Saline Lakes,* Special Publication No. 50, ed. R. W. Renaut and W. M. Last (Tulsa, Okla.: Society for Sedimentary Geology, 1994), pp. 239-47.

[29]Whitcomb and Morris, *The Genesis Flood;* Morris and Morris, *Science, Scripture and the Young Earth;* Roth, *Origins: Linking Science and Scripture;* Brown, *In the Beginning.*

[30]Brown, *In the Beginning,* p. 143.

[31]C. C. Reeves, *Introduction to Paleolimnology* (Amsterdam: Elsevier, 1967); A. Matter and M. E. Tucker, eds., *Modern and Ancient Lake Sediments* (Oxford: Blackwell Scientific, 1978); P. A. Allen and J. D. Collinson, "Lakes," in *Sedimentary Environments and Facies,* ed. H. G. Reading, 2nd ed. (Oxford: Blackwell Scientific, 1986), pp. 63-94; P. Anadon, L. I. Cabrera and K. Kelts, eds., *Lacustrine Facies Analysis* (Oxford: Blackwell Scientific, 1991); E. H. Gierlowski-Kordesch and K. R. Kelts, eds., *Lake Basins Through Space in Time,* Studies in Geology 46 (Tulsa, Okla.: AAPG, 2000).

[32]P. E. Sullivan, "Annually-Laminated Lake Sediments and the Study of Quaternary Environmental Changes—A Review," *Quaternary Science Reviews* 1 (1983): 245-313.

The Green River Formation varves may not always represent annual cycles. Although many varve couplets appear to represent annual cycles, Buchheim has presented clear stratigraphic evidence that the varve count within a tightly defined interval—the Lower Sandwich Bed—increased from the lake-margin facies to the central, deep-lake facies in the Fossil Butte member of the Green River Formation.[33] He suggested that the carbonate laminae of this member responded to fluvial events that occurred on a scale of more than one per annum.

Regardless of the annual nature of the varve couplets in the Green River Formation, they have by now been analyzed inside and out and their lacustrine origin is established. Criteria for distinguishing the Green River beds as lacustrine have in fact been well established for decades.[34] The acceptance of the lacustrine nature of the Green River sediments by some of the new generation of Flood geologists (discussed in chap. 9) is welcome.

Acknowledgment of the existence of ancient lakes by Flood geologists does not make the problem of the Green River lakes or other ancient lakes vanish. The Green River Formation achieves its maximum thickness in the Uinta Basin of northeastern Utah, where it is as much as 22,000 feet (four miles). Mainstream stratigraphers believe that the lake lasted between fifteen and twenty million years during the early Cenozoic. Catastrophists who believe that this great lake was a post-Flood phenomenon are forced to take the position that 20,000+ feet of sediment were deposited over a time span of a few years to a few thousand years. Still more damaging to the Flood hypothesis is the existence of well-documented lacustrine sequences that lie in the *middle* of the stratigraphic column, such as those of the well-studied Triassic Rift lakes of eastern North America. Globally, there are many suites of lacustrine sediments that lie in the middle of long sequences of strata that catastrophists are forced to classify as deposits from Noah's Flood.

SUMMARY AND POSTSCRIPT

During the nineteenth century, long before radiometric dating techniques were developed, stratigraphers came to the realization that extensive vertical sequences of rocks implied the passage of great stretches of time for their formation. These vertical sequences contain great packets of sediment that

---

[33]H. P. Buchheim, "Paleoenvironments, Lithofacies and Varves of the Fossil Butte Member of the Eocene Green River Formation, Southwestern Wyoming," *University of Wyoming Contributions to Geology* 30 (1994): 3-14.

[34]M. D. Picard, "Criteria Used for Distinguishing Lacustrine and Fluvial Sediments in Tertiary Beds of Uinta Basin, Utah," *Journal of Sedimentary Petrology* 27 (1957): 373-77; M. D. Picard and L. R. High, "Sedimentary Cycles in the Green River Formation (Eocene), Uinta Basin, Utah," *Journal of Sedimentary Petrology* 38 (1968): 378-83.

testify to protracted periods of uniform conditions, punctuated by climatic or tectonic changes.

The separate packets of sediment and entombed biota provide numerous clues as to their formation. The packets (formations and their subdivisions) maintain distinctions in grain size, color, texture, physical sedimentary structures, trace fossils, cements and bounding surfaces that relate to clear physical, chemical and ecological parameters under which these rocks formed. To ascribe all the varieties and their circumstances of formation to a catastrophic Flood is to turn such a Flood into a magical device that can accomplish anything that one desires.

During the past thirty years, many mainstream geologists, including many Christians, have attempted to draw the attention of Flood geologists and their lay followers to the kinds of evidences outlined in this chapter.[35] Some Flood geologists have reacted in a charitable and thoughtful manner, but others have brushed aside extensive sedimentological evidence.

---

[35]E.g., Wonderly, *God's Time-Records in Ancient Sediments;* Davis A. Young, *Creation and the Flood: An Alternative to Flood Geology and Theistic Evolution* (Grand Rapids: Baker, 1977); C. C. Albritton, *The Abyss of Time* (San Francisco: Freeman, Cooper, 1980); Davis A. Young, *Christianity and the Age of the Earth* (Grand Rapids: Zondervan, 1982); Michael R. Johnson, *Genesis, Geology, and Catastrophism: A Critique of Creationist Science and Biblical Literalism* (Exeter: Paternoster, 1988); D. M. Raup, "The Geological and Paleontological Arguments of Creationism," in *Scientists Confront Creationism,* ed. L. R. Godfrey (New York: W. W. Norton, 1983), pp. 147-62; B. F. Glenister and B. J. Witzke, "Interpreting Earth History," in *Did the Devil Make Darwin Do It?* ed. D. B. Wilson (Ames: Iowa State University Press, 1983); Don Stoner, *A New Look at an Old Earth* (Eugene, Ore.: Harvest House Publishers, 1997); Wonderly, *Neglect of Geologic Data;* Young, "The Discovery of Terrestrial History"; D. U. Wise, "Creationism's Geologic Time Scale," *American Scientist* 86 (1998): 160-73.

11

OF TIME, TEMPERATURE
AND TURKEYS

Clues from the Depths

SINCE THE LATE NINETEENTH CENTURY, geologists have typically classified most rocks into one of three categories: igneous, sedimentary and metamorphic. Most discussions of geology by young-Earth creationists emphasize sedimentary rocks such as sandstone, limestone, shale, coal and evaporites. They stress that these rocks were deposited extremely rapidly all over the globe during the Genesis Flood. The layered sedimentary rocks exposed so spectacularly in the Grand Canyon are frequently presented as a prime example of materials deposited by the Flood.[1] Young-Earth creationists, however, have said relatively little about igneous and metamorphic rocks.[2] Such

---

[1]S. A. Austin, ed., *Grand Canyon: Monument to Catastrophe* (Santee, Calif.: Institute for Creation Research, 1994).

[2]Important exceptions are Andrew A. Snelling and John Woodmorappe, "The Cooling of Thick Igneous Bodies on a Young Earth," in *Proceedings of the Fourth International Conference on Creationism 1998 Technical Symposium Sessions,* ed. R. E. Walsh (Pittsburgh: Creation Science Fellowship, 1998), pp. 527-45; and Andrew A. Snelling, "Regional Metamorphism within a Creationist Framework: What Garnet Compositions Reveal," in *Proceedings of the Third International Conference on Creationism,* ed. R. E. Walsh (Pittsburgh: Creation Science Fellowship, 1994), pp. 485-96.

rocks, however, provide compelling evidence that the Earth is considerably more than a few thousand years old.

Igneous rocks form by solidification to glass and/or crystals from molten rock material called *magma*. Lava is magma that is erupted onto the surface of the Earth. Magma originates deep within Earth's crust or upper mantle a few miles below the surface. Igneous rocks display wide variation in chemical and mineralogical composition, texture and mode of occurrence. Most igneous rock types are unusual and rare, but a few types are so abundant and widespread that even nongeologists are familiar with them. Virtually everyone has seen granite tombstones, building façades and kitchen countertops.[3] Many people have seen samples of lava, most of which is composed of the rock type basalt.

Metamorphic rocks form in the solid state by recrystallization of preexisting rocks at various depths under extreme heat and stress. In many cases, rocks that undergo metamorphism originated on Earth's surface either as sedimentary rocks or lava flows. In other cases, rocks formed at depth, such as granite, may later undergo episodes of heating that convert them into metamorphic rocks. Slate is a metamorphic rock familiar to most people. Schist is a widely used building stone in Philadelphia and New York City where it occurs in great abundance. Gneiss is a banded and folded rock that makes spectacular countertops and building facing stone.

Most of the discussion in this chapter concerns igneous rocks. We conclude with a brief consideration of metamorphic rocks.

WHAT ARE IGNEOUS ROCKS?

Geologists distinguish two broad categories of igneous rocks: *extrusive* (volcanic) and *intrusive* (plutonic). *Extrusive igneous rocks* have been extruded onto the surface of the Earth by volcanic eruption. Basalt, andesite, rhyolite and obsidian as well as welded tuff, volcanic ash and volcanic bombs are examples of common extrusive igneous materials. The Hawaiian Islands, for example, consist almost entirely of extrusive rocks. The overwhelmingly dominant Hawaiian rock type is *basalt* in the form of lava flows. Some other places where relatively recently erupted lava and/or volcanic ash can easily be seen within the United States are Lava Beds National Monument in northern California, Crater Lake National Park in Oregon, and Mount St. Helens National Volcanic Monument in southwestern Washington. Older lava is present at the surface

---

[3]Countertops that are not composed of marble are typically referred to in the industry as "granite" countertops. In reality, "granite" countertops include a wide range of igneous rocks such as syenite, gabbro, anorthosite, larvikite and granite as well as metamorphic rocks such as gneiss and migmatite.

Figure 11.1. Cross-cutting basalt in St. Francois Mountains, Missouri. Photo by R. F. Stearley.

in many localities throughout the western United States.

Intrusive igneous rocks form by injection of magma into preexisting rock at depth. Magma is typically injected into fractures or along bedding planes or faults in these preexisting *wall* or *country rocks.* In the process of intrusion, magma solidifies underground, in some cases several miles beneath the surface. Many intrusive rock bodies are presently exposed at the surface because the rock that was originally above them was eroded away when the igneous bodies were uplifted to the surface. Yosemite, Kings Canyon and Sequoia National Parks in the Sierra Nevada of eastern California feature enormous volumes of granite and granite-like igneous rocks that crystallized a few miles below the surface prior to uplift and exposure. West of Colorado Springs, Pikes Peak and its surroundings are composed largely of coarse-grained red granite. Vast tracts of granite make up large portions of the Canadian shield in Ontario.

EVIDENCE FOR THE IGNEOUS CHARACTER OF INTRUSIVE ROCKS

Anyone can readily understand that solidified lava flows are igneous rocks. It is often possible to see flows form from hot lava during an eruption of Kilauea on the island of Hawaii or at Mount Etna in Sicily. But nobody has seen an intrusive igneous rock form. So how do geologists know that intrusive rocks form from magma beneath the surface? How do we know, for example, that granitic rocks in the Sierra Nevada are intrusive igneous rocks?[4]

Bodies of intrusive igneous rock commonly display a *cross-cutting relationship* to the country rocks that they have intruded, suggesting that they were injected into fractures within those rocks (fig. 11.1). Because not all cross-cutting rock bodies are frozen remnants of magma, the cross-cutting relationship alone is not sufficient to prove magmatic origin for a rock, and therefore other evidence must be added.

In many cases, large bodies of intrusive magmatic rock show *chill zones* (fig. 11.2). A chill zone forms within a cooling igneous rock body immediately adjacent to its contact with the wall rock into which it was intruded. The chill zone consists of exceptionally fine-grained rock compared to that in the interior of the igneous body. Geologists attribute the fine-grained nature of the rock to very rapid chilling and crystallization of magma as it comes into contact with wall rocks that are much cooler than the magma. Basaltic magmas commonly have temperatures around 1100° to 1200° C and granitic magmas commonly have temperatures around 750° to 1000° C, depending on the amount of H_2O dissolved in them, whereas the rocks into which they are intruded may be as "cold" as 200° to 300° C. Magma in the interior of an igneous body loses heat much more slowly than magma at the margin. Consequently, crystals that make up the rock in the interior of the igneous body have time to grow larger than minerals in the chill zone next to the contact.

If rocks solidified from molten magma underground, heat escaping into the much cooler wall rocks ought to cook those rocks and produce some form of alteration. Very commonly igneous intrusions are surrounded by thick sequences of layered limestone, sandstone and shale. Near their contact with an igneous rock body, these sedimentary rocks commonly show obvious alteration, including changes in grain size, crystalline character and mineral content. In some cases, the original rocks have been altered into *contact metamorphic* rocks. Such altered rocks may contain a variety of calcium-bearing (and in some cases aluminum-bearing) silicate minerals such as garnet, epidote, pyroxene and wollastonite that were formed by the reaction of calcite (calcium carbonate) in the original limestone, quartz (silicon dioxide) in the original sandstone, and aluminum-bearing clay minerals in the original shale. In some

---

[4]The reader who is interested in learning more about igneous rocks should consult a general geology textbook. Those who know some of the fundamentals of geology and mineralogy might wish to consult introductory textbooks on igneous petrology such as John D. Winter, *An Introduction to Igneous and Metamorphic Petrology* (Upper Saddle River, N.J.: Prentice Hall, 2001); and Myron Best, *Igneous and Metamorphic Petrology*, 2nd ed. (Oxford: Blackwell, 2003). For a history of ideas about igneous rocks, including their magmatic origin, see Davis A. Young, *Mind over Magma: The Story of Igneous Petrology* (Princeton, N.J.: Princeton University Press, 2003).

Figure 11.2. The chill zone at the base of Salisbury Crags adjacent to the underlying sedimentary rocks is marked by uniform fine grain in comparison to coarser grain higher in the sill. As hot magma of the sill came in contact with the underlying cold sedimentary rocks at depth, heat was rapidly lost to the wall rocks. As a result the magma crystallized so quickly that crystals had insufficient time to grow to a large size. Photo by D. A. Young.

cases, silica was derived from silica-rich fluids expelled by the cooling magma instead of the wall rocks. Minerals such as garnet, epidote and pyroxene are stable at the very high temperatures that would likely be reached near the contact with an intruded magma. The presence of such high-temperature minerals in rocks adjacent to an igneous body provides a strong indication that the rocks were heated by the cooling magma.

In many igneous rock bodies, fragments of the rock into which the magma was intruded have been incorporated into the igneous rock, implying that those fragments were torn from the walls as the magma was emplaced (fig. 11.3). These out-of-place fragments are termed *xenoliths* (Greek for "strange rocks" or "foreign rocks"). In many cases, xenoliths in bodies that are believed to be igneous show evidence of chemical reaction with, or being melted by, the invading magma.

Another line of evidence of magmatic origin is the presence of textures (that is, the grain-to-grain relationships) within the rock that are consistent with crystallization from a liquid over a range of decreasing temperatures.[5]

---

[5]Igneous rock textures such as those observed in chill zones are examined in very thin slices of rock under a petrographic microscope equipped with polarizing lenses. The interested reader

Minerals and textures found in igneous rocks are consistent with results from thousands of high-temperature laboratory experiments to determine the pressure and temperature stability of a wide range of minerals.[6] In many cases, the rocks surrounding an igneous rock body may be intensely folded or faulted, suggesting that the country rocks were deformed concomitantly with emplacement of viscous magma under pressure.

Normally no one criterion is sufficient by itself to persuade a geologist that a given rock body has an intrusive magmatic origin. More often than not, several lines of evidence are present, and the combined weight of evidence persuades contemporary geologists that there are thousands of examples of intrusive igneous rocks worldwide now exposed at Earth's surface.

THE RELEVANCE OF IGNEOUS ROCKS TO THE QUESTION OF EARTH'S ANTIQUITY

So, what is the relevance of igneous rocks to the antiquity of Earth? No individual body of igneous rock is as old as Earth itself, but a vast array of igneous rock bodies all over the globe bear evidence of antiquity far exceeding a few thousands of years. First, let's think about extrusive rocks. Many volcanic terranes contain enormously thick piles consisting of hundreds of lava flows stacked on top of one another. The Columbia-Snake River plateau of western Idaho, northeastern Oregon and eastern Washington, for example, consists of about 60,000 square miles of basaltic lava flows as much as 12,000 feet thick and averaging 3,000 feet thick.

Great shield volcanoes like Mauna Loa and Mauna Kea on the island of Hawaii or Haleakala on Maui are enormous structures that were built up from

---

is invited to consult William S. Mackenzie, C. H. Donaldson and C. Guilford, *Atlas of Igneous Rocks and Their Textures* (New York: John Wiley & Sons, 1982) for beautiful color photographs of the textures of common igneous rocks.

[6]Although experimental studies on rock-forming minerals are presently carried out at dozens of major universities throughout the world, the pioneering work in calibrating the high-temperature scale, in constructing suitable apparatus for performing high-temperature experiments and in determining the stabilities of a wide range of common igneous rock minerals was done at the Geophysical Laboratory in Washington, D.C., beginning as early as 1905. For insight into some of the work done at the Geophysical Laboratory, see Hatten S. Yoder Jr., *Centennial History of the Carnegie Institution of Washington,* vol. 3: *The Geophysical Laboratory* (Cambridge: Cambridge University Press, 2004); and Davis A. Young, *N. L. Bowen and Crystallization-Differentiation: The Evolution of a Theory* (Washington, D.C.: Mineralogical Society of America, 1998).

For some helpful compilations of experimental results on pressure-temperature stability of minerals and geologically important liquids, see W. Gary Ernst, *Petrologic Phase Equilibria* (San Francisco: W. H. Freeman, 1976); Stearns A. Morse, *Basalts and Phase Diagrams* (New York: Springer-Verlag, 1980): Paul C. Hess, *Origins of Igneous Rocks* (Cambridge, Mass.: Harvard University Press, 1989); and W. Johannes and F. Holtz, *Petrogenesis and Experimental Petrology of Granitic Rocks* (New York: Springer-Verlag, 1996).

the floor of the Pacific Ocean by eruption after eruption. Mauna Loa began by eruption of basaltic lava directly onto the sea floor approximately 18,000 feet below sea level, but the summit of Mauna Loa is now 13,680 feet above sea level. The mountain is higher from its base on the floor of the ocean to its top than is Mount Everest, and in total volume Mauna Loa is overwhelmingly

Figure 11.3. Angular blocks (xenoliths) of country rock incorporated into light-colored igneous rock at Kinsman Creek, New Hampshire. During injection at depth, viscous magma pried off fragments of country rock that were then frozen in place in the mass of cooling magma. Photo by D. A. Young.

larger than Mount Everest. Mauna Loa is a pile of lava flows at least 32,000 feet thick. If each flow is about a couple of feet thick and a single eruption does not spread out over a great distance in all directions at the same time, it is evident that it takes a while for even a thickness of one foot of lava flows to cover an area of several tens of square miles in the vicinity of the volcano. Even if a particular area on one side of the volcano were covered by a two-foot-thick lava flow erupted at the rate of one flow per year, it would take 16,000 years to accomplish just that. But the lava flows do not all accumulate in one small area, and they are by no means erupted every year. As a result, far more than 16,000 years was needed to build up the volcano. Mauna Loa is considered an active volcano today even though it has not erupted since 1984. In South Africa, Karoo basalts, locally as much as 40,000 feet thick, once covered an area on the order of 750,000 square miles. Similar examples occur in India and Brazil.

Great thicknesses and volumes of basalt in such volcanic provinces imply that a long period of eruptive activity was involved. The evidence for the passage of vast amounts of time is made even more persuasive by the fact that lava flows in many localities are separated from one another by soil horizons. On the islands of Kauai and Maui in the Hawaiian Islands or on the Isle of Skye, for example, one- to three-foot-thick basalt lava flows are commonly capped by thoroughly weathered soils. In some cases, the soil consists of *laterite*, a dark reddish weathering product from which nearly all major chemical constituents except iron and aluminum oxides have been leached out. Laterite formation requires deep, thorough weathering of iron-bearing bedrock over hundreds of years in a humid climate. In portions of the Columbia River Plateau in Oregon, basaltic volcanic layers are commonly separated by layers of fossiliferous sedimentary rocks. These rocks, too, indicate the passage of a considerable amount of time between eruptions of individual lava flows.

Some of the most persuasive evidence for the antiquity of Earth, however, comes from intrusive igneous rocks, particularly very large bodies such as the Sierra Nevada batholith in California or the Bushveld Igneous Complex in South Africa, a gigantic layered igneous intrusion whose dimensions are on the order of 300 miles long, 150 miles wide, and 24,000 feet thick. The passage of a lot of time during formation of a large intrusive body is necessitated by the time required to melt a substantial quantity of liquid from rocks deep within the crust or upper mantle; to accumulate that melt into a large volume; to bring magma through the crust to the surface or to the place where it will finally be emplaced; to emplace the magma; to crystallize the magma; and to cool the newly crystallized igneous body to the ambient temperature of the surrounding rocks. In addition, many intrusive igneous bodies that originally crystallized at considerable depth are now exposed at Earth's surface. It also took time for the igneous rock body to get to the surface from the subsurface location where it originally crystallized. We focus here primarily on the time of melting of magma, ascent time, time of emplacement, and time of crystallization and cooling.

THE SOURCE AND COALESCENCE OF MAGMA

The formation of an igneous rock body begins with the generation of magma. The process begins with a deep-seated volume of rock with a chemical and mineralogical composition that lends itself to melting under the appropriate range of temperatures and pressures that exist in the crust. The minerals in that rock volume must absorb sufficient heat for melting to begin, typically at the boundaries between the mineral grains. The total volume of melt generated depends on the amount of rock that is subjected to the appropriate temperature-pressure conditions. Very likely, the temperature is not high enough to melt the rock completely. The rock only partially melts, and the percentage of rock that melts depends on temperature. Small quantities of melt accumulate at grain boundaries, but these small volumes must migrate through the rock to coalesce with other small melt volumes. Migration and coalescence typically do not occur until a critical percentage of rock has melted. The ability of melt to migrate and coalesce is related to the chemical composition of the magma. Granitic melts, by virtue of their high viscosity, migrate and coalesce with greater difficulty than gabbroic magmas. Not until significant volumes of melt have segregated from the rock being melted can magma ascent take place. We simply note that to melt and collect a mass of magma with a volume of 1,000 cubic miles (10 miles on a side) requires very much more time than it takes to form a body of magma with a volume of 1 cubic mile (1 mile on a side), and this in turn takes more time to form than a magma body with a volume of 1 cubic foot (1 foot on a side).[7]

THE ASCENT OF MAGMA

The determination of the amount of time that it takes for magma, after it has accumulated, to rise through the crust is not easy, although reasonable estimates are possible. Igneous rocks commonly contain xenoliths that were torn from the walls of the fractures through which magma ascended. In some instances, these xenoliths have been transported from the upper mantle as much as 35 miles below the surface of continents. Upper mantle xenoliths typically are composed predominantly of the minerals olivine and pyroxene, both iron-magnesium silicates with characteristic densities on the order of 3.5 grams per cubic centimeter (3.5 times the density of water). In contrast, basaltic lava that commonly contains these xenoliths has a density that is around 2.5 grams per cubic centimeter. If a mass of basaltic liquid were standing stagnant in a fissure and not moving upward, the denser xenoliths would sink through the liquid and would never rise, let alone reach the surface! It is possible to make rough calculations of the rate at which magma must move toward the surface to carry the xenoliths upward. Most such estimates are on the order of 4 inches to 3 feet per second. The rate of ascent of the xenoliths would vary depending on the density and viscosity of the lava in the fracture, the density of the xenoliths, the width of the fracture and the shape of the xenoliths. Generally, the

---

[7]The amount of source rock that needs to be melted to initiate segregation of melt into larger mobile masses varies depending on the type of rock being melted, temperature, presence of fluid and deformation. In most cases at the very least 10 to 20 percent of the rock must be melted to initiate segregation.

rate at which basaltic magma rises through the crust is rather rapid.

Before the 1970s, geologists generally thought of granitic magma rising as large balloon or tear-drop shaped masses called *diapirs*. Upward movement of a buoyant diapir would be a very slow process, perhaps requiring tens or hundreds of thousands of years.[8] More recently, a number of geologists have advocated the idea that granitic magma rises through planar fractures in the crust to its site of emplacement.[9] Nick Petford, now at Bournemouth University, has calculated that granitic magma could rise through a 19-foot-wide fracture at a rate of about 0.4 inch per second. At that rate, a small packet of magma could ascend 18 miles through the fracture from its source to the final emplacement site in as little as 41 days. If the fracture were not so wide, however, the rate would be considerably less and the amount of time to ascend would increase. If granite bodies were indeed emplaced through feeder dikes, Petford's result would be very encouraging to young-Earth creationists. There is more to the story, however.

EMPLACEMENT OF IGNEOUS BODIES

Magma generally will rise, provided it has unimpeded access, as in a large fracture, toward the surface. Depending on the distribution of stress within Earth's crust, magma that is still underground may spread out laterally as a sheet along bedding planes of sedimentary rocks into which it is being intruded. In other instances, magma may balloon outward, shouldering aside rocks already present. Such processes are referred to as magma emplacement. Upon emplacement, the cooling magma may form a sheet-like body of rock called a *sill* that is parallel to the layers into which it is intruded; a bulbous mass with a flat bottom called a *laccolith;* a cross-cutting sheet called a *dike;* a roughly cylindrical pipe-like mass called a *stock;* or a gigantic lenticular mass of rock, in many cases a combination of smaller intrusions, called a *batholith.* How long does it take to emplace magma into an igneous body that may vary in size from a few cubic feet to tens of thousands of cubic miles?

As an example, Petford calculated how long it would take to fill the volume occupied by the Cordillera Blanca batholith, one of hundreds of smaller granitic intrusions that comprise the huge Coastal batholith of central Peru.[10] This intrusion has an estimated volume of 1,350 cubic miles. At an ascent rate

---

[8]Bruce D. Marsh, "On the Mechanics of Igneous Diapirism, Stoping, and Zone Melting," *American Journal of Science* 282 (1982): 808-55.

[9]N. Petford, A. R. Cruden, K. J. W. McCaffrey and J.-L. Vigneresse, "Granite Magma Formation, Transport and Emplacement in the Earth's Crust," *Nature* 408 (2000): 669-73.

[10]For calculations pertaining to the Cordillera Blanca batholith of Peru, see Nicholas Petford, R. C. Kerr and J. R. Lister, "Dike Transport of Granitoid Magmas," *Geology* 21 (1993): 845-48.

of 0.4 inch per second through a 19-foot-wide fracture about six miles long, the batholith would be formed in about 350 years.

This calculation was greeted with enthusiasm by two young-Earth creationists, Andrew Snelling and John Woodmorappe, who suggested in light of Petford's calculations that igneous rock bodies could have been emplaced and cooled off within the time frame of a 10,000 year-old world.[11] Some cautions are in order, however. Not all geologists are persuaded that all granitic bodies were emplaced along dikes rather than diapirs. Assuming that such is the case, however, there are still additional factors that need to be taken into account. Snelling and Woodmorappe overlooked Petford's statement that "this time of active flow [i.e., the 350 years of filling] is likely to be divided into a number of smaller events interspersed by periods of source recharge."[12] In other words, filling of a batholith chamber depends not only on how fast magma can rise through a channel but also on how fast granitic magma can be generated in the source area. After a batch of magma is generated, it may be expelled upward through a dike system in a more or less continuous pulse until the melt in the source area is largely exhausted.

The problem is that the source may produce only a fraction of the total amount of magma in a single melting episode. As a result, only a portion of the final volume may ascend in a single pulse. More time must elapse until additional magma is generated at the source. After a critical amount of melt has been produced, a new pulse of magma may rise, and so on for several intermittent episodes. Thus, the time required for complete emplacement of all the magma may far exceed the 350 years, which represents only the time involved in actual magma rise, not the "dead time" in between pulses. For example, it might actually take 10,000 years to emplace the batholith with most of the time representing hiatuses between pulses of active magma rise. One must also consider that the filling time would be increased if the fracture through which the magma ascended were narrower than 19 feet and also shorter than six miles.

There are other considerations. Large, ready-made empty cavities do not exist several miles below the surface just waiting to be filled by any magma that happens to come along. Space needs to be made for any ascending magma. A very important question is how quickly room can be made available for emplacement of rising magma. The space is generated by the localized application of stress fields within Earth's crust as is the case, for example, within strike-slip fault zones like California's San Andreas fault or within tensional environments where the crust is being stretched. As space becomes available,

[11]Snelling and Woodmorappe, "The Cooling of Thick Igneous Bodies on a Young Earth."
[12]Petford, Kerr and Lister, "Dike Transport of Granitoid Magmas," p. 847.

magma can then move into it. Although estimates of the amount of movement on faults that could provide the kind of space needed are fraught with pitfalls, these estimates suggest that space is made available at a much slower rate than magma can rise, another line of evidence that is consistent with the notion that magma rises in pulses separated by time. For example, measurements along the San Andreas fault suggest long-term average movement of around 1.5 inches per year and estimates in tensional environments such as the Great Basin of the western United States suggest as much as two inches per year.

At first blush, the finding that a moderate-sized batholith might be filled in 350 years seems to be consistent with the theory of a young Earth, but that finding might be consistent with an Earth that is only ten thousand years old provided that all bodies of igneous rock are rather small and provided that they were all emplaced within a relatively small time interval. Many bodies are small, and their total volume could be filled up by ascending magma in a matter of years assuming availability of space and continuous magma production at the source and ascent in a single pulse of magma. There are, however, much larger bodies of igneous rock than the Cordillera Blanca batholith. Many of these had to form in separate episodes of magma rise. Consider the giant Sierra Nevada batholith of California (see chap. 13), or the even larger Coast Range batholith of British Columbia, or the entire Coastal batholith of Peru. These batholiths are composite igneous intrusions composed of dozens to hundreds, if not thousands of individual plutons (intrusive igneous rock bodies), many of which have a volume of several hundreds to thousands of cubic miles. Geologic field evidence indicates that individual bodies within such batholiths were not emplaced all at the same time but rather in a sequence. One body cuts across another that had already been emplaced and solidified. Moreover, the chemical compositions of the individual intrusions are not all the same, suggesting that the composition of the magma source area had changed or that different sources were being tapped at different times. Even if an individual pluton were emplaced in a matter of hundreds of years or even less, it would doubtless require several thousands to hundreds of thousands of years to emplace a large composite batholith consisting of hundreds of smaller intrusions.

THE COOLING OF MAGMA BODIES

The ascent of a given pulse of magma in a fracture may be relatively rapid, but emplacement of magma to form a sizeable body of igneous rock surely takes considerably longer. What really takes time, however, is the crystallization and further cooling of a mass of intruded magma. We are all familiar with everyday examples of the kind of thing we are talking about. In the United States, on Thanksgiving Day, most families relish the consumption of a large turkey.

To cook the turkey properly the oven must be heated to around 325° F, and the turkey needs to be in the oven for four to five hours. The time required to cook the turkey is related to the initial temperature of the turkey (it was probably in the freezer until just a few days before Thanksgiving and may still be very cold the night before the big day), the shape, size and weight of the turkey, and the heat-conducting properties of the lately departed bird. After the turkey has been cooked, it is placed on the table and it begins to cool. Thanks to the fact that hungry eaters require that the turkey be sliced, it generally takes an hour or so for it to cool nearly to room temperature. If the turkey were not sliced at all, it would take considerably longer to reach room temperature.

A large volume of emplaced magma behaves like a gigantic Thanksgiving turkey, but it will take considerably longer to cool off than a turkey. The body of magma is a tad larger than the turkey, for one thing, but the temperatures involved are also considerably different. The turkey needs to cool from 325° F to room temperature, but the magma needs to cool from initial temperatures between 750° and 1100° C (about 1382° and 2000° F), depending on what kind of magma it is. Unlike lava flows, a magma body also cools while it is still in the underground "oven." We can't suddenly take the magma and put it "on the table" at the surface to cool off quickly as we do a turkey. It remains well underground surrounded by other rocks that are also very hot in comparison to Earth's surface. Many magma bodies were emplaced into wall rocks at depths where the temperatures were on the order of 200° to 600° C or even higher, well above the boiling temperature of water at Earth's surface. Under such circumstances the magma body will take a long time to cool and crystallize.

HOW IS HEAT TRANSFERRED FROM A COOLING BODY?

There are several ways in which a cooling body of magma can lose heat. One way is by heat *conduction,* in which heat diffuses from a region of higher temperature to a region of lower temperature in accord with the second law of thermodynamics. We are all familiar with conduction. If we place a metal poker into glowing embers in a fireplace, heat will be so efficiently conducted through the poker that it will quickly become much too hot to pick up with our bare hands. The hot embers have transferred heat to the cool poker.

In the case of cooling magma, heat diffuses from the margins of the hot magma body into the cooler wall rock. Heat is also conducted from the interior of the magma body to its cooler margins. The rate at which conduction occurs depends on the dimensions and shape of the igneous body. Generally speaking, the larger a magma body is the longer it will take to cool off. Cooling rates also depend on the initial temperature of the magma. If the initial temperature of one magma body is 1100° C and the initial temperature of a second identical body is 800° C, the hotter body will cool to a given temperature more slowly than the cooler body if all other factors are equal. The rate of cooling also depends on the initial temperature of the surroundings. Suppose we have two identical bodies of magma at 900° C. One magma body is injected into wall rocks at 100° C, and the other into rocks at 400° C. The magma injected into the cooler wall rocks will cool more quickly than the one injected into the warmer wall rocks. The chemical composition of the magma is very important because the heat-conducting properties of magmas are functions of composition. Some liquids are capable of storing and conducting heat more rapidly than are others. Finally, we need to consider the thermal conductivity of the wall rocks. If the wall rocks are extremely poor heat conductors, then heat will not be lost readily from the magma even though the magma might be a good heat conductor. As an analogy, consider what happens when hot coffee is placed into a thermos bottle. It stays hot a long time.

A second way by which heat is transferred is *convection*. As soup heats up on a stove burner, the soup begins to circulate spontaneously. The heated soup at the bottom of the pot expands, and its density decreases. The hot, less dense soup becomes buoyant and rises toward the top of the pot. Heat is transferred toward the top of the pot by the rising, hotter soup. The rising soup also displaces colder soup at the top of the pot. The colder soup has a higher density than the rising soup and, therefore, begins to sink toward the bottom where it, in turn, displaces hot, less dense soup. These changes in the temperature and density of the soup result in spontaneous, continuous overturning (convection) of the soup. As a liquid, magma can also convect. If convection in a magma chamber is relatively rapid, heat can be efficiently transferred from the center of a hot magma body toward the slightly cooler margins. But the magma body will be unable to cool unless the heat is also transferred from the magma body into the wall rocks. Because solid wall rocks do not overturn (convect), however, convection is unimportant for transferring heat away from cooling magma bodies. If heat diffuses away from the contact of a cooling igneous body into wall rocks by conduction, then convection *within* the magma body may play an important role in transferring heat to the margins when it can then be removed.

Third, heat is transferred by *radiation*. Think of how pleasantly warm you feel on a beautiful sunny day. That warmth is clearly coming from the Sun, but how does that heat get from the Sun to Earth? It does not get to us by conduction, because there is virtually nothing but a few isolated atoms between Earth and the Sun to conduct heat. If there is virtually no matter to conduct heat from the Sun, then there is also nothing to transfer heat by convection. Hence, we are left with radiation as the heat-transfer mechanism.

Radiation is the heat-transfer mechanism that is operative in our microwave ovens. Heat is transferred from the Sun to the rest of the Solar System by electromagnetic waves such as radio waves, x-rays, gamma rays and microwaves. Is radiation of any importance in transferring heat from a cooling body of magma? The answer is *yes* if magma is on Earth's surface in the form of lava. Anyone standing close to a lava flow with a temperature of 2000° F will definitely feel very warm thanks to the blistering radiant heat emanated by the glowing lava. Because heat from a body of magma buried deep within rocks cannot escape effectively by radiation, the amount of radiant heat is considerably less than the amount of heat that is conducted away from the cooling magma body.

A fourth type of heat transfer known as *advection* is very important in geologic situations. Advection occurs where a fluid circulates through porous, permeable wall rocks. In this case, heat may be conducted away from a cooling body of magma through the surrounding solid rock, but heat is also removed by the fluid that is circulating through the pore spaces in that rock. Normally the percolating fluid is H_2O, a substance that has a very high heat capacity, a measure of the ability of a substance to absorb heat. One gram of H_2O can absorb much more heat than one gram of rock. However, at temperatures above 374° C and pressures above 219.5 atmospheres, H_2O exists as a *supercritical fluid* in which the distinction between liquid water and water vapor is blurred. Only at temperatures below 374° C can H_2O exist as liquid water, provided the pressure is sufficiently high. As H_2O in contact with hot magma is heated, it expands and rises toward the surface through any available interconnected pores or fractures. Cool water near the surface is denser than the rising aqueous fluid and sinks toward the magma body through pore spaces. Circulation of H_2O in the wall rocks is continuously driven by the heat loss of the magma.

Magma bodies, therefore, can lose heat in four ways. If a magma is injected into dry rocks, heat will generally be lost by conduction through those dry rocks. If magma is injected into "wet" rocks containing pore fluids, heat will be lost by a combination of conduction through the wall rocks and advection via circulating fluids. The relative importance of these two processes will be determined primarily by the permeability of the wall rocks and the availability of fluids. In many cases, the magma body itself may emit aqueous fluid into the wall rocks. Wall rocks with high permeability are capable of transmitting large quantities of fluid. In these situations, advection is probably the dominant heat transfer process. Wall rocks with low permeability are capable of transmitting lesser quantities of fluid and heat, and in this case conduction will likely be the dominant process of heat loss.

EARLY STUDIES OF HEAT LOSS FROM COOLING MAGMAS

When geologists first began to consider the problem of the amount of time that it takes for a body of magma to crystallize and cool to the temperature of the surrounding rocks, they considered the problem strictly in terms of heat conduction. Given the mathematical complexity of the calculations, it made sense to begin with as simple a model of heat loss as possible to make the problem tractable. In addition, the role that is played by circulating fluids during magmatic heat loss was not yet fully appreciated.

Early calculations were based on the heat conduction equation formulated by French mathematician Jean Baptiste Joseph Fourier (1768-1830) in the early nineteenth century.[13] To obtain the cooling time of a magma body from the heat conduction equation, it is necessary to know, or estimate, such variables as the initial temperature of the magma body; the temperature of the rocks into which the magma was intruded; the dimensions of the magma body; the depth of intrusion; and properties of the magma and wall rock such as thermal conductivity, density and heat capacity. If the magma is assumed to crystallize, then it is also necessary to know the latent heat of fusion of magma, that is, how much heat is released per unit volume or mass of liquid in the process of freezing. A number of estimates based strictly on heat conduction were published throughout the twentieth century.

One estimate was that of Esper S. Larsen Jr. (1879-1961), a professor of petrology at Harvard University throughout much of the first half of the twentieth century.[14] Larsen noted that the Southern California batholith (now known as the Peninsular Range batholith) is more than 370 miles long and about 60 miles wide. The batholith consists of numerous smaller intrusions. Field evidence indicates that the earlier intrusions were nearly crystallized prior to intrusion of the later plutons. For the sake of simplicity, therefore, Larsen assumed that the entire batholith was intruded in one pulse. He assumed an initial temperature of 820° C, an initial temperature of the wall rocks of 120° C, and a temperature of complete crystallization at 620° C. He assumed loss of heat by conduction through vertical walls. He calculated that, if the batholith approximated the shape of a huge dike 60 miles wide, then it would require 70 million years to crystallize to its center. Thinner dikes would require less time to crystallize.

---

[13]For those interested in the details of heat conduction, see the thorough treatment by H. S. Carslaw and J. C. Jaeger, *Conduction of Heat in Solids*, 2nd ed. (Oxford: Clarendon, 1959). Their treatment includes some geological examples, including the age of Earth. Also important is J. C. Jaeger, "Cooling and Solidification of Igneous Rocks," in *Basalts: The Poldervaart Treatise on Rocks of Basaltic Composition*, ed. H. H. Hess and Arie Poldervaart (New York: John Wiley & Sons, 1968), 2:503-36.

[14]E. S. Larsen Jr., "Time Required for the Crystallization of the Great Batholith of Southern and Lower California," *American Journal of Science* 243A, Daly vol. (1945): 399-416.

More than likely, the top of the batholith would have been covered by a stack of sedimentary rocks through which heat would be conducted to the surface. Thus, Larsen also calculated the effect of intrusion into stacks of sedimentary rocks of varying thicknesses. He calculated, for example, that beneath a cover of three miles of sedimentary rock, the batholith would have cooled to 600° C to a depth of eight miles after seven million years; crystallized to a depth of five miles in only 2.9 million years; and crystallized to a depth of 2.3 miles below the top of the batholith in 700,000 years. All of these numbers are vastly greater than the approximated 10,000 years for the age of Earth entertained by young-Earth creationists. Given that the batholith was intruded into a cover of metamorphosed sedimentary rocks, there is no way that the crystallization of the batholith could be accounted for in terms of the action of a Flood of a single year's duration. To make matters worse for young-Earth creationist theories, recall that Larsen just calculated the times required to crystallize the batholith completely at 620° C. By the time the batholith had been exposed *at the surface*, it had to cool from 625° C to 25° C, and that additional cooling process would clearly require a substantial additional amount of time. The fact that the batholith was emplaced as a sequence of smaller intrusions rather than as a single large pulse also indicates passage of even more time.

As another example, in a classic study of the Stillwater Igneous Complex, a layered intrusion in the Beartooth Mountains of southern Montana that is 27,000 feet thick, Harry Hess (1906-1969) of Princeton University evaluated the thermal relations of the intrusion.[15] He estimated the initial magma temperature at 1225° C and the initial wall rock temperature at 25° C. From mineralogical evidence, Hess estimated that the temperature within the intrusion was around 1125° C at the time when crystallization began, and he calculated that crystallization would have been approximately 60 percent completed by the time the intrusion cooled to 1100° C. Hess further calculated that approximately 50,000 years were required to crystallize 60 percent of the complex, and that means that even more time was required to crystallize the entire complex. Even at the completion of crystallization the temperature would still have been extremely high. To cool the entire intrusion to room temperature would have required still more lengthy time intervals.

EARLY STUDIES INVOLVING ADVECTION

Young-Earth creationists Andrew Snelling and John Woodmorappe have suggested that the calculations of people like Larsen and Hess, based on heat loss

---

[15]Harry H. Hess, *Stillwater Igneous Complex, Montana: A Quantitative Mineralogical Study*, GSA Memoir 80 (New York: GSA, 1960). For the discussion on thermal relations of the intrusion, see pp. 144-46.

from magmas purely by conduction, are geologically unrealistic and yield unreasonably high time estimates. They claimed that magmas in the vicinity of hydrothermal systems should be able to cool in only a few thousand years. These fluids of such systems, commonly referred to as *hydrothermal fluids*, may consist of groundwater already present in the surrounding rocks, or the fluids may have been released by the magma as it cooled. These hot fluids circulate through the pores and fractures of wall rocks. Aqueous fluid circulating through permeable rock can transfer heat toward the surface much more rapidly than solid rock can conduct heat. The result is that an igneous intrusion surrounded by permeable wall rocks containing aqueous fluid is likely to cool much more rapidly than an intrusion that loses heat through the wall rocks by conduction alone.

The more complex models of heat transfer from cooling magmas in the last thirty years have incorporated the effects of circulating fluids. Earlier studies of this type entailed generalized mathematical models of hypothetical plutons of arbitrarily specified dimensions. Larry Cathles of Kennecott Copper Corporation and Cornell University and Denis Norton of the University of Arizona conducted some of the more important studies.[16] In essence, they calculated cooling times for hypothetical intrusions with rectangular cross-sections of specified dimensions that had been injected into water-saturated, fractured wall rocks of varying permeability. Neither calculated the time of crystallization, but they did calculate cooling times of very hot, already crystallized igneous rocks. They typically obtained values in excess of 100,000 years for cooling times to 200° or 300° C. The lower cooling times that they calculated provide little comfort to young-Earth proponents. These calculated times are still a lot greater than 10,000 years despite the fact that advection of fluid promotes more rapid cooling of igneous bodies. Moreover, the sizes of the hypothetical plutons considered by Cathles and Norton, although typical for many bodies, are far smaller than many granitic batholiths such as the one considered by Larsen. One also needs to include the time required for the earlier crystallization of the plutons from magma.

The question remains as to the time of cooling required for a very large batholith. As one example, Norton and Knight modeled a batholith-sized pluton with a width of about 33 miles. The height of the pluton was set at 2.5 miles and the top at 2.5 miles below the surface. The initial temperature assigned was 920° C. Their calculations showed that after 160,000 years, most of the batholith still remained above 600° C, and after 1.2 million years most of

---

[16]See Larry M. Cathles, "An Analysis of the Cooling of Intrusives by Ground-Water Convection which Includes Boiling," *Economic Geology* 72 (1977): 804-26; and Denis Norton and J. Knight, "Transport Phenomena in Hydrothermal Systems: Cooling Plutons," *American Journal of Science* 277 (1977): 937-81.

the entire batholith was still above 400° C. Not only do these large intrusions require long periods to cool off, but the hydrothermal systems of circulating hot water associated with them may also remain at relatively constant high temperatures for hundreds of thousands of years before returning to the ambient temperature determined by the ordinary geothermal gradient.

An analysis of heat loss in a real intrusion was undertaken later by Norton in conjunction with Hugh Taylor Jr., of California Institute of Technology and an expert in the geological applications of oxygen-isotope distribution.[17] Norton and Taylor studied the cooling history of the Skaergaard Intrusion, one of the world's most thoroughly studied bodies of igneous rock.[18] This intrusion, located along the southeastern coast of Greenland, is a funnel shaped mass, about 8,250 feet thick, that was intruded into gneiss capped by basalt lava flows. The interior of the funnel consists of a thick sequence of layered igneous rocks composed primarily of the common minerals olivine, pyroxene and plagioclase. Taylor used the distribution of oxygen isotopes in the rocks of the intrusion and of the wall rocks invaded by the intrusion to estimate the extent of isotopic exchange between the magma and the hydrothermal fluids circulating around the intrusion. Norton and Taylor then constructed several possible mathematical models for cooling of the intrusion based on varying values of permeability of the intrusion and of the surrounding basalt and gneiss. They concluded from computer calculations that the intrusion required approximately 130,000 years just to crystallize from an initial temperature of about 1150° C. Even after complete crystallization, a very long time was required to cool the hot rock to a temperature of 270° C, roughly that of the wall rocks at the depth of original emplacement. The total cooling times to 270° C varied from 350,000 years to 900,000 years depending on the values of wall-rock permeability that were chosen. The model that fit best with the oxygen isotope data yielded a cooling time of approximately 600,000 years to 270° C. Obviously such values are far greater than the 10,000 years that young-Earth creationists propose for the age of Earth. Moreover, the Skaergaard Intrusion is not an exceptionally large body of igneous rock.

RECENT STUDIES INVOLVING ADVECTION

One very recent thermal modeling effort incorporating advection was developed by Daniel O. Hayba and Steven E. Ingebritsen of the United States Geo-

---

[17]Denis Norton and Hugh P. Taylor Jr., "Quantitative Simulation of the Hydrothermal Systems of Crystallizing Magmas on the Basis of Transport Theory and Oxygen Isotope Data: An Analysis of the Skaergaard Intrusion," *Journal of Petrology* 20 (1979): 421-86.

[18]For a review of some of the ideas about the formation of the Skaergaard Intrusion, see Young, *Mind over Magma*, pp. 319-26.

logical Survey.[19] They refined earlier models to account for changes in fluid density and the effects of the coexistence of water and steam, thus making the heat transfer models geologically even more realistic. Fortunately, the much more complex mathematical expressions required to evaluate such situations can be handled with the aid of the current generation of extremely high-speed, high-power computers. In their study, Hayba and Ingebritsen simulated the cooling history of small hypothetical intrusions and plotted the rate of cooling for each situation. As an example, they simulated heat loss from a hypothetical pluton 1.2 miles wide and 2.4 miles high whose top was placed at a depth of three miles beneath the surface. They assumed "instantaneous" intrusion of this rather small pluton at 900° C. Calculations showed that the upper portions of the pluton were still in the 300° to 500° C range after 50,000 years.

In a more recent study, Cathles and colleagues calculated the thermal history of sills that are very rich in olivine and pyroxene.[20] They assumed a 1.2 mile-thick sill with a diameter of 24 miles and an initial temperature of 1650° C intruded at ten miles beneath the surface into host rocks with a fairly high permeability. Fluids hotter than 200° C reached the surface after about 50,000 years and were sustained for 260,000 years. The sill did not cool below 200° C until at least 300,000 years had elapsed. In a second case they assigned a lower permeability to the wall rocks. In this case, fluids took 300,000 years to reach the surface and remained active for about 830,000 years. The sill was still hotter than 200° C after 1.1 million years. Because they chose values of permeability to achieve maximum longevity of the hydrothermal system, they concluded that hydrothermal systems with much greater permeability would maintain 200° C surface temperatures for less than 100,000 years and that hydrothermal circulation would cool even a very large intrusion in a few tens of thousands of years at most, provided that the hydrothermal system was generated by a single-pulse intrusion.

Young-Earth creationists might find some solace in the results of calculations on hydrothermal systems associated with cooling igneous intrusions, but again there are caveats. Cathles and his colleagues pointed out that longer-lived hydrothermal systems may result from a group of intrusions injected in a sequence, especially if the wall rocks have low permeability. In addition, many intrusions, especially those injected at great depth in the crust have no associated hydrothermal systems because the wall rocks have very low permeability. Many deep-seated magmas also contain little dissolved fluid. Such intrusions

---

[19]D. O. Hayba and S. E. Ingebritsen, "Multiphase Groundwater Flow Near Cooling Plutons," *Journal of Geophysical Research* 102 (1997): 12235-52.
[20]L. M. Cathles, A. H. J. Erendi and T. Barrie, "How Long Can a Hydrothermal System Be Sustained by a Single Intrusive Event?" *Economic Geology* 92 (1997): 766-71.

are unlikely to have cooled significantly by advection.

Let's look at the results of some recent calculations for another real igneous intrusion, the Bushveld Complex near Johannesburg, South Africa, the world's largest known *layered* igneous intrusion.[21] The total volume of rock in this intrusion is an estimated 75,000 to 120,000 cubic miles. The lower part, sometimes designated as the Rustenberg Layered Suite, contains extensive layers of anorthosite, pyroxenite, dunite, gabbro and chromitite that vary in thickness. These layers are divisible into a lower zone, a critical zone, a main zone and an upper zone. These zones are distinguishable from one another of the basis of the mineralogical variations. The mineralogy and chemical composition of the rocks indicate that the pronounced layering resulted primarily from accumulation of crystals of olivine, pyroxene, and plagioclase on the floor of the magma chamber and that through time the floor gradually rose as crystals accumulated on it. The details of the geology also suggest that not all the magma has been preserved in the present intrusion but that some of it was probably erupted onto the ancient land surface and ultimately eroded away. Estimates of the total pre-eruption volume of Bushveld magma are on the order of 200,000 cubic miles.

Detailed study of chemical variations in the minerals strongly favors a reconstruction of the history of the Bushveld Complex involving several pulses of magma injection. Grant Cawthorn and Feodor Walraven concluded from a detailed analysis of the cooling history of the intrusion(s) by computer modeling that the injection of the successive pulses of magma occurred throughout a period of approximately 75,000 years.[22] In other words, it took approximately 75,000 years just to fill the magma chamber. Fresh pulses of hot magma whose original temperature was around 1300° C were injected into earlier sheets that had cooled to perhaps 1200° C. Cawthorn and Walraven calculated that the entire intrusion had cooled only to 900° C after 180,000 years. At that temperature, the magma would have crystallized almost entirely. But because the freshly crystallized igneous rocks were still extremely hot, much additional time elapsed as the intrusion cooled to the temperature of the surrounding wall rocks. Yet more time was needed to cool to the temperature at the surface where it ultimately became exposed.

A subsequent study of the Bushveld Complex examined the role of hydrothermal fluid flow in relation to cooling.[23] Robb and coworkers looked at the

[21]H. V. Eales and R. G. Cawthorn, "The Bushveld Complex," in *Layered Intrusions*, ed. Richard Grant Cawthorn (Amsterdam: Elsevier, 1996), pp. 181-229.

[22]R. Grant Cawthorn and Feodor Walraven, "Emplacement and Crystallization Time for the Bushveld Complex," *Journal of Petrology* 39 (1998): 1669-87.

[23]L. J. Robb, L. A. Freeman and R. A. Armstrong, "Nature and Longevity of Hydrothermal Fluid Flow and Mineralization in Granites of the Bushveld Complex, South Africa," *Transactions of*

question of how long a hydrothermal fluid system could be sustained in terms of heat loss. They simulated the cooling history of the Lebowa Granite Suite (LGS), the upper granitic portion of the Bushveld Complex. They treated it as a slab of granite 1.2 miles thick that was injected at a depth of 2.4 miles. Heat was lost by conduction alone. Their model calculated that, after 50,000 years, the intrusion would still have had a temperature above 500° C. After half a million years, the intrusion would have cooled to around 200° C, low enough that any hydrothermal activity would finally have ended. The authors suggested, however, that a more realistic calculation would be based on the assumption that a 4.3-mile-thick slab of hot mafic magma would be lying directly beneath the intruded LGS, and its top would be three miles deep. The presence of the hot mafic rock below the LGS would keep the temperature of the LGS above its final crystallization temperature much longer in this case. Under these circumstances, the temperature of the entire complex would have cooled to 200° C only after about 4 million years.

Conclusions from Thermal Modeling Studies

The most recent simulations indicate that cooling times of igneous rock bodies are certainly shorter when we take into account the role of fluid in comparison to cooling times of igneous bodies that cooled entirely by conduction. Such models will undoubtedly be further improved to simulate real geological situations more accurately. The transfer of heat from cooling magma is affected by the release of fluids directly from the cooling magma chamber; by the presence of dissolved constituents in such magmatic fluids and in circulating ground water; and by chemical reactions that take place between those dissolved constituents and the wall rocks through which they circulate. Thermal models of the future will take these factors into account.

Although our knowledge of the cooling of igneous intrusions will improve, evidence already at hand refutes the contention that Earth is only a few thousand years old. Without question, very small igneous intrusions could easily have crystallized and cooled to surface temperatures in a matter of a few years to hundreds of years. Even larger intrusions could have cooled in a few thousand years if hydrothermal fluids removed heat. Large intrusions like the Bushveld Complex and the Sierra Nevada batholith in California appear to have required hundreds of thousands to a few millions of years to crystallize and cool to surface temperatures, especially if they lacked enormous hydrothermal systems. The problem for Flood geologists becomes much more acute because, in many situations, a large volume of igneous rock actually consists of a series

the Royal Society of Edinburgh: Earth Science 91 (2000): 269-81.

of smaller intrusions, each one of which was intruded only after the preceding one had effectively crystallized. Moreover, cooling of magma intruded into very hot, recently crystallized rock takes far longer than cooling of magma intruded into cool country rocks. In other words, if the complete cooling of each of ten different intrusions required at least fifty thousand years, but each of the ten intrusions was intruded only after the complete cooling of all of the previous intrusions, the total time elapsed would be at least 500,000 years.

To make matters still more complicated, field mapping demonstrates that not all igneous intrusions in the world were formed at the same general time. Geologic evidence clearly indicates that intrusions of various sizes were injected into previously existing rocks at many different times throughout geologic history. For example, geologic field relations show that the Pikes Peak batholith in Colorado was intruded well before intrusion of the Sierra Nevada batholith.

If the evidence of cooling of igneous bodies is incompatible with an Earth only a few thousands of years old, that evidence is even less compatible with the theory of Flood geology. Many igneous bodies are situated in previously existing sedimentary rock. Flood geology proponents claim that these fossiliferous, sedimentary rocks are deposits of the Flood. Thus, if we find a body of igneous rock that clearly intrudes fossiliferous sedimentary rocks and is overlain by hundreds or thousands of feet of fossiliferous sedimentary rocks, it must be the case that the igneous body should have cooled off substantially in the course of less than one year (the Flood year) to support a very thick overlying mass of sedimentary rocks. The cooling times of large igneous intrusions, however, simply cannot be reduced to one year! Even if one claims that the cooling of the intrusions could have extended after the Flood year, there still is not enough time. Large igneous intrusions, such as the Sierra Nevada batholith, that were injected into sedimentary rocks have obviously been exposed at the surface and have been cold, hard rock for quite a long time. We are left, at most, with two or three thousand years to cool igneous rock bodies completely on this generous scenario. But big intrusions simply can't lose their heat that fast. Snelling and Woodmorappe have not made a convincing case that the rates of cooling of igneous intrusions are compatible with a young Earth. They are, in effect, asking a hot, roasted turkey to stay in the oven and cool to room temperature in a few seconds. It doesn't happen.

COOLING OF LAVA LAKES

If one refuses to accept the heat-loss calculations for intrusive rocks, we still have direct empirical evidence of cooling rates of magma bodies from *lava* lakes. On many occasions, craters on the flanks of Kilauea volcano on the island of Hawaii have filled with lava (fig. 11.4). The cooling rates of some of the resulting lava lakes were measured. Alae, Makaopuhi and Kilauea Iki craters were drilled, and thickness and temperature at various depths of the lakes were measured. Alae was a small crater along the East Rift Zone of Kilauea that erupted for three days in August 1963.[24] A lava lake formed in the crater that was approximately 45 feet deep. Drilling indicated that crystallization of the lake was complete within 13 months and that the maximum temperature within the lake was lower than 100° C by August 1967.

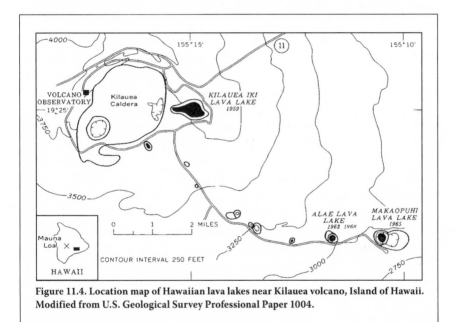

Figure 11.4. Location map of Hawaiian lava lakes near Kilauea volcano, Island of Hawaii. Modified from U.S. Geological Survey Professional Paper 1004.

The Alae crater was completely covered by lava erupted from nearby Mauna Ulu in February 1969.

Eruptions occurred at the much larger Makaopuhi crater between March 5 and 15, 1965, producing a lava lake in the western portion of the crater.[25] The lava lake had a depth of approximately 270 feet. A permanent crust formed on the lake on March 19. Lava from the nearby Mauna Ulu eruption of February 1969 covered the Makaopuhi lava lake, the upper crust of which had achieved

[24]Thomas L. Wright and Dallas L. Peck, *Crystallization and Differentiation of the Alae Magma, Alae Lava Lake, Hawaii,* Professional Paper 935-C (Washington, D.C.: USGS, 1978), pp. 1-20.

[25]Thomas L. Wright and R. T. Okamura, *Cooling and Crystallization of Tholeiitic Basalt, 1965 Makaopuhi Lava Lake, Hawaii,* Professional Paper 1004 (Washington, D.C.: USGS, 1977), pp. 1-78.

a thickness of about 58 feet. After a period of nearly four years, the lava lake was nowhere near completely solidified.

At Kilauea Iki, drilling and monitoring continued for more than twenty years after the original eruption.[26] Kilauea Iki, a relatively small crater about 3,600 feet by 160 feet, is located on the northeast side of the main summit caldera of Kilauea (fig. 11.4). In 1959, for a period of 38 days, Kilauea Iki erupted spectacularly. A vent on the southwest side of the crater fountained to heights of 1,900 feet, and lava filled the crater to a depth of 350 feet above the old crater floor. The fountaining produced a large cinder cone, Pu'u Puau, on the southwest corner of the crater, and large tracts of land were covered by cinders downwind from the cone. Current visitors to Hawaii Volcanoes National Park can walk the Devastation Trail through the cinder-covered area to the cinder cone and to an overlook into Kilauea Iki crater. For the more ambitious tourist, a four-mile-long trail leads along the north rim of the crater, down into and across the crater floor. Hiking this trail is well worth the effort, as it displays magnificent flows of *pahoehoe* lava and spectacularly fragmented, thick slabs of lava.[27] One gets a strong sense of the undulating nature of the surface of the lava lake by the manner in which it froze. Steam rising from the crater floor is still visible, and with patience one can still locate the drill-hole sites. After more than twenty years, the entire lake had not yet solidified completely. Down-hole temperatures were still extremely high. The lava lake, 350 feet thick, had not solidified and was nowhere near room temperature.

The Kilauea Iki lava lake was erupted onto Earth's surface. Had this body of magma been emplaced well below the surface of Earth, it would have taken much longer to solidify because heat would have escaped more slowly from its walls, roof and floor. Bodies of similar mineral and chemical composition like the Skaergaard Intrusion in east Greenland that is at least 8,000 feet thick or the Bushveld Complex in South Africa that is locally as much as five miles thick are not only considerably thicker but also vastly greater in total volume. We have every reason to expect that they would have taken far longer to crystallize than Kilauea Iki. A magma body the size of Kilauea Iki lava lake, had it been intruded underground, would not have solidified during the year of the Flood.

---

[26]Rosalind T. Helz, "Crystallization History of Kilauea Iki Lava Lake as Seen in Drill Core," *Bulletin Volcanologique* 43-44 (1980): 675-701.

[27]*Pahoehoe* is a term applied to a very widespread form of lava flow found on the Hawaiian Islands and in many other volcanic provinces around the world. Pahoehoe is characterized by a smooth, glassy surface that commonly includes parallel arc-shaped rope-like features on the surface that are produced by the flow of the lava.

EVIDENCE FROM METAMORPHIC ROCKS

The processes of heating and cooling are also involved in the formation of metamorphic rocks. In general, metamorphic rocks form by the application of extreme heat and/or pressure to previously existing rock. In many cases, the rock in question originates at Earth's surface as layers of sedimentary rock or lava flows that are subsequently buried deeply become warmer and are subjected to pressure exerted by the mass of overlying rock. In other cases, rocks undergoing metamorphism form underground and are never at the surface until long after their metamorphism. For example, a body of granitic magma may crystallize and cool while underground and subsequently be buried more deeply where it is reheated.

In *Creation and the Flood,* Young drew attention to the fact that some metamorphic rocks contain deformed fossils of brachiopods or other organisms.[28] On the grounds that death entered the world only after the entrance of human sin, young-Earth proponents generally maintain that fossils represent remains of organisms that died after the rebellion of Adam and Eve in the Garden of Eden. As a result, fossil-bearing sedimentary rocks had to form on Earth's surface sometime after the Fall, most likely during the Flood. They then had to be buried to great depths to become metamorphosed. The mineral content of the fossiliferous metamorphic rocks provides indications of the temperatures and pressures experienced by such rocks, because the temperature and pressure stability ranges of a wide variety of rock-forming minerals have been experimentally determined. Depths of burial can be reasonably estimated from these pressure-temperature determinations. The mineralogy of such brachiopod-bearing metamorphic rocks from New Hampshire suggests that they were buried to depths on the order of twelve miles and heated to around 600° C.[29] Thus, on the young-Earth view, the rocks had to form at the surface, be buried to a depth of twelve miles, be heated and recrystallized, and then uplifted and exhumed until they arrived at the surface, in the space of two or three thousand years at most. If the entire process of burial, metamorphism and uplift was accomplished during Noah's Flood, it all had to happen in a single year.

Since publication of *Creation and the Flood* in 1977, an entirely new field, designated as ultra-high pressure (UHP) metamorphism, has developed thanks

---

[28]Davis A. Young, *Creation and the Flood: An Alternative to Flood Geology and Theistic Evolution* (Grand Rapids: Baker, 1977), pp. 193-97.

[29]See A. J. Boucot, G. J. F. MacDonald, C. Milton and J. B. Thompson Jr., "Metamorphosed Middle Paleozoic Fossils from Central Massachusetts, Eastern Vermont, and Western New Hampshire," *Bulletin GSA* 69 (1958): 855-70; and A. J. Boucot and J. B. Thompson, "Metamorphosed Silurain Brachiopods from New Hampshire," *Bulletin GSA* 74 (1963): 1313-34.

to the discovery of numerous terranes consisting of large tracts of rocks that contain such rare minerals as coesite, majorite and diamond.[30] These minerals have typically been found as tiny grains included within the major minerals such as omphacite (a very high pressure pyroxene), epidote, garnet, zircon and kyanite (the high pressure form of aluminum silicate). Coesite, majorite and diamond all form at extremely high pressures (see below). *Coesite* is a very high pressure form (polymorph) of SiO_2, the chemical composition of the extremely common mineral quartz. Prior to its discovery in these UHP rocks, coesite was recognized only from rocks at known meteorite impact sites such as Meteor Crater, Arizona, where the extremely high pressure that converted quartz to coesite was generated by the nearly instantaneous shock of meteorite impact.

Majorite is a very high-pressure variety of garnet, a magnesium-, iron- and aluminum-rich silicate mineral. Because majorite forms in very high-pressure laboratory experiments, it is probably an important constituent in the upper mantle. Prior to its discovery in some UHP rocks, majorite, too, was known only from meteorite impact sites such as Coorara Crater in Western Australia.

Lastly, *diamond* is the high-pressure form of carbon which, in near-surface rocks in the Earth's crust, occurs as the mineral graphite, the material used in so-called lead pencils. Diamond does occur as small crystals in relatively uncommon although widespread, pipe-shaped igneous rock bodies of small volume called *kimberlite*.[31] Kimberlite represents the crystallization product of extremely volatile-rich magma that originated in the upper mantle and was rapidly and explosively impelled toward the surface. Diamonds have been recognized in UHP metamorphic rocks only within the last couple of decades.

Because the minimum pressures at which coesite, majorite and diamond become stable are well established from laboratory studies, it has been possible to estimate, in conjunction with information about the stabilities of the major minerals in UHP rocks, the pressures and temperatures at which the UHP rocks were formed. As examples, UHP metamorphic rocks from east Greenland are estimated to have formed at a pressure around 36 kilobars (ap-

---

[30]On UHP metamorphic rocks, see G. T. Roselle and M. Engi, "Ultra High Pressure (UHP) Terrains: Lessons from Thermal Modeling," *American Journal of Science* 302 (2003): 410-41; and D. Rumble, J. G. Liou and B. M. Jahn, "Continental Crust Subduction and Ultra-high Pressure Metamorphism," in *Treatise on Geochemistry*, vol. 3: *The Crust*, ed. Roberta Rudnick (New York: Elsevier, 2003), pp. 293-319.

[31]Kimberlite was named after the town of Kimberley, South Africa, site of the world's largest known diamond-bearing kimberlite pipe. The diamonds were mined in an open pit. In the United States, the largest diamond-bearing kimberlite is at Murfreesboro, Arkansas, where visitors may search for diamonds for a fee. In early 2006, one family discovered a large pale yellow 4-carat diamond of gem quality at Murfreesboro.

proximately 36,000 times the pressure exerted by Earth's atmosphere) and a temperature around 970° C; from the Erzgebirge in Saxony, Germany, at a pressure greater than 40 kilobars and a temperature greater than 900° C; from the Dabie-Su-Lu region of eastern China, pressures ranging from 27 to 50 kilobars and temperatures between 600° and 900° C; and from Rhodope, Greece, to pressures between 31 and 39 kilobars and temperatures between 600° and 900° C. One estimate called for a pressure as high as 60 kilobars.[32]

If we know the pressure at which the minerals in a rock formed, we can estimate the depth at which the rocks crystallized. Roughly speaking, given the approximate densities of many of the rocks in which the UHP minerals occur and of rocks that overlie them, a pressure of one kilobar is generated by an overlying column of such rock that is a minimum of 1.8 miles thick. A pressure of 40 kilobars would be exerted by a column of rock at least 70 miles thick. In the extreme case, a pressure of 60 kilobars would be exerted by a column of rock a minimum of 110 miles thick.

We have, then, a number of situations in which we encounter rocks, now exposed at the surface, that crystallized at depths as much as 70 miles, or even more, below the surface! On the young-Earth hypothesis, therefore, we must find a way to get these rocks to the surface from around 70 miles deep, and we must do it in just a very few thousand years and probably much less, perhaps even one year, if it all happened during the Flood.

But the situation is far more problematic for the young-Earth hypothesis. The UHP rocks are found in layered rock sequences that consist predominantly of quartzite and marble, rocks that are formed only by metamorphism of quartz-rich sandstone and limestone, interlayered with mica schist and gneiss, probably formed by metamorphism of shale and graywacke, another type of sandstone. These metamorphic rocks are formed from ordinary sedimentary rocks that originated at Earth's surface. The layers of eclogite, a rock composed mostly of pyroxene and garnet and the most likely to contain the UHP minerals, might represent metamorphosed basalt lava flows or sills. In addition, studies of both hydrogen and oxygen isotopes in a wide range of UHP rocks indicate that these rocks were altered by waters precipitated in cold climates, another indication that the rocks originated at Earth's surface.

The problem for the young-Earth hypothesis becomes very acute, because its proponents must now explain how ordinary sedimentary rocks deposited at Earth's surface were buried to depths of 70 miles or more, where they were thoroughly recrystallized, and then uplifted and exhumed back to the surface in an extremely short time! There is little doubt that these rocks have been

---

[32]A pressure of one kilobar equals one thousand bars. One bar is approximately equal to the pressure exerted by the Earth's atmosphere at the surface.

at the surface for the past three or four thousand years because the world's geography was undoubtedly much the same in 2000 B.C. as it is now. So, if the world is only 6,000 years old and has been stable for the past 4,000 years, we have to accomplish this feat of burial, recrystallization and uplift of solid rocks in 2,000 years. And if the sedimentation, burial, recrystallization and uplift occurred because of the Flood, we have essentially one year to accomplish the feat. One must keep in mind, too, that the uplift of a large volume of rock from a depth of 70 miles entails the complete removal by erosion of the overlying 70-mile-thick mass of rock, and that thickness must then be distributed and deposited on Earth's surface. Given occurrences of UHP in Greece, Germany, France and China, places that have been occupied by humans for a long time, one would suspect that someone should have noticed the enormous rate of uplift and erosion and left a record of the event.

Modern geology accounts for UHP terranes by burial of the sedimentary rocks along subduction zones that occur at the margins of continental land-masses and subsequent uplift in active mountain-building zones associated with subduction zones. Although the mechanism of the processes is not fully understood, time offers no difficulty for the standard geologic explanation. Given our knowledge of rates of subduction and mountain building today, there is no compelling reason to doubt that rocks can be dragged 70 miles downward and uplifted to the surface over the course of millions of years. And, of course, we have been showing all along that the Earth is billions of years old. Flood geology, on the other hand, cannot account for occurrences of the remarkable UHP metamorphic rock associations.

12

Time and the Stratigraphy of the Michigan Basin

A Case Study

Now that we have discussed some of the major lines of geologic evidence that point to the vast antiquity of Earth, let's investigate a few details of the geology of specific regions. The evidence that emerges in close examination of geological details leads to the irrefutable conclusion that the planet is far greater than a few thousand years old. If we carefully consider the astounding geologic complexity of virtually any region on Earth we are compelled to concede that such regions have experienced histories that far exceed a few thousands of years.

Young-Earth creationists are very fond of discussing the geology of the Grand Canyon in northern Arizona. They claim that both the thick succession of sedimentary rock layers exposed in the walls of the canyon and also the carving of the canyon itself are most satisfactorily explained in terms of catastrophic processes, the centerpiece of which allegedly was Noah's Flood. Young-Earth creationist groups have conducted field trips into the canyon for Christian tourists to inform them of the supposedly superior Flood theory of canyon geology. In addition, several books provide a catastrophic interpretation of the geology of the Grand Canyon.[1]

Although field trips and attractive books with color photographs of the canyon will undoubtedly convince some Bible-believers who know little about geology that the Deluge was responsible for most of the world's sedimentary rocks, we insist, along with virtually all other professional Christian geologists, that the geology of the Grand Canyon can in no way be accounted for legitimately in terms of a global Flood. One of us has previously written briefly about Grand Canyon geology to show the inadequacy of the Flood geology hypothesis.[2] More important, two of the many professional geologists who have spent a lifetime investigating the geology of the Grand Canyon and its environs, Stanley Beus and Michael Morales of the University of Northern Arizona, have recently edited a second edition of their excellent book, *Grand Canyon Geology*.[3] Readers who wish to learn more about legitimate Grand Canyon geology are encouraged to consult that book.

The canyon region consists essentially of a stack of horizontal layers of sedimentary rocks that are piled on top of each other (fig. 2.2, 8.1). Although the geology of the Grand Canyon region is considerably more complex than it looks at first glance, it is *relatively* simple in comparison with the geology of most mountainous regions on Earth. Perhaps that relative simplicity accounts for its appeal to lay people. Because the geology of the Grand Canyon has often been a focal point in the debate between young-Earth and old-Earth proponents, we have decided to introduce here detailed case studies of the geology of two different regions that have received scant attention in the literature pertaining to Earth's antiquity. One case study deals with layered sedimentary rocks in a context that is not quite so familiar to our readers, the Michigan Basin. Like that of the Grand Canyon, the geology of the Michigan Basin is *relatively* simple. A major difference is that much of our knowledge of Michi-

---

[1]Among books espousing a catastrophic interpretation of the Grand Canyon are Steven A. Austin, ed., *Grand Canyon: Monument to Catastrophe* (Santee, Calif.: Institute for Creation Research, 1994); Larry Vardiman, *Over the Edge: Thrilling, Real-Life Adventures to Grand Canyon* (Green Forest, Ark.: Master Books, 1999); and Tom Vail, ed., *Grand Canyon: A Different View* (Green Forest, Ark.: Master Books, 2003). This last book, a coffee-table book featuring a Colorado River guide who converted to Christianity, caused quite a stir both in the scientific community and in the National Park Service over its placement in the bookstore in Grand Canyon National Park!

[2]Davis A. Young, "The Discovery of Terrestrial History," in Howard J. Van Till, Robert E. Snow, John H. Stek and Davis A. Young, *Portraits of Creation: Biblical and Scientific Perspectives on the World's Formation* (Grand Rapids: Eerdmans, 1990), and "Making Mysteries Out of Missing Rock," in Howard J. Van Till, Davis A. Young and Clarence Menninga, *Science Held Hostage: What's Wrong with Creation Science and Evolutionism* (Downers Grove: InterVarsity Press, 1988), pp. 93-124.

[3]Stanley S. Beus and Michael Morales, eds., *Grand Canyon Geology*, 2nd ed. (Oxford: Oxford University Press, 2002).

gan Basin geology has been derived from subsurface studies. Our second case study is that of a mountain range with extremely complex geology, the Sierra Nevada of California.

INTRODUCTION TO THE MICHIGAN BASIN

In this chapter we develop a regional case history of the Michigan Basin.[4] Geographically, this means that we examine the Lower Peninsula and the eastern Upper Peninsula of Michigan, plus portions of neighboring Wisconsin, Illinois, Indiana, Ohio and Ontario. We chose this particular example because we have each studied and/or taught geology in this region for more than twenty years and have had the pleasure of introducing many students of all ages to its fascinating geological history.

The Phanerozoic strata of the Michigan Basin, that is, those that lie above the Precambrian rocks, provide a handy set of sedimentary formations illustrating the detection of elapsed time in the rock record. We do not attempt to discuss the whole of these strata but dwell on particular formations as vignettes of ancient depositional environments. A few of these stratigraphic units have already been discussed in chapters eight and ten. On occasion we refer back to these sections.

In their overall aspect, these Phanerozoic strata of Michigan, like those of the upper Midwest in general, are relatively undisturbed. They have not been severely deformed by major tectonic events. Thus, the interpreter does not have to worry about geometrical techniques necessary to "remove" any large overprint of such deformation in historical reconstructions. Although Michigan strata are relatively undisturbed, they have been subjected to mild warping and episodic exposure and erosion.

The Phanerozoic strata of Michigan, as well as bordering areas of neighboring states, uniformly dip toward the center of the Lower Peninsula. This large-scale structure is termed the "Michigan Basin." The basin formed in response

---

[4]For more detailed expositions of Michigan Basin geology, refer to J. A. Dorr Jr. and D. F. Eschman, *Geology of Michigan* (Ann Arbor: University of Michigan Press, 1970); H. B. Willman et al., eds., *Handbook of Illinois Stratigraphy*, Bulletin 95 (Urbana: Illinois State Geological Survey, 1975); R. M. Feldman, A. H. Coogan and R. A. Heimlich, *Field Guide: Southern Great Lakes* (Dubuque: Kendall/Hunt, 1977); R. T. Lilienthall, *Stratigraphic Cross-Sections of the Michigan Basin*, Report of Investigation 19 (Lansing: Geological Survey of Michigan, 1978); J. H. Fisher, M. W. Barratt, J. B. Droste and R. H. Shaver, "The Michigan Basin," in *Sedimentary Cover—North American Craton, U.S., The Geology of North America*, vol. D-2, ed. L. L. Sloss (Boulder, Colo.: GSA, 1988), pp. 361-83; P. D. Howell and B. A. van der Pluijm, "Early History of the Michigan Basin: Subsidence and Appalachian Tectonics," *Geology* 18 (1990): 1195-98; P. A. Catacosinos and P. A. Daniels, eds., *Early Sedimentary History of the Michigan Basin*, Special Paper 256 (Boulder, Colo.: GSA, 1991); G. L. LaBerge, *Geology of the Lake Superior Region* (Phoenix: Geoscience Press, 1994).

to a crustal instability in the Precambrian basement rocks that episodically causes the center of the basin to sag. In many cases the strata significantly thicken toward the center of the Lower Peninsula. The degree of thickening toward the central axis of the basin is one indicator of the timing of events of downward sag throughout the basin history.

Although the overall vertical thickness of the Michigan Basin strata in many places exceeds two miles, it achieves thicknesses in excess of 17,000 feet near the axis of the basin. Diagrams are drawn deliberately to exaggerate the vertical dimension of this basinal structure. Although the central portion of the basin has sagged more than two miles in relation to the outer portion, the basin is hundreds of miles wide. Thus, the strata form a series that resembles a set of stacked tea saucers, each of which thickens toward the middle. Erosion has planed this nest of saucers such that if one could gaze from space on this region with the uppermost 300 feet of loose soil and glacial till removed, one would see a bull's-eye pattern of bedrock layers (fig. 12.1).

General Regional Context

All Michigan Basin strata are epicontinental, that is, they were deposited from the sea over continental crust rather than sea-floor crust. The center of the ancient North American continent was lower in elevation than it is today. At intervals, sea level was also higher than it is today, especially during times when there were no polar ice caps. The combination of low elevation plus higher sea level favored invasion of a shallow sea over the interior of the continent. Such incursions of the sea are termed *transgressions* by geologists. At intervals, water-lain sediments were deposited over the middle of the North American continental landmass during a time of higher relative sea level, but water depths would never have been very deep. During Phanerozoic time, there was no nearby tectonic activity, other than the downward flexure of the Michigan Basin structure, or volcanism that affected the nature of sediments entering the Michigan Basin. These environmental parameters set limits on the types of sedimentary rocks that could form in the region. Sedimentary layers, with minor exceptions, consist of pale, quartz-rich *sandstones;* finely laminated gray *shales* composed of clay particles; pure calcium carbonate *limestones* with marine fossils; chemically altered limestones termed *dolostones;* relatively pure layers of *evaporites* such as halite or gypsum; and thin layers of *coal* consisting of concentrated, carbonized plant remains.

Michigan strata are parts of larger, continuous blankets of sediment deposited across the entire midwestern region. These can be read backwards in time as one progresses downward in the stack of layers, much as one could interpret a multiple layer cake by removing one layer at a time. In fact, geologists often refer to the stratigraphy as "layer-cake stratigraphy."

Michigan rocks can be observed and sampled directly at localities where rocks are exposed at the surface. These exposures are termed *outcrops.* The eastern end of Michigan's Upper Peninsula is only poorly covered with soil and thus provides many good exposures of some of the lower strata in the Michigan basin—Cambrian, Ordovician and Silurian rocks (fig. 12.1). Good

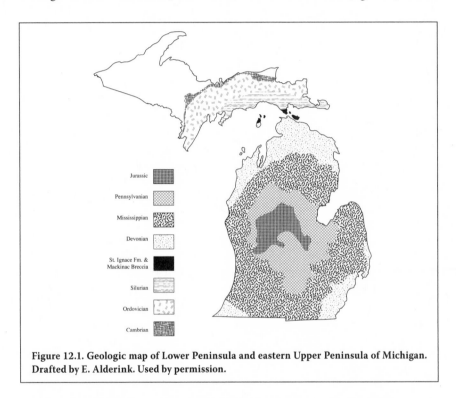

Figure 12.1. Geologic map of Lower Peninsula and eastern Upper Peninsula of Michigan. Drafted by E. Alderink. Used by permission.

outcrops of Devonian rocks are located around the northern perimeter of the Lower Peninsula. Rivers in areas inland of the Great Lakes coast have cut through and exposed layers of Mississippian age. Exposures of Pennsylvanian rocks can be located in the central portion of the Lower Peninsula, especially near the town of Grand Ledge, which is named after the prominent sandstone ledges adjacent to the Grand River.

Much of Michigan's Lower Peninsula is covered with a soil that is between 100 and 300 feet deep, obscuring the bedrock below. Another means of accessing the rock directly is through mining and/or quarrying. Large amounts of limestone are quarried from near-surface exposures near the northern edge of the Lower Peninsula and also near Detroit. These rocks are extracted for the manufacture of cement. Salt is mined in large quantity from deep mines near Detroit. Mississippian rocks are mined for gypsum along the central-western edge and the east-central edge of the Lower Peninsula. During the first half of the twentieth century, coal was mined in the center portion of the state.

Michigan's oil and gas resources have stimulated the charting of its subsurface through drilling. In fact, the oldest oil well in North America was

Streamlined Composite Michigan Stratigraphic Column
wavy lines indicate major uncomformities

Era	Period	Groups	Representative Formations
Cenozoic	Pleistocene		Pleistocene glacial tills and pond deposits
Mesozoic	Jurassic		Ionia Formation
	Pennsylvanian		Saginaw & Grand River Formations
	Mississippian		Michigan Formation
			Marshall Formation
			Antrim Shale
		Traverse Group	Alpena Limestone plus others
	Devonian		Dundee Limestone
		Detroit River Group	
Paleozoic			Mackinac Breccia
		Bass Islands Group	St. Ignace Formation
		Salina Group	Pt. Aux Chanes Shale (north) various evaporite formations (south)
	Silurian	Niagara Group	Lockport Formation plus others
		Cataract Group	
		Richmond Group	
	Ordovician		Trenton plus Black River Formations
			St. Peter Formation
		Prarie Du Chien Group	
	Cambrian		Munising Formation & Mt. Simon Sandstone
Pre-Cambrian			

Figure 12.2. Streamlined composite Michigan stratigraphic column. R. Stearley and E. Alderink.

drilled into the Michigan Basin structure at Sarnia, Ontario. As of this writing, more than 53,000 oil and gas wells have been drilled, with an average depth of 4,000 to 5,000 feet. Approximately 400 wells have been drilled to a depth of 10,000 feet or more; the deepest well drilled is more than 17,000 feet deep.[5] Of the millions of feet that have been drilled in the Michigan Basin, more than 300,000 linear feet of drill core have been preserved. For a great number of wells in which cores were not obtained, we have records of the rock through well cuttings and/or geophysical wireline logs. This means that the three-dimensional structure of the subsurface and the character of the strata are extremely well known.

We now turn to an examination of the stratigraphic record of the Michigan Basin and its inferred history. This is presented from the bottom up; we begin with the lowest Phanerozoic layers and then work upward toward the surface. A composite section of the stratigraphy is presented in figure 12.2.

Cambrian Sandstones

The lowest Phanerozoic rocks of the Michigan Basin are sandstones ascribed to the Cambrian Period. These sandstones lie under nearly two miles of other sedimentary strata in the center of the Michigan Basin but slowly flex toward the surface on its flanks. The Mount Simon Sandstone, the lowest of these units, thickens toward the west-central portion of the Michigan Basin, until it reaches a maximum thickness of more than 1,400 feet, implying sagging in the Michigan Basin during the interval in which it was deposited. The Mount Simon Formation and overlying Upper Cambrian sandstones are exposed at the surface in southern Wisconsin, where they form scenic buttes and escarpments, like those exposed at The Dells of the Wisconsin River.[6]

In the Upper Peninsula of Michigan, Cambrian sandstones correlated to the units overlying the Mount Simon are known as the Munising group strata (named for the town of Munising, Michigan, the so-called *type locality*). Along the Lake Superior shoreline, these sandstone layers form picturesque cliffs that are preserved at Pictured Rocks National Lakeshore. Inland, they form prominent ridgelines or cuestas that are scoured by streams into a series of waterfalls (fig. 12.3). Upper Tahquamenon Falls is one of North America's largest after Niagara Falls.[7]

---

[5]Data on Michigan Basin oil and gas production were obtained from William Harrison, director of the core repository at Western Michigan University. Data can be accessed at www.wmich .edu/geology/corelab/Michigan_oil_gas_facts.htm.

[6]P. A. Catacosinos and P. A. Daniels, "Stratigraphy of Middle Proterozoic to Middle Ordovician Formations of the Michigan Basin," in Catacosinos and Daniels, *Early Sedimentary History of the Michigan Basin*, pp. 89-100; LaBerge, *Geology of the Lake Superior Region.*

[7]Dorr and Eschman, *Geology of Michigan.*

Figure 12.3. Upper Tahquamenon Falls, eastern Upper Peninsula, Michigan. The rock exposed is Upper Cambrian sandstone. Photo by R. F. Stearley.

The Mount Simon Formation and the Munising Group of formations are pale, quartz-rich sandstones. Although fossils are generally lacking in these sandstones, rare trilobites demonstrate that at least some of the units were deposited by marine waters. Some of the sandstone units present elegant sedimentary features such as low-angle tabular cross-bedding and ripple marks. Altogether, these features indicate that a shallow sea encroached over the North American continent during Upper Cambrian time.

Although these Upper Cambrian sandstones contain some marine fossils such as trilobites, they are mostly devoid of fossils. There are no corals, for example. Nor are there any remains of vascular plants or vertebrates whatsoever, including fishes. This situation, of course, accords with the global biostratigraphic pattern discerned by Smith, Cuvier, Brongniart, Sedgwick and others at the beginning of the nineteenth century. The biotas of successive layers of Michigan Basin strata conform to the general order of the global pattern discovered nearly two hundred years ago, as we shall see.

Ordovician Sandstones, Limestones and Dolostones

In the Lower Peninsula of Michigan and neighboring portions of Illinois, Wisconsin and Indiana, the Cambrian sandstones grade upward into earliest Ordovician sandy dolostones with marine fossils. A large, regional unconformity separates these lowermost Ordovician strata from overlying Ordovician formations. This regional unconformity can be correlated to a continent-scale unconformity, marking the retreat of the shallow Upper Cambrian sea and the readvance of marine waters over central North America in Middle Ordovician time.[8]

The lowest unit of the Middle Ordovician strata is a widespread sandy rock that is well known throughout the Midwest, the St. Peter Formation (or St. Peter Sandstone), named after the type locality in central Minnesota (fig. 8.9). Residents of northern Illinois know this striking formation from its exposures at Starved Rock State Park. The sandstone is pale with cross-bedding. In popular literature, the St. Peter is commonly characterized as pure quartz sandstone that originated in a beach setting. Detailed petrography reveals many gradations within the St. Peter Formation. In parts of the formation, many of the quartz sand grains are "frosted," indicating a phase of (a)eolian activity. In other portions of the St. Peter, layers contain sand that has a significant percentage of abraded feldspar in addition to the quartz. Some layers are cross-bedded. Minor beds contain skeletons of marine ostracods, brachiopods and trilobites. Other minor beds are composed of gypsum. In its totality, however, the St. Peter is reasonably interpreted as having formed in a migrating arid coastline environment, with shallow marine shoreface sands overlying beach sands with dunes and evaporitic ponds.[9] Although the St. Peter is typically a few hundred feet thick over the Midwest, in the east-central Michigan Basin it achieves a thickness of 1,200 feet.[10]

Flood geologists have claimed that the St. Peter Sandstone is a result of a catastrophic slurry process.[11] They repeatedly point to the widespread geography of this layer as supposed evidence of its catastrophic formation.[12] How-

---

[8]L. L. Sloss, "Sequences in the Cratonic Interior of North America," *Bulletin GSA* 74 (1963): 93-114; H. B. Willman and T. C. Buschbach, "Ordovician System," in Willman et al., *Handbook of Illinois Stratigraphy,* pp. 47-87; L. L. Sloss, "Tectonic Evolution of the Craton in Phanerozoic Time," in *Sedimentary Cover—North American Craton, U.S., The Geology of North America,* vol. D-2, ed. L. L. Sloss (Boulder, Colo.: GSA, 1988), pp. 25-51; W. J. Frazier and D. R. Schwimmer, *Regional Stratigraphy of North America* (New York: Plenum Press, 1987).

[9]Willman and Buschbach, "Ordovician System"; Catacosinos and Daniels, "Stratigraphy of Middle Proterozoic to Middle Ordovician Formations of the Michigan Basin"; D. A. Barnes, C. E. Lundgren and M. W. Longman, "Sedimentology and Diagenesis of the St. Peter Sandstone, Central Michigan Basin, United States," *AAPG Bulletin* 76 (1992): 1507-32; G. LaBerge, *Geology of the Lake Superior Region.*

[10]We follow the suggestion of Barnes, Lundgren and Longman by including the Bruggers sandstone with the St. Peter Formation.

[11]S. A. Austin (as S. E. Nevins), "Stratigraphic Evidence of the Flood," in *Symposium on Creation III,* ed. D. C. Patten (Grand Rapids: Baker, 1971), pp. 33-65; H. M. Morris, ed. *Scientific Creationism,* 2nd ed. (El Cajon, Calif.: Creation-Life Publishers, 1982).

[12]See p. 227 in chap. 8 for Brown's quote.

ever, it is well understood that the St. Peter Formation is time-transgressive, that is, it was deposited over a sloping, irregular surface and represents a body of sediment that was produced by a migrating zone of deposition as the sea encroached on the land. As Walt Brown notes, the sand is "similar to sand on a white beach." Size-sorting of the sand, internal structures, fossil content and evaporite minerals all provide further evidence that the sand was *indeed* deposited on a beach; the beach moved laterally over the continental surface. In contrast to Brown's assertion, it is doubtful that a global catastrophic Flood would produce such a pure and consistent sand deposit with shallow current structures.

Ordovician strata above the St. Peter Formation are mostly limestones or limey shales and are highly fossiliferous.[13] The Upper Ordovician shaly strata, correlated to the Cincinnatian Series rocks of the Indiana-Ohio area, are far less resistant than other Upper Ordovician and lowest Silurian strata and form prominent lowlands (e.g., Green Bay, Wisconsin).[14] A prominent erosional unconformity tops the Ordovician rocks of the Michigan Basin.

Silurian Coral Reefs and Evaporites

Lower Silurian rocks of the Michigan Basin are mostly limestones and dolostones. These carbonate units host many hundreds of isolated carbonate skeletal framework structures: bioherms and reefs. The reefs have been subject to numerous studies—both for their intrinsic interest and as structures hosting oil and gas.

The Thornton reef and associated Silurian reefs of the Racine Formation were described in chapter ten. Silurian carbonate strata in the area average slightly more than 300 feet thick; most of this thickness is contributed by the Racine Dolomite. Many individual reefs are scattered throughout the Racine Formation in the northeastern Illinois/southeastern Wisconsin area. Reefs of the Moccasin Springs Formation, a southern and central Illinois equivalent to the Racine Formation but not dolomitized, extend toward the southwest and into Missouri, marking the position of an ancient Silurian archipelago. Silurian rocks thicken in central Illinois to 600 feet or more.[15]

---

[13]R. C. Hussey, "The Richmond Formation of Michigan," *Contributions from the Museum of Geology* 2, no. 8 (1926): 113-87; R. C. Hussey, *The Middle and Upper Ordovician Rocks of Michigan*, Publication 46, Geological Series 39 (Lansing: Michigan Geological Survey, 1952); R. V. Kesling, *Revision of Upper Ordovician and Silurian Rocks of the Northern Peninsula of Michigan*, Papers on Paleontology 9 (Ann Arbor: University of Michigan Museum of Paleontology, 1975); Catacosinos and Daniels, "Stratigraphy of Middle Proterozoic to Middle Ordovician Formations of the Michigan Basin."

[14]LaBerge, *Geology of the Lake Superior Region.*

[15]Willman et al., *Handbook of Illinois Stratigraphy;* C. J. Schuberth, *A View of the Past: An Intro-*

The Silurian reef trend follows the contours of the Michigan Basin architecture from Chicago to the east, sweeping in a long arc through northern Indiana and southernmost Michigan, wrapping to the north up the Bruce Peninsula of Ontario, then along an east-west trend through the southern margin of the upper peninsula, and turning south, forming the backbone of the Door Peninsula of Wisconsin (fig. 12.4). The strata were given different regional names by various workers but are subsumed under the Niagara Group. The Niagara Group is named for Niagara Falls and Niagara Gorge, where Silurian rocks are prominent. Millions of visitors to Niagara Falls have observed the tough Lockport Dolostone forming the vertical face at the lip of the Falls. Stout Niagaran dolostones like the Lockport undergird a massive ridge of rock, the Niagaran Escarpment, familiar to residents of southern Ontario. The Niagaran Escarpment resisted erosion by the Pleistocene glaciers and stands a hundred or more feet higher than its surroundings. The Escarpment is visible from space satellites, because it forms the Bruce Peninsula and Manitoulin Island of Ontario, the southern lip of Michigan's eastern Upper Peninsula as well as the Garden Peninsula of Lake Michigan, and the Door Peninsula of Wisconsin. To

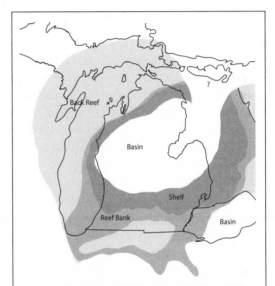

Figure 12.4. Map of Mid-to-Late Silurian shallow marine environments in the Michigan Basin region. Boundaries outside Michigan are left indeterminate. Redrawn from G. M. Friedman and D. C. Kopaska-Merkel, "Late Silurian Pinnacle Reefs of the Michigan Basin," in *Early Sedimentary History of the Michigan Basin*, ed. P. A. Catacosinos and P. A. Daniels, Special Paper 256 (Boulder, Colo: GSA, 1991).

a large degree, the modern topographic relief of the central Great Lakes region is controlled by the circular architecture and character of the strata of the Michigan Basin![16]

---

duction to Illinois Geology (Springfield: Illinois State Museum, 1986).

[16] A. W. Grabau, *A Guide to the Geology and Paleontology of Niagara Falls and Vicinity*, Bulletin 45, vol. 9 (Albany: New York State Museum, 1901); G. D. Ells, *Michigan's Silurian Oil and Gas Pools*, Report of Investigations 2 (Lansing: Michigan Geological Survey, 1967); Dorr and

The circular pattern of reef distribution is thought to relate to the fundamental structure of the Michigan Basin. During Silurian time, the central portion of the basin experienced a major downward flexing, stimulated by compression along what is now the Appalachian region. Eventually, the central basin accumulated more than 3,000 feet of Silurian sediment. The center of the basin sagged until it lay below the well-lit zone. Growth of organisms that had strong dependence on light, such as algae and corals, was inhibited in this area.[17]

In the shallow margins surrounding the deeper waters of the central basin, "pinnacle reefs" flourished. These reefs are exposed by erosion in the Wabash Valley of northern Indiana; by quarrying across northern Indiana and northwestern Ohio, along the Bruce Peninsula, Michigan's eastern Upper Peninsula and the Door Peninsula; and by plentiful drilling through Michigan's Lower Peninsula. Their ecological structure has been delineated by numerous studies; successive upward stages of colonization by waves of organisms (similar to those of the Thornton Reef discussed in chap. 10) are well documented. These structures, often several hundred feet high, are the remains of intact ecological reefs and, like the Thornton Reef, cannot be interpreted as mounds of debris swept and collected during a massive Flood.[18]

Later during Silurian time, the region underwent periods of sea level lowering. During these intervals, water circulation was restricted. Some of the reef tops evidence erosion and karstification. The reef limestones were chemically altered by saline brines, causing dolomitization. The center of the basin experienced prolonged periods of limited contact with open marine waters. The

Eschmann, *Geology of Michigan;* G. M. Ehlers, *Stratigraphy of the Niagaran Series of the Northern Peninsula of Michigan,* Papers on Paleontology No. 3 (Ann Arbor: University of Michigan, 1973); Fisher, Barratt, Droste and Shaver, "The Michigan Basin"; G. M. Friedman and D. C. Kopaska-Merkel, "Late Silurian Pinnacle Reefs of the Michigan Basin," in Catacosinos and Daniels, *Early Sedimentary History of the Michigan Basin;* LaBerge, *Geology of the Lake Superior Region.*

[17]Ells, *Michigan's Silurian Oil and Gas Pools;* Howell and van der Pluijm, "Early History of the Michigan Basin"; Friedman and Kopaska-Merkel, "Late Silurian Pinnacle Reefs of the Michigan Basin."

[18]The literature on Silurian reefs in the Great Lakes is extensive. For summaries and bibliographies, refer to the Thornton reef references in chap. 10; See also P. H. Heckel and G. D. O'Brien, eds., *Silurian Reefs of Great Lakes Region of North America,* AAPG Reprint Series 14 (Tulsa, Okla.: AAPG, 1975); A. M. Johnson, R. V. Kesling, R. T. Lilienthal and H. O. Sorenson, *The Maple Block Knoll Reef in the Bush Bay Dolostone (Silurian, Engadine Group), Northern Peninsula of Michigan,* Papers on Paleontology No. 20 (Ann Arbor: University of Michigan, 1979); M. E. Johnson and G. T. Campbell, "Recurrent Carbonate Environments in the Lower Silurian of Northern Michigan and Their Inter-regional Correlations," *Journal of Paleontology* 54 (1980): 1041-57; Friedman and Kopaska-Merkel, "Late Silurian Pinnacle Reefs of the Michigan Basin"; R. H. Shaver, "A History of Study of Silurian Reefs, Michigan Basin Environs," in Catacosinos and Daniels, *Early Sedimentary History of the Michigan Basin,* pp. 101-38; D. G. Mikulic and J. Kluessendorf, *Classic Silurian Reefs of the Chicago Area,* Michigan Basin Geological Society Field Excursion Guidebook, 1998.

isolated basin waters grew briny. Thick deposits of bedded marine evaporites, collectively referred to as the Salina Group, were deposited in the center of the basin, thinning towards the margins. Sedimentary structures in the bedded salts of the Salina Group indicate that the evaporites typically crystallized subaqueously, from shallow mineral-saturated waters that were recharged periodically by seawater overtopping the reefs. There are several cycles of evaporite precipitation; during the later cycles, many reefs were covered by salt.[19] Large volumes of Silurian salt are mined commercially in the Detroit area and in northern Ohio.

Drilling into Silurian reef structures for oil and gas began in the late 1800s. During the 1950s, the possibility that subsurface reef structures in Michigan and Ontario might produce large volumes of oil and gas created a hydrocarbon exploration boom that lasted through the 1970s. Oil and gas are still being extracted in significant volumes from Silurian structures in this region. As noted in chapter eight for industrial stratigraphy in general, the particular stratigraphic analysis of these regional rocks has resulted in large economic gains. Fossil fuel exploration and development has in turn funded large numbers of field studies and graduate theses on Michigan Basin Silurian rocks.

Mackinac Breccia

Along the margins of the Straits of Mackinac and on Mackinac Island, a bizarre, massive limestone formation, termed the Mackinac Breccia, is exposed.[20] The term *breccia*, derived from Italian for "broken," is a conglomerate made from angular broken debris. The clasts or fragments comprising the breccia range in size from small fragments to very large blocks. The largest blocks range from a few feet to hundreds of feet in diameter, and are termed *megabreccia*. The highly angular clasts obviously formed by breakage of previously solid rocks. The breccia occupies vacancies in underlying uppermost Silurian strata. The irregularly shaped blocks can be clearly identi-

---

[19]Ells, *Michigan's Silurian Oil and Gas Pools*; Dorr and Eschmann, *Geology of Michigan*; K. J. Mesolella, J. D. Robinson, L. M. McCormick and A. R. Ormiston, "Cyclic Deposition of Silurian Carbonates and Evaporites in Michigan Basin," *AAPG Bulletin* 58 (1974): 34-62; K. R. Cercone, "Evaporative Sea-level Drawdown in the Silurian Michigan Basin," *Geology* 16 (1988): 387-90; P. Sonnenfeld and I. Al-Aasm, "The Salina Evaporites in the Michigan Basin," in Catacosinos and Daniels, *Early Sedimentary History of the Michigan Basin*, pp. 139-54; C. L. Satterfield, R. K. Lowenstein, R. H. Vreeland and W. D. Rosenzweig, "Paleobrine Temperatures, Chemistries, and Paleoenvironments of Silurian Salina Formation F-1 Salt, Michigan Basin, U.S.A., from Petrography and Fluid Inclusions in Halite," *Journal of Sedimentary Research* 75 (2005): 534-46.

[20]Dorr and Eschmann, *Geology of Michigan*; K. K. Landes, G. M. Ehlers and G. M. Stanley, *Geology of the Mackinac Straits Region and Sub-Surface Geology of Northern Southern Peninsula*, Publication 44, Geological Series 37 (Lansing: State of Michigan Geological Survey Division, 1945).

fied as fragments of some of the latest Silurian rocks and some of the overlying earliest Devonian rocks. Elsewhere in the Michigan Basin, solution-work
and collapse features can likewise be demonstrated in drill cores through
the uppermost Silurian rocks.[21]

During early Devonian time, a series of caverns developed in the uppermost
Silurian strata. Large collapse features developed as the caverns were enlarged
and failed. Of great import is the fact that the collapsed boulders are intact
blocks that were clearly lithified prior to the collapse.

The boulders, of small to very large size, are cemented together to form the
Mackinac Breccia. The unit is overlain unconformably by one of the Lower
Devonian limestones of the Michigan Basin, the Dundee Formation. The
Dundee Formation itself is not fractured or brecciated and smoothes over the
preexisting karst topography.

Thus the Mackinac Breccia provides an amazing look into a series of events
implying a long epoch of elapsed time. During the time when the Lower Devonian rocks were forming, solution-work (paleokarst) developed in the uppermost Silurian rocks below, leading to the formation of an elaborate system of paleo-caverns and subsequent collapse. Most of the dissolved rock was
originally Silurian salt of the Point aux Chenes Formation, a component of
the uppermost Salina Group. The collapse occurred *after the lithification* of
the Lower Devonian rock strata. Subsequently, shaly lime sediments were deposited over the entire region and were lithified, burying and preserving the
historical record of solution and collapse.

Devonian Coral Reefs

Following the formation of the collapse breccia of the Mackinac Formation,
Devonian sedimentation continued in the Michigan Basin. Midcontinent
stratigraphers and paleontologists working with Devonian rocks and fossils
are treated to abundant surface exposures in Iowa and New York, as well as
Michigan, Ohio and Ontario. These units have been sampled by thousands of
professional paleontologists, geology students and amateur rockhounds during the past 150 years.

Overlying the Dundee Formation are several Middle-to-Upper Devonian
carbonate layers with some shaly interbeds (fig. 12.2). The carbonate rocks,
mostly limestones, are well exposed along the northeast and northwest coast
of Michigan's Lower Peninsula. Several very large quarries have been opened
into these strata for excavation of limestone for cement. At Alpena, Michigan, the LaFarge Corporation quarries limestone and manufactures cement.

---

[21]Ells, *Michigan's Silurian Oil and Gas Pools.*

In terms of production, this is the largest cement plant in the western hemisphere (see figs. 10.4, 10.5).[22]

Several of Michigan's Upper Devonian limestones are clustered in the Traverse Group, named after Grand Traverse Bay along the Lake Michigan coast. Traverse Group limestones, like Silurian layers a thousand or more feet below, host large coral buildups. In this case, the major framework organisms resemble those of the Silurian reefs but include some new ecological actors. The Silurian reefs were built on a stout framework of laminated stromatoporoid sponges and tabulate corals, with pentamerid brachiopods and pelmatozoan echinoderms as major ecological associates. The Devonian reefs are built from

Figure 12.5. Coral, *Hexagonaria,* from Traverse Group limestone (Devonian) exposed near Charlevoix, Michigan. From Calvin College geology collections, D. A. Young, collector.

stromatoporoid sponges, tabulate corals and abundant rugose corals (figs. 10.5 and 12.5). Pentamerid brachiopods are all but missing, but abundant spiriferid and atrypid brachiopods occupy many of the reef interstices. There is a turnover in minor ecological actors, such as trilobites, as well. Rock and fossil collectors of the Great Lakes region are very familiar with the coral *Hexagonaria,* the trilobite *Phacops* and the brachiopod *Mucrospirifer.* The cast of ecological characters can be sampled in abundance at several quarries in the northern

---

[22]Don Oliver, LaFarge Corporation geologist, personal communication.

Lower Peninsula, but quarries around Toledo, Ohio, and in southern Ontario are famous among local rockhounds as well.[23]

In addition to the above-mentioned contrasts to the reef communities exposed in Silurian strata, Devonian shallow-marine strata include prominent fish remains, unlike their Silurian counterparts. By far the vast majority of Devonian fish remains obtained from the Great Lakes region are those of the Placoderms, an extinct tribe of fishes equivalent in biologic diversity and presence during the Devonian to that of sharks today.

Like their Silurian counterparts, the Devonian reefs represent intact ecological assemblages. There are consistent differences between the two fossil assemblages in taxonomic composition. One does not find Devonian-style community assemblages in Silurian strata and vice versa. A catastrophic planetary Flood envisaged by most Flood geologists, capable of transporting whole reefs in place, would surely have dismembered the reefs and mingled their biotas, as well as mingling nonmarine taxa with them. These reef communities clearly experienced water turbulence and major storms whose effects can be discerned, but not catastrophic transport into their current locations!

Mississippian Shales and Evaporites

Near the end of the Devonian Period, carbonate deposition gave way to the deposition of clay-rich shales. Latest Devonian shales in Michigan actually straddle the stratigraphic boundary between the Devonian and Mississippian Periods, and shale deposition continued through much of the Mississippian Period. Mississippian strata of the Michigan Basin include a prominent sandstone layer, the Marshall Formation, and a shale with interbedded evaporites, the Michigan Formation. Total thickness of Mississippian rocks near the cen-

---

[23]E. C. Stumm, "Devonian Bioherms of the Michigan Basin," *Contributions from the Museum of Paleontology, University of Michigan* 22 no. 18 (1969): 241-47; Dorr and Eschmann, *Geology of Michigan;* G. M. Ehlers and R. V. Kesling, *Devonian Strata of Alpena and Presque Isle Counties, Michigan* (Lansing: Michigan Basin Geological Society, 1970); R. V. Kesling, R. T. Segall and H. O. Sorenson, *Devonian Strata of Emmet and Charlevoix Counties, Michigan,* Papers on Paleontology No. 7 (Ann Arbor: University of Michigan, 1974); R. V. Kesling and R. B. Chilman, *Strata and Megafossils of the Middle Devonian Silica Formation,* Papers on Paleontology No. 8 (Ann Arbor: University of Michigan, 1975); R. V. Kesling, A. M. Johnson and H. O. Sorenson, *Devonian Strata of the Afton-Onaway Area, Michigan,* Papers on Paleontology No. 17 (Ann Arbor: University of Michigan, 1976); R. M. Feldman, A. H. Coogan, R. A. Heimlich, *Field Guide: Southern Great Lakes* (Dubuque: Kendall/Hunt, 1977); R. C. Gutschick, "Devonian Shelf-basin, Michigan Basin, Alpena, Michigan," in *GSA Centennial Field Guide—North Central Section,* ed. D. L. Biggs (Boulder, Colo.: GSA, 1987), pp. 297-302; R. C. Gutschick and C. A. Sandberg, "Upper Devonian Biostratigraphy of the Michigan Basin," in Catacosinos and Daniels, *Early Sedimentary History of the Michigan Basin,* pp. 155-79; R. M. Feldman and M. Hackathorn, eds., *Fossils of Ohio,* Bulletin 70 (Columbus: Ohio Division of Geological Survey, 1996).

ter of the Michigan Basin averages between 1,500 and 2,100 feet.[24]

The Michigan Formation contains several subdivisions, most of which are shales. In its uppermost portion, it includes three prominent seams of gypsum and anhydrite, both evaporites composed of calcium sulfate. The three seams of calcium sulfate are commonly known as the "Triple Gyp" by miners; gypsum has been mined at shallow depth in western Michigan and quarried from surface outcroups in eastern Michigan near the "thumb." Although production of gypsum from Michigan is now at a very low level, Michigan was a major producer of gypsum for plaster and sheet rock in the past. Shallow underground mines in the Michigan Formation near Grand Rapids, Michigan, are now used for restricted-access, climate-controlled natural storage.

Although sedimentation continued through the Devonian-Mississippian boundary without interruption, a large regional unconformity exists between Mississippian rocks and Pennsylvanian rocks over Michigan and the entire Midcontinent.

Pennsylvanian Coal Beds

Pennsylvanian strata form one of the uppermost "saucers" of sedimentary rock in the stack of Phanerozoic strata. Because erosion has planed the set of saucers, the lateral extent of this upper unit is restricted (fig. 12.1). Pennsylvanian strata belonging to the Saginaw Group crop out in several places near the center of Michigan's Lower Peninsula. In other places, the soil cover has been removed so that open-pit coal mining or quarrying of shale (for brick manufacture) could be undertaken. The Saginaw Group constitutes a complex series of interbedded thin sandstones, limestones, shales and thin, discontinuous coal seams.[25]

Near Lansing, Michigan, the Saginaw Group rocks lie about 3,500 feet above the Silurian reef limestones and salt and 1,500 feet above the Devonian strata.[26] The Saginaw Group exposed here includes dark gray shales with marine fossils, nonfossiliferous sandstones interpreted as forming in beach and barrier-island environments, and coal beds.[27] Some of the sandstones adjacent to coal beds contain extensive root systems that correspond to the kinds of

---

[24]Dorr and Eschmann, *Geology of Michigan;* Lilienthal, *Stratigraphic Cross-Sections of the Michigan Basin;* J. A. Harrel, C. B. Hatfield and G. R. Gunn, "Mississippian System of the Michigan Basin: Stratigraphy, Sedimentology, and Economic Geology," in Catacosinos and Daniels, *Early Sedimentary History of the Michigan Basin,* pp. 203-20.

[25]R. Vugrinovich, *Lithostratigraphy and Depositional Environments of the Pennsylvanian Rocks and the Bayport Formation of the Michigan Basin,* Report of Investigations 27 (Lansing: Michigan Geological Survey, 1984); Fisher, Barratt, Droste and Shaver, "The Michigan Basin."

[26]Lilienthal, *Stratigraphic Cross-Sections of the Michigan Basin.*

[27]Dorr and Eschmann, *Geology of Michigan;* R. L. Milstein, "The Ledges of the Grand River, Michigan," in *GSA Centennial Field Guide—North Central Section,* ed. D. L. Biggs (Boulder, Colo.: GSA, 1987), pp. 311-14.

plant fossils located in the coals (figs. 10.1, 10.2, 10.3).

The plant fossils located in coals and associated sandstones and shales of the Saginaw Group comprise a stereotypical Carboniferous "coal-age" botanical suite (fig. 12.6). Major plant components of this suite are large "scale trees," such as *Lepidodendron,* belonging to the Lycophyta, a phylum that includes the modern ground-hugging vascular plants commonly termed "club mosses"; large and small representatives of the phylum Sphenophyta (e.g., *Calamites*) that includes only a handful of modern species, commonly known as "horsetails," all of which are less than four feet in height; ferns of the phylum Filicinophyta; archaic fern-like plants that bear seeds rather than spores, the "seed-ferns" of the phylum Pteridospermophyta; and early gymnosperm trees of the poorly understood group Cordaitales.[28] Notably, there are no flowering plants (angiosperms) nor any of our modern coniferous trees. The Michigan Pennsylvanian flora is identical in composition to similar floras known from North America at this time; all lack angiosperms and conifers. There is every reason, including the presence of preserved root traces, to conclude that this association represents the standard lowland swamp ecological community of its time.

The global Flood model for stratigraphy cannot explain how such widespread accumulations of plant debris retain such stereotypical associations, rather than mingling modern angiosperms and other phyla like cycads with these archaic plants. Similarly, Flood models cannot explain why plant debris is not mixed into Early Paleozoic strata.

Exposures of the Saginaw Formation in quarries and ledges in central Michigan reveal small-scale vertical successions in rocks in which gray clays bearing marine fossils are succeeded by sandstone layers with intact plant root traces and then coals. Vertical successions similar to this are seen throughout the North American mid-continent and were long ago termed *cyclothems,* cyclic sedimentary sequences.[29]

---

[28]C. A. Arnold, "Fossil Flora of the Michigan Basin," *Contributions from the Museum of Paleontology, University of Michigan* 7, no. 9 (1949): 131-269; C. A. Arnold, "Fossil Plants in Michigan," *The Michigan Botanist* 5 (1966): 3-13; J. R. Jennings, *Guide to Pennsylvanian Fossil Plants of Illinois,* Educational Series 13 (Champaign: Illinois State Geological Survey, 1990); Feldman and Hackathorn, *Fossils of Ohio.*

[29]See discussions in J. M. Weller, "Cyclic Sedimentation of the Pennsylvanian Period and Its Significance," *Journal of Geology* 38 (1930): 97-135; H. R. Wanless and F. P. Shephard, "Sea Level and Climatic Changes Related to Late Paleozoic Cycles," *Bulletin GSA* 47 (1936): 1177-206; P. H. Heckel, "Sea-Level Curves for Pennsylvanian Eustatic Marine Transgressive-Regressive Depositional Cycles Along Mid-continent Outcrop Belt, North America," *Geology* 14 (1986): 330-34; Frazier and Schwimmer, *Regional Stratigraphy of North America;* G. deV. Klein and D. A. Willard, "Origin of the Pennsylvanian Coal-bearing Cyclothems of North America," *Geology* 17 (1989): 152-55; G. Einsele, W. Ricken and A. Seilacher, eds., *Cycles and Events in Stratigraphy* (Berlin: Springer-Verlag, 1991). See also contemporary sedimentology textbooks.

Figure 12.6. Reconstruction of Pennsylvanian-age lowland forest. Jody Brown Zylstra, artist. Used by permission.

Mainstream sedimentary geologists interpret Pennsylvanian cyclothems as the products of regular advances and retreats of the global ocean, affecting low-lying coastal areas in many regions and on many continents. In eastern North America, the Appalachian mountains were beginning to achieve their final shape at this time; large volumes of sand and clay were eroded off the rising mountains and carried away, forming low-lying coastlands in the sea-covered continental interior.[30] The small-scale rise and fall of sea

[30]C. B. Trask and J. E. Palmer, "Structural and Depositional History of the Pennsylvanian System in Illinois," in *Paleoenvironmental and Tectonic Controls in Coal-Forming Basins in the United States,* Special Paper 210, ed. P. C. Lyons and C. L. Rice (Boulder, Colo.: GSA, 1986).

level then caused fluctuations of the shoreline, which are reflected as cyclic sequences of coastal swamp/delta development. The cycles of sea level that affected these coastal lands are globally correlated, and they are thought to reflect advances and retreats of large continental ice sheets over the south pole at that time.[31]

Flood geologists have attempted to mesh the phenomenon of coal-bearing cyclothems with a global planetary catastrophe.[32] However, the presence of rooted zones below coals testifies to elapsed time during which these roots grew. Whitcomb and Morris, and others since, have hypothesized that these roots were transported in the great Flood along with the vegetation they accompanied. At Grand Ledge and at many, if not most, Carboniferous coal-bearing cyclothems, there is no indication of transport of these roots. Furthermore, the presence of numerous "ironstone" concretions in the rooted zone indicates oxidation reactions in the rooted zone—a typical chemical phenomenon accompanying active root biology.

Mesozoic Red Beds

A profound erosional unconformity lies above the Carboniferous strata of the Michigan Basin. In the very center of the basin, rare outcrops plus drilling have demonstrated the presence of layers of red sandstone and shale, plus minor evaporite.[33] These have not been well studied to date because they are poorly exposed. Plant microfossils are typical Mesozoic plants, so they are ascribed to that geologic era.

Pleistocene Glacial Deposits

What is now the Great Lakes region was overridden by large ice sheets during the Pleistocene Ice Age.[34] The ice sheets eroded rocks and, when they retreated, left behind substantial deposits of rock rubble. A careful search in

---

[31]Wanless and Shephard, "Sea Level and Climatic Changes Related to Late Paleozoic Cycles"; J. J. Veevers and C. McA. Powell, "Late Paleozoic Glacial Episodes in Gondwanaland Reflected in Transgressive-Regressive Depositional Sequences in Euramerica," *Bulletin GSA* 98 (1987): 475-87; Einsele, Ricken and Seilacher, *Cycles and Events in Stratigraphy;* A. D. Miall, *Principles of Sedimentary Basin Analysis* (Berlin: Springer-Verlag, 2000).

[32]J. C. Whitcomb Jr. and H. M. Morris, *The Genesis Flood: The Biblical Record and Its Scientific Implications* (Philadelphia: Presbyterian & Reformed, 1961); Austin (as Nevins), "Stratigraphic Evidence of the Flood"; Walter Brown, *In the Beginning: Compelling Evidence for Creation and the Flood,* 7th ed. (Phoenix: Center for Scientific Creation, 2001).

[33]Dorr and Eschmann, *Geology of Michigan.*

[34]Ibid.; J. Imbrie and K. P. Imbrie, *Ice Ages: Solving the Mystery* (Hillside, N.J.: Enslow Publishers, 1979); P. F. Karrow and P. E. Calkin, *Quaternary Evolution of the Great Lakes,* Special Paper 30 (Ottawa: Geological Society of Canada, 1985); B. G. Andersen and H. Borns, *The Ice Age World* (Oslo: Scandinavian University Press, 1994); Laberge, *Geology of the Lake Superior Region.*

gravel pits of Michigan can turn up several dozen recognizable rock types. Some very distinct rock types can be traced to bedrock outcrops in southern Ontario or to the Upper Peninsula of Michigan—so they may have traveled hundreds of miles!

The Pleistocene ice sheets served as the equivalent of a gigantic plane or sanding device. They scraped across the Michigan Basin strata, leaving the bedrock configuration as seen in figure 12.1. They very probably removed more than one stratum! On top of the bedrock, the retreating ice sheets left a blanket of rock debris known as "till." In places, the till forms prominent regional ridges that mark the sites of temporary pauses in the glacial retreat.

After glacial retreat, plants occupied the new landscape, and animals followed these into the Great Lakes region. These plants have left a record in the form of pollen grains in the bottoms of ponds and lakes. The pollen record gives us a picture of the early vegetated landscape of Michigan following the Ice Age: a mixed plant community of northern pines and spruces intermingled with plains-style grasslands.

The early postglacial landscape of Michigan was colonized by a large number of extravagant (by contemporary standards) mammals.[35] They included mastodons, mammoths, peccaries, caribou, muskoxen and beavers much larger than our present-day beavers (fig. 12.7). Some of these creatures remain alive today, but many of these species are extinct. The riddle of the extinction of the large mammalian fauna of North America following the retreat of the Pleistocene glaciers remains a fascinating problem to be solved by future paleontologists.[36]

Development, movement and retreat of glacial ice sheets are related to long-term climatic changes and indicate the passage of significant time. As a well-studied example, existing ice sheets on Greenland have been cored and dated by counting visible yearly layers and calibrating these to radiometric dates. The annual layers form from variation in snowfall and melting between summer and winter seasons. At least 150,000 years of accumulation can be documented in these ice layers.

---

[35]J. A. Holman, *In Quest of Great Lakes Ice Age Vertebrates* (East Lansing: Michigan State University Press, 2001); J. A. Holman, ed., "Michigan's Ice Age Behemoths," *Michigan Academician* 34, no. 3, Special Issue (2002).

[36]B. Kurtén and E. Anderson, *Pleistocene Mammals of North America* (New York: Columbia University Press, 1980); D. K. Grayson, "Late Pleistocene Mammalian Extinctions in North America: Taxonomy, Chronology and Explanations," *Journal of World Prehistory* 5 (1991): 193-232; P. S. Martin, *Twilight of the Mammoths: Ice Age Extinctions and the Rewilding of North America* (Berkeley: University of California Press, 2005); P. L. Koch and A. D. Barnosky, "Late Quaternary Extinctions: State of the Debate," *Annual Review of Ecology, Evolution and Systematics* 37 (2006): 215-50.

Figure 12.7. Mastodon *(Mammut americanum)* excavation, Cascade, Michigan, 1999. Photo by R. F. Stearley.

CONCLUSION

The Michigan Basin structure has preserved a record of past Earth environments that existed in central North America over a protracted period of time. Like other sedimentary basins, it did not record a universal, global record of Earth history but preserved an extensive regional history when the basin was accumulating sediment. Generalized histories of entire continents must of necessity involve the synthesis of several such regional histories through correlation of stratigraphic units. Hence, criticisms by Flood geologists that the histories of the Earth, as schematized in generalized textbook stratigraphic columns, are artifacts of geologists' imaginations, and hence deliberately misleading, are incorrect.

The Michigan Basin structure can actually be discerned from orbiting space satellites! Thanks to the waters of the Great Lakes, which fill in low-lying regions formed from eroded softer strata, the rims of more durable layers, particularly upper Ordovician and Silurian dolostones, stand out as a concentric series of islands and peninsulas. Although loose soil covers much of the bedrock, the strata can be sampled in adequate surface outcrop and in many quarries and mines. The subsurface structure and character of the strata are extremely well understood through thousands of exploratory punctures, many of these quite deep.

The Michigan Basin strata are unified packages of sediment that retain internal consistency in composition, texture, structures and fossils. Coal is not mixed into coral reefs. Devonian coral reefs include placoderms whereas Silurian reefs do not. Thick layers of homogenous rock salt or gypsum, chemically precipitated from seawater, are not contaminated with sand grains, cobbles or boulders, nor with the skeletons of land-dwelling vertebrates. Catastrophic Flood models of regional sedimentation must posit a single agent that can simultaneously move and deposit two to three miles of sediment, yet retain these internal consistencies.

Within each stratum lie subtle timepieces. Like the hourglass depicted in chapter ten, these clocks do not require sophisticated knowledge of radioactive decay equations or sampling procedures for their understanding. Structures such as rooted zones, burrows excavated by marine invertebrates, upward growth frameworks in coral reefs, and erosion surfaces all attest to lengthy intervals of time during the accumulation of their respective sedimentary packages.

Over the past century, sedimentary geologists have elucidated principles for understanding sediments and sedimentary rocks. These principles include basic mechanics, which dictate the strength of water energy or wind energy driving a given body of sediment; aquatic chemistry, which explains under what sorts of conditions salt or gypsum or carbonate could be precipitated; and ecological or biological principles, which constrain the temperature and light conditions under which coral reefs or forests could flourish. When applied to stratigraphic packages such as those described for the Michigan Basin, these principles yield consistent pictures of ancient environments, not descriptions of a violent chaos.

Study of Michigan Basin rocks is not a dry, impractical enterprise. Surface and near-surface outcrops have yielded abundant limestone and gypsum plus small amounts of coal. Deep mines have recovered large volumes of rock salt. Substantial volumes of oil and gas are produced daily from deep wells. The subsurface strata of the Michigan Basin have provided more than 1.3 billion barrels of oil and nearly 6 trillion cubic feet of natural gas since 1920. These industries provide thousands of jobs and millions of dollars in state revenues each year. As stressed in chapter eight, regional stratigraphic investigations like those for the Michigan Basin yield valuable economic and cultural resources. Economic stratigraphy has prospered as a human enterprise through its realization that details of history could be inferred in a straightforward manner by application of the principles of normal physics, chemistry and ecology acting over time.

Regional histories like those of the Michigan Basin, admittedly incomplete due to an incomplete record, can be synthesized into a fascinating history of past environments and creatures on the Earth. We believe that this past testifies to the creativity and majesty of a Creator. To denigrate this past by dismissing it as imaginary is an exercise in ingratitude.

13

ILLUMINATION FROM THE
RANGE OF LIGHT

The Sierra Nevada

THE SIERRA NEVADA (literally, "snowy mountain range" in Spanish) occupies a large area along the eastern margin of California.[1] Although the overall upward slope on the western edge of the range is gentle, the slope of the eastern face is extremely steep, reflecting uplift and tilting along active faults. In fact, the Sierra Nevada peaks continue to be uplifted. Thus, the highest point in the lower forty-eight states, Mount Whitney, has become measurably higher between the time this chapter was written and the time you are reading it! The present phase of active tectonism and uplift is, however, merely a small chapter in a long and exceedingly complex history that has resulted in an extravagantly beautiful landscape.

In overview, this region of east-central California has experienced the following major episodes of geological activity: a long period of accumulation of sedimentary rocks during Paleozoic and early Mesozoic time; tectonism and extensive volcanism, recorded locally by emplacement of large amounts

---

[1]For an overview of the geology of California, see Robert M. Morris and Robert W. Webb, *Geology of California*, 2nd ed. (New York: John Wiley & Sons, 1990). On the Sierra Nevada, see chap. 3, pp. 62-125.

of granitic rock in the form of deep-seated intrusions into the early rocks during the Jurassic and Cretaceous Periods; metamorphism of the surrounding
sedimentary rocks and emplacement of metallic ore deposits associated with
metamorphism; Mesozoic and Cenozoic uplift and erosion, yielding sheets of
sediment spread into subsiding basins to the west of the Sierra Nevada; renewed uplift along high-angle faults during the latest Cenozoic Era; sculpting
by several episodes of Pleistocene glaciation; and very recent volcanism. The
U.S. Geological Survey maintains an ongoing program to monitor the motions
of magma below the eastern margin of the Sierra Nevada, anticipating future
volcanic activity in the region.

Each of these major episodes could easily be analyzed in separate chapters.
We present here only a bare outline of that complicated history and hope that
geologists familiar with the region will kindly overlook our efforts at simpli-
fication. We begin this abbreviated discussion in the middle of the history by
considering the highly aesthetic granitic rocks exposed at Yosemite National
Park and elsewhere in the Sierras.

THE SIERRA NEVADA BATHOLITH

Yosemite National Park is the best known and most beloved part of what famed
naturalist John Muir (1838-1914) termed the "Range of Light."[2] From Glacier
Point, 3,000 feet above the floor of Yosemite Valley, one takes in awesome,
jaw-dropping, breath-taking vistas of light gray granitic rocks. To the north
is the spectacular cleaved mass of Half Dome. Beyond that are Clouds Rest
and gray masses of rock in Tenaya Canyon. To the east, best viewed at nearby
Washburn Point, are the smooth glaciated slopes over which the waters of the
Merced River glide with exquisite grace and power at the 572-foot Nevada Fall
(fig. 13.1). Driving along the Tioga Pass Road in the northern part of the park
to Tenaya Lake at the head of Tenaya Canyon, one likewise sees vast tracts of
light gray granitic rocks that extend in every direction. Everywhere within
the Sierra Nevada, particularly in Yosemite National Park, there is abundant
evidence of past glacial activity in the form of highly smoothed and polished
rock exposures, erratic boulders sitting atop the rounded granitic masses, and
scoured out valleys (fig. 13.2). By looking across Yosemite Valley and pondering the action of glaciers on granitic rock, one gains a sense of the power and
scale of the events that were involved in shaping the landscape.

A motorist driving south about 150 miles to Sequoia and Kings Canyon
National Parks would still be in the heart of seemingly endless expanses of
granitic rocks. The rocks of Yosemite National Park, therefore, form only a

---

[2]John Muir, *The Mountains of California* (New York: The Century Company, 1894).

Figure 13.1. Washburn Point, Yosemite National Park, California, Nevada Fall right of center. Virtually all rock visible from this vantage point consists of granitic rocks rich in quartz, feldspars, biotite and hornblende. Such rocks are characteristic of virtually the entire Sierra Nevada batholith. Photo by D. A. Young.

very small portion of an enormous mass of granitic rock approximately 350 x 60 miles in area that geologists term the Sierra Nevada batholith (fig. 13.3). Geologists apply the term *batholith* to any intrusive mass of igneous rock that has an exposed area of at least 40 square miles.

The Sierra Nevada batholith is only one among a great series of batholiths that extends from southeastern Alaska through British Columbia into the western United States, Baja California, and on into the Andes Mountains of South America all the way to its southern terminus. Most of these west coast batholiths are extremely large in contrast to the numerous batholiths in the lowlands of the Piedmont of the southeastern United States (fig. 13.4).

Like most of these exceptionally large batholiths along the rim of the Pacific Ocean, the Sierra Nevada batholith does not consist of one continuous mass of granite that was injected as magma all at once. The batholith is an example of what geologists term a *composite batholith* because it consists of dozens, if not hundreds, of smaller individual intrusions termed *plutons* that were emplaced at different times. These plutons range in size from a little less than one square mile to more than 350 square miles.[3] For example, the Mount Givens Grano-

---

[3]Paul C. Bateman, *Plutonism in the Central Part of the Sierra Nevada Batholith, California*, Pro-

diorite, located about 20 to 80 miles southeast of Yosemite National Park, is approximately 520 square miles in exposed area; the Lamarck Granodiorite, located about 45 to 85 miles southeast of Yosemite, is about 220 square miles; and the El Capitan Granite and closely related rocks, a small portion of which are so beautifully exposed in Yosemite Valley, has an area greater than 110 square miles. All of these intrusions are large enough to be considered small batholiths in their own right.

Figure 13.2. Glacial features at Olmstead Point, Yosemite National Park. Note the highly smoothed and polished rock surface as well as isolated erratic boulders lying on outcrop. Photo by D. A. Young.

Field geologists distinguish the individual plutons from one another on the basis of differences in mineral composition, grain size and texture. The plutons are commonly separated from one another by sharp contacts. The order in which plutons were intruded can normally be determined on the basis of a variety of field criteria. Where two plutons are in contact, the younger one may contain fragments, called *xenoliths,* of the granitic rock that was intruded by the younger magma (chap. 11).[4] As the magma was injected into fractures in the existing crystallized pluton, pieces of the older rock were dislodged and rafted into the magma where they were eventually trapped as the new magma cooled

fessional Paper 1483 (Washington, D.C.: USGS, 1992), p. 25.
[4]Ibid.

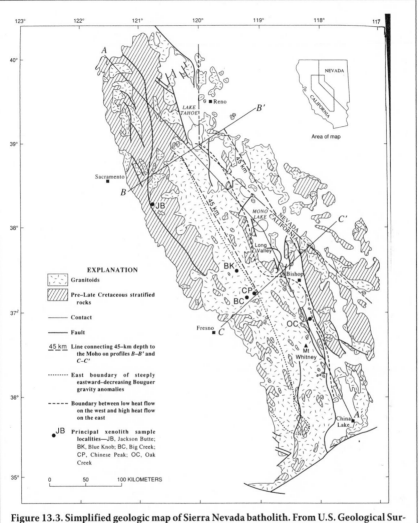

Figure 13.3. Simplified geologic map of Sierra Nevada batholith. From U.S. Geological Survey Professional Paper 1483.

and crystallized. Older plutons are also commonly cut by dikes that emanate from plutons that were intruded into them (fig. 13.5). Magma that formed the younger pluton sometimes worked its way along a fracture in the older pluton and then froze to form a tabular body of rock called a *dike* (chap. 11).

In addition, many of the plutons in the Sierra Nevada display preferred orientation of mineral grains, especially dark minerals like biotite and hornblende, that is roughly parallel to the contacts of the pluton. This preferred alignment, referred to by geologists as *foliation,* developed by flow of the partly crystallized magma within the cooling pluton. Another way in which one can determine the relative age of two plutons is that the contact between the two plutons may cut across structures within the older pluton, such as foliation. In some cases, it may not become evident that the contacts of a given pluton cut across an older pluton until their regional relations are plotted on a geological map (fig. 13.6).

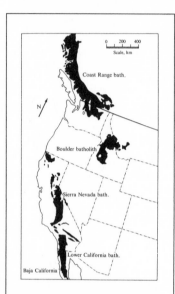

Figure 13.4. Map of distribution of major batholiths of western Canada and United States. Reproduced by permission of The McGraw-Hill Companies from Ian S. E. Carmichael, Francis J. Turner and John Verhoogen, *Igneous Petrology* (New York: McGraw-Hill, ©1974), p. 577.

Within the Sierra Nevada batholith, individual plutons have been grouped into so-called *intrusive suites* on the basis of their field relations. In the vicinity of Yosemite National Park, geologists from the U.S. Geological Survey who have been mapping in the region for the past several decades have assigned the several individual plutons of varying sizes and mineralogical compositions to a number of intrusive suites.[5] Within the central part of the batholith, the Fine Gold Intrusive Suite, the intrusive suite of Yosemite Valley, the Shaver Intrusive Suite, the intrusive suite of Buena Vista Crest, the intrusive suite of Merced Peak, the intrusive suite of Washburn Lake, the Tuolumne Intrusive Suite, the John Muir Intrusive Suite, the Scheelite Intrusive Suite and the Palisade Crest Intrusive Suite have been named.[6] Numerous plutons, both small and large, have not yet been assigned to any intrusive suite.

The assignment of plutons to intrusive suites has been made on the basis of field mapping of spatially associated, genetically related and sequentially intruded plutons. By way of example, the Tuolumne Intrusive Suite that makes up much of the eastern part of Yosemite Valley includes the granodiorite of Kuna Crest, the Half Dome Granodiorite, the Cathedral Peak Granodiorite and the Johnson Granite Porphyry (fig. 13.6). The intrusive suite of Yosemite Valley that crops out in the central and

---

[5]Ibid., pp. 26, 121-76.

[6]Formal names of intrusive suites use capital letters, whereas informal names (more loosely used and not precisely defined) use lower-case letters.

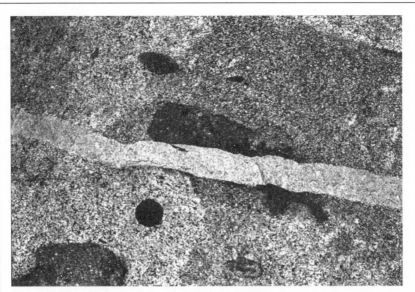

Figure 13.5. Dike cutting through an inclusion (possibly a xenolith) incorporated into a granitic intrusion. By the principle of cross-cutting relationships, dikes are younger than the rocks they cut through. If the dike emanated from a nearby intrusion, then that intrusion must be younger than the intrusion that is cut by the dike. The inclusion, in turn, is probably older than the intrusion that is cut by the dike. Lens cap for scale. Photo by D. A. Young.

western part of Yosemite Valley includes the El Capitan Granite, the granite of Rancheria Mountain and the Taft Granite. In both of these examples, the first-named pluton is the oldest within the suite, and the last-named pluton is the youngest within the suite. The age relations of intrusive suites can also generally be inferred from cross-cutting relations or deformation of rocks of an earlier suite by emplacement-related processes associated with a younger suite. Hence, the Tuolumne Intrusive Suite is younger than the intrusive suite of Yosemite Valley, and the John Muir Intrusive Suite is younger than the Shaver Intrusive Suite. Like individual plutons, intrusive suites also vary in size. The Tuolumne Intrusive Suite, for example, is approximately 450 square miles in size.

The plutons are composed of a wide variety of igneous rocks with granite, granodiorite and quartz diorite predominating. Virtually all of these rocks are composed of varying combinations of the minerals quartz, potassium feldspar, plagioclase feldspar, biotite mica and hornblende. Although different names are given to the rocks depending on the relative abundances of minerals, collectively the rocks of the batholith are loosely considered as "granitic" or "granitoid." All of these granitic plutons consist of very coarse-grained

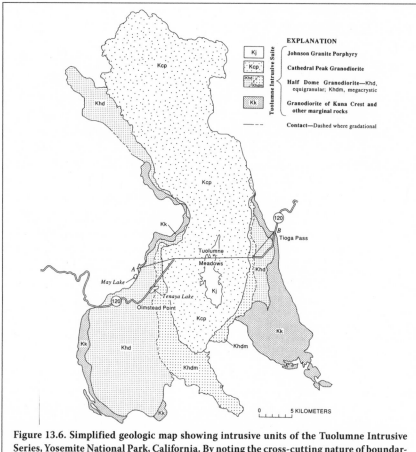

Figure 13.6. Simplified geologic map showing intrusive units of the Tuolumne Intrusive
Series, Yosemite National Park, California. By noting the cross-cutting nature of boundaries between intrusive units, geologists can ascertain that the Cathedral Peak granodiorite
is younger than the Half Dome granodiorite which, in turn, is younger than the Kuna Crest
granodiorite. From U.S. Geological Survey Professional Paper 1483.

crystalline rocks. The major minerals are typically a few millimeters to more
than one centimeter in diameter. Feldspar crystals as much as one inch long
are very common within the granitic rocks of the Sierra Nevada batholith. The
coarse-grained character of these igneous rocks indicates that they formed far
beneath the surface and cooled very slowly. Experimental studies of melts with
compositions like those of the granitic rocks of the Sierra Nevada suggest that
the liquids from which these rocks crystallized originally existed at temperatures on the order of 750° to 825° C.[7]

---

[7]W. Johannes and F. Holtz, *Petrogenesis and Experimental Petrology of Granitic Rocks* (Berlin:
Springer-Verlag, 1996); and O. F. Tuttle and N. L. Bowen, *Origin of Granite in the Light of Experi-*

As noted in chapter eleven, the four processes of magma melting, ascent, emplacement and crystallization typically require tens to hundreds of thousands of years, depending on size, depth of intrusion, initial magma temperature and other factors. But that's just for one pluton! Now consider that there are more than one hundred plutons in the Sierra Nevada batholith. Conceivably some of them were intruded around the same time but in very different parts of the batholith. But in any one region of the batholith there is a sequence of perhaps dozens of plutons, each of which required a substantial amount of time for the four-stage process to occur. Add to that the fact that there were probably extended stretches of time that elapsed between cooling of an older pluton and injection of the magma of the next pluton in the sequence. In fact, field evidence indicates that, in most cases, the individual plutons had crystallized almost completely before the magma of the next pluton was injected. Angular inclusions of rock from the earlier plutons are commonly embedded in newer plutons near the contacts; dikes that solidified from the magma of the younger plutons commonly occupy fractures in the older plutons; and sharp, well-defined boundaries rather than gradational boundaries that might indicate magma mixing near the contact between the two plutons are strong indicators of virtually complete crystallization of older plutons at the time of intrusion of the next pluton. Moreover, dikes from an older pluton injected into a younger pluton are typically lacking. Such dikes would indicate mobilization of small quantities of magma that had not yet crystallized in the older plutons. In other words, many, if not most, of the plutons certainly had cooled well below the temperature of complete crystallization prior to injection of the next batch of magma.

Considered together, the evidence overwhelmingly indicates that the formation of all the plutons of the Sierra Nevada batholith could easily have taken hundreds of thousands to millions of years to accomplish. The batholith alone offers compelling evidence that Earth is far more than a few thousand years old. And note that the evidence presented here is based on field observations, theoretical calculations based on the known thermal properties of rocks and laws of heat transfer, and the physics and fluid mechanics of magma ascent and emplacement. We have made no appeal whatever to radiometric dating. Had methods of radiometric dating never been developed, the evidence from these igneous rocks undeniably favors passage of a vast amount of time during construction of the Sierra Nevada batholith far in excess of the restricted time demands of young-Earth creationism.

In summary, field evidence tells us about relative time relationships and

mental Studies in the System $NaAlSi_3O_8$—$KAlSi_3O_8$—SiO_2—H_2O, GSA Memoir 74 (New York: GSA, 1958).

may lead to rough estimates of how much time elapsed. Radiometric dating gives us the absolute ages. Now that we have established that the great antiquity of the batholith can readily be demonstrated on grounds other than radiometric dating, let's take a brief look at some of the results of radiometric dating on rocks of the Sierra Nevada.

Dozens of ages have been determined by U-Pb dating on separates of the mineral zircon that commonly occurs in the granitic rocks (see chap. 14). Moreover, data have been obtained by K-Ar dating of separates of hornblende and biotite from the granites, and Rb-Sr whole-rock isochron ages have also been obtained.[8] The overwhelming majority of ages from the batholith are around 160-170 million years (Jurassic) and 80-120 million years (Cretaceous).

It is striking that the radiometric ages are consistent with relative ages that have been worked out on the basis of field relationships. For example, field relationships in the central Sierra Nevada indicate that the Fine Gold Intrusive Suite is cut by the Shaver Intrusive Suite to its east and is, therefore, older. In turn, rocks of the Shaver Intrusive Suite are cut by the gigantic Mount Givens Granodiorite pluton and other plutons of the John Muir Intrusive Suite on the eastern side of the Sierra Nevada and is, therefore, older. The relative age sequence of the intrusive suites from oldest to youngest is, therefore, Fine Gold, Shaver and John Muir. Let's see if the radiometric dates are consistent with that conclusion derived from field relations. U-Pb ages on 13 samples of the Bass Lake Tonalite, the largest pluton of the Fine Gold Intrusive Suite, and possibly a composite pluton, range from 124 to 105 million years and average 114 million years. A U-Pb age of 104 million years was obtained for the Dinkey Creek pluton in the Shaver Intrusive Suite, and a U-Pb age of 102 million years was obtained for the younger granite of Shuteye Peak, also part of the Shaver Intrusive Suite. Two samples of Mount Givens Granodiorite, a pluton of the John Muir Intrusive Suite, yielded U-Pb ages of 93 and 88 million years, and the Lamarck Granodiorite, also a member of the John Muir Intrusive Suite, gives a U-Pb age of 90 million years. *The radiometric dating sequence is consistent with the field relationships:* Fine Gold Intrusive Suite (124-105 million years); Shaver Intrusive Suite (104-102 million years); and John Muir Intrusive Suite (93-88 million years).

Let's consider an additional example from the central Sierra Nevada. Field relationships indicate that the intrusive suite of Yosemite Valley is older than the intrusive suite of Buena Vista Crest which, in turn, is older than plutons

---

[8]Bateman, *Plutonism in the Central Part of the Sierra Nevada Batholith*, pp. 63-67; see also T. W. Stern, P. C. Bateman, B. A. Morgan, M. F. Newell and D. L. Peck, *Isotopic U-Pb Ages of Zircon from the Granitoids of the Central Sierra Nevada*, Professional Paper 1185 (Washington, D.C.: USGS, 1981).

of the Tuolumne Intrusive Suite. The El Capitan Granite, part of the intrusive suite of Yosemite Valley, yields U-Pb ages of 103-102 million years. The granodiorite of Illilouette Creek, part of the intrusive suite of Buena Vista Crest, has a U-Pb age of 100 million years. Incidentally, the intrusive suite of Buena Vista Crest intrudes the Shaver Intrusive Suite (104-102 million years). Finally, the U-Pb ages for intrusions of the Tuolumne Intrusive Suite range from 86 to 91 million years. *The radiometric dates are consistent with the field relationships.*

Radiometric ages for the Sierra Nevada batholith indicate that most of the various magma bodies were injected during the Jurassic and Cretaceous Periods that lasted from 200 million to 65 million years ago. But let's note that even without the evidence of radiometric dating, the field and heat conduction evidence alone leads to the conclusion that very substantial amounts of time were required for this complex batholith, consisting of more than one hundred plutons, to crystallize entirely.

Pre-Batholithic Sedimentary Rocks

These large volumes of granitic magma had to be intruded into *something* that was already there! They were not intruded into a gigantic empty void. As we see later, mineralogical evidence in rocks that are now at the surface suggests that the magmas originally crystallized around three to four miles below the surface. There must have been some preexisting rocks lying above the magmas when they were intruded.

Geologists have a good idea of what those rocks were. Scattered throughout the Sierra Nevada are large blocks of metamorphosed sedimentary and volcanic rocks that are several miles in length and width (fig. 13.7). These blocks, termed *roof pendants,* are embedded within the upper part of the batholith and are remnants of the overlying roof rocks that were injected by rising magmas (fig. 13.8). In addition, other remnants of rock, termed *septa,* consist of elongated strips of metamorphosed sedimentary rocks that were intruded by the magmas. Some septa may be roof pendants, but others may represent wall rocks of plutons rather than overlying rocks. These septa characteristically appear at the boundaries between adjacent plutons or intrusive suites.

Because the rocks within the roof pendants and septa that are above and/or adjacent to the batholith were intruded by the granitic magmas, they are older than the granitic rocks of the Sierra Nevada batholith. These roof pendants represent only small remnants of what were much larger volumes of sedimentary rock that covered the rising masses of magma and that have been substantially eroded.

Detailed field studies show a consistent succession of layered sedimentary rocks in several of these roof pendants along the eastern side of the batholith that are assigned to the so-called Morrison block.[9] This stratigraphy occurs in the Bishop Creek roof pendant, through the Pine Creek and Mount Morrison roof pendants, and includes the Gull Lake and Log Cabin Mine and parts of Northern Ritter Range roof pendant (fig. 13.7). Less metamorphosed sedimentary rocks similar to the rocks in the roof pendants are also well preserved in the White and Inyo Mountains just east of the Sierra Nevada.

The sequences of layered rocks within these roof pendants consist of limestone, chert, quartz sandstone, mudstone and conglomerate. Some rocks have been mildly metamorphosed to slate, hornfels and marble. There is a succession of ten formations within the Morrison block from the basal Mount Aggie Formation to the uppermost Bloody Mountain Formation. Several of these formations contain fossils of such marine organisms as conodonts, graptolites, radiolarians, crinoids, corals and brachiopods. The aggregate thickness of this succession is approximately 20,000 feet.

Overlying this sedimentary rock succession within portions of the Mount Morrison, Ritter Range and Saddlebag Lake roof pendants is a thick accumulation of volcanic and sedimentary rocks assigned to the Koip Group. This group of rocks is thought to be separated from underlying strata by an unconformity. Deposition of the Koip Group would have begun only after extensive downward erosion. Many rocks at the base of the Koip Group are conglomerates in channel fills that contain cobbles and pebbles of the underlying rocks, another indication that there had been elevation and erosion of the underlying rocks prior to deposition of the Koip Group rocks. Estimates of the thickness of the rocks belonging to the Koip Group range from 16,000 to 36,000 feet.

Thus, the Sierra Nevada contains remnants of a very thick succession of sedimentary rocks that were deposited prior to the intrusion of the granitic batholiths. The composite thickness of the formations in the roof pendants may far exceed 36,000 feet (seven miles) and may be as much as 55,000 feet (ten miles).

Let's consider how we might apply a Flood geology model to these rocks. In the Flood geology scenario, fossil-bearing sedimentary rocks are commonly attributed to the year-long Deluge of Noah. Some adherents of Flood geology, however, might assign some fossiliferous rocks to the period between the conclusion of creation week and the onset of Noah's Flood. Flood geologists have also suggested that some fossil-bearing rocks might have been deposited in catastrophic events subsequent to the Flood, but because we are discussing geologically older

[9]See C. H. Stevens and D. C. Greene, "Stratigraphy, Depositional History, and Tectonic Evolution of Paleozoic Continental-Margin Rocks in Roof Pendants of the Eastern Sierra Nevada, California," *Bulletin GSA* 111 (1999): 919-33; and Calvin H. Stevens and David C. Greene, "Geology of Paleozoic Rocks in Eastern Sierra Nevada Roof Pendants, California," in *Great Basin and Sierra Nevada*, GSA Field Guide 2, ed. D. R. Lageson, S. G. Peters and M. M. Lahren (Boulder, Colo.: GSA, 2000), pp. 237-54.

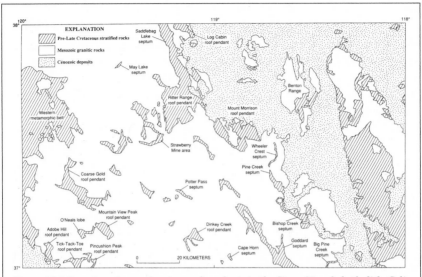

Figure 13.7. Location of several major roof pendants in the Sierra Nevada batholith, California. From U.S. Geological Survey Professional Paper 1483.

Figure 13.8. View of Mount Morrison roof pendant looking west from U.S. 395 near Crowley Lake, California. Note the pronounced layering of the sedimentary rocks in the pendant. These represent rocks that were intruded by the granitic magmas of the Sierra Nevada batholith. Photo by D. A. Young.

parts of the Sierra Nevada, we need not consider that last explanation. According to young-Earth proponents, all fossiliferous rocks were formed after the Fall of Adam and Eve, on the grounds that there was no death in the world until then. The rocks in the roof pendants do contain a wide variety of fossils. If these rocks were deposited between the Fall and the Flood, then only 1,656 years were available on the most strictly literal reading of the Genesis genealogies. Many of the rocks contain evidence that they were deposited in deep, generally tranquil water far from shore. The question is whether minimally seven miles of fine-grained sediments and volcanic rocks accumulated in only one and a half millennia. We would be talking about an average sedimentation rate of about 20 feet per year for 1,656 years! If these rocks were all deposited during a one-year planetary Flood, however, then the sedimentation rate was seven miles or at least 36,000 feet per year! Do Flood geologists really expect anyone to believe that?

These hypothetical calculations, however, assume that deposition of the sedimentary rocks was uninterrupted. We already noted that sedimentation must have been interrupted to account for the unconformity that separates the Koip Group from the underlying strata. In addition, uplift and erosion presume prior cementation of the underlying sediment pile to consolidated sedimentary rock. Consolidation, uplift and erosion must all be factored in to any attempt to estimate the amount of time involved. There's more, however!

One also needs to ask why these sedimentary rocks contain absolutely no fossils of fish or other contemporary organisms if they were formed from water during the past few thousand years. Where is the evidence for fossil trees, flowering plants, reptiles and mammals? Surely some of these organisms should have been preserved in the rocks of the roof pendants if they were deposited within the past few thousand years.

The rocks now exposed in the roof pendants were also strongly folded. Some geologists believe that the rocks in the roof pendants experienced at least two or, in some cases, three distinct episodes of intense folding. In the roof pendants of the Morrison block, there is at least one well-defined set of folds, indicating a period of intense deformation of the entire succession of rocks. Associated with this set of folds are several high-angle reverse faults that probably formed at about the same time in response to the same set of stresses within the crust. Because field mapping demonstrates that the batholith truncates these folds and faults, the period of deformation must have occurred prior to invasion of the granitic magmas. But we are still not done! There is a long strike-slip fault known as the Tinemaha Fault that separates rocks of the Morrison block from their equivalents in the nearby Inyo and Owens Mountains. This fault also is clearly truncated by the granitic rocks of the batholith, and, therefore, is older than the batholith.

In summary, prior to emplacement of vast volumes of granitic magma that make up the batholith, the Sierra Nevada region experienced a complex sequence of events that involved subsidence, deposition of a sediment pile several miles thick, cementing of the sediment, some volcanism, at least one major episode of folding and more than one episode of faulting. In the first place, at least 36,000 feet of sediments accumulated locally. The sedimentary rock types of the various formations and within individual formations differ from one another, indicating that there were changing environments of deposition, including marine and stream environments. The existence of the unconformities indicates that some portion of the sedimentary rock stack was eventually deformed by folding and then tilted, uplifted, weathered and eroded before being depressed and covered by new layers of sediment, most likely under water. The rocks were also faulted under nearer-surface conditions prior to invasion of magmas because the faults do not penetrate into the igneous granites. And then large volumes of granite magma intruded the stacks of sedimentary rocks and cut across the folds and faults. Are we expected to believe that all these events could have occurred within 6,000 years? Within 1,656 years from creation to the Flood? Within one Flood year? But there is yet more!

Contact Metamorphism of the Roof Pendant Rocks

Not only is a lengthy pre-batholith history indicated by the sedimentary rock sequences that are preserved in the roof pendants and septa, but these rocks were metamorphosed during intrusion of the granitic magmas, calling for yet another lengthy process. Because intense heat was applied to the original pre-batholithic rocks adjacent to the granitic magmas being intruded, many of these rocks were subjected to intense contact metamorphism. Substantial amounts of time were necessary to convert the older sedimentary rocks into slate, schist, marble, hornfels and skarn. Original shale and clay-rich sandstone were commonly converted into fine-grained *hornfels* that include, among other minerals, the diagnostic minerals andalusite and cordierite. Experimental laboratory studies on the stability of andalusite and cordierite indicate that those rocks were subjected to temperatures between 500° and 600° C at a pressure of around 1.5 to 2 kilobars.[10] Pressures of this magnitude are generated at depths around three to four miles.

---

[10]D. M. Kerrick, "Contact Metamorphism in Some Areas of the Sierra Nevada, California," *Bulletin GSA* 81 (1970): 2913-38. See also D. M. Kerrick, "The Genesis of Zoned Skarns in the Sierra Nevada, California," *Journal of Petrology* 18 (1977): 144-81; and John M. Ferry, Boswell A. Wing and Douglas Rumble III, "Formation of Wollastonite by Chemically Reactive Fluid Flow during Contact Metamorphism, Mt. Morrison Pendant, Sierra Nevada, California, USA," *Journal of Petrology* 42 (2001): 1705-28.

This figure for depth of metamorphism gives us an appreciation of the depth at which magmas were intruded and provides a very good estimate for how much uplift had to occur to bring the crystallized igneous rocks to the surface where they are presently exposed. Beyond that, these experimental data imply that the metamorphosed rocks not only had to be elevated from that depth to the surface where we now see them, but also were depressed from the surface to that depth prior to their metamorphism. These sedimentary rocks, originally formed at the surface, must have been buried under another several miles of rock to generate the requisite pressure.

However, these overlying rocks have been eroded away. Intuitively, one can sense that the burial of a mass of rock under a thickness of another four miles of rock and then the reemergence of that mass of rock is bound to take a lot of time. Reinforcing that conclusion is the fact that metamorphic minerals like andalusite and cordierite, especially the former, are notoriously slow to crystallize and react. The very fact that andalusite and cordierite persist in these rocks at the surface well out of their fields of stability is an indication of how sluggish these minerals are. In other words, even if we could imagine a way of bringing the sedimentary rocks to a depth of four miles and returning them to the surface very quickly, the reality is that a considerable amount of time is needed for mineralogical reactions to occur. Not only does pressure need to increase, but, far more important, the temperature of the sedimentary rocks must rise from near surface temperatures to between 500° and 600° C in response to intrusion of the granitic magmas. Again, we are faced with the time required for heat transfer from the magma into the wall rocks. As we saw in chapter eleven, large stretches of time are needed for such heat transfer even if fluid is present. The metamorphism of these large volumes of sedimentary rock, coupled with their burial and uplift, undoubtedly required, at a barest minimum, thousands of years.

Scattered throughout much of the Sierra Nevada are more than 150 bodies of tungsten-bearing rocks formed by metamorphism of limestone and shaly limestone accompanied by introduction of metal-bearing fluids from the batholith. Mineralogical evidence for temperatures of metamorphism of these rocks is consistent with the evidence of the andalusite-cordierite-bearing rocks.

POSTBATHOLITHIC EVENTS—UPLIFT AND EROSION

But there was also much geologic activity after the batholith was intruded. Because the batholith was formed several miles below the surface and is at the surface today, there must have been an enormous amount of uplift and erosion of the sedimentary rock cover and the upper part of the batholith. To the west of the Sierra Nevada lies the Great Valley of central California. The valley is the topographic expression of two major linked sedimentary basins, the Sacramento and San Joaquin basins, both of which contain many thousands of feet of sediment and poorly consolidated sedimentary rock derived from the Sierra Nevada to the east as well as the Coast Range to the west. Along the eastern margin of the Great Valley a variety of sedimentary rocks rests on eroded basement consisting of batholithic granitic rocks and metamorphic rocks belonging to the Sierra Nevada province. These sedimentary deposits consist predominantly of various kinds of sandstones and reworked volcanic materials. For example, the Ione Formation contains abundant rocks that are rich in quartz and the clay mineral kaolinite, an indicator that the source material was thoroughly weathered. The Ione Formation grades toward the east into several sinuous, finger-like deposits of gold-bearing gravel that occupy channels in the underlying granite and metamorphic rocks. These gravel deposits have been traced many miles up the gentle western slope of the Sierra Nevada. Geologists maintain that these channels were formed by rivers prior to the present-day drainage of the San Joaquin, Merced, Tuolumne, Stanislaus, Sacramento, Feather and other rivers. These ancient rivers transported clay, sand, gravel and gold from the rising mountain range. Much of the finer-grained sediment was carried out into the Great Valley where it formed the sheet-like Ione Formation.

Eventually parts of the Ione Formation, including the channel gravels, were covered by the Valley Springs Formation, consisting primarily of reworked volcanic ash deposits (tuff). Given that the prior drainage was covered by this formation, new drainage patterns developed. Evidence of a later set of river channels is preserved on some of the "table mountains" along the western foothills of the Sierra Nevada, such as at Oroville. One striking example is Table Mountain, exposed in Calaveras and Tuolumne Counties from Knights Ferry and Jamestown parallel to the course of the present-day Stanislaus River toward Sonora Pass (fig. 13.9).[11] The flat cap of Table Mountain consists of lava flows with a sinuous outcrop pattern. The lava flows filled a deep valley, no longer existing, that was excavated into the volcanic rocks of the Relief Peak Formation which, in turn, covers the Valley Springs Formation (fig. 13.10). The flows are about 45 to 295 feet thick. The topography and geology clearly indicate that thick lava flowed down the course of another, younger river prior to the present-day Stanislaus River. The lava is presently exposed as a cap along the top of the "ridge" because the surrounding rocks of the Relief Peak Formation are softer and more easily eroded than the lava flow cap. Regional

---

[11]Dallas D. Rhodes, "Table Mountain of Calaveras and Tuolumne Counties, California," in *Centennial Field Guide*, vol. 1: *Cordilleran Section*, ed. Mason L. Hill (Boulder, Colo.: GSA, 1987), pp. 269-72.

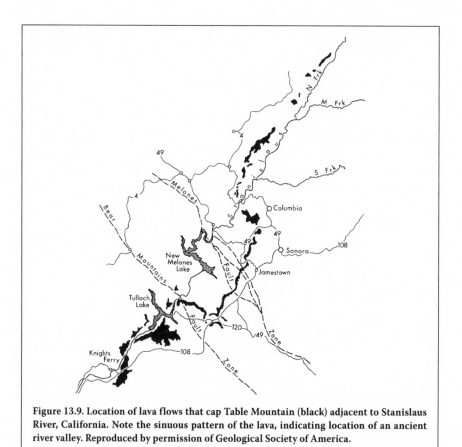

Figure 13.9. Location of lava flows that cap Table Mountain (black) adjacent to Stanislaus River, California. Note the sinuous pattern of the lava, indicating location of an ancient river valley. Reproduced by permission of Geological Society of America.

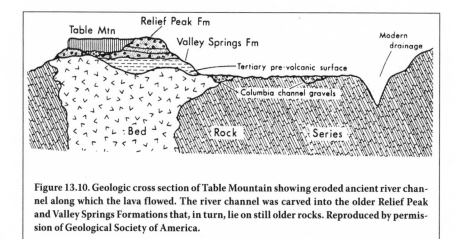

Figure 13.10. Geologic cross section of Table Mountain showing eroded ancient river channel along which the lava flowed. The river channel was carved into the older Relief Peak and Valley Springs Formations that, in turn, lie on still older rocks. Reproduced by permission of Geological Society of America.

evidence indicates that this latite lava flow originated at a source near present-day Sonora Pass and that some lava also flowed down the east side of the Sierra Nevada. Clearly, following eruption of this flow, a period of erosion removed sedimentary bedrock back down to the batholith basement. The modern river drainage developed since that episode of erosion and accompanied the period of uplift resulting in the present-day Sierra Nevada.

After these sedimentary rocks, gravels and volcanic rocks were deposited along the entire western side of the Sierra Nevada, the entire range was uplifted as much as 1.8 miles and tilted toward the west along a system of great normal faults that bounds the eastern flank of the range. This episode of uplift and tilting is the westernmost manifestation of rifting and block faulting throughout the Basin and Range province of Nevada and western Utah.

RECENT VOLCANIC AND GLACIAL ACTIVITY

During and after this final, and ongoing, stage of uplift and tilting of the Sierra Nevada, several centers of volcanic activity developed along the eastern margin of the rising range. From north to south these are the Mono-Inyo Craters in the vicinity of Mono Lake, the Long Valley Caldera, the Big Pine volcanic field and the Coso volcanic field. As an example, we'll consider the Long Valley Caldera in the vicinity of Mammoth Lakes, California.[12] Prominent among the early volcanic features are extensive lava flows and rubbly domes of thick, pasty rhyolite at Glass Mountain as much as 3,200 feet thick. Between Mammoth Lakes and Bishop is a widespread sheet of volcanic ash called the Bishop Tuff that formed in a cataclysmic eruption of the gigantic Long Valley caldera (10 x 20 miles) just east of Mammoth Lakes. So much material was erupted from the magma chamber that the land collapsed to form a steep-sided caldera, presently the site of Mammoth Lakes ski area. The older lava flows and the Glass Mountain rhyolite were both truncated by the caldera collapse. Although the Bishop Tuff forms thick deposits in the immediate vicinity of the caldera, the erupted ash was blown across much of the western United States and can be identified as far east as Nebraska. Hot magma yet resides below the Mammoth Lakes region; the U.S. Geological Survey monitors the magma, anticipating a future eruption!

In the vicinity of Mammoth Lakes, the Bishop Tuff locally rests on a glacial deposit called the Sherwin Till (fig. 13.11). One well-known, excellent road cut on U.S. 395, 23 miles north of Bishop, beautifully displays the superposition of the tuff on glacial till. Pumice and tuff at the base of the Bishop Tuff

[12]Roy A. Bailey, "Long Valley Caldera, Eastern California," in *Centennial Field Guide*, vol. 1: *Cordilleran Section*, ed. Mason L. Hill (Boulder, Colo.: GSA, 1987), pp. 163-68.

have both been dated by means of K-Ar dating of feldspar crystals at 730,000 years.[13] Extensive weathering of the underlying glacial deposits also indicates that these deposits are quite a bit older than the Bishop Tuff. That the interval of weathering of the glacial till was extensive can be deduced from the fact that granitic boulders as much as 25 feet below the contact with the overlying tuff are deeply disintegrated. In addition, a soil layer that is 1.5 feet thick caps the till just beneath the pumice deposit. The entire sequence is overlain by 10 feet

Figure 13.11. Fine-grained Bishop Tuff, erupted during explosive collapse of Long Valley Caldera, rests on top of thoroughly weathered glacial material, the Sherwin Till. Change in color marks the boundary. The entire sequence is cut by small dikes of tuff that were injected into fractures. Roadcut on U.S. 395, 23 miles north of Bishop, California. Photo by D. A. Young.

of gravel of probable fluvial origin. These gravels consist of very smooth, small round pebbles and boulders.

After the cataclysmic eruption that produced the Bishop Tuff, later volcanic rocks developed within the caldera. First, a large dome consisting of rhyolite built up near the center of the caldera. Further rhyolite eruptions occurred

[13]Robert P. Sharp, "Big Pumice Cut, California: A Well-Dated, 750,000-Year-Old Glacial Till," in *Centennial Field Guide*, vol. 1: *Cordilleran Section*, ed. Mason L. Hill (Boulder, Colo.: GSA, 1987), pp. 161-62; and G. Brent Dalrymple, "K-Ar Ages of the Friant Pumice Member of the Turlock Lake Formation, the Bishop Tuff, and the Tuff of Reds Meadow, Central California," *Isochron/West*, no. 28 (1980): 3-5.

around the outside of the central dome. For a time, lake sediments accumulated within the caldera. These in turn were succeeded by extensive outputs of trachybasalt and quartz latite lavas. Superimposed on the most recent volcanic deposits are a series of small volcanic cones belonging to the Mono-Inyo Craters. The eastern flank of the Sierra Nevada has experienced numerous successive episodes of volcanic activity since the most recent mountain uplift, in itself suggestive of a passage of a considerable amount of time.

After the Sierra Nevada was sufficiently elevated, and during episodes of volcanism associated with the Long Valley caldera and other volcanic centers on the east side of the mountains, glaciation affected much of the range. In particular, the valleys and canyons that extend from the high eastern crest of the range contain abundant evidence of several episodes of carving by glaciation. The evidence includes U-shaped valleys and cirques as well as terminal, lateral and recessional moraines that display cross-cutting relations and varying degrees of weathering.[14] The most recent of these glaciations is termed the Tioga stage, represented by moraines at the mouths of several valleys on the east side of the Sierra Nevada. Tioga deposits can normally be recognized by the fact that the till is much fresher than the till in other glacial deposits. For example, the granite boulders within the Tioga till are only slightly weathered, and granite boulders are also much more abundant relative to finer-grained material, indicating that they had not yet broken down into smaller fragments by the weathering process. Tioga deposits also can be seen locally overlying other glacial deposits. An older glacial episode is the Tahoe episode. Between these two is the Tenaya stage. Tahoe deposits display more extensive weathering of the granite boulders and the finer-grained material than do the Tioga deposits. In addition, the number of granitic boulders in a given volume of till is smaller in the Tahoe till, indicating their greater degree of weathering, as do also measurements of the overall size of the boulders in the Tahoe till, which is less than in Tioga glacial deposits. Still other glacial deposits predate the Tahoe glacial deposits and postdate the Sherwin till.

West of Bishop, the flanks of some of the valleys are lined with large ridges of Tahoe glacial deposits called *lateral moraines.* These moraines formed when an advancing glacier left behind deposits of rubble between the ice and the sides of the valley. As the glacier retreated during a phase of warming, the moraine was left behind as evidence of the glaciation. On the floor of the valley between the lateral moraines of Tahoe age on either side of the valley are a series of smaller arcuate ridges of till which are convex downslope. These are attributed to the Tioga phase of glaciation. Not only do they show a lesser

---

[14]More highly weathered tills typically contain a relatively low ratio of boulders to finer-grained matrix as well as relatively small boulders in comparison with less weathered tills.

degree of weathering than the materials of the Tahoe glaciation, but their superior position indicates that they are younger. One can envision that after the lateral moraines had been formed by an earlier glacier that had subsequently retreated, a newer glacier flowed down the same valley, flowing between the two lateral moraines, no doubt disturbing them and possibly adding some additional material to them (unless it was a smaller glacier than the Tahoe glacier). As the ice margin of the newer glacier began to retreat with gradual local warming and melting, rubble was left behind at the terminal margin of the glacier. If the glacier retreated in an episodic fashion, then every time the glacier margin remained somewhat stable, an accumulation of rubble was built up at the margin to form an end moraine. If the glacier retreated in fits and starts, then a sequence of end moraines would be left behind, and that is exactly what we find in the instance mentioned.

Now let's consider the implications of the glaciation, apart from all the volcanic activity going on more or less simultaneously. For glaciers to form, the Sierra Nevada had to achieve a significant elevation (remember that sedimentary deposits on top of the batholith appear to have been formed at a relatively low elevation) so that it would have a significant supply of precipitation but also be cold enough for snow, lots of it, to be precipitated, enough to form ice owing to the weight of overlying snow. There must have been passage of a substantial amount of time to allow for the uplift of the mountain range, the passage of many winters to accumulate sufficient snow to form a glacier, plenty of time for the glacier to flow several miles from high in the mountains down valleys toward low land where they would ultimately melt. Then there had to be an extended period of warming such that the glacier would retreat and leave the moraines behind. Then the process had to begin all over again—cooling, precipitation, glacier formation, ice flow, deposition of moraines and so on. Conditions favorable to significant glacier formation recurred more than a dozen times, in each instance resulting in a well-documented stage of advance and retreat on the eastern flank of the Sierra Nevada. Moreover, a substantial amount of time elapsed in between each glaciation cycle to allow the moraines to weather extensively.[15] The fact that the Tioga moraines have not yet weathered anywhere near as much as the Tahoe deposits indicates that a long time is needed for the weathering to take place.

IMPLICATIONS OF THE SIERRA NEVADA GEOLOGY

The Range of Light sheds a brilliant and clear light on the nature of God's ac-

---

[15]For a summary of the various glacial stages in the Sierra Nevada, see Norris and Webb, *Geology of California*, pp. 94-99.

tion in the physical world. We need only open our eyes to see it. The evidence from the Sierra Nevada (and we have presented only the barest outline of a much more complex history) makes it plain that God, in his wisdom, chose to form that part of the world through a fascinatingly complicated set of geologic processes that took staggeringly long stretches of time. The Sierra Nevada is not unique in this regard. In looking closely and carefully at the geology of virtually any region on the Earth in detail, we should all understand that God took plenty of time to make his world. But then time is nothing to the Creator who is from everlasting to everlasting (Psalm 90:2).

14

RADIOMETRIC DATING

Part One

IN THE LAST SIX CHAPTERS WE DISCUSSED a wide range of geologic evidences indicating that Earth has undergone a long, complex and dynamic history. We have shown that geology does not support the idea that our home planet is only a few thousands of years old. Although the information gleaned from the rocks in the field demands vast amounts of time, none of the evidence we have discussed thus far indicates precisely how much time geologic processes may have taken. For example, we can make reasonable estimates for cooling times of igneous rock bodies, but these estimates still do not tell us exactly when an igneous body was formed. We might state that a certain body of granite is younger than the surrounding sedimentary rocks on the basis of cross-cutting relationships, but field data do not tell us how long ago the granite was intruded. Or we might establish that a layer of volcanic ash is younger than sedimentary rock layers below it and older than sedimentary rock layers above it on the basis of the principle of superposition, but this field evidence does not tell us when the volcanic ash was erupted and deposited. To answer the question of precisely when, geologists need a set of methods capable of providing accurate absolute ages.

The task of assigning ages to geologic events pertains to the discipline of *geochronology*. The need for absolute dating of geologic events is addressed by *radiometric dating,* a large set of scientific methods and procedures that involve measurements of the extent of radioactive decay for a variety of chemical elements within geologic samples. Because radiometric dating necessarily entails very technical scientific reasoning, we present the basics of radiometric dating as simply as possible. Young-Earth creationists do not hesitate to use technical arguments invoking mathematics in their support of a young Earth. This practice may lend an air of authority and believability to some of their claims in the eyes of those who are not scientists. The fallacies of their claims must be refuted with technical arguments. We'll try to make it as painless as possible.

The claim that radiometric dating is the only support for the idea that Earth is old is not valid. As a result, attempts to discredit radiometric dating are pointless if the goal is to support the notion that Earth is really young. Long before the development of radiometric dating methods, there was abundant geologic evidence for an ancient world. Even if young-Earth creationists could somehow discredit all radiometric dating methods, the conclusion that Earth is very old is firmly established on solid scientific grounds such as those we laid out in the preceding chapters. What would be lost if radiometric dating were totally invalidated would be a more accurate determination of the age of Earth and of geologic events.

In this chapter we present a brief synopsis of radiometric dating methods.[1] Fasten your seat belts! Put on your thinking caps! Hang on. Here we go!

CHEMICAL ELEMENTS, ATOMS AND ISOTOPES

All methods of radiometric dating are based on the recognition that some atoms of a particular chemical element spontaneously change into atoms of a different chemical element. Such atoms are *unstable,* and we must begin our discussion with an explanation of why these atoms are unstable.

The universe consists of approximately one hundred different chemical elements. Many chemical elements, such as oxygen, carbon, iron, copper, zinc, gold and sulfur, are familiar to everyone. Other chemical elements are not so well known. Among those that are not likely to become part of dinnertime

---

[1]Those who desire a more detailed treatment of the general principles of radiometric dating and of specific methods should consult Gunter Faure and Teresa M. Mensing, *Isotopes: Principles and Applications,* 3rd ed. (Hoboken, N.J.: Wiley, 2005); and Alan P. Dickin, *Radiogenic Isotope Geology* (Cambridge: Cambridge University Press, 1997). Also of great value are the summaries of radiometric dating in G. Brent Dalrymple, *The Age of the Earth* (Stanford, Calif.: Stanford University Press, 1991); and G. Brent Dalrymple, *Ancient Earth, Ancient Skies: The Age of Earth and Its Cosmic Surroundings* (Stanford, Calif.: Stanford University Press, 2004). These works include references to highly technical works devoted to detailed discussion of individual methods.

conversation anytime soon are hafnium, dysprosium, neodymium, rhenium and bismuth. Each chemical element consists of atoms that are distinguishable from the atoms of all other chemical elements by the number of *protons,* extremely small particles that possess a positive electrical charge, that are contained within their nuclei. For example, all atoms of hydrogen in the universe, *by definition,* contain one and only one proton within their nuclei. All atoms of calcium contain 20 protons within their nuclei. All atoms of uranium contain 92 protons within their nuclei. The number of protons within the nucleus of an atom is termed the *atomic number.* Thus, each chemical element is characterized by its own atomic number. For example, the atomic number of hydrogen is 1, that of carbon is 6, that of oxygen is 8, and that of iron is 26.

Typically, atoms of various chemical elements also possess *electrons* orbiting about a nucleus. The number of electrons is normally the same as, or very close to, the number of protons in the nucleus. Electrons are tiny particles that weigh only 1/1,836 as much as a proton and possess a negative electrical charge that is exactly as strong as that of the proton. If a proton and an electron interact they may combine to form a *neutron,* a particle that is slightly more massive than a proton but lacking an electrical charge. Atoms of most chemical elements contain a variable number of neutrons in their nuclei. Most hydrogen atoms, however, contain no neutrons. The total number of protons plus neutrons within the nucleus, called the *atomic mass,* of the vast majority of hydrogen atoms is, therefore, one. However, a small percentage of hydrogen atoms do contain a single neutron in their nuclei along with the one proton. Because there are two major particles in the nuclei of these hydrogen atoms, their atomic mass is two. This form of hydrogen is called *deuterium* or "heavy hydrogen." There is yet a third form of hydrogen known as *tritium* that contains two neutrons in its nucleus along with the single proton. Therefore, the atomic mass of a tritium atom is three, because there are two neutrons and one proton in the nucleus.

These different kinds of hydrogen atoms are referred to as *isotopes.* Hydrogen has three isotopes: "normal" hydrogen atoms with just a single proton and no neutrons, deuterium and tritium. In the universe, as it turns out, the majority of chemical elements possess more than one kind of isotope. So, the isotopes of a particular chemical element are distinguished from one another by the number of neutrons within the nucleus, that is, by their different atomic masses. Oxygen provides another example. Oxygen atoms have eight protons in their nuclei, by definition. The most abundant isotope of oxygen also contains eight neutrons, and its atomic mass is 16. Some oxygen atoms, however, contain nine neutrons in their nuclei so that the atomic mass is 17. And a somewhat greater percentage of oxygen atoms contain ten neutrons in their nuclei, and their atomic mass is 18. These three isotopes of oxygen are sometimes referred to as oxygen 16 (^{16}O), oxygen 17 (^{17}O) and oxygen 18 (^{18}O).

The isotopes of oxygen are all naturally stable. That is, an oxygen 16 (^{16}O) atom always remains an oxygen 16 atom unless we bombard it with some other particle in a high-energy particle accelerator. But the isotope is stable in the sense that the atom does not spontaneously alter into some other kind of isotope. The same is true for oxygen 17 and oxygen 18. In other words, these three oxygen isotopes are not unstable, that is, not radioactive. This fact is rather comforting because one fifth of the air that we breathe is oxygen, and inhaling radioactive oxygen atoms on a regular basis would likely cause damage to our lungs. Because of the stability of oxygen atoms, radiometric dating methods cannot involve oxygen.

In the case of a *radioactive* isotope, however, its atoms are unstable. Given sufficient time, an atom of a radioactive isotope will spontaneously change or "decay" into an atom of a different chemical element. There are dozens of naturally-occurring radioactive isotopes, and they decay in a variety of distinct ways.

RADIOACTIVE DECAY MECHANISMS

One kind of radioactive decay is illustrated by rubidium 87 (^{87}Rb). There are two isotopes of rubidium. The more abundant isotope rubidium 85 (^{85}Rb), which contains 37 protons and 48 neutrons, is stable. In contrast, rubidium 87 (^{87}Rb), an unstable isotope, contains 37 protons and 50 neutrons. Sooner or later, an individual atom of ^{87}Rb will spontaneously transform into an atom of the isotope strontium 87 (^{87}Sr), an atom that contains 38 protons and 49 neutrons in its nucleus. In going from ^{87}Rb to ^{87}Sr a proton was gained and a neutron was lost. The process that was involved is termed *beta decay*. In beta decay, in effect, one of the neutrons in the ^{87}Rb atom is transformed into a proton and an electron. The net result is that a proton is produced, making the ^{87}Rb atom into ^{87}Sr, and a neutron is lost. Because one nuclear particle was gained while one was lost, the atomic mass (87) remains the same. The electron is ejected from the atom as a high-energy *beta particle*. Many isotopes disintegrate by *beta decay*.

Another kind of decay is illustrated by samarium 147 (^{147}Sm). Samarium, with 62 protons, has the atomic number 62. There are also 85 neutrons in the nucleus of this isotope, just one of many samarium isotopes. Sooner or later, a samarium 147 atom will spontaneously decay, and, when it does, it effectively releases two protons and two neutrons from its nucleus so that the number of protons is reduced to 60 and the number of neutrons is reduced to 83. The result is that the new atom is neodymium 143 (^{143}Nd). There are four fewer major particles in the nucleus of ^{143}Nd than there are in ^{147}Sm. The emitted combination of two protons and two neutrons is referred to as an *alpha particle*, and the type of decay is called *alpha decay*. An alpha particle is equivalent to the nucleus of a helium atom. As a result, during radioactive decay of samarium 147 and other isotopes that decay by the alpha process, helium may accumulate in the radioactive sample.

A third kind of decay is called *electron capture*, illustrated by potassium 40 (^{40}K). Potassium is a very abundant element, and, although potassium 39 (^{39}K) is the most abundant isotope of potassium, there is plenty of ^{40}K in very common minerals like feldspar and mica. Potassium 40 can decay in two ways. One way is by the beta process, but some potassium 40 also decays by electron capture, a process that is somewhat the reverse of beta decay. In electron capture, an electron that orbits close to the nucleus is captured by the nucleus and, in effect, interacts with one of the protons to produce a neutron. The end result is that a proton is lost and a neutron is formed. Thus, potassium 40 with 19 protons and 21 neutrons ends up being converted into argon 40 (^{40}Ar) with 18 protons and 22 neutrons.

Last, some isotopes decay by means of a long chain of decay events until they are finally converted into their ultimate stable end-product isotope. A very important example is the decay of uranium 238 (^{238}U). Uranium 238 is a very heavy isotope with 92 protons and 146 neutrons. When it first decays, ^{238}U emits an alpha particle and is converted into thorium 234 (^{234}Th), consisting of 90 protons and 144 neutrons. But thorium 234 is also radioactive and converts into protactinium 234 (^{234}Pa) by beta decay. ^{234}Pa is also radioactive. Beginning with ^{238}U, a long series of decays involving eight alpha decays and six beta decays eventually results in the production of the stable isotope lead 206 (^{206}Pb), consisting of 82 protons and 124 neutrons.

MATHEMATICAL DESCRIPTIONS OF THE RADIOACTIVE DECAY PROCESS

The nature of the radioactive decay process and the basic mathematics describing that process were established and refined very early in the twentieth century, primarily by Rutherford and Soddy.[2] Radioactive decay must be modeled statistically because we have no way of predicting when a specific individual atom of some isotope is going to disintegrate spontaneously. We could observe an individual atom for ten billion years, and it would be completely unpredictable when that atom would decay. The atom might decay after only ten minutes or it might well wait the full ten billion years before disintegrating.

---

[2]See especially E. Rutherford and F. Soddy, "The Cause and Nature of Radioactivity, Part I," *Philosophical Magazine* 4, ser. 6 (1902): 370-96; and "Radioactive Change," *Philosophical Magazine* 5, ser. 6 (1903): 576-91.

On the other hand, we can predict very accurately the rate at which an extremely large number of radioactive atoms of a specific isotope within a sample would decay. Imagine that we have a pure sample of a compound such as rubidium chloride (RbCl) in which the rubidium consists entirely of the radioactive isotope [87]Rb. A sample that has a volume of only one cubic inch contains trillions of atoms. The decay of a very large number of atoms will follow an *exponential decay law* that is expressed mathematically in a manner exactly analogous to that of growth and decay phenomena widespread in the natural world.[3]

In simple terms, after a specified amount of time that we call a *half-life*, half the original number of radioactive atoms in the sample will decay into their end-products. For example, suppose our sample contains exactly one trillion atoms of rubidium 87 as the radioactive "parent" isotope. By the end of one half-life, half of the original amount of rubidium 87 will decay into the "daughter" isotope strontium 87 ([87]Sr). We now have only 500 billion atoms left in our sample, and 500 billion atoms of strontium 87 have been produced. After a lapse of a second half-life, only half of the 500 billion [87]Rb atoms, namely, 250 billion atoms, remain. After a third half-life only half of the 250 billion atoms, that is, 125 billion atoms are left. After a fourth half-life, only 62.5 billion atoms remain, but there are now 937.5 billion atoms of daughter [87]Sr, and so on. Thus, after each half-life has elapsed, only half the amount remaining at the conclusion of the preceding half-life is left in the sample. This decay behavior is illustrated in the graph shown in figure 14.1. The theory of radioactive decay indicates that this is how radioactive isotopes should behave, and measurements of decaying isotopes indicate that they do in fact behave that way. Geochronologists just won't be able to predict *which* [87]Rb atoms will decay.

Figure 14.1 indicates that we have a basis for determining the age of a sample of radioactive isotope. For example, if a sample now contains an equal number of [87]Rb isotopes and of [87]Sr isotopes, and the sample originally contained no [87]Sr atoms whatever, then our sample is one half-life old. If [87]Rb and [87]Sr present in a ratio of 1:3, the sample is two half-lives old, and so on.

Of course, a big question is, What *are* the values of the half-lives of the various radioactive isotopes? If the values of the half-lives of radioactive isotopes have not been determined, the age of any radioactive material cannot be calculated. We must know the rate at which radioactive isotopes decay. In

---

[3]The exponential decay law for radioactive substances is $N = N_o e^{-\lambda t}$, where N is the amount of radioactive isotope in a sample; N_o is the amount of radioactive isotope in a sample at the time of its formation; λ is the decay constant of the isotope; and t is the age of the sample. e is the base of the natural logarithm and has the value e = 2.7183.

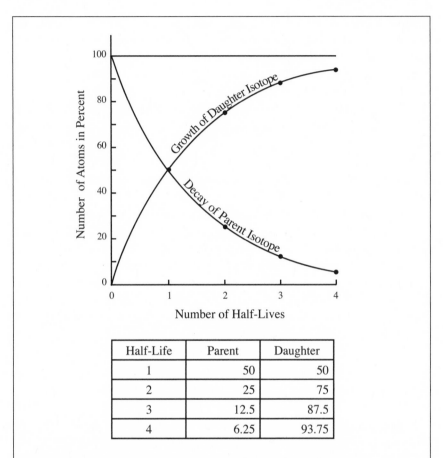

Figure 14.1. Diagram of radioactive isotope half-life: the diagram illustrates the decrease in amount of parent radioactive isotope and increase in amount of daughter isotope with time. Note that for each half-life that has elapsed the amount of parent isotope decreases by one-half.

fact, the half-lives have been measured by means of a variety of laboratory experiments. For some isotopes, more than one method has been used to determine the half-life. For a few isotopes it is difficult to measure the half-life. The half-life, represented by the symbol $\tau_{1/2}$, is related mathematically to a quantity known as the *decay constant* by means of the simple mathematical expression $\tau_{1/2} = 0.693/\lambda$.[4] The decay constant, always a very tiny number, represents that

---

[4]$N = \frac{1}{2}N_0$ when $t = \tau_{1/2}$. In other words, $\tau_{1/2}$ is the *half-life*, the age of the sample when the amount of radioactive isotope in the sample is one-half the original amount. By substituting these two expressions into the radioactive decay law equation, we obtain $\frac{1}{2}N_0 = N_0 e^{-\lambda\tau_{1/2}}$ which by rearrangement yields $e^{-\lambda\tau_{1/2}} = \frac{1}{2}$ and ultimately $\tau_{1/2} = 0.693/\lambda$.

probability of an individual atom's decay per year. The decay constants and half-lives are generally well known for a wide variety of isotopes. Table 14.1 lists the half-lives and decay constants of some of the more useful geological radioactive isotopes.

Table 14.1. Decay Constants and Half-Lives of Some Geologically Important Radioactive Isotopes

Radioactive Isotope	Decay Constant (years^{-1})	Half-Life (years)
^{10}Be	4.62 x 10^{-7}	1.5 x 10^6
^{14}C	0.1209 x 10^{-3}	5730
^{26}Al	9.60 x 10^{-7}	7.20 x 10^5
^{36}Cl	2.3 x 10^{-6}	3.08 x 10^5
^{40}K	0.581 x 10^{-10}	1.19 x 10^{10}
^{87}Rb	1.42 x 10^{-11}	48.8 x 10^9
^{147}Sm	0.654 x 10^{-12}	1.06 x 10^{11}
^{232}Th	4.9475 x 10^{-11}	14.01 x 10^9
^{235}U	9.8485 x 10^{-10}	0.7038 x 10^9
^{238}U	1.55125 x 10^{-10}	4.468 x 10^9

The value of the half-life of a particular isotope is extremely important in determining whether that isotope is useful for measurement of the age of geologic materials. For example, the isotope beryllium 7 (^{7}Be) has a half-life of 0.146 years, approximately two months. This means that most of a quantity of beryllium 7 will have decayed to its daughter product within a year. Some isotopes decay even more rapidly than ^{7}Be. The likelihood of finding measurable quantities of ^{7}Be in the crystal structure of a mineral specimen that is 100,000,000 years old or even only 10,000 years old would be nil. Virtually all of the isotope would have decayed in that period of time. As a result, ^{7}Be is not used for dating old minerals. By the same token, if the half-life is sufficiently large, the value of the half-life gives an indication of the range of ages of geologic materials that can be obtained.

If we wish to date materials that are probably only hundreds or thousands of years old, we might employ carbon 14 (^{14}C) with a half-life of 5,730 years. To date minerals and rocks that are a few million to a few tens of millions of years we might utilize ^{40}K with a half-life of 1.31 x 10^9 years. On the other hand, if our rock is several hundred million to a few billion years in age we might employ a method involving ^{238}U with a half-life of 4.468 billion (4.5 x 10^9) years or ^{147}Sm with a half-life of 106 billion (1.06 x 10^{11}) years. If we use the ^{14}C method on a very old sample, the amount of ^{14}C remaining in the sample would be so extremely small that the chances for large errors in measurement would result, and if we tried to date an extremely young rock by a ^{238}U-based method, a negligible amount of the daughter isotope lead 206 (^{206}Pb) would have accumulated, again resulting in large errors in the measurement of lead.

ARE RADIOACTIVE DECAY RATES CONSTANT?

If it can be demonstrated that radioactive decay rates are not constant, then the whole enterprise of radiometric geochronology is undercut. The dating of a geologic object will be reliable if and only if the decay constant or the half-life of a given isotope really has a constant value. In other words, a dating method would yield unreliable results if the decay constant changed over time. In that case, the decay constant would not be a constant! Over the years, young-Earth creationists have tried all manner of ways to discredit radiometric dating.[5] In recent years, the Institute for Creation Research and the Creation Research Society have cosponsored a technical team of young-Earth creationists known as the RATE (Radioisotopes and the Age of The Earth) group. The RATE group has moved beyond appeals to creation of apparent age and beyond some of the fallacious arguments employed in the past against radiometric dating.[6] They are now promoting the idea that there were major events in which all radioactive decay constants were changed so drastically that rates of radioactive decay increased by several orders of magnitude. The outcome would be that isotope ratios suggesting an old event were really produced in a much briefer period of time. The ages of minerals and rocks are all anomalously too old. Members of the RATE group speculate that the main events during which nuclear decay rates speeded up in spectacular fashion were creation week and the Genesis Flood.[7]

Young-Earth physicist D. Russell Humphreys of Sandia National Laboratories and, more recently, the Institute for Creation Research, is one of the members of the RATE group. He has claimed that the Bible implies that such an

---

[5]Davis A. Young, *Christianity and the Age of the Earth* (Grand Rapids: Zondervan, 1982) summarized several previous efforts to invalidate radiometric dating by Harold Slusher, Robert Whitelaw, D. O. Acrey, Melvin Cook, and Whitcomb and Morris.

[6]For advocacy of the concept of apparent age, see J. C. Whitcomb Jr. and H. M. Morris, *The Genesis Flood: The Biblical Record and Its Scientific Implications* (Philadelphia: Presbyterian & Reformed, 1961), pp. 232-39 and 344-57. Whitcomb and Morris, for example, claimed that some radiometric dates provided apparent ages only because God had originally created isotope ratios in minerals that looked as if they had resulted from a great amount of radioactive decay.

[7]The theory of changes in decay rates has been advocated in Larry Vardiman, Andrew A. Snelling and Eugene F. Chaffin, eds., *Radioisotopes and the Age of the Earth: A Young-Earth Creationist Research Initiative*, 2 vols. (El Cajon, Calif.: Institute for Creation Research and Creation Research Society, 2000, 2005).

event occurred.[8] Humphreys cited biblical texts, such as Deuteronomy 32:22 and Psalm 18:7-8, which refer to fires and the foundations of the earth, and he suggested that such texts have nuclear processes in mind and support the idea of accelerated radioactive decay, especially during the Genesis Flood. Unfortunately for Humphreys, the scriptural texts actually refer, as they state, to fires and the foundations of the earth. Humphreys has seriously misused Scripture by pressing the biblical text into uses for which it was not intended. Scripture was not given to provide us with all sorts of cryptic, arcane clues to the inner workings of the cosmos. The Bible is a book of salvation through Jesus Christ (John 20:31). As the Westminster Shorter Catechism states, Scripture principally teaches us what we are to believe concerning God and what duty God requires of us. The message of the Old Testament was originally addressed to the ancient Israelites in language adapted to their cultural situation, including the use of imagery suited to them. The Israelites were a prescientific people who needed a saving message from God, not subtle hints about twenty-first-century physics!

Apart from the fallacious appeal to Scripture, the claim that radioactive decay rates might have been speeded up poses insurmountable problems from a scientific viewpoint. Geologists and geophysicists, knowing that the validity and utility of radiometric dating hinges on the constancy of the half-lives or decay constants of radioactive isotopes, have not blindly assumed that the decay constant cannot be changed in time. To assess whether decay constants really are constant, atomic physicists have conducted a wide range of laboratory experiments in an effort to change the rate at which radioactive decay takes place. As early as the first decades of the twentieth century, Rutherford heated radon gas to 2500° C and subjected it to pressures of 1,200 atmospheres and found no change in activity. Since then, physicists have attempted to force changes in decay rates. A variety of radioactive isotopes have been subjected to extremely high and low temperatures; high pressures; bombardment by high-energy particles; irradiation with x-rays and gamma rays; electrical and magnetic fields of varying intensity; and very high accelerations of gravity. Radioactive isotopes have been also placed in a variety of chemical compounds. Experimentalists have successfully altered nuclear decay rates of a small number of radioactive isotopes such as ^{7}Be by as much as 1.5 percent by altering chemical and physical environments.[9] Most of the experimentally induced changes in decay constants, however, amounted to only a few tenths of one

---

[8]D. R. Humphreys, "Accelerated Nuclear Decay," in Vardiman, Snelling and Chaffin, *Radioisotopes and the Age of the Earth,* 1:351-57.

[9]Chih-An Huh, "Dependence of the Decay Rate of ^{7}Be on Chemical Forms," *Earth and Planetary Science Letters* 171 (1999): 325-28.

percent. Isotopes that decay by electron capture appear to be the most susceptible to very slight variations in the decay constant.[10] *To date, no evidence for perturbation in the decay constant of any geologically important radioactive isotope has been found.*

Even if by some as yet unknown physical mechanism the decay rates of geologically important radioactive isotopes such as ^{238}U, ^{87}Rb, ^{147}Sm or ^{40}K could be shown to vary by as much as, say, five percent, young-Earth creationism would have little cause for rejoicing. Although the method on which modern geologists depend for obtaining reliable ages would no longer yield ages with as high a degree of accuracy, there would be no challenge whatever to the concept of an extremely ancient Earth. Given a 5 percent variability in decay constants, a rock that yields an age of one billion years might be as much as 1.05 billion years or as little as 0.95 billion years. But let's face it; a date of 950 million years is not appreciably closer to a 6,000-year-old Earth than is a date of 1,050 million years. But recall that so much variation in the decay constants of geologically important radioactive isotopes has never been confirmed experimentally.

So how do young-Earth advocates deal with this situation? Even Humphreys granted that the observed changes in geologically unimportant decay constants are "minuscule compared to the million-fold or greater acceleration of decay rates which is required by the evidence for a young earth." Not discouraged by such inconvenient realities, Humphreys suggested that "we should not be surprised if we find evidence that God has supernaturally intervened, either directly or indirectly, to produce such large changes."[11] The decision of Humphreys to invoke a divine miracle to support his view is a clear indication that his hypothesis is in trouble. He is making a tacit admission that there is no credible, experimentally confirmed, scientifically viable mechanism as to why radioactive decay rates suddenly increased more than a million times at the onset of the Genesis Flood.

However, if rates of radioactive decay were in fact speeded up a million-fold, enormous amounts of heat would be generated. Every time a radioactive atom disintegrates in a mineral there is release of energy and consequent production of heat. Suppose that radioactive decay occurred as rapidly as suggested by Humphreys: an enormous quantity of heat would have been released in a very short time. The heat that would normally have been dissipated slowly and

---

[10]For reviews of experimental studies of changes in rates of radioactive decay, see G. T. Emery, "Perturbation of Nuclear Decay Rates," *Annual Review of Nuclear Science* 22 (1972): 165-202; and H.-P. Hahn, H. J. Born and J. I. Kim, "Survey on the Rate Perturbation of Nuclear Decay," *Radiochimica Acta* 23 (1976): 23-37.

[11]Humphreys, "Accelerated Nuclear Decay," p. 334.

gradually, say, over a period of several hundred million years would have been emitted within the one year in which the Genesis Flood occurred! The release of enormous quantities of heat in such a short amount of time would have resulted in extensive melting of rocks, boiling of the ocean, and heating of the atmosphere to such an extent that life would have been wiped out. Talk about global warming! If Humphreys were correct, no one would be on Earth to talk about nuclear decay rates. The proposal of vastly speeded up decay rates is completely untenable.

Humphreys attempted to get rid of the excess heat by invoking spectacular increases in the rate of expansion of the universe during the creation week and during the Flood year. Appealing to the phenomenon of energy loss of radiation traveling through space in an expanding universe, that is, a universe in which space itself is stretching and in which the wavelengths of radiation are lengthening (the so-called red shift), Humphreys proposed that a miraculous acceleration of the expansion of the universe could absorb the enormous amount of heat energy released by catastrophic radioactive decay. However, he was left with trying to explain why the universe should suddenly expand at an unbelievable rate during creation and the Flood year. Humphreys reasoned that the rapid expansion of the universe would profoundly affect forces between particles in atomic nuclei so that decay constants would be altered. Although trying his best to come up with a scientific explanation for catastrophic changes in radioactive decay rates, in the end, Humphreys unwittingly conceded that his proposals are purely conjectural, quasi-scientific stretches with no evidential support whatever. Thus, he made statements such as "I prefer to start with theories in which God would make a very precise change, in order to limit the effects to just those which would accomplish His purposes."[12] In response to questions that might be asked about his theory such as "How would God change the boson masses?" Humphreys said that "if we could follow this chain (or any other) backwards far enough, we might find a point where God has intervened supernaturally."[13]

---

[12]Ibid., pp. 364-65.

[13]Ibid., p. 366. It should further be noted that the RATE group, realizing that an unrealistically large quantity of heat would be generated if nuclear decay rates were drastically increased, now claims to have found evidence that extremely rapid cooling also accompanied the alleged accelerated nuclear decay so that temperature on Earth would not necessarily have gotten out of hand (see D. R. Humphreys, "Young Helium Diffusion Age of Zircons Supports Accelerated Nuclear Decay," in Vardiman, Snelling and Chaffin, *Radioisotopes and the Age of the Earth*, 2:68). Volume 2 also contains forays into the effects of altering various physical constants. Because response to the arguments from physics advanced by members of the RATE team are beyond the scope of this book and the expertise of its authors, we recommend that interested readers see a critical review by a Christian physicist, Randy Isaac, "Assessing the RATE Project," *Perspectives on Science and Christian Faith* 59 (2007): 143-46.

In the absence of explicit biblical assertions to the effect that God miraculously speeded up radioactive decay rates or miraculously speeded up the
expansion of the universe during the time of creation and the Deluge, Humphreys's appeal to miracle is completely ad hoc. Furthermore, in the absence
of an evidentially supported, coherent scientific explanation for a dramatic increase in radioactive decay rates and universe expansion rate during creation
week and the Flood year or at *any* time, for that matter, we are also inclined
to dismiss his scientific speculations. The scientific community is highly unlikely to place any credence whatsoever in Humphreys' speculations given the
overwhelming experimental support for the essential stability of nuclear decay
rates. To sum up, although we totally agree that God has performed miracles
and is free to act as he chooses, we are very suspicious of the validity of any position that needs to invoke miracles arbitrarily to prop up a theory that, while
desperately wanting to be considered scientific, not only lacks evidential support but also flies in the face of abundant scientific evidence that demolishes
the theory in question.

FUNDAMENTAL ASSUMPTIONS IN THE APPLICATION OF
RADIOACTIVITY TO DATING GEOLOGIC MATERIALS

The physics on which radiometric dating is based is well established. Moreover, the decay constants appear to be constants indeed. Thus, if we want
to determine the age of a sample, we next develop some suitable analytical
methods for measuring the abundances of the parent and daughter isotopes
in the sample. As noted in chapter five, mass spectrometers have been used
for decades to provide accurate, precise measurements of isotope or isotope
ratio abundances. For analysis by mass spectrometer, a mineral or rock sample
is typically crushed and dissolved in various chemical reagents to produce a
solution of the relevant isotopes. Solutions are fed into the mass spectrometer
as ionized gases, and the instrument magnetically separates and counts the
atoms of the isotopes of differing masses. In more recent years, a new instrument known as the Sensitive High-Resolution Ion MicroProbe (nicknamed
SHRIMP) has been developed for the isotope analysis of individual mineral
grains, particularly those of zircon. These grains do not need to be crushed
and dissolved in this technique. Atoms sputtered from the mineral surface
by an ion beam are fed through a mass spectrometer and analyzed. These
analytical instruments have been thoroughly tested and refined and routinely
produce results of extremely high accuracy and precision.

So, the fundamental, physical basis for radiometric dating is perfectly sound
and can, in principle, be successfully applied, provided that we have a suitable
means of isotope analysis. But now we need to place radioactive decay into

geologically realistic contexts. If significant errors occur in radiometric dating, they are most likely to enter in because of geologic factors. Minerals and rocks commonly have complex geologic histories. Chemical alteration of a rock or mineral during its long history may cause leakage of parent or daughter isotopes. Geochemists have developed protocols and techniques to assess problems of leakage and reject specimens or ages that are suspect. In some cases, as we shall see, geochemists can actually turn element migration to their advantage to wring more information from the rocks! However, upon superficial examination, the work of discriminating between good and bad ages can appear to be arbitrary to one unfamiliar with geochemistry. Unfortunately, this situation has been exploited by young-Earth creationists.

Young-Earth creationist John Woodmorappe has vigorously challenged the validity of radiometric dating in numerous publications, the most important of which is *The Mythology of Modern Dating Methods*.[14] Woodmorappe has alleged that three fallacies characterize the claims of those who defend radiometric dating. The first fallacy he called CDMBN (Credit Dating Methods for "ostensible" successes but Blame Nature for failures). He has charged proponents of radiometric dating with a "heads I win, tails you lose" mentality that exempts dating from rational criticism and protects it from falsification. What Woodmorappe and others failed to appreciate is that the basis of radiometric dating is a matter of physics. If he wants to discredit or falsify the methods, he must find a way to discredit the physics of radioactivity and to show either that the decay equation is false or that the decay constant is not constant. In fact, writing before the RATE conclusions were published, Woodmorappe expressed disapproval of attempts by fellow creationists to find ways to change decay constants! It would also be perfectly legitimate for Woodmorappe to point out that a particular method almost never yields a satisfactory result and so should be discarded, or else that the method yields less satisfactory results than some other method. That is exactly what geologists have been doing throughout the history of radiometric dating. As we will see, geochronologists have largely abandoned the Rb-Sr mineral dating method, not because it is wrong in principle, but because we now have developed superior methods that overcome difficulties in successful application of the method.

Much the same can be said for Woodmorappe's second alleged fallacy: ATM (Appeal To Marginalization). Here he criticized geochronologists for attempting to account for poor results by appealing to unusual geologic circumstances. Well, exactly! It's the complexities of geology that make the assessment of analytical results both challenging and fun. It is the *business* of

---

[14]John Woodmorappe, *The Mythology of Modern Dating Methods* (El Cajon, Calif.: Institute for Creation Research, 1999).

geochronologists to evaluate their own work and to discard problematic ages. Later in this chapter we provide an example of a situation in which geologic circumstances would almost certainly render an age suspect. But as long as the basic physics remains uncontested, the fundamental dating methods remain valid in principle.

Woodmorappe's third fallacy he called ATT (Appeal To Technicalities). As an example, he mentioned that geologists might reject an "unwelcome date" because of a technicality like "the host rock showed traces of alteration." But that's just another instance of the complexity of geologic conditions that needs to be taken into account. All three of Woodmorappe's alleged fallacies essentially amount to the same thing, namely, that radiometric dating methods should be discredited and thrown out because bad or meaningless ages are sometimes obtained.[15] But the reason that bad, discrepant, meaningless, puzzling or unexpected ages are obtained has nothing to do with the established physics and mathematics of radioactive decay. Such unusable ages come about because of geologic factors, pure and simple (unless the analyst did a bad job). "Explaining away" such ages is not, as Woodmorappe charged, blaming nature: it is recognizing the profound complexities of nature. It is taking into account the complexity of geologic processes and history. Sometimes, a "bad" age for a rock can alert the geologist to a geologic process or event that otherwise might have been overlooked. Again, if Woodmorappe really wants to discredit the fundamental method of radiometric dating, he must show the falsity of the simple equation (see footnote 3) $N = N_0 e^{-\lambda t}$ with respect to the radioactive decay of a large quantity of a radioactive isotopes. Then, *and only then*, might geologists have reason to abandon radiometric dating. Even then, they would try to redevelop a whole new set of dating techniques on the basis of whatever new equation is established.

Now let's explore the role of geologic factors in radiometric dating. If we want to apply the methods of radiometric dating to mineral and rock specimens successfully, then certain geologically relevant conditions need to be met. These conditions are extremely important, and we must pay close attention to them. We have already noted the physically relevant condition that the decay constant really is a constant, and we have seen that there is no evidence of sufficient weight to challenge the validity of that condition.

---

[15]For example, Steven A. Austin, "Testing Assumptions of Isochron Dating," in Vardiman, Snelling and Chaffin, *Radioisotopes and the Age of the Earth*, 2:325-92. As have mainstream geochronologists before him, Austin conducted a detailed study of the behavior of radiometric systems in a couple of suites of rocks from different localities and discovered some discordances in ages. On that basis, Austin concluded that the assumption of constant decay rates might be invalid. There is, of course, a far cry between the possibility of small-scale changes in decay rates and the kind of change envisioned by members of the RATE group.

A very important geologically relevant condition is that neither parent nor daughter isotope has migrated into or out of the specimen that is being dated during its history.[16] If we are to obtain an accurate age for the time of formation of a sample, for example, we need to know that the mineral or rock being dated has behaved as a *closed system* since it was first formed. If either parent or daughter isotope escaped from a specimen after it had formed, the isotope measurements would probably yield ages that were either too large or too small, depending on exactly what happened. We must do our homework regarding the geologic context, however. Some isotopes are relatively mobile when the rocks in which they occur are heated. For example, helium and argon may easily be lost from radioactive minerals during heating. Rubidium and potassium are capable of migrating in intergranular fluids during heating of rock specimens.

How can we avoid such possibilities? One way is to be extremely careful in collecting the specimens that are to be analyzed. Severely fractured rock specimens would likely provide avenues of access by fluids that could easily promote the mobility of elements like rubidium. Weathering processes likewise provide opportunity for migration of some soluble chemical elements. As a result, geochronologists normally collect specimens that are as fresh as possible, that is, that show no signs of chemical weathering.[17] Whether a rock has been chemically altered can normally be determined by very careful examination of thin slices of rock under a petrographic microscope.[18] Geochronologists also generally avoid the collection of specimens of fractured rocks or rocks that are located within fault zones unless they are specifically interested in the effects of fracturing and weathering on presumed radiometric dates. It turns out, too, that methods have been developed in which the assumption is made that isotopes *have* been lost during a heating event or disturbance. These methods can tell us how much material has been lost and when it was

---

[16]The closed system condition can be stated mathematically as $N + D = N_o + D_o$, where N and D are the present amounts of parent and daughter isotopes respectively in a sample, and N_o and D_o are the amounts of parent and daughter isotopes respectively at the time of formation of the sample. If we rearrange the equation as $N_o = N + D - D_o$ and substitute into the decay equation $N = N_o e^{-\lambda t}$, we obtain $N = (N + D - D_o) \cdot e^{-\lambda t}$. Rearranging that equation yields an expression for calculation of the age of a closed system specimen, $t = (1/\lambda) \cdot \ln [(N + D - D_o)/N]$.

[17]Chemical weathering is not a problem in the radiometric dating of lunar rocks. In view of lack of an atmosphere and water on the moon, lunar rocks are completely free of chemical weathering. The thin sections are stunningly beautiful because of their lack of iron stain and fine-grained clay minerals. Because sedimentary rocks are composed of materials derived from the weathering of other rocks, in many cases, component minerals cannot be dated meaningfully.

[18]The petrographic microscope is a standard geologic tool for examining very thin slices of rock in polarized light. Complex interference of cross-polarized light produces spectacular interference colors that serve as a powerful aid in the identification of minerals and analysis of textural features in the rocks.

lost. We later review some instances of dating rock specimens in which there has been lead loss, for example.

Another extremely important condition that needs to be addressed is one that we set aside when we were talking about the use of half-life to calculate an age for a sample of a pure radioactive isotope. Our introduction of the closed system condition raises the issue of how much daughter isotope was present in a sample at the time of its formation. When we introduced our discussion of calculating the time of a radioactive specimen, we assumed, without stating as much, that our radioactive sample contained no daughter product at the time when it was formed. But we cannot always safely make that assumption.

Geologists must assess the validity of the assumption that no daughter product was present in a sample at the time of its formation every time they undertake a radiometric dating procedure. If a chemist prepares a sample of a pure radioactive compound such as RbCl in a laboratory under carefully controlled conditions, the assumption that no daughter product, in this case ^{87}Sr, is present at the time of formation of the compound is a perfectly reasonable one because the chemist took great pains to eliminate strontium and other elements in the preparation of the compound. In some cases, the assumption of no initial daughter isotope may also be reasonable from a geologic point of view.

In most cases, however, there is every reason to believe that there already was some daughter product in our mineral or rock at the time of its formation. If a geochronologist were to analyze such a sample, the calculated age would be greater than its actual age, if he or she assumed that none of the daughter element had been present. This poses a serious problem. In most instances, no one was present at the time of formation of minerals and rocks to analyze whether there was any daughter isotope in a specimen. So how can one ever hope to obtain an accurate age? An absolutely necessary condition for obtaining a meaningful age is that geochronologists need to devise some logical means for determining the amount of the daughter element that was present at the time of formation. Or perhaps they need to devise a method in which they don't need to worry about original daughter isotopes. In fact, geochronologists *have* devised various means for dealing with the problem of original daughter isotopes in radioactive samples. We discuss examples of such procedures below in our review of specific kinds of dating methods.

RADIOMETRIC METHODS

Now let's look at some of the methods that are useful for determining the ages of geological events. Most readers have probably heard about radiocarbon dating. As it turns out, radiocarbon dating is of relatively limited usefulness in geology because the method is not capable of providing reliable ages that are more than a few tens of thousands of years. Radiocarbon dating has been most useful in archeological work and in providing ages for events, such as glaciation and lake formation, that occurred toward the end of the Pleistocene Ice Age. As a result, we ignore radiocarbon dating and, instead, devote a great deal of discussion to several geologically more relevant dating techniques. Some methods that were used in the past have been discarded because they eventually proved not to be as reliable as hoped. Either they were subject to too many geologic variables or, better, more accurate methods were developed that provided the same information.

There are a very large number of radioactive isotopes. What isotope a geochronologist decides to use for dating a geological object is determined largely by what geologic event he or she is attempting to date. One very important kind of geologic event is the crystallization of a body of igneous rock like granite, diorite or gabbro. When did crystallization occur? Another geologic event might be the cooling of a body of igneous rock to a temperature at which argon (Ar) atoms might be retained in minerals. When did that happen? Some methods are used to determine when solidification of lava occurred. What was the timing of a metamorphic event? Other methods must be used to determine when a sediment layer was deposited on the bottom of a lake or the ocean. Still other methods are used for figuring out how long a mass of rock has been exposed to the atmosphere. There are methods for calculating the amount of time elapsed since a meteorite fragment broke off from its parent object. Some methods give us information about rates and times of uplift by recording the time at which a particular mineral had cooled sufficiently to retain a gas like helium (He) in its structure. Other methods record the time at which lead was lost from a rock. Thus, it is extremely important in any discussion of radiometric dating to take into account precisely what geologic event is being dated, what method is being used and what geologic factors can influence the results. Because coverage of all the available methods is well beyond the scope of this book, we discuss a few important representative methods.

DIRECT DATING OF MINERALS

Most older methods of radiometric dating, some of which are still in use, entail analysis of individual minerals such as sanidine, orthoclase and microcline (all of which are forms of potassium feldspar), biotite (a mica), hornblende (an amphibole) and zircon that have been extracted from their host rocks. Biotite and potassium feldspar have commonly been used to determine the time of eruption and formation of layers of solidified volcanic ash known as tuff by means of potassium-argon (K-Ar) dating. In this case, large crystals of biotite and/

or sanidine are separated from the fine-grained matrix of the tuff and analyzed for ^{40}K and ^{40}Ar. Although the biotite or feldspar may have crystallized in magma while it was in the throat of a volcano or even deeper in the volcanic conduit, it is reasonable to assume that the time of eruption was not significantly later than the time of original crystallization of the mineral. Moreover, the time of deposition and solidification of the volcanic ash is virtually the same as the time of eruption.

The advantage of the K-Ar mineral dating method is that there is unlikely to be a significant amount of the daughter element argon (Ar) in the mineral crystals at their time of crystallization. Argon is an inert gas element that does not bond chemically with the other elements that make up the crystal structure of biotite or feldspar. The crystal structure of feldspar also has little room for an Ar atom to be trapped. Such sites do exist in biotite, but Ar can also easily escape from them. There is little reason, therefore, to expect that these minerals contain a significant amount of Ar when they crystallize. Moreover, such crystals from ash deposits that were erupted within the last few years, hence with essentially zero age, typically contain negligible amounts of initial Ar.[19]

Another characteristic of the K-Ar method is that the ^{40}Ar that accumulates within the mineral by decay of ^{40}K is readily lost until the mineral cools below a critical temperature called the *closure* or *blocking temperature.* For example, the closure temperature for Ar retention in biotite is around 325° C, and around 550° C for hornblende. As a result, a K-Ar date on biotite, hornblende or feldspar readily indicates the time at which the mineral cooled below the closure temperature. Moreover, because hornblende retains Ar more effectively than biotite, the K-Ar age of a biotite sample will typically be less than the K-Ar age of hornblende from the same rock specimen. For volcanic ash, however, the time of cooling below the closure temperature is essentially identical to the time of eruption because the mineral crystals cool in a matter of hours after the eruption takes place. A matter of hours between the time of eruption and the time of cooling to the closure temperature makes little difference if the mineral is, say, 5 million years old!

K-Ar ages are also obtained on minerals from plutonic igneous rocks such as granite. There are several such ages for the granitic rocks of the Sierra Nevada batholith in California, for example. In these cases the ages are almost invariably lower than those obtained by other methods such as U-Pb dating. The reason is that in this geologic situation, we are likely determining the time at which the mineral cooled sufficiently to begin retaining Ar. Given that granites form far underground, they take much longer to cool than a lava flow

---

[19]Lava that erupted through prior lava flows might inherit Ar from the older flow, thus rendering anomalously old ages.

or volcanic ash that was erupted onto Earth's surface. It may take thousands to hundreds of thousands of years for an igneous rock body to cool to 325° C, at which point biotite begins to retain Ar. This geological reality is reinforced by the fact that most of the granitic rocks in the Sierra Nevada were later intruded by other very hot magmas before they had a chance to cool below closure temperatures. Intrusion of a hot mass of granitic magma next to a cooling granite body could easily keep the already crystallized granite substantially above the closure temperature of biotite for a very long time. In these cases, then, the K-Ar data may provide some indication of how long it took a given body of igneous rock to cool below the closure temperatures of various minerals.

A method that is rarely used anymore is $^{87}Rb/^{87}Sr$ mineral dating. Ions of the chemical element Rb have the same electrical charge as those of potassium, and they are only slightly larger than potassium ions. As a result, Rb readily substitutes for K in the crystal structures of biotite, muscovite and feldspar. Because Rb is much less abundant than K, it makes up only a small fraction of the minerals. For this dating method, the amounts of ^{87}Rb and its daughter ^{87}Sr in the mineral of interest are measured, typically in the form of ratios. The ratio that is measured for Sr is $^{87}Sr/^{86}Sr$. ^{86}Sr is used as the denominator in the ratio because it is neither radioactive nor the daughter product of radioactive decay. In other words, it is a stable isotope, the amount of which should not change through time if the mineral is a closed system. Unlike the K-Ar method, there is bound to be some Sr already present in the mineral at the time it forms. Either biotite or sanidine that crystallized in a volcanic ash deposit will incorporate Sr that is present in the magma. Sr readily substitutes for either Ca or K in these minerals, particularly the former. But how is a geochronologist to know how much ^{87}Sr was incorporated at the time of crystallization? Another way to phrase the question is to ask what was the initial ratio of $^{87}Sr/^{86}Sr$ in the mineral. In the past, one way that geochronologists addressed that problem was to assume a value based on the fact that minerals formed at present have $^{87}Sr/^{86}Sr$ ratios ranging mostly from 0.703 to 0.715. So a value of 0.704, commonly measured in modern lava flows, or some similar value was plugged into the operative equation and a date was calculated. The problem was that quite different results were obtained depending on the initial Sr isotope ratio chosen. This should, however, be of no consolation to young-Earth proponents, because no matter which value was selected, very large ages were obtained. But geochronologists desired a more dependable method, and so $^{87}Rb/^{87}Sr$ dating of minerals is rarely, if ever, employed today.

ISOCHRON DATING: THE WHOLE-ROCK METHOD
Today geochronologists commonly employ methods that yield so-called model

ages. The validity of these ages depends on the validity of a particular model or picture of how certain geologic situations affect the behavior of different kinds of isotopes. Model ages characteristically require isotope analyses of several rock or mineral samples from a given body of rock. The ages obtained from mathematical or graphical treatment of the analyses can yield various kinds of information, for example, time of crystallization of magma, time of metamorphism or time of lead loss from the rocks.

We consider two different and widely used sets of model dating methods: the isochron and concordia methods. We begin with the *Rb-Sr whole-rock isochron* method. This method is superior to the Rb-Sr mineral dating method because geologists do not have to make any assumptions regarding the value of initial Sr; instead, the model reveals mathematically what the initial amount of ^{87}Sr was in rocks when they were originally formed in terms of the ratio $^{87}Sr/^{86}Sr$. The whole-rock method is also sensitive to the possibility of diffusion of Sr or Rb into or out of the samples. That is, from the results, the geochronologist can normally determine whether diffusion has occurred. The whole-rock method has been applied to a variety of rock types but is used primarily for the determination of the crystallization age of an igneous rock body.

To see how the method works, note that geologists first construct a model or picture of the geologic process they wish to date. Let's construct a model for the crystallization of magma. Imagine that a hot magma exists somewhere within Earth's crust a few miles below the surface. The magma is composed of a great variety of atoms, most of which are common elements like potassium, sodium, calcium, iron, magnesium, aluminum, silicon, oxygen, hydrogen and titanium, but there will also be trace amounts of other chemical elements like scandium, barium, uranium, nickel, rubidium and strontium. Atoms of several Sr isotopes will be present in the magma. Of these Sr isotopes, we are especially interested in ^{87}Sr and ^{86}Sr. ^{87}Sr differs from ^{86}Sr only in having an extra neutron in its nucleus. This makes ^{87}Sr a slightly heavier atom than ^{86}Sr. In spite of the slightly greater mass of ^{87}Sr atoms, the ratio $^{87}Sr/^{86}Sr$ should normally be essentially uniform throughout the magma. There is no evidence to indicate that ^{87}Sr atoms preferentially migrate to one portion of a body of magma and ^{86}Sr to another to the extent that the $^{87}Sr/^{86}Sr$ ratio would vary drastically throughout the body. Although a geochronologist does not know what this initial ratio $^{87}Sr/^{86}Sr$ in the magma was at the outset of the investigation, he or she will be able to determine that ratio eventually. It does not matter that the investigator does not know how much ^{87}Sr is present at the time of crystallization, and it doesn't matter that this ^{87}Sr is not derived by later decay of ^{87}Rb after crystallization has begun. This ratio $^{87}Sr/^{86}Sr$ is termed the *initial strontium isotope ratio* and is expressed mathematically as $({}^{87}Sr/{}^{86}Sr)_o$ or sometimes as $({}^{87}Sr/{}^{86}Sr)_I$.

Now suppose that the model magma with a uniform initial strontium isotope ratio $({}^{87}Sr/{}^{86}Sr)_o$ begins to crystallize as it cools. Several different minerals begin to form as the temperature declines, each of which has a different internal atomic crystal structure.[20] If the magma is a granitic magma, then the major minerals to form are feldspar and quartz and typically substantial amounts of hornblende or mica minerals like biotite or muscovite. Suppose that the magma crystallizes to a biotite-bearing granite. The crystal structure of quartz has little tolerance for most chemical elements apart from silicon and oxygen. As a result, quartz is virtually pure silicon dioxide (SiO_2). Ions of chemical elements that are of interest to geochronology like Rb and Sr are excluded from the quartz structure because they have electrical charges that are too different from that of silicon and because they are much too large to substitute for the major atom (silicon) in that structure. Some Rb and Sr, however, are incorporated into such minerals as biotite and feldspar. Trace amounts of Rb and Sr substitute for K in the biotite structure. Rb generally substitutes for K in the feldspar structure, and Sr generally substitutes for calcium (Ca) because the charges and sizes of these ions are similar.

After the granite has crystallized, the ${}^{87}Sr/{}^{86}Sr$ ratio is essentially constant throughout the body of rock reflecting the uniformity of that ratio in the magma. During initial crystallization relatively little ${}^{87}Rb$ has yet decayed to produce a significant amount of new ${}^{87}Sr$. However, at the time of formation some samples of the granite are relatively enriched in Rb and others are relatively enriched in Sr because of the variation in proportions of minerals. For example, samples of granite with greater amounts of biotite and K feldspar have higher ratios of Rb/Sr than samples of the granite with higher contents of quartz and plagioclase feldspar and lesser amounts of biotite and K feldspar. This is so because the Rb of the granite is concentrated in the K-rich minerals like biotite and K feldspar, and the Sr is concentrated in Ca-bearing minerals like plagioclase. Thus, the ratio of Rb/Sr and, therefore, of ${}^{87}Rb/{}^{86}Sr$ varies from one sample of the granite to another. If a geologist could sample several chunks of granite just after it had crystallized, and if the ${}^{87}Sr/{}^{86}Sr$ and ${}^{87}Rb/{}^{86}Sr$ ratios in those samples were analyzed and the data were plotted on a graph, our geologist would obtain a diagram like the horizontal line (t=0) portrayed in figure 14.2.

None of the above reasoning is wild guessing. It is soundly based on the

---

[20]The crystal structures of minerals, well-defined and well-ordered geometrical arrays of different types of atoms, are determined largely by the sizes, electrical charges and bonding characteristics of the atoms composing the minerals.

empirical behavior of Sr and Rb isotopes in mineral structures and is noncontroversial. The straight line on the diagram in figure 14.2 is called an *isochron* and the slope of the line is a function of the age of the rock. For an igneous rock of zero age, that is, one that has just crystallized, the slope of the isochron is zero and is, therefore, a horizontal line.

But what will this diagram look like for a granite that has aged somewhat? How does the position of the isochron vary as the granite gets older? As the granite becomes older, some of the ^{87}Rb atoms spontaneously decay by beta emission into ^{87}Sr atoms in accordance with the radioactive decay law. This means that, through time, the ratio of ^{87}Rb/^{86}Sr of any sample of granite gradually decreases. The more time that elapses, the more the ratio decreases be-

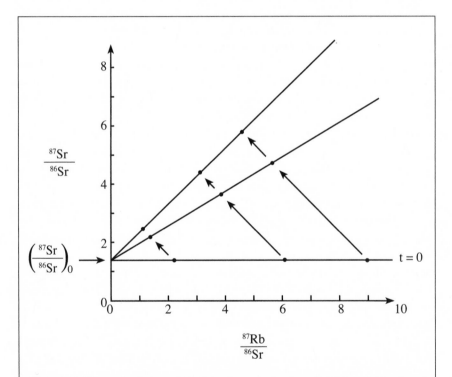

Figure 14.2. Schematic whole-rock Rb-Sr isochron. At the time of initial crystallization of a body of magma, the initial ^{87}Sr/^{86}Sr isotope ratio is essentially constant, no matter what the Rb/Sr ratio. As time elapses, ^{87}Rb in the crystallized igneous rocks decay to ^{87}Sr. As a result, over time, the values of ^{87}Rb and ^{87}Sr in each individual sample follow paths with a slope of -1 as indicated by the arrows with negative slopes. Therefore, through time, the isochron line pivots around the point representing the initial Sr isotope ratio. The slope of the isochron is a function of time, and the steeper the slope, the greater the age of the samples represented by the isochron.

cause ^{87}Rb is continuously lost from the sample by decay. At the same time, the ^{87}Sr/^{86}Sr ratio of the sample increases, because ^{87}Sr is added to the rock by decay of ^{87}Rb. Consequently, a data point on the isochron diagram shifts its position through time as shown in figure 14.2. The older the sample becomes, the farther along the arrows the data point is shifted.

All the points shift along paths with the same slope as indicated in figure 14.2. In a given amount of time, points with the highest ^{87}Rb/^{86}Sr ratios shift the most because they have higher ^{87}Rb contents, and thus a greater amount of ^{87}Rb is converted into ^{87}Sr by decay. As a result, the isochron line pivots around the point of $(^{87}$Sr/^{86}Sr$)_o$ through time. The slope of the isochron becomes steeper as the granite ages. This change in slope of the isochron through time is illustrated in figure 14.2. In the process, it is specified that no Rb or Sr has been added to or removed from samples other than by radioactive processes. In other words, the samples of granite have acted as closed systems.

Mathematically, the isochron is expressed by the equation

$$^{87}\text{Sr}/^{86}\text{Sr} = {^{87}\text{Rb}}/^{86}\text{Sr} \cdot (e^{\lambda t} - 1) + (^{87}\text{Sr}/^{86}\text{Sr})_o$$

where λ is the decay constant and t is the age of the samples. This equation has the form of the equation of a straight line: $y = mx + b$. Thus $(^{87}$Sr/^{86}Sr$)_o$ is the y-intercept (b), and $(e^{\lambda t} - 1)$ is the slope (m) of the isochron on a plot of ^{87}Sr/^{86}Sr (y-axis) versus ^{87}Rb/^{86}Sr (x-axis).[21]

To obtain an isochron for a body of the granite it is necessary to collect several large samples of the granite with variable mineral contents to insure a wide range of Rb/Sr ratios. The ^{87}Sr/^{86}Sr and ^{87}Rb/^{86}Sr ratios in the rocks at present are measured analytically by mass spectrometry and plotted on the diagram. These analytical data points commonly define a straight line, the slope of which gives the age of the crystallization of the rock body. If the samples had been seriously affected by the diffusion of Sr and/or Rb out of or into the rock, then the data points would commonly be scattered. Presumably the data points of rocks affected by a small degree of Rb or Sr diffusion would show small enough scatter of points that they would still indicate a clear linear trend. Data points for rocks affected by a great deal of diffusion should be expected to show very strong scatter of points, in some cases so much that it is difficult to detect any clear-cut straight-line trend. More specifically, if the rocks had been affected only by Sr diffusion out of the granite body, ^{87}Rb/^{86}Sr values would all be greater than what they would have been had no diffusion

---

[21]The equation just presented in the text is derived as follows: The equation for the decay of ^{87}Rb is $^{87}\text{Rb} = {^{87}\text{Rb}}_o\, e^{-\lambda t}$. The equation for the closed system condition is $^{87}\text{Rb} + {^{87}\text{Sr}} = {^{87}\text{Rb}}_o + {^{87}\text{Sr}}_o$. By rearrangement of the closed system condition equation and substitution into the decay equation we obtain $^{87}\text{Rb} = (^{87}\text{Rb} + {^{87}\text{Sr}} - {^{87}\text{Sr}}_o)\, e^{-\lambda t}$. By rearranging that expression we obtain $^{87}\text{Sr} = {^{87}\text{Rb}}\,(e^{\lambda t} - 1) + {^{87}\text{Sr}}_o$. Finally, by dividing each side of that equation by ^{86}Sr $(= {^{86}\text{Sr}}_o)$ we obtain $^{87}\text{Sr}/^{86}\text{Sr} = {^{87}\text{Rb}}/^{86}\text{Sr}\,(e^{\lambda t} - 1) + {^{87}\text{Sr}}_o/^{86}\text{Sr}_o$.

taken place, and the data points would be displaced to the right of a true isochron for such rocks. If the data points do yield a well-defined straight line, a geochronologist may safely infer that the granite body has not been affected by significant Rb or Sr diffusion throughout its history. In such cases, the isochron diagram also yields the initial $^{87}Sr/^{86}Sr$ ratio of the magma from the intercept of the straight line on the y-axis. In effect, the whole-rock isochron is capable of indicating what the original amount of daughter product in the rock we are dating was, and it does so in the form of the $^{87}Sr/^{86}Sr$ ratio. Figure 14.3 illustrates an actual whole-rock isochron for Algoman granite from Canada.

If the body of crystallizing granite behaved as described in our model scenario we should obtain a well-defined isochron. If the granite did not behave in accord with the simple scenario, which is entirely possible, a geochronologist probably would not obtain a reliable age. Failure to obtain a reliable age is *not* a failure of the method, which is both mathematically sound and geologically reasonable. Failure to obtain a reliable age is not *blaming* nature, as Woodmorappe claimed; it is simply an indication that the geologic conditions required for satisfying the conditions of the model did not exist. That the method does not give a good isochron in some instances comes as no surprise to geologists. That the method results in good isochrons as often as it does is encouraging and exciting. The fact that data points line up as a straight line on an isochron diagram in and of itself is not a guarantee that the line is a true isochron telling us something about age. Other age-independent geologic processes can also lead to straight-line behavior of the data points. We consider these alternatives shortly, but for now note that alternative explanations for straight lines on isochron diagrams are relevant only in rare instances and that most straight lines are true isochrons.

Owing to the fact that both Rb and Sr can be remobilized during heating events, the Rb-Sr whole-rock method does not always yield reliable isochrons, especially for rocks with complex histories. In recent years, therefore, geochronologists have increasingly turned to whole-rock isochrons based on the alpha decay of ^{147}Sm to ^{143}Nd. The relevant Sm-Nd isochron equation is exactly analogous to that for Rb and Sr. All one needs to do is substitute ^{147}Sm for ^{87}Rb, ^{143}Nd for ^{87}Sr, and ^{144}Nd for ^{86}Sr, and substitute the decay constant of ^{147}Sm for that of ^{87}Rb. The Sm-Nd isochron plot, therefore, is a plot of the ratio $^{143}Nd/^{144}Nd$ vs. $^{147}Sm/^{144}Nd$. ^{144}Nd is the stable isotope of Nd. Both Sm and Nd are so-called rare-earth elements (REE) and are considerably less abundant than are Rb and Sr. Sm and Nd each have a charge of +3, and Sm is a slightly larger ion than Nd. Geochemically, therefore, Sm and Nd, unlike Rb and Sr, behave in a similar manner. This means that both Sm and Nd tend to enter into the same sites in crystal structures of the same minerals. They are particularly likely to enter into *accessory minerals* like zircon, apatite, titanite and monazite, all of which are relatively common in typical igneous rocks, especially granite and syenite.[22] Somewhat lesser concentrations of Sm and Nd also occur in minerals such as pyroxene, biotite and hornblende or other amphiboles where they may substitute for elements like Ca. Because Sm and Nd tend to enter into the same minerals, it is more difficult to obtain a suite of rocks in which a sizeable spread of $^{147}Sm/^{144}Nd$ ratios occur. Nevertheless, the difference in mass between ^{147}Sm and ^{144}Nd is sufficiently large that some fractionation occurs. Very high precision and accuracy of analyses are required to make the method effective.

One of the main advantages of the method in comparison with the Rb-Sr isochron method is that rare-earth elements are relatively immobile during geologic disturbances. If fluids are percolating through a body of granite that

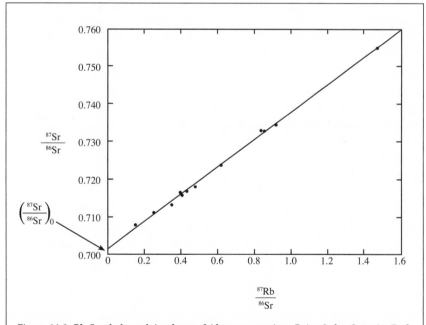

Figure 14.3. Rb-Sr whole-rock isochron of Algoman granites, Rainy Lake, Ontario. Each data point represents the present-day analyzed isotopic values of ^{87}Rb, ^{87}Sr and ^{86}Sr expressed as ratios. The analytical data points fall on a well-defined straight line whose slope indicates an age of 2.54 billion years.

---

[22]Syenite is a very coarse-grained igneous rock that is closely allied to granite but lacks a significant amount of quartz. The predominant minerals in syenite are alkali feldspar and one or more iron-magnesium silicate minerals such as hornblende, biotite or pyroxene.

is undergoing metamorphism, there is a very good chance that there will be some redistribution of Sr and Rb during the event, thus disturbing the Rb-Sr system and reducing the chances of obtaining a good isochron. In contrast, rare-earth elements generally remain within the minerals in which they were originally incorporated. The result is that Sm-Nd systems are generally not significantly disturbed during metamorphism, and information pertaining to the original age of crystallization of a granitic intrusion is likely to be preserved even though that body may have been subsequently metamorphosed. This characteristic feature of the system means that Sm-Nd whole-rock isochron dating is especially important for the dating of exceptionally old rocks in the one to four billion year range. Although reliable ages in this range can be obtained by means of Rb-Sr isochron dating (fig. 14.3), there are still cases in which subsequent mobility of the ions has reset the original ages.

Several additional isochron methods have been developed more recently, based on the beta decay of lanthanum 138 (^{138}La) to cerium 138 (^{138}Ce), of rhenium 187 (^{187}Re) to osmium 187 (^{187}Os), and of lutetium 176 (^{176}Lu) to hafnium 176 (^{176}Hf). Although none of these methods is likely to be as widely useful as Sm-Nd or Rb-Sr, they do offer promise in the investigation of igneous rocks rich in iron-magnesium minerals, meteorites or certain ore samples. They do, therefore, find application to geologic materials that are not so readily datable by the Rb-Sr and Sm-Nd isochron methods.

Thus far, we have applied the whole-rock isochron methods to the time of crystallization of a cooling mass of magma. In some cases, however, whole-rock isochrons may yield the age of a metamorphism event. Space does not allow for a detailed explanation of the model, but suffice it to say that during an intense, long-lasting episode of high-temperature metamorphism, Sr or Nd isotopes may be sufficiently redistributed throughout the mass of rock being metamorphosed that a more or less uniform ^{87}Sr/^{86}Sr or ^{143}Nd/^{144}Nd ratio may prevail throughout the rock mass. Pervasive fluid should aid in the redistribution of the Sr or Nd isotopes. If the Sr or Nd isotopes are reset to a constant value throughout the rock body, the radiometric clock would be reset to zero so that an isochron measured today would yield the age of resetting, namely, the time of metamorphism. Rocks that have well developed preferred orientation of mineral grains, especially gneisses, rocks that form at high metamorphic temperatures, are the most likely candidates for obtaining metamorphic crystallization ages. What is needed is collection of several whole-rock samples from the mass of gneiss that, as in the case of igneous rocks, have a wide range of ^{87}Rb/^{86}Sr or ^{147}Sm/^{143}Nd values brought about by mineralogical variations.

15

RADIOMETRIC DATING

Part Two

IN ADDITION TO MINERAL DATING METHODS and whole-rock (iso-chron) methods presented in chapter fourteen, several other methods of ra-diometric dating have been developed to overcome various problems and to determine the age of a rock mass more accurately.

MINERAL (INTERNAL) ISOCHRONS

In some cases where metamorphism has not sufficiently redistributed Sr iso-topes throughout a large volume of rocks, a *mineral* or *internal isochron* may yield a metamorphic crystallization age. Generally, metamorphism of an older igneous rock results in redistribution of Sr isotopes at least on the scale of a large hand specimen so that the individual minerals within that hand speci-men have their Sr isotope ratios reset to a common value. Although there are exceptions, Nd isotopes are generally less mobile than Sr, and Nd isotope ra-tios are not so easily homogenized during a metamorphic event. As a result, the method that we are describing is typically less applicable to Sm-Nd dating than it is to Rb-Sr dating. Redistribution of Sr isotopes may have homogenized and reset the Sr isotope ratio within a small, fist-sized sample, while not hav-ing been sufficiently pervasive to reset the Sr isotope ratio throughout an en-tire rock mass to the same value.

In this case, geochronologists collect one whole rock sample (or possibly several), separate the individual minerals from the whole rock sample, analyze their isotope ratios, and plot the results on the isochron diagram. Different minerals characteristically plot at different locations on an isochron. Biotite, for example, readily accommodates Rb in its crystal structure and typically has, therefore, a relatively high $^{87}Rb/^{86}Sr$ ratio. In contrast, plagioclase feldspar preferentially incorporates Sr more than Rb and has a much lower $^{87}Rb/^{86}Sr$ ratio than biotite does (see figure 15.1). In some cases, analyses of mineral separates from several different whole rocks yield parallel isochrons, presumably providing the age of metamorphism.

In some instances, mineral isochrons can also be determined where only a single rock sample is available and construction of a whole-rock isochron from

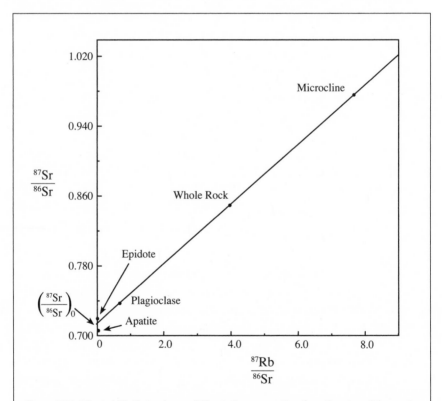

Figure 15.1. Mineral Rb-Sr isochron of Giants Range granite of northeastern Minnesota. Data points represent present-day analyzed values of ^{87}Rb, ^{87}Sr and ^{86}Sr expressed as ratios in individual minerals (plagioclase, apatite, epidote and microcline) in a large whole-rock sample of granite as well as analyzed values of the whole-rock sample. The data points fall on a well-defined straight line whose slope indicates an age of 2.44 billion years.

several related rock samples from a given rock mass is out of the question. Meteorites, for example, consist of only a single fragment of rock, many of them fist-sized or smaller. In such cases, various minerals (olivine, pyroxene, plagioclase, apatite, iron oxides) are extracted from the meteorite, analyzed individually, and the resulting data are plotted on an isochron diagram. The major drawback is that the isochron plot may be determined by data from a very small number of minerals. In contrast, many whole-rock isochrons are well-defined by data points from as many as a dozen or more whole-rock samples. Despite the limitations, Rb-Sr, Sm-Nd, Re-Os, and Pb-Pb mineral isochron methods have been applied with striking success to meteorites.

Numerous meteorites that originated in the asteroid belt between Mars and Jupiter have been studied by this method, and the slopes of the mineral isochrons indicate a striking consistency of ages in the 4.4 to 4.6 billion-year range, no matter what method is used. For example, the Caddo County iron meteorite has a Sm-Nd age of 4.53 ± 0.02 billion years; the Estherville stony iron meteorite yields a Rb-Sr age of 4.54 ± 0.20 billion years; the Angra dos Reis stony meteorite gives a Sm-Nd age of 4.55 ± 0.04 billion years; the Juvinas stony meteorite has a Rb-Sr age of 4.50 ± 0.07 billion years; and the St. Severin stony meteorite has a Sm-Nd age of 4.55 ± 0.33 billion years. The original Sr isotope ratios for meteorites determined by Rb-Sr internal isochrons are invariably very close to 0.699. The major exception to this range of ages comes from a group of meteorites with unusual texture and mineral content that probably originated from the surface of Mars rather than in the asteroid belt, and their ages are generally about 1 to 1.5 billion years.[1]

The interpretation of meteorite ages is not always straightforward. For many meteorites, internal isochron ages likely refer to the time of original crystallization, but in some cases, the isochron age may indicate the time of a subsequent metamorphism. The texture of the meteorite, as observed in thin sections under a polarizing microscope, provides clues as to whether the meteorite experienced metamorphism. Evidence of preferred mineral alignment or reaction among minerals is also suggestive of a metamorphic event.

Before we leave the topic of meteorite dating altogether, let's note that it is possible to produce whole-rock isochrons for these objects. We already stated that the mineral-isochron methods have been applied to meteorites because we don't have access to a gigantic meteorite to which we can apply the whole-rock method. So how does one go about generating a whole-rock isochron for meteorites, a method that requires sampling and analysis of a large number of samples from a large body of material? Obviously NASA is not ready to send

---

[1]For meteorite ages, see chap. 7 in G. B. Dalrymple, *Ancient Earth, Ancient Skies: The Age of Earth and Its Cosmic Surroundings* (Stanford, Calif.: Stanford University Press, 2004).

astronauts to the asteroid belt to sample a large asteroid, but fortunately, a sufficient number of meteorites have been collected that specialists can classify them in distinctive groups (for example, carbonaceous chondrites, ordinary chondrites, enstatite chondrites, basaltic achondrites and irons). Each group has its own characteristic mineral composition, textural features in thin sections, and major and trace chemical element composition.

Because these distinctive types exist, it is possible to construct a whole-rock isochron by analyzing the isotope composition of samples of several representatives of each group. For example, we might analyze samples of several different basaltic achondrites to obtain a whole-rock isochron of the basaltic achondrite group. The analytical data for each specific group characteristically lie on remarkably well-defined straight lines with very little scatter. Moreover, the ages calculated from these straight lines of the various groups all agree, with most ages falling between 4.4 and 4.6 billion years. As one example, a Rb-Sr whole-rock isochron based on analysis of 23 samples of different basaltic achondrites yields an age of 4.53 ± 0.19 billion years. The initial isotope ratios of the different groups are also remarkably similar. In addition, the ages obtained from Rb-Sr isochrons agree closely with ages obtained from Sm-Nd isochrons, Pb-Pb isochrons and other methods. Moreover, they agree with the ages obtained from internal isochrons.

THE CONCORDIA METHOD

The concordia method is a powerful technique developed in 1956 by George Wetherill of the Department of Terrestrial Magnetism at the Carnegie Institution of Washington. This method addresses the problems of lead loss in uranium-bearing minerals and of discordance between different dates based on the decay of uranium (U) to lead (Pb).[2] There are a variety of methods for dating minerals using the radioactive chemical elements uranium and thorium (Th). For example, it is possible to determine the age of the same mineral by analyzing ^{238}U and its ultimate daughter product ^{206}Pb, by analyzing ^{235}U and its daughter ^{207}Pb, and also by analyzing ^{232}Th and its daughter ^{208}Pb. In addition, there is a lead-lead method that provides another calculation that is based on the ratio of radiogenic ^{207}Pb to radiogenic ^{206}Pb. There are also four independent methods for calculating the time of crystallization of a zircon, for example, in an igneous rock body such as granite. The difficulty is that most U-Pb analyses of the same sample yield discordant ages. In other words, the four calculated ages do not agree. One well-known example concerns the

---

[2]George W. Wetherill, "Discordant Uranium-Lead Ages I," *Transactions of the American Geophysical Union* 37 (1956): 320-26.

Boulder Creek batholith just west of Boulder, Colorado, in the Front Range of the Rocky Mountains.[3] A sample of quartz diorite from the batholith yielded the following ages: $^{207}Pb/^{206}Pb$, 1,705 million years; $^{207}Pb/^{235}U$, 1,410 million years; $^{206}Pb/^{238}U$, 1,220 million years; and $^{208}Pb/^{232}Th$, 1,005 million years. A sample of granodiorite from the same batholith yielded the following ages: $^{207}Pb/^{206}Pb$, 1,720 million years; $^{207}Pb/^{235}U$, 1,635 million years; $^{206}Pb/^{238}U$, 1,575 million years; and $^{208}Pb/^{232}Th$, 1,520 million years. A sample of quartz monzonite from the same batholith yielded the following ages: $^{207}Pb/^{206}Pb$, 1,715 million years; $^{207}Pb/^{235}U$, 1,535 million years; $^{206}Pb/^{238}U$, 1,405 million years; and $^{208}Pb/^{232}Th$, 1,285 million years.

One can see that, for these three samples, the results for a given sample vary greatly for the four methods and also that the $^{207}Pb/^{206}Pb$ method typically yields the highest age and the Th-Pb method yields the lowest age. This result is characteristic for U-Th-Pb dates of almost any igneous intrusion. Although there are exceptions, the lead-lead ($^{207}Pb/^{206}Pb$) method generally yields the highest age, and the thorium-lead ($^{232}Th/^{208}Pb$) method generally yields the lowest age on the same sample. Geochemists normally attribute the discordant ages to loss of lead, either by slow, continuous leakage or in discontinuous episodes, from the uranium-bearing mineral that is being analyzed. The $^{207}Pb/^{206}Pb$ method should be least affected by loss of lead because any lead lost from a mineral sample that one wants to date should have essentially the same $^{207}Pb/^{206}Pb$ ratio as that of the mineral itself. As a result, the measured Pb/Pb age is far less likely to be reduced than ages obtained by the other methods. In those methods, if lead is lost at all, no matter what the lead isotope ratio, the mineral will appear to be younger than it really is. So, for the Boulder Creek batholith, one may surmise from the Pb/Pb age that the probable age of crystallization of the rocks is around 1,705 to 1,710 million years. The concordia method was devised as one means for circumventing the lead loss problem.

The concordia method is used in conjunction with a graphical plot of radiogenic $^{206}Pb/^{238}U$ versus radiogenic $^{207}Pb/^{235}U$ (fig. 15.2). The concordia diagram includes a curve that shows how the values of these two ratios change continuously over time for a hypothetical mineral sample that has been completely closed to U and Pb and that contained no ^{206}Pb or ^{207}Pb at the time of formation so that all of the Pb in the hypothetical specimen is radiogenically derived from U decay. Now let's suppose that a uranium-bearing mineral is analyzed for U and Pb. A correction is made for any original daughter Pb isotopes, and the appropriate Pb/U ratios are plotted on the diagram. Next let's

---

[3]On the Boulder Creek batholith, see Thomas W. Stern, George Phair and Marcia F. Newell, "Boulder Creek Batholith, Colorado Part II: Isotopic Age of Emplacement and Morphology of Zircon," *Bulletin GSA* 82 (1971): 1615-34.

suppose that the data point plots exactly on the concordia curve. A reasonable initial conclusion would be that the point represents the true crystallization age of the U-bearing mineral. But that often doesn't happen. Rather, data points representing analyses of several mineral separates from samples at different locations within a large body of igneous rock commonly align on an essentially straight line that intersects the concordia curve at two points. The upper intercept (fig. 15.2) is said to represent the original time of crystallization, and the lower intercept of the straight line with the concordia curve is said to represent a time of disturbance when lead was lost. The argument is that the straight line results from a specific episode of lead loss with some zircon separates losing more lead than others, and in which some of the zircons lost all their lead at the time of the lead-loss episode. If Pb is lost from the zircons, the $^{207}Pb/^{206}Pb$ ratios of lead lost from the various zircon samples will all be virtually identical because the masses of the two Pb isotopes are so similar to each other that there would be negligible fractionation upon lead loss. In addition, if lead is lost from individual zircon separates at a particular time, then their Pb/U ratios will decrease. The more lead they lose, the lower their Pb/U ratios become. If the minerals were analyzed at the time of lead loss, the data points would fall on a line passing through the origin where the ratios of Pb/U are both zero. After the episode of lead loss, radioactive decay of U continues, Pb continues to accumulate in the zircons, and the Pb/U ratios increase. Through time, the straight line rotates as shown in figure 15.2.

The concordia method is highly powerful because it takes advantage of the phenomenon of lead loss and circumvents the problem of discordant U-Pb dates. Despite lead loss, geochronologists can still recover the crystallization age of the zircons, but they also have the potential added benefits of detecting when lead loss occurred and the relative amounts of lead lost from each sample. There are other instances in which lead is lost continuously or in more than one specific lead-loss episode. These cases are much more complicated and are not treated here. Suffice it to say, the concordia method demonstrates that geochronologists need not know ahead of time that a mineral has lost a daughter element to make use of it for dating purposes.[4] A concordia diagram for the Boulder Creek batholith is presented in figure 15.3.

The sharp-eyed reader may have noticed that we breezed right past another potential problem. We said that the concordia curve is calculated on the basis that the hypothetical sample contains no daughter product at the time of formation. In other words, all the daughter product lead is radiogenic. In reality, geochronologists plot *radiogenically* produced Pb/U ratios. The concordia curve,

---

[4]On concordia and various other U-Th-Pb methods, see chaps. 10 and 11 in G. Faure and T. M. Mensing, *Isotopes: Principles and Applications*, 3rd ed. (Hoboken, N.J.: John Wiley, 2005).

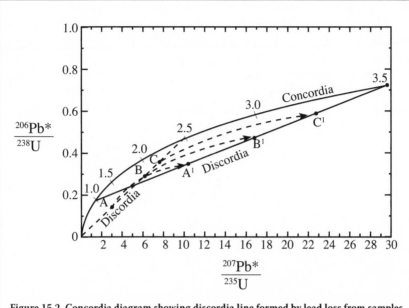

Figure 15.2. Concordia diagram showing discordia line formed by lead loss from samples. When the time of lead loss is over, the position of the discordia line rotates from points A, B and C to A', B' and C'. The upper intercept of discordia with the concordia curve yields the age of crystallization, and the lower intercept yields the probable age of the episode of lead loss.

therefore, presents an idealized situation. They want, however, to analyze real minerals like zircon, plot their compositions on the diagram, and evaluate their relation to the concordia curve. But how can they be certain that there was no radiogenic lead in these zircon samples when they were first formed?

In cases where there might have been some Pb incorporated into the zircons or another U-bearing mineral at their time of crystallization, geochronologists may determine how much *common lead* is present in the specimens. Common lead includes some radiogenic lead 206, lead 207, and lead 208, as well as lead 204 (^{204}Pb), a stable isotope of lead. The abundance of ^{204}Pb in a mineral does not increase over time because of the radioactive decay of some other isotope. If a mineral incorporated any lead at all during its original crystallization, some of that lead will be ^{204}Pb. In most cases, zircons contain negligible amounts of ^{204}Pb, an indication that a negligible amount of any kind of lead was incorporated at the time of its crystallization. If so, one may reasonably assume that virtually all of the radiogenic lead in the zircon was formed after crystallization by decay of uranium and thorium. The reason that a negligible amount of lead is incorporated into a zircon during its crystallization from magma, for example, is that Pb^{2+} ions are typically too large and have too small an electrical charge to substitute for the Zr^{4+} ions in the crystal structure of zircon.

But suppose that the zircon or some other U-bearing mineral does contain common lead that was incorporated at the time of crystallization. How then do we determine how much radiogenic lead was in the mineral at the time of crystallization? Consider that during the crystallization of granite, for example, some lead is incorporated into crystals of feldspar, a major constituent of

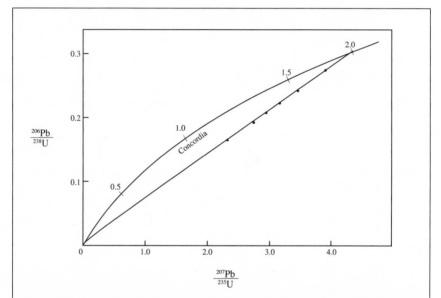

Figure 15.3. Concordia plot of zircon samples from the Boulder Creek batholith, Colorado. Data points from analyses of U and Pb isotopes in several zircon samples fall on a well-defined straight line (discordia). The upper intercept of the discordia line intersects concordia at about 1.725 billion years, indicating the probable crystallization age of the zircons. The lower intercept intersects concordia at 65 million years, indicating a probable time of lead loss from the zircons.

all granites. The lead (Pb^{2+}) ions substitute readily for K^{1+}, an ion of similar size and reasonably similar electrical charge. In feldspar, the ratios of all the lead isotopes do not change over time because virtually no radiogenic Pb is generated in the feldspar on account of the fact that feldspar contains virtually no uranium or thorium—ions that do not fit well into feldspar structure. At the time of formation, therefore, feldspar crystals incorporate lead with certain ratios of $^{208}Pb/^{207}Pb/^{206}Pb/^{204}Pb$ that remain unchanged through time. Because these lead isotopes have very similar masses, it is reasonable to assume that a

U-bearing mineral crystallizing from the same granitic magma as the feldspar incorporates these four Pb isotopes in virtually the same proportions as those in the feldspar, if it incorporates any Pb at all. It is then a simple matter of making a correction to the analyzed values of Pb ratios in a radioactive mineral like uraninite or possibly zircon. This is done by subtracting the measured ratio of lead isotopes in adjacent feldspar crystals. In some cases, the mineral galena (lead sulfide, PbS) may occur in veins associated with granite, and lead ratios in the galena can also be used to determine the likely original Pb ratios in the minerals being analyzed for use with the concordia method.

The concordia method is especially useful for zircon, but has also found application with titanite and apatite. Although zircon is present in small amounts and tiny crystals, it is an extremely abundant mineral that occurs in a wide range of igneous and metamorphic rocks.

Cosmogenic Isotopes and Exposure Ages

Among more recently developed radiometric methods are several that yield information about the length of time that a rock surface has been exposed to the atmosphere or that a meteorite has been exposed to cosmic rays as it travels through space from the asteroid belt to the Earth. These so-called *exposure ages* are based on the abundances of both radioactive and stable isotopes that are produced by interaction of cosmic "rays," particularly high-energy neutrons, with atoms that are present in the atmosphere or in common minerals occurring in rocks exposed at the surface.[5]

Cosmic "rays" are not really rays like light but are tiny particles, typically protons, neutrons or alpha particles originating in the Sun and in distant sources outside the Solar System. Generally, galactic cosmic rays that originate beyond the Solar System have considerably higher energies than solar-derived cosmic rays and are, therefore, more likely to produce cosmogenic isotopes during collisions with atoms in the atmosphere or in rocks. Among the more important cosmogenic nuclides are beryllium 10 (^{10}Be), aluminum 26 (^{26}Al) and chlorine 36 (^{36}Cl), all three of which are radioactive. ^{10}Be is produced in the atmosphere by fragmentation *(spallation)* of stable isotopes of oxygen and nitrogen in the air when impacted by cosmic rays. Such ^{10}Be atoms in the atmosphere adhere to tiny dust particles and are quickly removed by precipitation and deposited in lake- and sea-bottom sediments. Being radioactive, ^{10}Be atoms attached to clay particles in the sediments gradually decay. The half-life

---

[5]T. E. Cerling and H. Craig, "Geomorphology and In-situ Cosmogenic Isotopes," *Annual Review of Earth and Planetary Sciences* 22 (1994): 273-317; and Lionel L. Siame, Didier L. Bourlès and Erik T. Brown, eds., *In Situ-Produced Cosmogenic Nuclides and Quantification of Geological Processes*, GSA Special Paper 415 (Boulder, Colo.: GSA, 2007).

of ^{10}Be is 1.5 x 10^6 years. As a result, the abundance of ^{10}Be in a sediment pile typically decreases with depth because deeper sediment layers are older and are buried beneath more recently produced sediment, and in older sediment layers beryllium has had more time to decay. Measurement of ^{10}Be in various layers of lake- or sea-bottom sediments can, therefore, provide information about the age and sedimentation rate of these sediments.

In polar regions ^{10}Be may adhere to snowflakes and accumulate in thickening sheets of snow that are ultimately converted into glacial ice. The measurement of ^{10}Be and other cosmogenic isotopes in various layers in the Antarctic ice sheet, therefore, can yield information about ages of ice layers and the rate of development of glaciers. Because half-lives of these nuclides are on the order of one million years (table 14.1), this method works most effectively for relatively recent sediments (and ice layers). The ages are much less than those obtained by methods discussed previously for the dating of rocks, but that is to be expected. After all, lake sediments, sea-bottom sediments and glacial ice all occur on top of the crystalline rocks that are dated by those other methods. Therefore, on the principle of superposition, we expect these sediments and glaciers to be younger than rocks beneath them. It should also be noted that studies of cosmogenic isotopes in sea-floor sediments have provided no indication of a spectacular rise in the rate of sedimentation a few thousand years ago, as might be expected during a catastrophic planet-wide Flood.

Because the great majority of rocks in Earth's crust are composed of silicate minerals such as quartz, feldspar, mica and olivine, the likelihood of production of ^{10}Be by interaction of cosmic rays with silicon and oxygen atoms in these minerals is great. In addition, aluminum 26 (^{26}Al) is produced by neutron bombardment of silicon atoms in quartz and other silicate minerals. Chlorine 36 (^{36}Cl) forms by the spallation of potassium and calcium atoms when bombarded by cosmic rays. Potassium and calcium are among the most abundant chemical elements in common rock types and are present in feldspar, pyroxene and mica. Cosmic-ray production of cosmogenic isotopes such as ^{10}Be, ^{26}Al and ^{36}Cl in minerals in rocks on Earth's surface raises the possibility that geochronologists can also determine the time at which a rock face was first exposed to the atmosphere.

How does the method work? First of all, geochronologists need to know the factors affecting the intensity of cosmic rays within the atmosphere. The distribution of cosmic rays within Earth's atmosphere is influenced by the geomagnetic field. Cosmic rays are deflected by Earth's magnetic field toward the poles. Therefore, the abundance of cosmic rays is affected by *geomagnetic latitude,* and the intensity of cosmic rays at a specific elevation above sea level is higher at higher latitudes, that is, closer to the north or south magnetic poles. Thus, production of cosmogenic isotopes by cosmic ray bombardment of minerals in the rock surface would be greater, all other factors being equal, at high latitudes, that is, closer to the magnetic poles. If, therefore, one wants to determine the age of sediments on a lake bottom or the length of time that a rock has been exposed to the atmosphere, he or she must take into account the latitude of the lake or of the rock surface.

As cosmic rays descend through the atmosphere they interact with atmospheric atoms and may ultimately be absorbed before interacting with a rock face at the surface. Therefore, the abundance of cosmic rays at Earth's surface is also a function of altitude, that is, the thickness of atmosphere through which a cosmic ray has traveled. As a result, production rates of ^{10}Be and other cosmogenic isotopes in a rock face are influenced by the *altitude* at which a rock is exposed. All other factors being equal, the production rate of cosmogenic isotopes is higher, the higher the elevation of the rock.

Geochronologists also need to know the orientation of a rock face being dated. For a constant flux of cosmic rays, a steeply tilted rock face experiences fewer interactions with cosmic rays per unit surface area than a gently tilted rock face, and a gently tilted rock face experiences fewer interactions with cosmic rays per unit surface area than a horizontal rock face. As a result, cosmogenic isotopes are produced more abundantly on a horizontal rock face per unit surface area than on a steeply tilted rock face.

Geochronologists must also know production rates of various cosmogenic isotopes under identical conditions for different minerals and rocks. For example, ^{26}Al is produced at more than twice the rate in quartz than in the mineral olivine for the simple reason that there are more silicon atoms in a given mass of quartz than there are in olivine, and it is the silicon atoms that are impacted by the cosmic rays to produce ^{26}Al.

Exposure ages are calculated from equations that take into account these and other factors as well as the decay constant and the amount of radioactive isotope present in the mineral sample. The variables that must be taken into account are generally more numerous and difficult to assess in the case of cosmogenic isotopes than with the methods discussed for determining the ages of rock crystallization. Consequently, exposure and sedimentation ages are not always considered as accurate as whole-rock isochrons, for example.

Exposure-age dating has been a very valuable tool for dating glacial deposits like *moraines.* The dating of glacial deposits was formerly dependent almost entirely on stratigraphy and paleontology, but development of methods involving the disintegration of cosmogenic isotopes has permitted determination of the time elapsed since a boulder was first exposed to the atmosphere after deposition at the margin of a melting glacier. Let's take a look at a specific example. One of the most studied localities for alpine glaciation within the United States is the Wind River Range in western Wyoming. Since the early twentieth century, recognition of a succession of superposed moraines led to the realization that there have been several episodes of glacial advance and retreat in the area, perhaps coinciding with the advances and retreats of the continental ice sheets. The most recent moraines, superposed on older moraines, are much less thoroughly weathered than the older moraines and contain large angular boulders. These moraines are assigned to the so-called Pinedale stage (or advance). An older set of moraines, characterized by a somewhat more weathered nature, are assigned to the Bull Lake episode. The oldest moraine in the area, assigned to the Sacagawea Ridge stage, is thoroughly weathered and contains only a few large boulders because originally large boulders disintegrated during weathering. Thus, stratigraphic superposition, correlation with nearby terraces, and extent of weathering indicate that moraines of the Sacagawea Ridge stage are older than the moraines of the Bull Lake stage, which in turn are older than moraines of the Pinedale stage.

Enter exposure-age dating. The dating of 56 samples of fragments of boulders using the ^{36}Cl method reveals that the Pinedale moraines are 15,000 to 23,000 years old, the Bull Lake moraines have a minimum age of 120,000 years, and the Sacagawea Ridge moraines have a minimum age of 232,000 years and are probably much older.[6] Several samples dated by the ^{10}Be method have ages consistent with those obtained by ^{36}Cl dating.

A young-Earth creationist might be encouraged by the fact that these ages are not all that old. In some cases they are only a little more than the approximately 6,000-10,000 year age of Earth that they prefer. These dates, however, have been derived from glacial deposits that are stratigraphically younger than the vast successions of layered, fossiliferous sedimentary rocks that are thousands of feet thick that cover much of Earth's surface. If one is inclined to attribute these fossiliferous rocks to the action of Noah's Deluge, then one must deal with the fact that glacial deposits are more recent than the Deluge. In fact, the Deluge of Noah had to take place more than 230,000 years ago on the basis of the age of the Sacagawea Ridge moraines. Glacial deposits elsewhere have yielded even higher exposure ages.

As an additional example, let's consider large rock fragments excavated by a meteorite impact at Meteor Crater (also known as Barringer Crater) in northern Arizona, near the town of Winslow and about six miles south of Interstate

---

[6]Fred M. Phillips, Marek G. Zreda, John C. Gosse, Jeffrey Klein, Edward B. Evenson, Robert D. Hall, Oliver A. Chadwick and Pankaj Sharma, "Cosmogenic ^{36}Cl and ^{10}Be Ages of Quaternary Glacial and Fluvial Deposits of the Wind River Range, Wyoming," *Bulletin GSA* 109 (1997): 1453-63.

40. At this site, a meteorite, or possibly a pair of meteorites, approximately 150 feet in diameter impacted the surface and excavated the crater that is presently about three-quarters of a mile in diameter and 550 feet deep. A great quantity of material was blasted out of the crater and is draped over the adjacent landscape. Horizontal bedrock layers of Coconino, Kaibab and Moenkopi Formations that underlie the area were tilted upward by the impact as material was ejected and by subsequent rebound.

One way to determine the time of impact is to obtain exposure ages from [10]Be and [26]Al isotopes embedded in fresh rock surfaces that face the atmosphere. Nishiizumi and his colleagues analyzed cosmogenic isotopes in quartz grains from four samples of Kaibab Formation collected from the crater walls and six samples taken from large blocks of ejected Kaibab Formation.[7] Because the Kaibab Formation occurs 30 feet or more beneath the surface at this location, these samples were sufficiently buried so as to be shielded from cosmic rays prior to meteorite impact. Therefore, cosmogenic nuclides detected in the samples were produced *after* excavation and exposure to the atmosphere. The samples yielded [10]Be ages ranging from 14,600 to 51,600 years, indicating the various times when ejecta material consisting of Coconino Sandstone was removed by erosion from the underlying rubble of Kaibab Formation, or else talus slopes on the crater walls were stripped away to expose the samples. The highest ages ranging from 44,700 to 51,600 years were obtained from the "summits" of large ejecta blocks. These dates are in excellent agreement with dates obtained by [26]Al dating. These dates range from 14,500 to 52,700 years. It is worth noting, too, that the same sample yielded the youngest age by both methods. Likewise the four samples with the greatest [10]Be ages also yielded the greatest [26]Al ages.

In another study of boulders of Kaibab Formation lying on the surface at Meteor Crater, cosmogenic [36]Cl was analyzed.[8] The investigators selected samples that they thought would have stood above the surface of the ejecta blanket immediately after impact. The [36]Cl ages of five samples ranged from 36,500 to 50,400 years. The average age of the samples with the four highest values was 49,000 years. They regarded that as a best estimate for the time of impact.

One final study of Meteor Crater employed a nonradiogenic method, namely that of thermoluminescence.[9] Quartz grains from four sandstone

[7]K. Nishiizumi, C. P. Kohl, E. M. Shoemaker, J. R. Arnold, J. Klein, D. Fink and R. Middleton, "*In Situ* [10]Be-[26]Al Exposure Ages at Meteor Crater, Arizona," *Geochimica et Cosmochimica Acta* 55 (1991): 2699-703.

[8]Fred M. Phillips, Marek G. Zreda, Stewart S. Smith, David Elmore, Peter W. Kubik, Ronald I. Dorn and David J. Roddy, "Age and Geomorphic History of Meteor Crater, Arizona, from Cosmogenic [36]Cl and ^{14}C in Rock Varnish," *Geochimica et Cosmochimica Acta* 55 (1991): 2695-98.

[9]S. R. Sutton, "Thermoluminescence Measurements on Shock Metamorphosed Sandstone and

samples ranged from 45,100 to 53,600 years old with a mean of 50,400 years. Quartz grains from four samples of dolomite ranged from 37,700 to 50,800 years old with a mean age of 46,000 years.

The reader will likely recognize the general consistency of ages obtained by four independent methods, one of which does not rely on radioactive decay. The consistency of the ages is remarkable considering that the variables necessary in cosmogenic nuclide-based dating are more difficult to assess than are the conditions for the methods that yield extremely old ages. Given this consistency, geochronologists generally accept that the impact at Meteor Crater occurred around 49,700 years ago. That result alone is sufficient to bring into question the idea that the world is only 10,000 to 20,000 years old. Lest one take comfort in the fact that 50,000 years is not that far from 20,000 years, again the reader must remember that Meteor Crater is the result of a relatively recent event that occurred *after* the thick stack of layers of rock that now make up the Colorado Plateau was deposited and cemented. Many of the layers that had been deposited were also subsequently stripped away prior to meteorite impact as indicated by the regional geology of northern Arizona. A lot of geologic history took place long before the meteorite impacted the desert.

OTHER METHODS

Many of the methods that we have mentioned thus far are useful for dating extremely old rocks and minerals with ages of hundreds of millions to billions of years. Among these are the Rb-Sr, K-Ar, Sm-Nd, U-Th-Pb, Lu-Hf, Re-Os and La-Ba methods. In many cases what is dated is the time of crystallization of an igneous intrusion, the time of metamorphism or the time of solidification of a lava flow or ash deposit. We have also discussed a few methods involving cosmogenic nuclides such as ^{10}Be, ^{26}Al or ^{36}Cl. Many of these methods are useful for dating materials or geologic features that are only a few years to a few thousands of years in age. Such methods can give us information about the age of a recent sediment, glacial ice or eroded surface.

Still, we have only begun to scratch the surface of the impressive field of radiometric dating. There are numerous other radiometric methods for which interested readers are invited to consult the literature cited in the footnotes. The ^{40}Ar/^{39}Ar method is another widely used method that is useful for dating crystallization and heating events hundreds of millions of years old.[10] Numer-

---

Dolomite from Meteor Crater, Arizona: 2, Thermoluminescence Age of Meteor Crater," *Journal of Geophysical Research* 90 (1985): 3690-700. For detail on thermoluminescence methods, see Martin J. Aitken, *Thermoluminescence Dating* (London: Academic Press, 1985).

[10]On argon-argon dating, see Ian McDougall and T. Mark Harrison, *Geochronology and Thermo-*

ous methods involving decay of radioactive members of the ^{238}U decay chain utilize situations involving disequilibrium in the uranium series.[11] These methods are based on the physical and chemical separation of a radioactive daughter product from its radioactive parent isotope in certain geologic situations. Ages are determined from the rate of decay of the daughter isotope or the build-up of a daughter isotope from the parent isotope. These methods are extremely useful for determining ages of materials that are tens to hundreds of thousands of years old. Thus, they provide a valuable set of techniques for assessing ages between those datable by Rb-Sr, Sm-Nd and U-Th-Pb methods and those datable by cosmogenic nuclides. Uranium-series dating is useful for determining ages of modern lake and oceanic sediments, coral reefs and glacial ice.

Another valuable method that has recently come into its own is the U-He method.[12] Although U-He dating was first used in the early twentieth century, its use was discontinued because of He diffusion from minerals in which it was produced by decay of the members of the uranium decay series. Recent studies of diffusion rates of He from minerals such as zircon, titanite, monazite and apatite, however, have made it possible to determine approximately the temperature at which He diffusion ceases to be a major concern. It is now possible, especially by analyzing U and He in zircons, to estimate the amount of time that has elapsed since a zircon cooled to a temperature sufficiently low to retain He in significant amounts. In conjunction with so-called fission-track dating, U-He dating has made it possible to reconstruct the uplift and cooling history of mountain belts.

Archeological studies have been aided by utilization of fission-track dating and several nonradiometric methods such as obsidian rim hydration and thermoluminescence dating.[13] At many archeological sites, a combination of these methods, along with cosmogenic nuclide dating, has proved very useful

---

chronology by the $^{40}Ar/^{39}Ar$ Method, 2nd ed. (New York: Oxford University Press, 1999).

[11]On uranium-series disequilibrium dating, see M. Ivanovich and R. S. Harmon, eds., *Uranium-Series Disequilibrium: Application to Earth, Marine, and Environmental Science*, 2nd ed. (New York: Oxford University Press, 1992); and B. Bourdon, Gideon M. Henderson, Craig C. Lundstrom and Simon P. Turner, eds., *Uranium-Series Geochemistry*, Reviews in Mineralogy and Geochemistry 52 (Washington, D.C.: Mineralogical Society of America, 2003).

[12]For a summary of U-Th/He dating, see Kenneth A. Farley, "U-Th/He Dating: Techniques, Calibrations, and Applications," in *Noble Gases in Geochemistry and Cosmochemistry*, ed. Donald Porcelli, Chris J. Ballentine and Rainer Wieler, Reviews in Mineralogy and Geochemistry 47 (Washington, D.C.: Mineralogical Society of America, 2002), pp. 819-44.

[13]On archeological dating, see R. E. Taylor and M. J. Aitken, eds., *Chronometric Dating in Archaeology* (New York: Plenum Press, 1997); Edward C. Harris, *Principles of Archaeological Stratigraphy*, 2nd ed. (London: Academic Press, 1989); and Michael J. O'Brien and R. Lee Lyman, *Seriation, Stratigraphy, and Index Fossils: The Backbone of Archaeological Dating* (New York: Kluwer Academic/Plenum, n.d.).

for determining ages of pottery, hearths and artifacts. Such methods have also proved useful for dating very recent geologic events a few hundreds to a few thousands of years old.

Is Radiometric Dating Valid?

Young-Earth creationists are clearly uncomfortable with the results of radiometric dating methods. Ages of millions to a few billions of years obviously are not compatible with the idea that Earth is only a few thousands of years old! Consequently, young-Earth creationists have gone to great lengths to refute a large array of specific dating methods as well as the whole concept of radiometric dating. These arguments are either alleged refutations of straw men, mutually conflicting, or simply spurious. Because rebuttal of all the arguments against radiometric dating put forward by young-Earth advocates would require an entire book in itself, we address only a few representative criticisms that young-Earth proponents have advanced against the various isochron methods.

The fundamental young-Earth objections to isochron methods consist in the two claims that rocks are subject to so many geologic complexities that any alleged age information on an isochron diagram is bound to be unreliable and that a straight line on an isochron plot does not necessarily have any age relevance whatever and, therefore, cannot be used to substantiate great antiquity of a rock or mineral specimen. Let's consider the first claim concerning geologic complexities such as diffusion of isotopes in the rocks being dated. John Woodmorappe has stated that loss of Rb from whole rocks during their history could lead to an apparent isochron that would yield an incorrect age. This claim is no big revelation; every geochronologist accepts the truth of that contention. They know that significant Rb diffusion would likely result in considerable scatter of data points on an isochron diagram. More to the point, however, is that isochron plots resulting from Rb diffusion would look quite different from how they actually look if Earth is only a few thousand years old.

Consider a body of magma with a more or less homogeneous initial Sr isotope ratio that rapidly crystallized underground, say, five thousand years ago during the great Flood, to form a body of plutonic igneous rock. Suppose that we could drill into the pluton to collect several whole rock samples of this igneous rock body immediately after the Flood had ceased, analyze their isotopes and plot the results on an isochron diagram. The data points will likely define a reasonably good straight line that will be either horizontal or very close to horizontal simply because of the approximately homogeneous original Sr isotope ratios (fig. 14.2). If the original ratios varied widely, then the data points would be scattered and would not define a very good straight line. There is no compelling geochemical reason to expect that the data points obtained from rocks just after they formed will define a straight line with a steep positive slope on the isochron diagram.

Now let's assume that this body of igneous rock is quickly exposed at Earth's surface because of incredibly rapid erosion of the overlying rock, and let's imagine that we could analyze the exact same samples of whole rocks today, some five thousand years after the igneous rock originally formed. Keep in mind that the rocks do contain radioactive isotopes, and they will begin to decay during the past five thousand years. However, the radioactive minerals in the igneous rock body will undergo such a small amount of radioactive decay during those five thousand years, given the extremely large half-life of ^{87}Rb of 48.8 billion years, that the originally horizontal isochron will have pivoted only imperceptibly around the y-intercept. The isochron plot will still look virtually the same as an isochron plot determined at the time of crystallization of the igneous rock. The only way to avoid that assertion is to invoke accelerated nuclear decay, a phenomenon that is totally unsubstantiated experimentally—the "findings" of the RATE group notwithstanding—and something that Woodmorappe, to his credit, is loath to do.

Now let's assume, as Woodmorappe asked us to do, that significant Rb loss in the igneous rock body also occurred during its 5,000-year history. The result of Rb loss is that the Rb/Sr ratios of all the samples that we collected for analysis will be displaced to the left because the ratios are decreasing (see the lower part of fig. 15.4). But moving data points to the left along a virtually horizontal line still results in the definition of a line with a virtually negligible slope. The slope of this false isochron will look nothing like the actual lines with steep positive slopes that routinely result from analysis of igneous rock bodies. Only if an igneous rock is already very old could significant Rb loss by diffusion produce a false isochron that yields a fictitious very old age. If igneous intrusions were produced during Noah's Flood only a few thousand years ago, young-Earth creationists must account for the complete absence of near horizontal whole-rock or mineral isochrons determined for plutonic igneous rock bodies. Failing that, they must provide a satisfactory geochemical reason why there would be a strong positive correlation between the initial strontium isotope ratios and Rb/Sr ratios of whole rocks.

Woodmorappe also suggested that Rb-Sr isochron dating had been undercut by the findings of geologists Timothy Lutz and Leeann Srogi who pointed out that the "assumption that all samples used to construct an isochron had the same initial ratio" might be violated by "the effects of heterogeneities in Sr

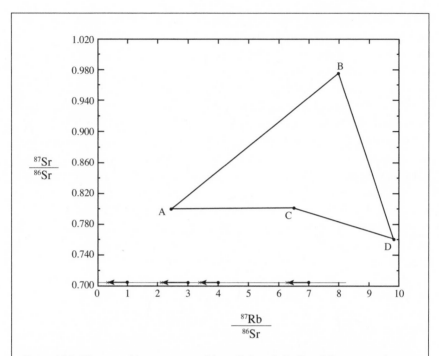

Figure 15.4. Diagram of isotope compositions: Points A, B, C and D represent isotope compositions of rocks from four different published isochrons. If all isochrons are really mixing lines, then, in principle, magmas with a very wide range of isotope compositions should have mixed, producing mixing lines like AB, which looks like a legitimate isochron, but also lines like AC, CD and BD. But note that if lines AC, CD and BD are projected to the $^{87}Sr/^{86}Sr$ axis, we would obtain extremely high "initial ratios" with values more than 0.800. Moreover, CD and BD have negative slopes. But no published isochrons based on analytical data ever have initial Sr isotope ratios much above 0.725 and never have negative slopes. These observations indicate that isochrons are not simply mixing lines. They really do provide age information.

At the bottom of the diagram is a horizontal isochron that represents a pluton of zero age. If the pluton becomes a few thousand years old, the isochron will still be virtually horizontal. If Rb diffuses from the pluton, represented by the left-pointing arrows, the data points representing isotope composition shift to the left and will still define a virtually horizontal isochron. Rb diffusion from a pluton in a very young Earth will not produce an apparent isochron with a steep positive slope. Hence, published isochrons cannot be explained by Rb diffusion from a pluton that is only a few thousand years old; they represent very old plutons.

isotopic composition created by hydrothermal alteration."[14] Such alteration, they claimed, is commonly initiated at the time of formation of igneous rocks

---

[14]Timothy M. Lutz and Leeann Srogi, "Biased Isochron Ages Resulting from Subsolidus Isotope Exchange: A Theoretical Model and Results," *Chemical Geology* 56 (1986): 63-71. For the quotation, see p. 64.

because water-rich fluids are generally concentrated in a cooling magma during the final stages of magmatic crystallization. Thus, there is likely to be some redistribution of Sr during the hydrothermal stage such that heterogeneities in Sr isotope composition are likely to be produced in the crystallizing rocks. If Lutz and Srogi are right, there should be some scatter of data points around a poorly defined straight line, and significant errors would be introduced into the interpretation of the actual age of the rocks.

Woodmorappe overlooked four points, however. In the first place, even where such hydrothermal alteration has taken place, the age of the misleading isochron is still going to be vastly greater than a few thousand years, even though it may not give an accurate determination of the time of crystallization. Second, Lutz and Srogi made "no claim that acceptance of inaccurate ages is a widespread problem."[15] Third, the problem of hydrothermal alteration is one that would primarily affect igneous plutons intruded at relatively shallow

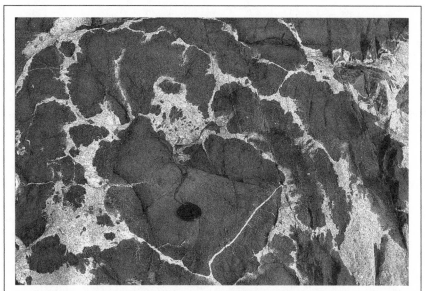

Figure 15.5. Portion of an igneous pluton along coastal Maine containing frozen remnants of "commingled" magmas that solidified to light-colored diorite and dark-colored basalt with very irregular outlines. Photo by D. A. Young.

levels of Earth's crust. Finally, Woodmorappe neglected to mention that Lutz and Srogi showed mathematically that samples with certain specific ranges of

---

[15]Ibid., p. 70.

geochemical characteristics involving Rb and Sr would have experienced less Sr introduction and would, therefore, demonstrate a minimum of age "distortion." Thus, by employing an appropriate sample selection strategy to the entire set of analyzed samples, a genuine isochron can still be obtained.

Let's now look at the second claim said to justify the rejection of isochrons. This is the claim that a straight line on an isochron diagram has no necessary age relevance whatever. Young-Earth creationists such as Arndts and Overn, Helmick and Baumann, Woodmorappe, and Brown have stated that the straight lines could just as well be the product of mixing of two end-members in varying proportions. We could, for example, take two magmas with contrasting chemical and isotopic compositions, mix them together in varying proportions, let them crystallize, analyze a suite of whole rocks, plot the data, and we would end up with a straight line on the isochron diagram. If such magmas were mixed a thousand years ago and analyzed last week, this straight line would have no age significance even though its slope might appear to indicate extremely old age. We would, indeed, have produced a pseudo-isochron.

However, the question is whether the straight lines on isochron diagrams are, in fact, due to mixing of magmas or other rock masses. Here is where firsthand knowledge of geologic factors is crucial. There are indeed situations in which hybrid magmas have been produced by mixing two other magmas with contrasting chemical compositions. More often than not, however, two magmas that encounter one another commingle. There is generally a minor amount of true mixing or solution of the magmas in one another. For example, the coexistence (commingling) of basalt or gabbro with granitic rock in the same dike or stock is relatively common.[16] The two contrasting igneous rock types occur side by side (fig. 15.5), an indication that two different magmas occupied the same dike or magma chamber at the same time. Field evidence in such situations, however, attests to the probability that, although some mixing of the two magmas produces a small quantity of homogenized hybrid or intermediate magma, complete mixing of the two end-member magmas in various proportions is obviously not even approached. Thorough mixing normally leaves detectable evidence in field relations or in thin sections of rock samples so that the nature of the two original magmas can be deciphered. It is uncommon for end-member magmas to become thoroughly hybridized.

---

[16]Commingled magmas refer to two or more magmas of contrasting chemical composition that occur in the same igneous rock body such as a dike or sill. Either the magmas were not soluble in each other, as is the case with gasoline and water, or the magmas had insufficient time to mix thoroughly and, therefore, retained their identity during solidification to physically distinct rock masses.

In cases where an isochron plot has been obtained for a stock or batholith of granite or another igneous rock, little if any field evidence exists for magma mixing on a large scale.

In the second place, Rb-Sr whole-rock isochrons (e.g., fig, 14.3) show that, if the mixing hypothesis is correct, one of the end-member magmas must have a high Sr isotope ratio. From almost any Rb-Sr whole-rock isochron plot it becomes evident that one of the magmas must have had a very high Sr isotope ratio if the isochron was produced by magma mixing. In some instances, the Sr isotope ratio exceeds 0.8. On the other hand, the left end of the isochron plot almost always has a Sr isotope ratio around 0.7 or a little higher.

Here's the problem for the young-Earth argument. If we analyze *any* modern day magma such as that erupted at volcanic centers like Hawaii or Mount St. Helens, we *never* encounter a magma with a Sr isotope ratio significantly more than 0.72. Basalt magmas erupted in the ocean basins today almost invariably have Sr isotope ratios around 0.705 or less, whereas more silica-rich magmas erupted on the continents today tend to have ratios as much as 0.72. But we just *never* find recently erupted magmas with Sr isotope ratios anywhere near as high as 0.8. If modern magmas never have these high ratios today, young-Earth creationists who advocate the mixing idea must explain why such magmas were seemingly so common in the geologic past. Where did these high Sr isotope ratio magmas disappear?

Moreover, if isochrons, each of which has its own slope, are mixing lines, there must have been many magmas with a wide range of high initial Sr ratios. And if there existed several magmas with high Sr isotope ratios, geochronologists should detect situations in which two magmas, both with relatively high Sr isotope ratios, were mixed together. But geochronologists *never* obtain an isochron that looks like the result of mixing of two magmas with high Sr isotope ratios. If such mixing did take place, the resulting mixing line would yield a high "initial" $^{87}Sr/^{86}Sr$ ratio isotope on the y-intercept (fig. 15.4). But isochrons almost always have initial Sr isotope values around 0.70-0.72. There is also no reason why we couldn't get two magmas with very similar Sr isotope ratios but different Rb/Sr ratios to mix. These would yield mixing lines with very gentle slopes that would give very low fictitious ages, but these are very rare. There is not even a reason why mixing lines with negative slopes could not occur, but we do not find those either.

In addition, one must take into account the fact that, in most cases, the isochron age is reasonably consistent with the age of the intrusion that had been previously estimated on the basis of field evidence. If isochrons are simply mixing lines, it is remarkable that the slopes of these alleged "mixing" lines generally increase with the relative ages of the intrusions. One could imagine a single instance in which a mixing line fortuitously gave a false isochron age consistent with the previously known geologic age, for example, Devonian or Cretaceous, but to expect such convergence on a large scale stretches credulity.

So far we have been talking only about igneous rocks. An even greater difficulty for adherents of the mixing hypothesis concerns the internal isochrons of meteorites. As noted earlier, the ages of meteorites are stunningly consistent at 4.4-4.6 billion years. Young-Earth (young-universe) advocates must address how to explain the isotope composition of individual minerals in meteorites in terms of a simple two-stage mixing process. The mixing hypothesis so beloved by young-Earth creationists cannot account for internal mineral isochrons for the simple reason that the individual minerals are not formed by mixing of end-member constituents from different fluids or by mixing of end-member minerals. Minerals like pyroxene, olivine, apatite or iron oxide have very specific chemical compositions that are not formed by blending. Whoever attempts to explain the Sr isotope composition by mixing must also explain the major chemical element composition in the various minerals by mixing. Even if young-Earth advocates could concoct an explanation for mineral isochrons by mixing, they must still explain why this mixing process always produces "false" isochrons with the exact same slopes and, therefore, the same apparent age. They must explain why the mixing process invariably produces straight lines that project to the Sr isotope axis at a value of 0.699 or close to it, and they must explain how whole-rock isochrons of stony meteorites can be the result of simple mixing.

Young-Earth creationists cannot appeal to mixing for meteorites in which there has been some degree of metamorphism, a process of *solid-state* crystallization. Rocks do not mix to produce rocks of hybrid composition very efficiently when they are already solid! Geochronologists who analyze the same meteorite by means of the Rb-Sr and Sm-Nd isochron and other methods obtain virtually the same age.[17] If the isochrons result from mixing, there is no reason to expect the remarkable consistency of meteorite dates. Nor is there any reason, on the mixing hypothesis, to expect that the supposedly initial strontium isotope ratios generated by the isochron lines would fall into such a very narrow range. The mixing explanation fails to account for all the evidence, is geologically unrealistic, and certainly provides no comfort for the young-Earth hypothesis.

The same observations apply to Sm-Nd isochrons. The mixing hypothesis cannot account for the consistency of initial Nd isotope ratios of isochrons, the existence of high Nd isotope magmas, the lack of negative isochrons, or internal isochrons.

---

[17]Dalrymple, *Ancient Earth, Ancient Skies.*

A specific example in which the mixing explanation for isochrons is advocated is a paper by Larry S. Helmick and Donald P. Baumann, professors of chemistry and biology, respectively, at Cedarville University. They wrote to "demonstrate the apparent superiority of the mixing model over the isochron method to explain rubidium-strontium isotopic data." They contended that

> whether the results are accepted or rejected is not determined by the accuracy and validity of the data itself or the inherent validity of the method, as is commonly thought, but by the validity of the researcher's presuppositions. If the results happen to agree with these presuppositions, they are arbitrarily accepted as valid. If they disagree, they are arbitrarily rejected as invalid. Consequently, the isochron dating method appears to add nothing of value to our understanding of the age of rocks. Furthermore, if the linear data is due to mixing, any age calculated from the slope would be erroneous.[18]

To illustrate their contention, Helmick and Baumann reported on a hypothetical exercise for their students. They provided students with beads of three different colors representing ^{87}Rb, ^{87}Sr and ^{86}Sr. They combined the beads into different groups with specified proportions of beads of each color to represent different values of the ratios of $^{87}Rb/^{86}Sr$ and $^{87}Sr/^{86}Sr$. Each group was plotted as a single data point on a diagram analogous to the isochron diagram. Students were asked to produce "mixtures" by combining numbers of groups of beads in various specified proportions. They might, for example, be asked to combine five of one group with ten of another group. Students plotted the values of bead proportions on the diagram. The bead "mixtures" that they made would plot on a diagram in a manner analogous to $^{87}Sr/^{86}Sr$ and $^{87}Rb/^{86}Sr$ with values intermediate to the values of the groups being combined. Two groups of beads could be combined in various proportions to yield "mixtures" that define straight lines on the diagram, analogous to isochrons. If all three groups of beads are combined in various proportions, then "mixtures" of the groups produce scattered plots.

Helmick and Baumann made the point that, by varying proportions of the various end members, both straight line and scatter diagrams can be produced, and that different kinds of straight lines can be produced. They reasoned that their bead model demonstration shows that mixing of magmas or other geologic materials can produce data plots with different possible outcomes, including mock isochrons. Thus, they felt that isotopic isochron plots really contain no meaningful age information.

---

[18]Larry S. Helmick and Donald P. Baumann, "A Demonstration of the Mixing Model to Account for Rb-Sr Isochrons," *Creation Research Society Quarterly* 26 (1989): 20-23.

The Helmick-Baumann "experiment" may persuade anyone who does not understand geology or the geologic basis for the isochron diagram. The bead exercise engages students in a purely abstract mathematical exercise that is totally divorced from geologic and geochemical reality. In the first place, mixing three different components may be important in producing some sedimentary rocks, but sedimentary rocks are very rarely dated by isochron methods. Simultaneous mixing of three different magmas to produce hybrid igneous rocks is extremely unlikely. Mixing of two magmas certainly occurs, but field and mineralogical evidence almost invariably indicates that such mixing has occurred. Moreover, Helmick and Baumann, neither of whom is a geologist, have not addressed real-world geologic problems with general application of the mixing hypothesis. All that they have demonstrated is that one can take beads of three different colors, combine them in different ways, plot the results on a diagram, and produce a straight line. They have not even begun to demonstrate the superiority of the mixing model as over against the isochron model. Absolutely no geochronologist would for a moment take their "demonstration" with the slightest amount of seriousness.

To conclude, we examine a paper by Robert Brown, long a critic of radiometric dating techniques.[19] Brown stated that "if uniformitarian scientists can comfortably resort to a mixing-line interpretation when an isochron interpretation gives an age inconsistent with the geologic time scale, there should be equal freedom to choose a mixing-line interpretation for data which give an isochron interpretation that contradicts a Biblically-based time frame."[20] To many readers, Brown's statement may sound eminently fair. But let's look more closely. In the first place, young-Earth creationists must consistently maintain that *every* isochron interpretation espoused by the geologic community contradicts what the young-Earth advocates take to be a "biblically-based time frame" of only a few thousands of years. Because isochron diagrams (and there are thousands of them) invariably yield ages in the range of tens of millions to billions of years, young-Earth creationists must conclude that all isochrons are false. Young-Earth creationists, therefore, must either invoke the mixing-line interpretation for all isochron diagrams; develop a satisfactory alternative interpretation of the diagrams, which they have not done; invoke speed-up of decay rates, for which there is no evidence; or invoke creation of apparent age, which many of them are hesitant to do. In any case, the mixing

---

[19]Robert H. Brown, "Mixing Lines—Considerations Regarding Their Use in Creationist Interpretation of Radioisotope Age Data," in *Proceedings of the Third International Conference on Creationism, Technical Symposium Sessions*, ed. R. E. Walsh (Pittsburgh: Creation Science Fellowship, 1994), pp. 123-30.
[20]Ibid., p. 126.

model cannot be based solely on a diagram. There must be field, mineralogical and geochemical evidence to support the idea of mixing to go along with every "isochron" plot.

Geologists, however, consider the mixing model only in special cases where independent empirical field and mineralogical evidence suggests the possibility of mixing. In cases where mixing is invoked to account for the origin of an igneous rock body, standard isochron plots are typically *not* involved. So-called pseudo-isochron plots may be used, however. For example, Brown referred to a study of granitic intrusions in central Idaho by Fleck and Criss.[21] He commented that "conventional geological considerations place the age of these formations in the vicinity of 80 million years." He then stated that "the 1640 m.y. Rb-Sr isochron obtained from these formations is accounted for . . . by proposed mixing of melted wall rocks with rising magma at the time of pluton formation."[22] Now the authors of the study did indeed argue that there is *evidence of mixing of the rising magma with partly melted wall rocks.* Brown's statements, however, are misleading. He gave the impression that a Rb-Sr isochron on the plutons yielded an age of 1,640 million years, seemingly in contradiction to the 80 million year age obtained by "conventional geological considerations." The reader may infer from Brown's statements that the Rb-Sr isochron age is badly out of agreement with the age based on the field geology, thus prompting geologists to consider mixing to account for the apparent isochron date.

But Fleck and Criss indicated that radiometric ages based on the K-Ar dating of hornblende and U-Pb dating yield ages of 75 to 95 million years. Well, these data are consistent with the field evidence of Late Cretaceous age around 80 million years ago. But what about the Rb-Sr isochron? No such whole-rock isochron was determined for these rocks. There was no report of an Rb-Sr isochron that yields an age of 1,640 million years in the paper by Fleck and Criss. Instead, the authors measured the Rb and Sr concentrations and the *present-day* $^{87}Sr/^{86}Sr$ ratios of several whole-rock samples from three different igneous rock suites. Already knowing the ages of the rocks from the K-Ar and U-Pb dating, they *calculated* the initial Sr isotope ratios of the igneous rocks at the time of formation from the present-day ratios. *Fleck and Criss did not obtain the initial Sr ratios graphically from an isochron plot.* They found that the calculated initial Sr isotope ratios increased systematically from <0.704 in Oregon to the west to >0.708 in central Idaho in the east. They plotted the values of the calculated *initial* Sr isotope ratios versus 1/Sr ratios (not the

[21]R. J. Fleck and R. E. Criss, "Strontium and Oxygen Isotope Variations in Mesozoic and Tertiary Plutons of Central Idaho," *Contributions to Mineralogy and Petrology* 90 (1985): 291-308.
[22]Brown, "Mixing Lines," p. 127.

^{87}Rb/^{86}Sr ratios) of the rocks and obtained a crude straight line. *This plot is not a standard Rb-Sr isochron plot, and it is not used to determine the ages of crystallization of igneous plutons.*

Fleck and Criss also plotted *initial* Sr isotope ratios versus Rb/Sr ratios, again not a standard whole-rock isochron plot. Data points on this diagram defined a crude "pseudo-isochron" of 1,640 million years. This "pseudo-isochron" is not designed to yield an age for the crystallization of the igneous rock bodies. On the other hand, the "pseudo-isochron" age is reasonably close to the known age of the wall rocks through which the magmas ascended. Fleck and Criss interpreted all their data in terms of mixing of high-Sr magmas ascending from the mantle and fluid-rich melt produced by partial melting of Precambrian schists and gneisses in the area. But there was no reinterpretation of a Rb-Sr whole-rock isochron that gave an age they didn't like.

Another paper to which Brown referred in which mixing was invoked likewise did not invoke mixing to explain Rb-Sr whole-rock isochrons, but rather utilized plots of initial Sr isotopes versus another variable, a very different kind of plot from a true isochron plot—to support their claim that mixing had taken place in the production of numerous magmas over a wide area in southeastern Australia. In no case was the mixing hypothesis invoked to explain a whole-rock isochron age for the time of crystallization of the magma bodies.[23]

To sum up, appeals to the mixing hypothesis by young-Earth advocates to account for isochron plots fail completely because of failure to understand the empirical geologic and geochemical factors that enter into the interpretation of the diagrams. The proponents of a young Earth have also failed to appreciate the distinction between a standard isochron plot in which the ordinate of a graph plots the Sr isotope ratio and other diagrams in which the ordinate plots the *initial* Sr isotope ratio. As it stands, isochron diagrams that include straight-line data plots are, in most instances, valid indicators of the age of some significant geological event, whether it is the time of crystallization or the time of metamorphism. And those indicated ages are vastly greater than young-Earth creationists care to hear.

THE ALLEGED PROBLEM OF DISCORDANT AGES

An oft-repeated criticism of radiometric dating leveled by young-Earth creationists is that different methods commonly yield discordant ages for the same rock or event.[24] Thus, the four U-Pb methods typically yield four differ-

---

[23]C. M. Gray, "An Isotope Mixing Model for the Origin of Granitic Rocks in Southeastern Australia," *Earth and Planetary Science Letters* 70 (1984): 47-60.

[24]For recent examples, see Steven A. Austin, "Do Radioisotope Clocks Need Repair? Testing the Assumptions of Isochron Dating Using K-Ar, Rb-Sr, Sm-Nd, and Pb-Pb Isotopes," and Andrew

ent ages. Moreover, a U-Pb mineral age obtained from zircon is often higher than a K-Ar age obtained from biotite in the same rock. And a K-Ar age on hornblende is generally higher than a K-Ar age on biotite from the same rock. If we constantly obtain all these discrepant ages for the same rock body, critics ask, then how can we trust any of them? Isn't there something faulty and contrived about the whole business? To many proponents of young-Earth catastrophism, this criticism seems fatal to the whole program of radiometric dating. However, if one understands how the different systems are affected by real-world geologic conditions, then discordant ages generally do not pose a serious problem. In many instances, the ages differ because the method is not measuring the same event. For example, because Ar is lost from minerals like biotite at high temperature, a K-Ar age determined from a biotite crystal within a granite intrusion is not going to tell us the time of crystallization of the granite. Instead, it tells the time at which the intrusion cooled to the temperature at which Ar ceases to diffuse effectively from the biotite and begins to accumulate within the biotite. In contrast, a whole-rock Sm-Nd age is far more likely to indicate the time of crystallization of the granite body. We should, therefore, normally expect that a K-Ar age on a biotite from a granite intrusion will be less than the age we might obtain by the Sm-Nd whole-rock isochron method on the same intrusion.

In another case, discordant U-Th-Pb ages result from substantial lead loss. Four different ages may be obtained from zircons in an igneous rock intrusion, none of which indicates accurately the time of crystallization of the intrusion and none of which accurately indicates the time of lead loss. One may often assume that the actual age of crystallization of the intrusion is greater than the discrepant ages and that the time of diffusion (lead loss) was less than at least three of the discrepant ages. The concordia method provides one way to address this problem head on and to gain extra information from the loss of lead.

At the very worst, all that discrepant ages indicate is that geologic conditions are so complex that radiometric dating methods are not as accurate as we might like. The fact that discordances do occur from time to time, however, should grant little solace to proponents of a young Earth for the simple reason that, almost invariably, the ages obtained by different methods for the same body of rock are on the order of hundreds of thousands to a few billions of years. Geochronologists generally do not encounter situations in which one

A. Snelling, "Isochron Discordances and the Role of Inheritance and Mixing of Radioisotopes in the Mantle and Crust," both in *Radioisotopes and the Age of the Earth: Results of a Young-Earth Creationist Research Initiative*, ed. Larry Vardiman, Andrew A. Snelling and Eugene F. Chaffin (El Cajon, Calif.: Institute for Creation Research and Creation Research Society, 2005), 2:325-92 and 2:393-524.

method yields an age of 57 million years for a given rock body, whereas an alternative method for the same body yields an age of 5,700 years, thus forcing us to choose between a young age and an old age. Geochronologists are more likely to encounter a situation where one method yields an age of 57 million years and another, 49 million years. Even rock samples with discordant ages are almost invariably all extremely ancient!

The complaints of young-Earth creationists about discrepant or discordant dates notwithstanding, the fact remains that there are a plethora of instances in which diverse methods produce remarkable agreement of ages. We already mentioned the astonishing concordance of many dozens of ages between 4.4 and 4.6 billion years on stony meteorites, iron meteorites, individual meteorites and groups of meteorites obtained from Rb-Sr and Sm-Nd internal isochrons, Rb-Sr, Pb-Pb, Lu-Hf, and Re-Os whole-rock isochrons, and by $^{40}Ar/^{39}Ar$ dating.

Good concordance also shows up for terrestrial samples. Consider some of the very oldest rocks known on Earth from the Isua Greenstone Belt in the Itsaq Gneiss Complex of western Greenland.[25] The Isua Greenstone Belt consists of a variety of metamorphosed sedimentary and volcanic rocks. Samples of conglomerate have been dated by U-Pb and Rb-Sr methods and have yielded ages ranging from 3.66 to 3.77 billion years. Much finer-grained sedimentary rocks in the Isua Greenstone Belt have been dated by Sm-Nd and give an age of 3.74 billion years. The volcanic rocks have been dated by Pb-Pb and Sm-Nd and are 3.71 to 3.81 billion years old. A banded iron formation has been dated by U-Pb and Pb-Pb and gives an age of 3.70 billion years. The reader should sense that these rocks are very close to being around 3.75 billion years old! The consistency of these ages is striking considering that rocks so old are likely to have experienced a complex geologic history. The fact that similar consistencies or concordances are not in the least unusual has contributed to the high degree of confidence placed in these methods by a scientific community that thrives on and depends on repeated critical challenges to scientific claims.

If radiometric dating cannot withstand legitimate, knowledgeable critique, then geologists will abandon its methods. They have already shown that they will abandon defective methods following compelling criticism by members of

---

[25]For further information on the Isua Greenstone Belt, see S. Moorbath, M. J. Whitehouse and B. S. Kamber, "Extreme Nd-isotope Heterogeneity in the Early Archean—Fact or Fiction? Case Histories from Northern Canada and West Greenland," *Chemical Geology* 135 (1997): 213-31; A. P. Nutman, V. R. McGregor, C. R. L. Friend, V. C. Bennett and P. D. Kinney, "The Itsaq Gneiss Complex of Southern West Greenland: The World's Most Extensive Record of Early Crustal Evolution (3900-3600 Ma)," *Precambrian Research* 78 (1996): 1-39; chap. 4 of G. Brent Dalrymple, *The Age of the Earth* (Stanford, Calif.: Stanford University Press, 1991); and chap. 5 of Dalrymple, *Ancient Earth, Ancient Skies.*

the scientific community. But several other methods have long withstood critical analysis and are thereby considered fruitful tools of geological investigation. The evidence obtained from a wide range of radiometric dating methods overwhelmingly indicates that the Earth is billions of years old. The clues that God has left in the rocks in the form of radioactive isotopes and their daughter products could not be much clearer.

Part Four

PHILOSOPHICAL PERSPECTIVES

16

Uniformitarianism, Catastrophism and Empiricism

In the last several chapters we demonstrated that geologic evidences reveal that Earth is extremely old, and we have shown that young-Earth creationists have misinterpreted the evidence in claiming that Earth is only a few thousand years old. Despite voluminous empirical evidences cited in opposition to young-Earth theories, most young-Earth creationists believe that modern geologists, including the authors of this book, are captive to a faulty, rationalistic philosophy of science. Specifically, young-Earth creationists have claimed that the picture of Earth's history advanced by modern geology has been constructed on the basis of a *principle of uniformitarianism* that is painted as anywhere from atheistic to anti-biblical. Young-Earth proponents imply that, by blindly holding to a dogma of uniformitarianism, geologists unwittingly misinterpret the geologic evidence pertaining to the antiquity of Earth, thus drawing incorrect conclusions. Advocates of a young Earth imply that geologists could be liberated from their false interpretations of the geologic record by invoking a biblical principle of *catastrophism.*

The Creationist Challenge to Uniformitarianism

Before defining the terms *uniformitarianism* and *catastrophism*, let's convey a sense of the importance of these two terms to young-Earth advocates. The heading of chapter four of *The Genesis Flood* (1961) is titled "Uniformitarianism and the Flood: A Study of Attempted Harmonizations." Chapter five, titled

"Modern Geology and the Deluge," includes sections titled "The Uniformitarian Interpretation of Geology," "The Inadequacy of Uniformity to Explain the Strata" and "Contradictions in the Uniformitarian System." Whatever uniformitarianism may be, it is not good, because it is characterized by inadequacy and contradictions. The section "The Uniformitarian Interpretation of Geology" asserted that "the uniformitarian assumption is certainly a reasonable assumption, provided there is no sufficiently valid evidence to the contrary, but it must always remain merely an assumption."[1] In the next section "The Inadequacy of Uniformity to Explain the Strata," Whitcomb and Morris attempted to substantiate that there might be "sufficiently valid evidence" contrary to uniformitarianism. They began with a more forceful claim that "it is not even true that uniformity is a *possible* explanation for most of the earth's geologic formations, as any candid examination of the facts ought to reveal."[2] With regard to the vast quantities of igneous rocks, including volcanic rocks, around the world, they asserted that "the principle of uniformity breaks down completely at this important point of geologic interpretation. Some manifestation of catastrophic action alone is sufficient."[3] Turning to the process of mountain building, the authors stated:

> it is here that the principle of uniformity would appear to be most inadequate. If it were valid, surely a feature of such prime importance in the interpretation of earth history as diastrophism and orogeny should be explainable in terms of some sort of present-day observable and measurable process which is now producing incipient earth movements of similar kind.[4]

In regard to continental ice sheets, they said that "the principle of uniformity is once again woefully inadequate to account for them" and that "the dogma of uniformity has thus far completely failed to account for this additional very important aspect of accepted geologic history."[5] As to various processes of sedimentation, they claimed that "the geologic dogma of uniformity has once again proved inadequate to explain the geologic data."[6] Furthermore, concerning the formation of coal, they wrote that "the fundamental axiom of uniformity, that the present is the key to the past, completely fails to account

---

[1]J. C. Whitcomb Jr. and H. M. Morris, *The Genesis Flood: The Biblical Record and Its Scientific Implications* (Philadelphia: Presbyterian & Reformed, 1961), p. 131.

[2]Ibid., p. 137 (italics in original).

[3]Ibid., p. 139.

[4]Ibid., p. 140. *Orogeny* is the technical term for "mountain building." *Diastrophism* is a term that is no longer used by geologists that refers to the formation of large structures within the Earth's crust by the application of stresses.

[5]Ibid., pp. 143-44.

[6]Ibid., p. 146.

for the phenomena."[7] They also asserted that "the doctrine of uniformity thus is supposed to render unnecessary any recourse to catastrophism, except on a minor scale."[8]

Whitcomb and Morris further charged that uniformitarian thinking stems from an unbiblical, false, unchristian philosophy. In support of this charge, they appealed to 2 Peter 3:3-10, a passage in which Peter warned that there would be scoffers in the last days who would say, "Where is this 'coming' he promised? Ever since our fathers died, everything goes on as it has since the beginning of creation." Peter then reminded his readers of the Flood judgment and of the coming final judgment. With regard to this passage Whitcomb and Morris said:

> Here again the Flood is used as a type and warning of the great coming world-wide destruction and judgment when the "day of man" is over and the "day of the Lord" comes. But the prophet is envisioning a time when, because of an apparent long delay, the "promise of his coming" is no longer treated seriously. It is to become the object of crude scoffing and intellectual ridicule. It will be obvious to "thinking men" in such a day that a great supernatural intervention of God in the world, as promised by Christ, is scientifically out of the question. That would be a miracle, and miracles contradict natural law.
>
> And how do we know that miracles and divine intervention contradict natural law? Why, of course, because our experience shows and our philosophy postulates that "all things continue as they were from the beginning of the creation"! This is what we call our "principle of uniformity" which asserts that all things even from the earliest beginnings can be explained essentially in terms of present processes and rates. Even the Creation itself is basically no different from present conditions, since these processes are believed to have been operating since even the "beginning of the creation." There is no room for any miracle or divine intervention in our cosmology; therefore, the concept of a future coming of Christ in worldwide judgment and purgation is merely naïve.[9]

We begin to see what young-Earth creationists understand uniformitarianism to be and why they so abhor it. In their understanding, the core of the concept of uniformitarianism is that the geologic past must be interpreted solely in terms of present-day causes operating at the slow, gradual rates of the present.

The supposed failure of uniformitarianism has been a continuing theme in the writings of Morris subsequent to 1961. Chapter five of his book *Scientific Creationism* (1974), is titled "Uniformitarianism or Catastrophism?" Much of

---

[7]Ibid., p. 165.
[8]Ibid., p. 131.
[9]Ibid., p. 452.

the chapter includes denigrations of so-called uniformitarian explanations of geologic phenomena. Like many other young-Earth creationists, Morris frequently invoked an intimate connection between evolution and uniformitarianism as a strategy for discrediting the latter, given the widespread hostility toward evolution on the part of so many Christians in the evangelical wing of the church. "It is true," Morris wrote,

> that the evolution model is fundamentally tied to uniformitarianism, since it assumes that present natural laws and processes suffice to explain the origin and development of all things. The creation model is fundamentally catastrophic because it says that present laws and processes are not sufficient to explain the phenomena found in the present world.

Morris also quoted several geologists who also "are seriously questioning or altering the traditional application of uniformitarianism to geology.[10] He then argued for rapid processes and opposed slow uniform processes in explaining geologic history.

This negative attitude toward uniformitarianism has persisted in young-Earth creationist literature to the present day. In 1975, Loren Steinhauer wrote that one aspect of uniformitarianism "is at variance with observation," whereas another aspect "leads to logical and philosophical contradictions."[11] In *Grand Canyon: Monument to Catastrophe* (1994), young-Earth geologist Steven Austin maintained that Young, one of the coauthors of our book, "adopts the uniformitarian framework, and does not acknowledge a legitimate alternative."[12] With reference to the strata of the Grand Canyon, Austin insisted that "it is imperative that we examine *both* uniformitarian and catastrophist frameworks." He linked the concepts of evolution and uniformitarianism by noting that "evolution and creation form different interpretive frameworks that geologists use to interpret Grand Canyon. . . . Evolutionists frequently make the uniformitarian assumption that strata of Grand Canyon formed during long ages, as oceans slowly advanced and retreated over the North American continent." In contrast, he said that "Creationists usually make the catastrophist assumption that a global flood was responsible for depositing a great thickness of Grand Canyon strata." These two very different explanations of the strata of the Grand Canyon exist because

> these interpretations of strata are determined by the presuppositions or bias of

---

[10]Henry M. Morris, *Scientific Creationism*, 1st ed. (El Cajon, Calif.: Master Books, 1974), p. 92.

[11]Loren C. Steinhauer, "Tracing the Past: Is Uniformity Meaningful?" in *Symposium on Creation V*, ed. Donald W. Patten (Grand Rapids: Baker, 1975), p. 89.

[12]Steven A. Austin, "Interpreting Strata of Grand Canyon," in *Grand Canyon: Monument to Catastrophe* (Santee, Calif.: Institute for Creation Research, 1994), p. 23.

the interpreter. If one begins with the uniformitarian model, that sedimentation occurred in calm and placid seas, an evolutionary interpretation of thick strata sequences will be derived. However, if one starts with a catastrophist framework, that sedimentation has occurred in response to catastrophic agents, a creationist interpretation is given to thick strata sequences. Conclusions, therefore, are derived from assumptions.[13]

Note that Austin claimed that "uniformitarians" *assume* that strata formed over long ages, whereas the "catastrophists" *assume* that strata were formed by a global flood. We will return to this point later.

In his book *In the Beginning* (1995), Walter Brown referred to an "anti-catastrophe principle" known as "uniformitarianism." This principle had "for more than 150 years" been "summarized by the phrase, 'The present is the key to the past.'" Brown understood this phrase to mean that "only processes observable today and acting at present rates can be used to explain past events."[14] Later he wrote that "the other foundational principle in evolutionary geology is the principle of uniformity. It states that all geological features can be explained by processes operating today."[15]

In writing about the origin of the Little River Canyon in Alabama, Alfred Jerry Akridge asked:

Are present process rates in geology sufficient to explain most geologic changes which have occurred throughout the history of planet Earth? Is an immensity of geologic time required? Uniformitarian geologists generally assume this [and here Akridge quoted from the 1955 fifth edition of a historical geology textbook by Carl Dunbar]. According to this doctrine, small amounts of moving water operating over vast periods of time can carve deep canyons through solid rock. . . . People who hold to the doctrine of catastrophism state that catastrophic or violent and sudden events which happened between periods of slow and gradual change in geologic history caused major changes in the Earth's features.[16]

Again we see the claim that geologists assume the dominance of "present process rates" in Earth history.

In his critique of radiometric dating, John Woodmorappe repudiated uniformitarianism because of its alleged negative effect on evangelism. To counter those who think that the promulgation of Flood geology is a hindrance to evangelism, he wrote that "compromise with the uniformitarian old-age

---

[13]Ibid., p. 22.

[14]Walter Brown, *In the Beginning: Compelling Evidence for Creation and the Flood*, 6th ed. (Phoenix: Center for Scientific Creation, 1995), p. 130.

[15]Ibid., p. 142.

[16]Alfred Jerry Akridge, "A Flood Based Origin of Little River Canyon Near Ft. Payne, Alabama," in *Proceedings of the Fourth International Conference on Creationism 1998*, ed. Robert E. Walsh (Pittsburgh: Creation Science Fellowship, 1998), p. 9.

system has the *precise opposite* effect of keeping the Gospel credible."[17] He also regarded the idea that "uniformitarian geochronology rests upon rigorous, objective methodology" as a myth to be exploded.[18]

Radioisotopes and the Age of the Earth (2000) is replete with references to uniformitarianism such as the "uniformitarian belief system," "uniformitarian radioisotopic dating," "uniformitarian process rates" and several mentions of the "uniformitarian timescale." Specific examples are found in chapters written by Eugene Chaffin and Andrew Snelling. The abstract of Chaffin's chapter "Theoretical Mechanisms of Accelerated Radioactive Decay" comments that "the data are examined in attempts to rid the discussion of the assumptions of the uniformitarian philosophy, and to formulate views of the relevant theories of physics which are not hampered by an anti-Biblical worldview."[19] In his chapter on radiohalos, Snelling states that "if a large amount of radioactive decay has actually occurred, then within the Biblical, young-earth creationist timeframe such large quantities of radioactive decay had to occur in a drastically shorter period of elapsed time than the constancy at today's decay rates assumed by uniformitarians would allow."[20]

Our final example comes from a recent paper titled "The Uniformitarian Stratigraphic Column—Shortcut or Pitfall for Creation Geology?"[21] The authors expressed apprehension over the use of the standard "uniformitarian" geological timescale by young-Earth creationists. One reason they gave for their concern is that

> these models of earth history are developed within the context of the Naturalist worldview and are thus at odds with Christianity on many levels. We believe there are significant differences between a radical approach of evaluating and reinterpreting data collected, analyzed, and published over many years by the uniformitarian establishment and introducing a "flood explanation" on top of an essentially uniformitarian interpretation.

The authors recommended that the role of the uniformitarian timescale

---

[17]John Woodmorappe, *The Mythology of Modern Dating Methods* (El Cajon, Calif.: Institute for Creation Research, 1999), p. 5 (italics in original).

[18]Ibid., p. 8.

[19]Eugene F. Chaffin, "Theoretical Mechanisms of Accelerated Radioactive Decay," in *Radioisotopes and the Age of the Earth: A Young-Earth Creationist Research Initiative,* ed. Larry Vardiman, Andrew A. Snelling and Eugene F. Chaffin (El Cajon, Calif.: Institute for Creation Research and Creation Research Society, 2000), 1;305.

[20]Andrew A. Snelling, "Radiohalos," in *Radioisotopes and the Age of the Earth: A Young-Earth Creationist Research Initiative,* ed. Larry Vardiman, Andrew A. Snelling and Eugene F. Chaffin (El Cajon, Calif.: Institute for Creation Research and Creation Research Society, 2000), 1:398.

[21]John K. Reed and Carl R. Froede, "The Uniformitarian Stratigraphic Column—Shortcut or Pitfall for Creation Geology?" *Creation Research Society Quarterly* 40 (2003): 90-98.

should be discontinued in creationist models "if it can be shown that the column rests, even in its supposedly empirical aspects, on presuppositions of evolutionary uniformitarianism, and deep time," because these presuppositions are "all inimical to creationism."[22] They charged that modern geologic interpretation of the Earth's crust encapsulated in the uniformitarian stratigraphic column "includes the rejection of the Christian worldview in favor of Naturalism—a worldview that replaces a reality founded on God's creation and governance of the universe with an impersonal, uncaring mechanism."[23]

The examples that we have cited represent only a tiny fraction of negative references to the concept of uniformitarianism that permeate young-Earth creationist literature.[24] The solution to the problem of the dreaded uniformitarianism, according to young-Earth creationists, is to reinterpret the data of geology in terms of the principle of catastrophe which involves the idea that the Earth has been subjected to global or very widespread catastrophic phenomena beginning with creation and extending to the Fall of Adam and Eve, and to times following the Fall, with emphasis on the Noachian Flood, believed to have inundated the entire Earth for about a year. Additional post-Flood adjustments are alleged to have been catastrophic. Flood geologists postulate planetary-scale erosion, with vast slurries of sediment washing back and forth. Simultaneously, catastrophic volcanism and vastly accelerated rates of mountain-building and plate-tectonic activity prevailed. The catastrophic philosophy is believed to offer a superior explanatory principle for accounting for regional stratigraphies, fossil preservation, fossil graveyards, polystrate trees, thick salt deposits, mountains, volcanoes, igneous intrusions, and the distribution of radioactive and radiogenic isotopes.

The Young-Earth Creationist Understanding of Uniformitarianism

The writings of young-Earth advocates invariably indicate that they view uniformitarianism as a method for the reconstruction of the Earth's past which *assumes* that geologic history can or must be explained almost entirely in terms of geologic processes that occur on the Earth at the present time and that have operated throughout geologic time at more or less the same slow, gradual rates that characterize the present day. Young-Earth creationists also typically claim that modern geology is founded on this conception of uniformitarianism. Young-Earth advocates seem to think that geologists postulate

---

[22]Ibid., p. 90.

[23]Ibid., p. 95.

[24]Several older citations pertaining to uniformitarianism may be found in the original version of Davis A. Young, *Christianity and the Age of the Earth* (Grand Rapids: Zondervan, 1982).

very slow process rates and a very peaceful Earth history in which there were virtually no catastrophes, inasmuch as modern processes like mountain building and sedimentation are supposedly extremely slow and anything but catastrophic.

Indeed young-Earth creationists seem to think that modern geologists reject a priori the very possibility of great large-scale or global catastrophes. They seem to think that uniformitarianism requires that an example of every kind of rock found in the geologic record must be forming somewhere in the world today if the present is the key to the past. They seem to think that, if we cannot find an example of a specific rock type forming in the world today, uniformitarianism is somehow contradicted. They seem to think that very rapid, violent processes are inconsistent with uniformitarianism. Room is allowed only for small-scale local catastrophic events.

Young-Earth creationists often claim that uniformitarianism is partnered with commitment to an evolutionary view of the universe that flows from hostility toward the Christian-theistic worldview presented in Scripture, a hostility that is directed particularly toward the concepts of divine creation and a catastrophic global Flood to exterminate sinful humanity. Before we conclude that one is faced with a stark choice between naturalistic uniformitarianism as allegedly adopted by modern geology and biblical, Christian-theistic catastrophism as allegedly adopted by young-Earth creationists, let's look at the concept of uniformitarianism from a historical point of view. Then let's examine how modern geologists understand the concept of uniformitarianism. Let's also see if young-Earth creationists have a beef with what modern geologists actually believe and do, and let's see if uniformitarianism is really unbiblical.

Uniformitarianism in History

Charles Lyell believed that his predecessors and contemporaries who studied the Earth appealed to catastrophic agents too readily in reconstructing the geologic past. He maintained that it isn't necessary to make such appeals, because he thought that ordinary modern geologic processes like glaciation, beach erosion, river sedimentation, volcanism and crustal movements explained the geologic record just fine, provided that these processes acted over very long periods of time. Lyell's contemporary, G. P. Scrope, for example, showed from a study of the volcanic deposits in the Auvergne region of south-central France that the topography of the area could be accounted for by successive eruptions of lava spreading over the terrain; later excavation of valleys into the lava flows, thanks to the work of down-cutting rivers; then further eruptions of lava that followed the river courses; reestablishment of rivers and further down-cutting into the lava in the valley; renewed volcanism and so on.

Scrope also pointed out that this succession of events, *empirically* supported by the geologic relations in the area, called out for time—Time—TIME.

Lyell compiled numerous examples of geologic relationships explicable in terms of ordinary processes acting gradually over long periods of time. As noted in chapter three, Lyell was by no means opposed to local catastrophic events. In fact, he insisted that they would occur from time to time, but he was unimpressed by the appeals of fellow geologists to large-scale catastrophes of which we have no experience. In his enthusiasm to make his point of the adequacy of contemporary processes as geologic causes, Lyell endorsed the idea that the intensity or energy of geologic causes *on a global scale* remained essentially constant throughout the Earth's history. In other words, although the intensity of volcanic activity might vary through time in a particular locale, when viewed on a global scale, volcanism remained more or less of the same intensity. In effect, Lyell denied that there was any kind of *direction* to the course of Earth history. His opposition to biological evolution for several decades—take note, readers who think that uniformitarianism is the handmaiden of evolution—flowed from his commitment to this steady-state conception of Earth history.

One of the most perceptive analyses of Lyell's geologic philosophy was that of William Whewell (1794-1866), one of his contemporaries. Whewell was a professor of mineralogy and also of moral theology at Cambridge, renowned for his work in mathematics and theology but especially for his massive volumes on both the history and the philosophy of the inductive sciences. It was Whewell who first put names on the geologic methodologies of the day. As early as 1832, in a review of the second volume of Lyell's *Principles of Geology*, Whewell observed that two opinions divided the geologic world into two "sects" which he dubbed "the Uniformitarians and the Catastrophists."[25] Later, in his work on the history of the inductive sciences, Whewell referred to the *doctrine of geological catastrophes* and the *doctrine of geological uniformity*.[26] He noted that geologists of England, France and Germany widely accepted the idea that "great changes, of a kind and intensity quite different from the common course of events . . . have taken place upon the earth's surface." This opinion, he said, "appeared to be forced upon men by obvious facts."[27] In particular, geologists who adhered to this doctrine of catastrophes were inclined to attribute the elevation of mountain ranges and the changes in life forms from one formation to the next to the action of causes that acted with much

[25]William Whewell, "Review of Lyell's *Principles of Geology*, v. 2," *Quarterly Review* 47 (1832): 103-32.

[26]William Whewell, *History of the Inductive Sciences* (London: John W. Parker, 1857), 3:506-20; see chap. 8 "The Two Antagonist Doctrines of Geology."

[27]Ibid., p. 506.

greater intensity than causes of the present. Among such geologists were Cuvier, Deluc and Élie de Beaumont (1798-1874). Catastrophists used terms like "paroxysms" and "great revolutions."

In opposition to the catastrophic school, Whewell mentioned those who adhered to the doctrine of geological uniformity like Hutton, some of the Italian geologists and particularly Lyell. According to Whewell, Lyell found that "proofs of catastrophic transition, the organical and the mechanical changes, failed at the same time." Lyell saw that earthquakes were capable of the elevation of mountain ranges "if allowed to operate for an illimitable time." Whewell understood Lyell to claim that "the operation of the causes of geological change may properly and philosophically be held to have been uniform through all ages and periods."[28] Lyell, of course, did not attribute the stratigraphic record to the biblical Flood, and he also believed that the Earth is quite old.

Whewell agreed with proponents of geological uniformity that one should not arbitrarily assume the existence of catastrophes. In his judgment, the degree of uniformity and continuity with which causes had acted during the course of Earth history had to be assessed "from the facts of the case." The effects teach us whether a geologic cause has been similar or dissimilar to causes acting at present. Lyell, therefore, erred because he rejected *ahead of time* "any difference between the intensity of existing and of past causes." According to Whewell, "we are in danger of error, if we seek for slow and shun violent agencies further than the facts naturally direct us, no less than if we were parsimonious of time and prodigal of violence."[29] Whewell also criticized Lyell for assuming that the present should be taken as the representative period in Earth history. We must not, he said, "select arbitrarily the period in which we live as the standard for all other epochs."[30] He concluded that "no confirmation of the doctrine of uniformity" had been found in the other sciences in the way that it had been maintained in geology.[31] Whewell maintained that any idea of the uniformity of nature ought to be used in a very broad sense to include "catastrophes and convulsions of a very extensive and intense kind."[32]

It is very significant that the vast majority of adherents of the doctrine of geological catastrophes in the late eighteenth and early nineteenth centuries did not attribute the bulk of the stratigraphic record to the biblical Deluge. Eminent catastrophists with significant field experience, like Cuvier and Buckland, restricted the symptoms of the Flood to surface deposits that are

[28]Ibid., p. 512.
[29]Ibid., p. 513.
[30]Ibid., p. 514.
[31]Ibid., p. 517.
[32]Ibid., pp. 514-15.

superposed on successions of fossiliferous rock strata. In addition, Whewell intimated that adherents of the doctrine of geological catastrophes typically thought that relatively long stretches of time intervened between the episodes of cataclysmic mountain building. The adherents of the doctrine of geological catastrophes were also not inclined to attribute the postulated catastrophes to supernatural interventions by the creator.

The "catastrophists" of whom Whewell spoke were *not* "Flood geologists" in the sense of the modern young-Earth creationists, nor did they confine themselves to belief in an Earth that is only a few thousand years old. They were as comfortable with a very old Earth as those who shied away from catastrophes. Adherents of both the doctrine of geological catastrophes and the doctrine of geological uniformity accepted an old Earth and denied the kind of Flood geology so near and dear to the hearts of contemporary young-Earth creationists.

Many geologists persisted in attributing mountain building to sudden, paroxysmal causes in spite of whatever appreciation for Lyell's ideas they might have entertained. By the late nineteenth century, however, geologists began to recognize that an increasing number of past geologic phenomena could be interpreted in terms of slower, more gradual processes akin to those acting today, and they tended to regard themselves as Lyellian uniformitarians even though many of them rejected his steady-state conception of Earth history. Geologists became more skeptical of appeals to great catastrophes and put great stock in the slogan that the present is the key to the past. Thus, despite occasional warnings against insistence on uniformity of rates through time by prominent geologists, the concept of continental drift met strong resistance from many geologists who perceived that the idea of lateral movement of continents was contrary to the stability of Earth's surface, implicitly regarded as a corollary of the idea of geologic uniformity. However, the ever-accumulating evidence for the theory of plate tectonics in the 1960s destroyed resistance to the idea of continental drift for all but a handful of geologists.

In the first half of the twentieth century, the geological community almost unanimously viewed with great suspicion the hypothesis of J Harlen Bretz (1882-1981) of the University of Chicago regarding the channeled scabland topography of eastern Washington. On the basis of giant gravel bars with braided geometry, Bretz proposed that the Channeled Scablands, characterized by large "coulees," had resulted from catastrophic flooding.[33] Subsequent discovery of sets of ancient shorelines on mountain slopes as

---

[33]Bretz proposed a catastrophic flood associated with glacial activity for the origin of the Channeled Scablands in J Harlen Bretz, "The Channeled Scablands of the Columbia Plateau," *Journal of Geology* 31 (1928): 617-49, and "The Spokane Flood Beyond the Channeled Scablands," *Journal of Geology* 33 (1930): 97-115.

well as giant ripples in southwestern Montana by J. T. Pardee led to the
proposal that so-called glacial Lake Missoula, formed by ice damming
of the Clark's Fork River, had experienced catastrophic outflow. In mid-
twentieth century, Bretz was finally vindicated when he demonstrated
that catastrophic draining of Lake Missoula provided the source of the
regional-scale, devastating floodwaters that produced the distinctive topography of the Channeled Scablands.[34]

In the early twentieth century, there was little interest in cratering processes on Earth. In fact, craters were often regarded as volcanic. However,
in the 1960s, planetary and lunar studies made possible by probes launched
throughout the solar system coupled with detailed investigations of large terrestrial craters such as Meteor (Barringer) Crater in Arizona confirmed that
meteorite bombardment had been a major process throughout solar system
history.[35]

By the 1980s, almost all geologists had become perfectly comfortable
with the role that catastrophic action had played in terrestrial history as
indicated by the widespread acceptance of the theory that the dinosaurs,
flying reptiles, marine reptiles, ammonites and many other life forms had
become extinct at the end of the Cretaceous Period 65 million years ago, in
the aftermath of a large meteorite impact on the Earth.[36] And, in the field
of stratigraphy, British stratigrapher Derek Ager drove home the idea that
many sedimentary formations, such as turbidites on continental slopes,
had been deposited very rapidly with his memorable statement that the
"history of any one part of the earth, like the life of a soldier, consists of

---

[34]J. T. Pardee published evidence for catastrophic draining of glacial Lake Missoula in western Montana, including giant current ripples where the lake had drained. See J. T. Pardee,
"Unusual Currents in Glacial Lake Missoula," *Bulletin GSA* 53 (1942): 1570-99. Subsequently
Bretz made many more field observations and synthesized data into a grand hypothesis
linking the distinctive topography of the Channeled Scablands to catastrophic outflow of
glacial Lake Missoula. See J Harlen Bretz, "Channeled Scabland of Washington: New Data
and Interpretations," *Bulletin GSA* 67 (1956): 958-1049. For a summary relating the geomorphic features to modern fluvial hydraulics, see Victor R. Baker, "Large-Scale Erosional and
Depositional Features of the Channeled Scabland," in *The Channeled Scabland: A Guide to
the Geomorphology of the Columbia Basin, Washington,* ed. V. R. Baker and D. Nummen-
dal, sponsored by Planetary Geology Program, Office of Space Science, NASA (Washington:
NASA, 1978), pp. 81-115.

[35]One of the leading proponents in explaining the origin of puzzling terrestrial craters by meteorite impact was Robert Dietz. See, for example, Robert S. Dietz, "Astroblemes: Ancient
Meteorite-Impact Structures on the Earth," in *The Moon, Meteorites, and Comets,* ed. Barbara M.
Middlehurst and Gerard P. Kuiper, The Solar System vol. 4, pt. 2 (Chicago: University of Chicago Press, 1963), pp. 285-300.

[36]The classic paper relating large-scale extinction to meteorite impact is L. W. Alvarez, W. Alvarez, F. Asaro and H. V. Michel, "Extraterrestrial Cause for the Cretaceous-Tertiary Extinction,"
Science 208 (1980): 1095-108.

long periods of boredom and short periods of terror."[37]

WHAT DO MODERN GEOLOGISTS UNDERSTAND BY UNIFORMITARIANISM?

A 1967 symposium on the uniformity of nature sponsored by the Geological Society of America highlighted the ambiguity of the concept of uniformitarianism. One contributor, noting the fundamental importance of the principle of uniformity to geologists, showed that a wide range of "nonequivalent" answers were likely to be given in response to the question as to "what, precisely, is the Principle of Uniformity."[38] Given this diversity of perception, a number of geologists and historians of science began efforts to clarify the meaning and significance of the principle of uniformity.

As an example, a young Harvard paleontologist who later gained fame as a popularizer of evolutionary theory and geology, Stephen J. Gould (1941-2002), distinguished between *methodological uniformitarianism* and *substantive uniformitarianism.*[39] The former concerned the uniformity of natural law, the presupposition that is basic to all natural sciences, including geology. Gould saw methodological uniformitarianism as amounting to "an affirmation of induction and simplicity" and said that it would be "subsumed in the simple statement: 'geology is a science.'"[40] On the other hand, substantive uniformitarianism concerned such matters as whether the kinds of processes that produce geologic effects have persisted throughout geologic time and whether rates of processes have been more or less constant, or gradual, as opposed to catastrophic. Gould saw substantive uniformitarianism as a "descriptive theory" that had "not withstood the test of new data and can no longer be maintained in any strict manner." He sensed the danger that "too rigidly held, this or any other testable theory is transformed into an a priori assumption, stifling to the formulation of new hypotheses which may better explain certain data."[41]

In a paper on catastrophism published in 1970, Dutch historian of science Reijer Hooykaas (1906-1994) of the Free University of Amsterdam stressed the distinction between "system" and "method."[42] He reasoned that the method employed in reconstructing the geologic past might be *actualistic* in that it in-

---

[37]Derek Ager, *The Nature of the Stratigraphical Record* (New York: John Wiley, 1973), p. 100.

[38]M. King Hubbert, "Critique of the Principle of Uniformity," in *Uniformity and Simplicity: A Symposium on the Principle of the Uniformity of Nature,* Special Paper 89, ed. Claude C. Albritton Jr. (New York: GSA, 1967), pp. 3-33. The quotation is on p. 4.

[39]Stephen J. Gould, "Is Uniformitarianism Necessary?" *American Journal of Science* 263 (1965): 223-28.

[40]Ibid., p. 227.

[41]Ibid., p. 226.

[42]Reijer Hooykaas, "Catastrophism in Geology, Its Scientific Character in Relation to Actualism and Uniformitarianism," *Mededelingen der Koninklijke Nederlandse Akademie van Wetenschappen* 33 (1970): 271-316.

voked the action of modern-day causes to account for past geologic effects, or that the method of procedure might be *nonactualistic* in maintaining that some ancient geologic causes were different in character from those of the present. In that case, the method of reconstructing the geologic past assumed that modern-day causes are insufficient to account for all past geologic effects.

Application of either the actualistic or the nonactualistic methods might result in a *system* or *historical description* that was *catastrophic*, in which case geologic evidence in a particular situation indicated that the cause producing a given effect acted with an intensity much greater than occurs at the present. Alternatively, application of either method might result in a system that was *actualist*, in which case geologic evidence in a particular situation indicated that the past geologic cause was just the same as a cause acting at present at essentially the same rate, intensity or energy as it does at present.

More recently, Rudwick, Gould and Austin, in analyzing "uniformitarianism," have detected four different senses in which the term might be used in geologic theorizing. Martin Rudwick, specifically addressing Lyell's uniformitarianism, said that one type of uniformity concerned the *theological status* of a past geologic cause.[43] Such causes might be *naturalistic*, in which case they were achieved by the action of secondary means, not excluding divine providential ordering, or they might be *supranaturalistic*, in which case a past cause was not attributed to secondary means but only to the primary act of a creative power. Rudwick stressed that, apart from the question of the origin of biological species, the geology and paleontology of the 1820s were thoroughly naturalistic, even though, for many practitioners, given a providentialist interpretation.

Closely related to Rudwick's first category is Gould's first sense of uniformitarianism, namely, the *uniformity of law*, which "is not a statement about the world" but an *a priori* claim of *method* that is necessary for the practice of all natural sciences that natural laws are invariant in space and time.[44] Without invoking this postulate, he argued, we would have no confidence in the validity of our inductive inferences.

Young-Earth geologist Steven A. Austin also published an assessment of

---

[43]Martin J. S. Rudwick, "Uniformity and Progression: Reflections on the Structure of Geological Theory in the Age of Lyell," in *Perspectives in the History of Science and Technology*, ed. Duane H. D. Roller (Norman: University of Oklahoma Press, 1971), pp. 209-27.

[44]For Gould's more recent writings on uniformitarianism, see Stephen J. Gould, "Catastrophes and Steady State Earth," *Natural History* 84, no. 2 (1975): 14-18, and "Toward the Vindication of Punctuational Change," in *Catastrophes and Earth History: The New Uniformitarianism*, ed. W. A. Berggren and John A. Van Couvering (Princeton, N.J.: Princeton University Press, 1984), pp. 9-34. For the quote, see Gould, "Catastrophes and Steady State Earth," p. 17.

the principle of uniformity.[45] He designated his first sense of uniformitarianism as *methodological uniformitarianism,* a procedure that assumes the continuity of the inherent properties of matter and energy through time. Like Gould, Austin concluded that the assumption of this kind of uniformity is not unique to geology but lies at the root of all sciences. He maintained that even though "the affirmation of the temporal continuity of the properties of matter and energy is vital to geological investigations," the term *methodological uniformitarianism* is superfluous because it is simply a part of the definition of an inductive science like geology.[46]

We concur with Rudwick, Gould and Austin that reliance on second causes and the uniformity of natural law and its corollary, the continuity of the properties of matter and energy through time, are fundamental postulates of all natural sciences, including geology. For example, we assume that Newton's law of gravitation holds good in all places and times so that we won't wake up one day to discover that the force of gravitational attraction between two objects has become proportional to the distance between them rather than inversely proportional to the square of the distance between those two objects as is now the case. In other words, we won't wake up to discover that the moon is suddenly moving closer and closer to Earth and that the Sun is moving toward Earth even more rapidly while objects on Earth's surface are beginning to float into the air.

Such law-like behaviors flow from the inherent physical properties of matter and energy that we assume always have been and always will be the same under the same set of conditions. For example, a piece of pure calcite ($CaCO_3$) will always have a hardness of 3 at room temperature and pressure because of its chemical composition and crystal structure and the bonding energies of the constituent atoms of calcium, carbon and oxygen. A piece of pure quartz (SiO_2) will always have a hardness of 7 at room temperature and pressure. However, if we change the conditions in which these minerals exist, for example, by raising the temperature by several hundred degrees Celsius, the hardness of both quartz and calcite may change. In addition, under surface conditions on the Earth, if calcite is exposed to hydrochloric acid, it will always react with it to form calcium chloride, carbon dioxide and water. We assume that calcite would have had the same behavior in the distant geologic past when subjected to the same set of conditions. All geologists in Lyell's time, including Christian geologists like Sedgwick and Buckland, assumed the uniformity of natural law and of material properties. All geologists today accept this postulate. All natural scien-

---

[45]Steven A. Austin, "Uniformitarianism—A Doctrine that Needs Rethinking," *Compass* 56 (1979): 29-45.

[46]Ibid., p. 37.

tists accept this postulate. Young-Earth geologist Austin accepts this postulate. We accept this postulate. And all practicing scientists, when doing science, appeal to second causes without necessarily rejecting divine providence.

Given the contemporary antipathy toward "naturalism" among conservative Christians, we emphasize that, although Rudwick used the term *naturalistic* in his analysis, the appeal to second causes alone as a foundational postulate of all natural science is not the same thing as espousing a purely naturalistic worldview. Assumption of this postulate in scientific practice *does not* mean that there can never be or never was a "supernatural" miracle. Acceptance of a miracle at a particular point in history does, however, mark a point at which a given phenomenon is no longer being accounted for in a scientific way.

For example, suppose that someone believes that there were no violent events during the time that God was creating the world because the initial creation was "very good," whether he did so in seven days or over millions of years. Suppose that this person discovers a layer in a stack of rock layers that looks exactly like a layer of modern volcanic ash. Being principially opposed to the occurrence of violent processes during the ancient past, however, our observer refuses to attribute the layer to a violent volcanic eruption, and so he or she maintains that the layer was put there by a miraculous action of God. Now it is conceivable that God did perform a miracle to create that layer.

The explanation of our observer may be correct, but it is not scientific, because natural science can function only in terms of continuity of natural laws, properties and behaviors. Our observer would have gone beyond science in advocating the supernatural explanation. There is no way that the supernatural hypothesis can be tested in a scientific manner by examining the properties of the ash layer, doing experiments on it or making predictions about the character of the set of rock layers. Scientists could not observe analogous examples of divine action either.

We suggest, however, that God is economical with miracles and that he has employed them mainly in the service of redemptive history. We find biblical miracles in conjunction with the exodus, the prophetic era, the ministry of Christ and the work of the apostles. All of these miracles were attestations of the word and revelation of God that were observed by human beings. Arbitrary unobserved miracles performed during the work of creation would have had absolutely no impact on people and would not serve to confirm the presence of God or the pronouncement of the word because no one was there to observe them.

Biblical miracles like the virgin birth, the resurrection or Jesus' walking on water were powerful signs to the observers to confirm the divinity of Christ, but such miracles have no bearing on the daily practice of scientific geology.

Such miracles have no effect on historical reconstructions of the Earth's past, nor do they affect the laws of physics or the course of chemical reactions. We see no reason why a Christian doing science cannot operate with the assumption of the uniformity of natural law and still accept the reality of biblical miracles. What would be a problem, however, is the introduction of arbitrary or capricious miracles with no compelling reason from the biblical text for assuming their existence. Assuming such miracles would make the pursuit of historical sciences more problematic.

In addition, the uniformity of law is a corollary of the biblical doctrine of creation, because God created an ordered cosmos. The Scripture repeatedly refers to God's ordinances, decrees and laws with reference to the natural world. The Old Testament, in particular, teaches that God is in a covenant relation with his creation by which he upholds the ordinances of day and night, the sun and moon, the boundaries of the sea, and so on. An implication of such a teaching is the idea of a uniformity of law and properties. Indeed the practice of science with its dependence on the regularity of nature is in part an outgrowth of Christian thought.

For Rudwick, the second type of uniformity concerned the *methodological status* of past geologic causes. Like Hooykaas, he noted that the status of a past geologic cause might be *actualistic* (uniform) or *nonactualistic*. In the former, a past cause might correspond to some cause now in operation; in the latter, a past cause, having no valid analogy to a present cause, had ceased to exist. Although research into the geologic past should undertake "actualistic comparison of past and present," Rudwick insinuated that the actualistic postulate could be assumed a priori to be "totally adequate for geological explanation." Rudwick also noted that Lyell's contemporaries were all actualists who differed among one another "only in the *extent* to which they believed the past could be interpreted by close analogy with present causes."[47]

For this second category, Gould referred to the *uniformity of process* or "actualism." Again, he asserted, this view of uniformity "is not an argument about the world; it is a statement about scientific procedure."[48] Gould had in mind the procedure of explaining past effects, if at all possible, as the result of causes that still operate on the Earth. Scientists, he said, generally "do not invent causes with no modern analogues when present causes can render the observed results." In effect, to do so is to employ the principle of simplicity, and to do that, in Gould's judgment, is to invoke "another a priori methodological assumption shared by all scientists and not a statement about the empirical world."[49]

---

[47]Rudwick, "Uniformity and Progression," p. 222 (italics in original).
[48]Gould, "Catastrophes and Steady State Earth," p. 17.
[49]Ibid., p. 11.

Austin called the second category *causal uniformitarianism*. He noted that geologists showed some difference in their understanding of causal uniformitarianism. Some geologists of earlier generations had interpreted this kind of uniformitarianism as a "statement about how geological processes must have operated." Austin properly considered this approach to be dangerous because it could easily become a doctrine that *only* the processes about which we know at the present are sufficient to account for all past geologic phenomena. Stated in this manner, the concept of causal uniformitarianism fell into the trap of assuming that geologists have already discovered every geologic process. Far better, Austin thought, to modify the idea of causal uniformitarianism to allow for unique kinds of ancient processes and as yet undiscovered present processes. A second category of geologists invoked causal uniformitarianism as a statement of procedure or "actualistic" method in which geologists associate "modern kinds of processes with their products to decipher ancient processes from their products preserved in the earth's crust."[50] Still another group of geologists, according to Austin, is willing to

> recognize unusual ancient processes, undiscovered processes, and inversions of actualistic reasoning as important problems for causal uniformitarianism. The geologist's technique in deciphering ancient processes, they affirm, relies not only on analogies with products of modern geological processes, but on analogies with products of similar ancient processes, on analogies with products from experimental replicas and other non-geological systems, and on logical deductions from theories or scientific laws.[51]

Austin viewed this approach as the most satisfactory one.

We concur with Rudwick, Gould and Austin that a sensible procedure in unraveling the geologic history of the Earth is to apply known causes to past effects and phenomena before invoking hypothetical, unknown causes. It makes sense to look at a layer of sedimentary rock and interpret it in terms of known processes like stream action, beach deposition, flooding, wind erosion and so on before conceding that none of those causes is satisfactory and initiating the search for a different cause. This is a matter of simplicity of procedure. This actualistic approach does not commit geologists to find, for every past geologic event, a present-day cause, but it does mean that this is where we start.

Sometimes geologists invoke a cause that doesn't seem to be operating today because they are not familiar with a present-day cause that satisfactorily explains a geologic phenomenon. For example, in the late eighteenth century Abraham Werner and his students explained the origin of the rock type basalt

---

[50]Austin, "Uniformitarianism," p. 38.
[51]Ibid., p. 39.

in terms of precipitation from a vast saline ocean. Although no one had ever seen the ocean actually precipitate basalt, Werner knew that the ocean does contain dissolved salts and does precipitate crystalline minerals upon evaporation. It was, therefore, reasonable to envision a time when the ocean had a much more complex solution chemistry and might have been capable of precipitating minerals like pyroxene and plagioclase feldspar, the main constituents of basalt. But Werner knew very little about volcanism. After decades of intense study of the products of present-day volcanoes, natural historians began to recognize that basalt resembled a volcanic product in many ways. When that happened, the Neptunist explanation of Werner gradually fell by the wayside and basalt was increasingly explained in terms of a modern-day process, volcanism, that had become much better recognized and understood.

Finally, adherence to an actualistic procedure in reconstructing the geologic past may well result in one geologic phenomenon being accounted for in terms of a modern cause whereas another phenomenon may have to be accounted for in terms of a cause unlike any operating at present.

Rudwick's third category concerned the rates at which past geological causes operated. He contrasted processes as either *gradualistic* (uniform) or *saltatory* (paroxysmal or catastrophic), but affirmed that these represent end members in a continuum of possibilities. According to Rudwick, the geologists of Lyell's era postulated saltatory or catastrophic events only when necessary against a backdrop of generally gradualistic change.

For Gould, the third and fourth senses of uniformity entail *substantive* assertions about the world. The third sense he called *uniformity of rate*, involving the idea that geologic change is typically slow and gradual, rather than paroxysmal or cataclysmic. Hence, the term *gradualism* is commonly invoked for this sort of uniformity. According to Gould, gradualism is an empirical claim about the world that must be tested, not assumed, to determine whether it is true or false. Writing even before development of the Alvarez hypothesis advocating demise of dinosaurs at the end of the Cretaceous Period due to a catastrophic meteorite impact, Gould suggested that, although Lyell's view had largely prevailed, his original "insistence on a near uniformity of rate was stifling to the imagination."[52]

Austin referred to *actional uniformitarianism* as the category corresponding to uniformity of process rates throughout time. After indicating that in some cases the uniformity of geologic process rates throughout geologic time had been assumed a priori, he noted several examples of geologic events that had been shown to occur at rates quite different from those of the present. "No

---

[52]Gould, "Catastrophes and Steady State Earth," p. 18.

theory of rates," Austin insisted rightly, "be it gradual or catastrophic, should be accepted without geological evidence."[53]

Young-Earth creationists repeatedly charge modern geologists with being uniformitarians in this third sense. They imply that we are all gradualists all the time. Young-Earth geologist Austin made it quite plain that he did not agree with this kind of uniformity, but neither Rudwick nor Gould agreed with this kind of uniformity either. And neither do we. We completely accept Gould's claim that uniformity of process rates through time is a testable claim. There is no question that the idea of pure uniformity of geologic process rates throughout time has failed the test. Rates have varied considerably, and in some cases, geologic effects have been produced quite catastrophically. We have already called attention to the channeled scabland floods of Bretz, the catastrophic meteorite impact hypothesis of Alvarez, and catastrophic turbidite deposition espoused by Ager. In chapter eight, we discussed several categories of large-scale, violent events in sedimentary rocks.

We reject now, and have always rejected, the idea that geologic processes have always acted at the same rates and that they have always acted in a slow, gradualistic manner. Yet we also are convinced that the geologic past has been characterized by many processes that acted in a very slow fashion. There are several regional packages of very fine-grained, finely laminated sediments that were deposited from large lakes that survived for hundreds of thousands, if not a few millions, of years. There are thick successions of fine-grained marine sediments that accumulated over millions of years of quiescent deposition. There are numerous examples of thick successions of lava flows, the tops of all or many of which were thoroughly weathered prior to deposition of the succeeding lava flow. These flows are also commonly intercalated with thin-bedded lacustrine sediments. There are numerous widespread formations of "fossil" sand dunes throughout the western United States.

Thick accumulations in excess of 95 percent pure quartz sand of uniform grain size *do not* result from catastrophic processes, but do require a lot of time. Reef structures, widespread throughout the stratigraphic record, take a lot of time to develop. An admission that catastrophic events have occurred in the geologic past hardly commits us to acceptance of a young Earth or a completely catastrophic interpretation of Earth history. A lot of catastrophes, a few maybe even of global extent, are bound to occur over 4.5 billion years of Earth history. But a world that is only 6,000 years old has little time for the accumulation of thick, fine-grained sedimentary deposits or thick sequences of pure salt that formed by slow deposition, and the rock record is full of exam-

---

[53]Austin, "Uniformitarianism," p. 40.

ples of precisely these sorts of deposits. When known processes explain such deposits, there is no need to wait for the discovery of an unknown process. Moreover, a violent, one-year deluge can never explain these deposits.

The final form of uniformitarianism in Lyell's thought, according to Rudwick, concerned the overall pattern of a geological cause traced over the course of Earth history. Here Rudwick contrasted a steady-state (uniform) Earth history as opposed to a directional, progressive or developmental pattern of history. Lyell strongly emphasized the former, envisioning an Earth history in which geologic processes, when averaged over the entire globe, acted at a relatively similar energy throughout the Earth's history. As a paradoxical result, Lyell denied the face-value implication of the fossil record as being directional, and he was skeptical of the claim that the globe is cooling. According to Rudwick, virtually all geologists of Lyell's time disagreed with Lyell over the steady-state character of Earth history and were directionalists.

Even Scrope, who, like Lyell, emphasized an actualistic approach, was a directionalist. Although he successfully pointed out that much of the geology of southern France could be accounted for in terms of an actualistic process, namely volcanism, acting in a gradualistic fashion over a lengthy stretch of time, Scrope still believed that during the course of Earth history, volcanism had decreased in intensity. Moreover, directionalism of volcanism could easily be linked to a perceived directionalism in terms of the cooling of the Earth. If the Earth has been cooling through time, it would make sense that volcanic activity might not be as pronounced at present as it was early on. As we saw earlier, Lord Kelvin strongly disputed Lyell's steady-state uniformitarianism on grounds of the Earth's cooling.

Gould mentioned *uniformity of conditions,* entailing a dynamic *steady-state* conception of Earth history, à la Lyell. On this view the Earth would be constantly changing within narrow limits. Its history would be characterized by unending cycles without trends or direction. Opponents of uniformity of conditions invoke a directional view of Earth history. Gould affirmed that uniformity of conditions is also a testable assertion about Earth's history that has proved largely incorrect.

Austin alluded to *configurational uniformitarianism,* the theory of Lyell that concerned the uniformity or continuity of geological conditions or environments throughout the course of Earth history. Austin, too, noted that Lyell's configurational uniformitarianism had been discredited.

We also reject, as most geologists have, Lyell's steady-state conception of Earth history. We are not uniformitarians in the sense that Lyell so strongly favored, because there is abundant evidence to indicate that there are directional characters to some aspects of the Earth's past. The fossil record documents a direction to the population of Earth from one-celled organisms to metazoans to vertebrates to mammals to human beings. Earth has cooled and has slowed in its rotation during its history. Meteorite bombardment was a much more significant phenomenon during the early stages of formation of the Solar System and Earth than it is now. Some sedimentary rock types are more prevalent in Precambrian terranes than in more recent ones, for example, banded iron formations. Some igneous rock types such as anorthosite were more prevalent in the geologic past than they are now. To sum up, neither we nor other modern geologists accept either the uniformity of geologic process rates or the uniformity of geologic conditions through time.

AUSTIN, THE GRAND CANYON AND UNIFORMITY

The reader should now have a pretty clear idea of how modern geologists perceive the concept of uniformitarianism, and we have stated that we are in essential agreement with our modern geologist colleagues. Even Austin, an advocate of a young Earth, seems to be in essential agreement with us and with modern geologists about the contemporary understanding of uniformitarianism in his paper on that concept. Austin also briefly discussed uniformitarianism in the book on the Grand Canyon that he edited. In his chapter on interpreting strata in the Grand Canyon, he discussed the principles for interpreting rock strata, pointing out that "strata, especially those so well exposed in Grand Canyon, need to be understood and interpreted as a historical record of sedimentary processes." He said that "over the years geologists have agreed on five major principles for interpreting strata."[54] The first four principles are the principles of superposition, original horizontality, original continuity and cross-cutting relationships (see chap. 2 and figs. 2.2 and 11.1). As we have seen, geologists constantly invoke these principles in reconstructing the historical sequences in which masses of diverse rocks developed. Austin fully agreed that the employment of these four principles is perfectly appropriate, and, in fact, he employed them in relation to the succession of rocks exposed in the Grand Canyon.

But now we turn to the fifth principle, which Austin called the *principle of process-product analogy*. He stated that *"geologists should prefer explanations for the origin of strata which are consistent with the kinds of geologic processes forming strata today."* He then developed the principle in detail, and we quote his statement in entirety:

> This principle directs geologists as to how to think about strata, not about how all strata are necessarily deposited. Geologists should seek analogies between

---

[54]Austin, "Interpreting Strata of Grand Canyon," p. 23.

the kinds of processes which form strata, not the *rates* of process which form strata. They should seek analogies with modern processes which obey natural laws. Water, wind, and ice today form strata. We should seek to explain ancient strata by reference to these agents, without assuming correspondence in the rate, scale, or intensity of modern and ancient processes. This does *not* make the extreme generalization that only known, modern processes, *operating at modern rates*, formed strata (uniformitarianism), or that only present natural laws explain all geologic features (naturalism).[55]

If geologists have agreed upon this fifth principle, then it follows that they do *not* adopt the "extreme generalization" that strata were formed solely by "known, modern processes, operating at modern rates" which Austin termed "uniformitarianism." Again, Austin could not have been much clearer that modern geologists *do not accept* the kind of uniformitarianism that insists on interpreting the geologic past solely in terms of modern processes operating at modern rates (see examples in chap. 8). Moreover, he again seemed to endorse the fifth principle.

If, however, Austin recognized that modern geologists have rejected the idea, as he does, that geologic processes must have occurred at relatively constant rates throughout geologic time, that the rates of those processes were pretty much the same as they are today, and that those processes have generally been slow, then it is very perplexing that, in the very same chapter on interpreting the strata of the Grand Canyon, Austin set up the typical contrast encountered so frequently in young-Earth creationist literature between uniformitarians and catastrophists. Why did he refer to one of the coauthors of this book (Young) as someone who adopts the "uniformitarian framework" when Young has *never* espoused the *necessity* for interpreting geologic history *solely* in terms of slow, gradual processes and when he is in complete agreement with Austin's fifth principle of stratigraphic interpretation? Why did he refer in that chapter to a "uniformitarian" (and evolutionary) model and contrast it with a catastrophic framework, implying that modern geologists somehow or other have an a priori bias against the notion that geologic rates can have changed and changed dramatically, too?

"If one begins with the uniformitarian model, that sedimentation occurred in calm and placid seas," Austin claimed, "an evolutionary interpretation of thick strata sequences will be derived." But what Austin did not say is that failure to adopt the extreme "uniformitarian" model as Austin described it in his fifth principle does not automatically compel a geologist to adopt the catastrophist framework. Most geologists would not interpret the strata of the Grand Canyon in a way in which they were committed ahead of time to "dis-

---

[55]Ibid., p. 24 (italics in original).

covering" that the rocks had been deposited slowly in environments analogous to modern-day environments. They would rather interpret the physical evidence in the strata in terms of processes, environments and rates that were most consistent with that evidence. And the fact is that geologists do not believe that the evidence in those rocks lends itself to some global catastrophic scenario. The empirical evidence in the rocks, not an a priori bias toward slow rates, compels acceptance of slow deposition of most of the Grand Canyon rocks. If the evidence indicated a catastrophic deposition, modern geologists would accept that conclusion. They have already shown themselves more than willing to adopt catastrophic hypotheses where the evidence points to it.

Not only is it puzzling that Austin posited a contrast between uniformitarian and catastrophist models, as if modern geologists accepted the uniformitarian model, despite his acknowledgment that modern geologists don't adhere to the uniformitarian model, but it is disturbing that young-Earth advocates persist in referring to modern geology as uniformitarian in the sense that it rejects the possibility of large-scale catastrophic events and insists on appealing to modern gradualistic processes to explain all geologic phenomena. We challenge young-Earth creationists to abandon the fruitless dichotomy between so-called uniformitarians and catastrophists. It is a false and useless dichotomy in which a straw man is erected for the purpose of demolition. We also challenge young-Earth creationists to desist from labeling modern geology as uniformitarian when they know full well that modern geologists repudiate any a priori commitment to slow, gradual process rates in the geologic past to the exclusion of all catastrophic events.

All geologists, whether committed to an old Earth or a young Earth, are uniformitarian in two senses and two senses only. One way in which all geologists are uniformitarians is in the sense that they believe that geology is a science that does its work under the assumption that the basic properties of material and laws of nature have remained constant. The only other way in which all geologists are uniformitarians is in the sense that they generally accept the actualistic procedure in which we first attempt to explain past geologic features in terms of geologic processes or causes that are either identical to or analogous *in kind,* but not necessarily in degree, to those of the present before proceeding to invoke unknown and unusual causes.

No geologist, whether committed to an old Earth or a young Earth, is a uniformitarian in the sense of Lyell's concept of a steady-state Earth. Everyone believes that at least some geologic processes have operated in a directional manner. Modern professional geologists believe that past meteorite bombardment of the Earth was more intense than it is now, whereas Flood geologists think that the intensity of flooding has decreased in the last few thousand years.

And, finally, no geologist, whether committed to an old Earth or a young Earth, is either a pure gradualist or a pure catastrophist. Young-Earth proponents are catastrophists in regard to Noah's Flood, but they do accept that many geologic processes have occurred in gradualistic fashion since the Flood. Old-Earth geologists allow for catastrophic action on a global scale in regard to large meteorite impacts and on a very large regional scale in regard to floods such as those that produced the Channeled Scablands of eastern Washington. Most geologists would agree with Ager that large portions of the stratigraphic record can be interpreted as long periods of very slow, gradual sedimentation, punctuated by short episodes in which sedimentation was extremely rapid as a result of large-scale regional floods, volcanic eruptions, turbidite flows, mudflows and the like. But if young-Earthers and old-Earthers both hold the same general conception of uniformity, and if they both accept the possibility of catastrophes on a global scale, why is there a dispute?

WHY THE BIG DIFFERENCE IN OPINION?

Austin stated that "evolutionists frequently make the uniformitarian assumption that strata of Grand Canyon formed during long ages" whereas "creationists usually make the catastrophist assumption that a global flood was responsible for depositing a great thickness of Grand Canyon strata."[56] He has said that evidence does not speak for itself and that rocks need to be interpreted. In effect, Austin attributed the differences between the two camps to differences in bias, in interpretive framework.

Although there is an element of truth in that suggestion, it is not the whole story. If we take the notion of divine creation seriously, then creation itself exerts a profound effect on the observer. No matter how strongly we may be committed to an interpretive framework, that framework can be modified or abandoned by the force of created and providentially sustained evidence. While geologic evidence must be interpreted, it has a way of challenging interpretations that are poorly founded. If our interpretive framework were the sole determinative factor in reconstructing geologic history, we could claim equal validity for all sorts of bizarre scenarios of the past.

But in the end interpretive frameworks have to answer to what is in the rocks. And that is exactly what happened a few centuries back. In the seventeenth century, the controlling framework was a young Earth and global Flood. By the mid-eighteenth century, that framework collapsed because the rocks "talked back." Natural historians kept making empirical discoveries in the rocks that just did not fit with the global Deluge hypothesis.

---

[56]Ibid., p. 22.

So let's return to Austin's assertion about the assumptions made by evolutionists and creationists. If we ignore the attempt to condemn modern geology by calling it "evolutionist," let it be known that modern geologists, as Austin stated, *do not make the assumption* that rock strata have formed over long ages. No one woke up one day and said, "I don't like the old interpretive framework that sees everything in terms of a global Flood only a few thousand years ago. I think I'm now going to assume that rocks formed over long ages and will begin to interpret them in light of that assumption." Rather, over the past three centuries, thousands of natural historians and geologists, both Christian and non-Christian, exploring rock masses around the world were gradually, imperceptibly persuaded by numerous lines of evidence, some of which we have discussed, that the Earth is much older than they had believed. They didn't think the Earth was ancient because they "wanted" it to be. They didn't think that the Earth is ancient because they wanted to discredit the Bible. They didn't think the Earth is ancient because they "rigged" the scientific game to come out with a desired result. They were persuaded by evidence in the rocks themselves of Earth's antiquity.

In contrast, Austin admitted that young-Earth creationists assume that strata were catastrophically deposited by the Flood. He is absolutely right on that score. They basically have no choice but to assume that. They have locked themselves into that desired outcome because of an a priori commitment to a particular way of reading the Bible that straight-jackets them into acceptance of a very young Earth. As a result, they cannot really approach rocks empirically because they already "know" in advance what the "answer" is supposed to be. They see only what they want to see.

Concluding Thoughts

Because they have put on blinders, young-Earth creationists are unwilling to accept the totality of the available geologic evidence. They are unwilling to abandon their young-Earth, global-Flood hypothesis even when the evidence shows it to be untenable. They have ignored or distorted a vast body of evidence that is contrary to their preconceived notion of what Earth history must have been like. They have focused only on data that, taken in isolation from geologic contexts, might be seen as favorable to their own theory. They claim continually to argue from the evidence of nature, but they have repeatedly ignored what is inconvenient for them. Although some of the phenomena of the sedimentary rock record might be interpretable in terms of a great Flood, most of the phenomena to which they appeal are far more satisfactorily explicable in terms of much smaller scale processes than a global catastrophic Flood. More important, young-Earth creationists have refused to accept the abundant evidence of glacial deposits, lake deposits, desert deposits, delta deposits, shore deposits, reef deposits and evaporite deposits in the rock record. Young-Earth creationists have refused to face the evidence from metamorphism, the kinetics of mineral formation and heat flow from cooling magmas. They have tried to make the evidence from radiometric dating say something opposite from what it does say. The attempt to find a way to have the decay constants of radioactive isotopes change in an unbelievably spectacular fashion is a desperate attempt to rescue their view of the world. To date, *all* physical evidence pertaining to decay constants indicates the virtual immutability of those constants. Although a tiny fraction of geologic evidence might suggest a global Flood if considered in complete isolation from the wealth of other evidence, the overwhelming totality of evidence argues mightily against a global Deluge.

In the end, the dogged persistence in holding on to a young Earth and a global Deluge has less to do with geology than with other concerns. Even some young-Earth creationists grant that the evidence at present does not support their view. Nelson and Reynolds, in an honest assessment, wrote in a philosophical and biblical defense of young-Earth creationism that "natural science at the moment seems to overwhelmingly point to an old cosmos," and they conceded that "it is safe to say that most recent creationists are motivated by religious concerns."[57]

But if this debate over the age of the Earth is not really about physical evidence, then what is it about? We believe that those who are most firmly committed to young-Earth creationism do so because they are convinced that a divinely inspired, infallible, inerrant Bible demands it. We admire young-Earth creationists for their total commitment to Scripture, because we are likewise committed. We are one with them in our total commitment to the gospel of Jesus Christ and rejection of the secular humanism of our day. Yet, as we pointed out, a firm commitment to the infallibility and inerrancy of Scripture does not require a Christian to believe the theory of a recent creation to which young-Earth creationists adhere. And certainly the gospel of Jesus Christ does not demand acceptance of a young Earth. Nor is the eternal salvation of anyone anywhere ever dependent on acceptance of a young Earth. It is the sacrifice of Jesus Christ that saves us from the wrath to come, not belief in a young Earth. The data of the Bible certainly do not demand that we hold to these views.

Christians need to relax and stop being afraid that some scientific evidence

---

[57]Paul Nelson and John Mark Reynolds, "Young Earth Creationism," in *Three Views on Creation and Evolution*, ed. J. P. Moreland and John Mark Reynolds (Grand Rapids: Zondervan, 1999), pp. 41-75; quotation p. 49.

will disprove the Bible or undermine Christianity. We should not be afraid of the evidence that God has put into his world. We do ask, however, that if the Bible convincingly teaches that the Earth is only a few thousand years old and if there was a geologically active global Flood of such cataclysmic proportions, why is it that the physical evidence in God's world, evidence that God put there, evidence interpreted by thousands of competent individuals, many of whom are themselves Bible-believing Christians, constantly points overwhelmingly against that idea and in the direction of an extremely ancient world?

The only recourse that flood catastrophists have to save their theory is to appeal to a pure miracle and thus eliminate entirely the possibility of historical geology. We think that would be a more honest course of action for young-Earth advocates to take. Young-Earth creationists should cease their efforts to convince the lay Christian public that geology supports a young Earth when it does not do so. To continue that effort is misguided and is detrimental to the health of the church and the cause of Christ.

17

CREATIONISM, EVANGELISM
AND APOLOGETICS

IN THIS BOOK WE HAVE ENDEAVORED TO SHOW that several purported scientific claims advanced by young-Earth creationists do not stand
up to scrutiny and fail to establish a young age for the Earth. These claims
are generally based on incomplete information, wishful thinking, ignorance of
real geologic situations, selective use of data and faulty reasoning.

God has placed a wealth of clues in the rocks that attest to great terrestrial
antiquity. From the abundant empirical evidence that has been extracted from
the rocks there is nothing that would remotely lead geologists to conclude that
Earth is anything other than extremely old. The hard fact that must be recognized and accepted by all Christians is that geologic investigations point
compellingly and overwhelmingly to the enormous antiquity of the Earth.

It is extremely improbable that future discoveries will lead the geologic
community to revive acceptance of a very young Earth. Yes, there have been
great revolutions in scientific thought in the past, and we should expect more
of them in the future. However, it is futile for proponents of a young Earth to
hope for such a revolution that would entail a complete reversal from acceptance of an old Earth that itself resulted from a lengthy scientific revolution.[1]

---

[1]We remind readers that the gradual abandonment of a young Earth in favor of an expanded
antiquity has been beautifully traced by Martin J. S. Rudwick, *Bursting the Limits of Time: The
Reconstruction of Geohistory in the Age of Revolution* (Chicago: University of Chicago Press,
2005).

Die-hard fans of a young Earth should no more expect a scientific revolution restoring mainstream acceptance of a very young Earth than they should expect a revolution restoring belief in geocentricity or the four elements of the ancient Greeks: earth, water, air and fire. Although some Christians might deny the evidence, turn a blind eye to the evidence or wish that the evidence would just go away because they find it very uncomfortable, and no matter how many Scripture verses they throw at the rocks, the evidence for Earth's vast antiquity is there—it is diverse, it is voluminous, and it will not vanish.

THE DANGERS OF CONTINUING TO PROMOTE A YOUNG EARTH

Unfortunately, despite the pleas of practicing Christian geologists, biologists and astronomers, many young-Earth creationists will undoubtedly persist in their efforts to promote their views as being in accord with Scripture and nature. We do well to ask, however, about the side effects of the crusade to convince the public that young-Earth creationism and Flood geology are valid theories. What are the unintended effects of attempts to introduce young-Earth creationism into the public schools of the nation? What are the unintended effects of the promulgation of young-Earth creationism in churches and Christian schools? We submit that persistent advocacy of young-Earth creationism and Flood geology by churches, Christian organizations and individual believers results in two extremely serious consequences that damage the cause of Christ.

The first consequence concerns the spiritual health of Christian youth. Currently, in hundreds to thousands of pulpits, Sunday schools, Christian schools and homes where children are home-schooled, Christian young people are being indoctrinated by well-intended pastors, Sunday school teachers, Christian school teachers and parents—few of whom have any competence in geology—to accept young-Earth creationism and Flood geology as legitimate science. Frequently, students are taught that the traditional six twenty-four-hour days interpretation of Genesis 1 is the only interpretation of the text that is consistent with belief in an inerrant Bible. Often they are also misleadingly taught that the tenets of young-Earth creationism stand on equal scientific footing with mainstream geologic views of an ancient Earth. Students are told that the data of geology really support the idea that the Earth is only a few thousands of years old and that Noah's flood was responsible for a substantial portion of the geologic record.

Many young Christians have been reared to believe that this concept of creation is a virtual article of faith that represents *the* biblical teaching. Those young Christians then go off to college, to a museum or to another source of knowledge where they may be exposed to legitimate geology and are stunned by the force of geologic evidence for Earth's antiquity. They have been personally confronted with an intellectual and spiritual fixed great gulf that is far wider than the Grand Canyon, between their newfound scientific understanding and the religious views of their youth. Not having been equipped to handle the resulting intellectual and spiritual stresses, they all too often conclude, because the geologic evidence is so persuasive, that what they were taught about creation must be incorrect. To them, the Bible now becomes a flawed book. Sensing that they have been misled about creation by the religious authorities of their youth, they lose confidence in the rest of their religious upbringing. Such students may suffer severe shock to their faith. They were not properly taught the truth about creation, nor were they equipped to deal with challenges to their faith. Christians who are professional scientists have all heard far too many accounts of individuals whose spiritual journeys sound much like the scenario just described. Let's have no shipwrecks of the faith of young, vulnerable, unprepared Christian youth that can be laid at the door of the pseudo-science promoted by Christians.

Although many Christian schools endorse and promote young-Earth creationism, other Christian schools that do not promote it still often lack the courage to teach a genuinely scientific understanding of Earth history for fear of alienating outspoken parents who are adherents of young-Earth creationism. Consequently, school administrators may be unwilling to make solid curricular decisions that might backfire in loss of enrollment. In addition, many science teachers in Christian schools, hired by Christian school boards and administrators who lack competence in assessing the scientific qualifications of the teachers they hire, have themselves been trained to adopt the young-Earth theory. As a consequence, young-Earth creationism receives favored status in Christian school science classrooms and is perpetuated in an ongoing cycle.

We challenge Christian school administrators and board members to introduce and endorse legitimate, mainstream, old-Earth geology into their science curricula. We challenge pastors and Sunday school teachers to omit endorsement of young-Earth creationism and Flood geology from sermons and classes. Finally, we challenge seminary professors who train the future pastors of the churches to learn some geology themselves, to encourage seminarians to learn some geology, to model for seminarians how to relate biblical teaching to valid science and to encourage students to abandon advocacy of young-Earth creationism as they embark on their pastoral work. Christian leaders need to flee from promoting scientific nonsense.

Likewise, continued aggressive advocacy of young-Earth creationism has negative consequences for evangelism and apologetics. Modern young-Earth creationism and Flood geology are not only useless apologetically but are counterproductive in evangelizing unbelieving scientists. Unfortunately, most people, even college graduates, have had little scientific training. Most know even less about geology than they do about chemistry or biology because of the low visibility of geology in many high school and college curricula—doubly so in the curricula of Christian schools and Christian colleges. Such people are vulnerable to young-Earth creationism because its claims can sound so plausible to laypersons.

We do not doubt for a moment that many people have been brought to faith in Christ through the ministry of churches and individuals committed to young-Earth creationism, and we acknowledge that specifically young-Earth creationist claims may have been part of the evangelistic presentation. Be that as it may, it is very improbable that a non-Christian scientist will succumb to young-Earth creationism in order to become a Christian. The overwhelming majority of practicing scientists know enough about geology to recognize that mainstream geology has reconstructed a long, dynamic history of Earth. Moreover, they sufficiently trust the abilities and credentials of the geologic community to accept the force of the evidence for an old Earth presented by geologists. In addition, the great majority of practicing natural scientists also know enough about the claims of young-Earth creationism and Flood geology to recognize that these claims are spurious.

When presented with the gospel, unbelieving scientists will reckon that, if it is an article of Christian faith that the world was created only a few thousand years ago and that most sedimentary rocks were deposited during Noah's flood, a religion that tolerates such bogus science is not worthy of further interest. By linking the gospel of Jesus Christ to young-Earth creationism, Christians place a serious barrier in the way of a person's acceptance of the gospel. In this sense, modern young-Earth creationism is a hindrance to evangelism.

Imagine the following scenario. A Christian has been witnessing to a neighbor who happens to be a scientist. At first, the Christian has simply focused on the person and work of Jesus Christ during his witness. The scientist has been intrigued by the biblical plan of salvation and is impressed by the case for the resurrection of Jesus. The scientist has become quite interested in the Christian faith. Suppose that the Christian invites the scientist to attend a church service, and the scientist agrees to come. The inquiring scientist is moved by the praise, prayers, singing, Bible readings and reverence of the worship service. Then comes the sermon, which happens to concern the doctrine of creation. The very likeable and clearly gifted pastor preaches from the Bible that God made the world and everything in it. He posits that a materialistic view of existence cannot satisfactorily explain the existence of anything—the properties of subatomic particles, the reality of physical law, the sense of right and wrong, the stability of the universe, consciousness and so on. So far, so good, the scientist thinks.

But then the pastor claims that God created the cosmos virtually instantaneously in a series of miracles over a period of six days, only a few thousand years ago. The pastor informs the congregation that the Christian cannot accept the big bang theory because it conflicts with Genesis and is, after all, only a speculation concocted by a scientific community that rejects miracles, is bound by a naturalistic philosophy, and is dominated by materialists who wish to expunge God from his universe. The pastor points out that creationist scientists have produced geological evidence that the Earth really is very young. Radiometric dating, he has learned, is completely unreliable because godly scientists have discovered that the decay constants of radioactive isotopes were speeded up during creation week and the Flood of Noah. In fact, the preacher continues, believing scientists have shown that the rocks of the Grand Canyon were not deposited millions of years ago but in a great watery cataclysm, quite possibly Noah's Flood. Take heart, dear believer, the pastor exhorts, the real facts of science, discovered by believing scientists who love God, far from demonstrating an old Earth, actually corroborate the infallible biblical account of creation in six days. In a flourish, he concludes the sermon by assuring the congregation that they may safely trust in the Lord because the ideas and schemes of men will come to naught, but the Word of our God endures forever.

Virtually any competent scientist—especially any competent geologist—who is investigating Christian faith, upon hearing such a sermon (or a talk on science and creation sponsored by a church) would almost certainly be turned off to further consideration of Christian faith. All such sermons, radio talks, public lectures and the like by young-Earth creationists will only alienate any knowledgeable non-Christian scientist in the audience. It is time for the evangelical world to realize that non-Christian scientists are not the devil's minions whose "false teachings" must be attacked to protect the Christian faithful. They are, like anyone else, image-bearers of God who need to be reclaimed for the Savior. To use missionary terminology, scientists are a "people group" who need evangelizing, just like Muslims, the Fulbe and Fulani tribes of northwestern Africa, the Jews of New York City, the secularized upper-class North Americans or Europeans, or the homeless of inner-city Chicago. Christians will have no success in claiming scientists for Christ, however, if the scientific profession is attacked or technical knowledge is ridiculed any more than Christians will have success in converting dentists by attacking the dental profession or ridiculing what dentists have to say about teeth and dental care.

Next time that you preach a sermon on creation, dear pastor, consider that there might just be a scientist in the congregation. Don't drive him or her away by associating the words "science" and "atheism," by using a negative or cynical tone if you mention science or by advocating scientific "theories" that are counter to mainstream science. If you do mention science, hold it up as an honorable profession to which God might be calling a young person in your congregation. Thank God for the wonderful gifts he has given scientists and for the amazing discoveries they have made about our world. You might even make the challenging suggestion that science is potentially useful for leading the church to improved understanding of the word of God!

The Christian faith has lost credibility in the public consciousness thanks to ill-advised attempts to introduce "scientific" creationism into the science classrooms of the public schools. Can we seriously expect non-Christian educational leaders to develop respect for Christianity if believers insist on teaching the brand of pseudo-science that young-Earth creationism brings with it? Will not efforts to impose modern young-Earth creationism on the public simply lend credence to the idea already entertained by so many intellectual leaders that Christianity, at least in its modern form, is sheer anti-intellectual obscurantism? We fear that it will. It already has. Let's not perpetuate the damage.

THE CHALLENGE OF AN ANCIENT EARTH

What then is the Christian to do if young-Earth creationism and Flood geology must be abandoned? Where does this state of affairs lead?

It can be an intellectual crisis for many Christians to face the idea that the Earth might be billions of years old. After all, many of them have been taught by Christian leaders for so long, some from Sunday school days, that the Bible teaches that the earth is very young. They have been told that Genesis 1 is a straightforward narration of facts that reinforce the young-Earth idea. In addition, such believers cannot understand why God should have taken so long to make the world. For them, the notion of a vast antiquity of the earth diminishes God's power and sovereignty. Frightening questions arise. Is the Genesis creation account reliable or is it only a fictional fantasy? Does the Bible present a flawed picture of the nature of the world?

Most frightening of all is the specter that if the Bible is wrong about creation then the whole story of salvation is placed in severe jeopardy and personal salvation is at stake. Fearing that their faith might be irreparably damaged, inasmuch as they have made a completely unnecessary connection between the great antiquity of Earth and the consequent untrustworthiness of the Bible and its message of salvation, some Christians steadfastly resist the conclusion that the Earth is old. They cannot bring themselves to consider that possibility. They distrust the conclusions of scientists and harbor the deep suspicion that geologists are engaged in a malicious conspiracy to doctor the evidence. Scientists are hostile to the Christian faith anyway, aren't they, so what else can we expect from such folk?

We have the greatest sympathy with Christians who are troubled by these questions and who feel threatened by science. We appreciate the struggles and anguish that believers may go through, because we have experienced some of those struggles ourselves. Obviously, many good Christians are not emotionally ready to accept an old Earth even if they realize intellectually that the scientific case for a young Earth has no merit. They may continue to cling to young-Earth creationism simply because they don't quite know how to integrate belief in an old Earth into their conception of the Bible and of Christian faith. We urge believers, however, to keep such an emotional attachment to themselves and to discontinue public support for young-Earth creationism.

"Proving" the Bible or Christianity with spurious scientific hypotheses does not honor God and can only be injurious to the cause of Christ.[2] We must not defend God's truth by arguing falsehood on its behalf. In fact, Christians must be very cautious in using even legitimate science as an apologetic device. We should not fall into the trap of thinking that Scripture is more reliable or trustworthy if it is backed up at every point by scientific evidence. Nor should we suspect that Scripture may be untrustworthy if science does not back it up at every point. Scripture stands on its own self-attesting authority.

The Unity of God's Word and Works

We have no doubt left some of our readers in a perplexed position in which they are still convinced that Genesis 1 teaches a recent creation in six twenty-four-hour days but, at the same time, are now ready to concede that the scientific claims of young-Earth creationism do not have a leg to stand on. Christians who find themselves in this predicament are bound to experience profound tension in their intellectual universe. How can we reconcile pre-

---

[2]It is with deep sadness that we read the following in Larry Vardiman, Andrew A. Snelling and Eugene F. Chaffin, eds., *Radioisotopes and the Age of the Earth* (El Cajon, Calif.: Institute for Creation Research, 2005), 2:768: "As the evidence accumulates, initial dissemination of these groundbreaking results should be made in creationist publications and to Christians in general to encourage them regarding the reliability of the Bible. Research on the age of the earth may, with God's help, be one of the most important methods for encouraging the church to work to return recognition and honor back to the Creator and Savior and away from naturalism." Conviction of the truth of Scripture is the work of the Holy Spirit and flows from Scripture's own self-attesting character. To be sure, extrabiblical evidence may provide powerful support for our belief in a reliable Bible, but how discouraging to see lay Christians being encouraged regarding the truth of God's Word because of utterly bogus "science."

sumed biblical teaching for a young earth with scientific evidence for an old Earth? Can we reconcile them? It is perfectly natural to want to blend the Bible with knowledge of nature. Deep within all of us is the sense that the whole of our experience and the world we live in and the realm of the divine all hangs together, all coheres, and we have an ineradicable craving to recognize that coherence. Nor is that sense or that craving misplaced. We believe that God has implanted that sense and craving in all of us. In that light, can we help those who believe that the Bible says creation took place in six twenty-four-hour days yet realize that they must abandon young-Earth creationism as a legitimate scientific theory?

Let's suppose then, just for the sake of our discussion, that the Bible and scientific evidence do not *seem* to agree in regard to the origin of the world, and let's see if we can understand why acceptance of the great antiquity of the Earth does not undermine Christian faith.

In studying Earth, geologists—whether they realize it or not—are investigating part of God's creation, a fragment of the universe that is replete with God-established facts. Minerals, rocks, fossils, volcanoes, mountains, strata, intrusions, faults, folds, lava flows, sedimentary structures, metamorphic features and all geologic phenomena are what they are by the will of God. Therefore, Christians (not to mention unbelievers) should handle geologic data reverently and worshipfully and not be afraid of the implications of the data. Even though human theoretical interpretations of aspects of the created world are invariably underdetermined by the data, these interpretations must, nevertheless, attempt to be as faithful to and consistent with the evidence as possible. After all, God made the evidence, and that evidence fits into his comprehensive, unified, coherent plan for the cosmos. Our Creator has providentially brought the world to its current state of existence, and therefore, the data of geology and all other sciences owe their existence to his sovereign counsel. Geologic data are just as much a reality, just as true, as any statement in the Bible. Any geologic fact is just as much a fact as the fact that you were born or that David was the king of Israel.

Obviously a geologic fact is far less important than the central facts of the Christian gospel. Probably no one's life will be significantly different because one is either aware or not aware of that geologic fact—unless one is a geologist, of course. Most of us do not consider the fact that small dikes of basalt cut through rhyolite flows in the St. François Mountains of southeastern Missouri to be a life-shaping or life-altering bit of information! But it is a fact, nonetheless. A geologic fact is very different from biblical statements, because the data of the Bible are expressed verbally whereas those in nature are not. The data of the Bible primarily tell human beings what to believe concerning God and what duty he requires of the sons and daughters of Adam and Eve. The data of the Bible are ethically normative for human lives; the data of nature are not. The Bible generally tells human beings what to do; nature generally does not. In the Bible and in nature, humans encounter different kinds of facts, but, in both cases, facts of divine origin.

Because Scripture and the created universe are both God-given, they cannot be in conflict. They form one comprehensive, unified, coherent whole that is an expression of the character and will of our Creator and Redeemer, who is the author of both. Nature and Scripture form a unity, because God himself is One. All his glorious attributes cohere perfectly. To God, the created universe and his Word are wholly transparent. Inasmuch as he created the world and is the author of Scripture, he comprehensively sees and knows all the details of his universe "at a glance." His interpretation of all the details coincides with his knowledge of them. God's "theology" and his "science" are identical with reality, for he created and sustains it. God also knows the interrelationships of the details as he knows the details by themselves, for he has created them all in accordance with his plan. All is light to him; there are no surprises for him. God makes no discoveries. Nothing is unknown to him. No data are unintelligible to God.

Unlike God, however, humans often express themselves in inconsistent and contradictory ways. Moreover, the fact that God's words and works form a perfect unity by no means indicates that humans can always see how they reflect that unity. In the act of interpreting the Bible, the world around us, ourselves or our teenage children, Christians constantly strive to grasp this unity by understanding the interrelationships. The interpretation of creation is carried out largely through the natural and social sciences, philosophical reflection and the arts, while the interpretation of Scripture is carried out largely by way of theology and biblical studies with the aid of archeology, history, linguistics, anthropology and other allied fields. Given their status as finite creatures, humans must construct their interpretations of reality by discursive thought that entails both induction from particulars and deduction from general principles. For humans, the details generally precede understanding of the interrelationships among those details both temporally and logically. We collect data and then think about their meaning, a process that leads us to further data collection and further reflection. As creatures, too, humans are not privy to all the information about creation. Our knowledge of details is incomplete. Moreover, because the entire created order has a history, a lot of information has been irretrievably eradicated while new facts come into being.

Then too, humans are sinners, and sin distorts the interpretation of God's truth. The human intellect has been affected by sin, and humans do resist, to some degree, dealing responsibly with God's works and words. Men, women and children do not always wholeheartedly accept the God-given evidence that confronts them, a particular danger in the human sciences like psychology and sociology and in theological interpretations of biblical data. At times, humans distort the meaning of evidence. Sometimes this distortion takes place unconsciously and sometimes more intentionally. The fact that humans must *interpret* the unified cosmos in which God has placed us means that they do not always perceive the unity. Humans do not always understand how everything fits together because of both creatureliness and sinfulness.

As a result, the Bible and nature sometimes seem to be unrelated to one another, in competition with one another or even in conflict with one another. Such disjunctions, however, lie not between the Bible and the created order, but rather between human understanding of the Bible and human understanding of nature. It is human interpretations of God-given data that lead into discrepancy, conflict and disagreement. Christians should not be dismayed by seeming conflicts between nature and the Bible. Such apparent conflicts are conflicts between interpretations drawn from natural science and the results of biblical exegesis.

THE UNITY OF THE BIBLE

Let's look more closely at why neither Christians nor non-Christians should become upset over presumed discrepancies between the Bible and the created order. Let's briefly focus on the Bible. Christians recognize that the Bible in itself is a unity. They believe that the Bible has one redemptive story to tell from beginning to end, from Genesis to Revelation. The story begins in the Garden of God with its rivers and life-giving fruit trees, and it ends there, too! The whole story is about intimate fellowship with our Creator God that was lost and is regained through Jesus Christ. Christians affirm that all parts of the Bible agree and are in harmony with one another, but they do not always hear this harmony fully.[3] There are statements and teachings in the Bible that Christians have difficulty fitting together. Although there are many examples of biblical difficulties that Christians cannot fully resolve to their satisfaction, they still recognize that the overarching message of the Bible rings true.

For example, there are a number of parallel accounts of the same event in

---

[3]If we insist on using musical metaphors in dealing with the relation of creation and Scripture, a more appropriate metaphor than harmony would be that of counterpoint. In the hands of master contrapuntalists like Bach or Mahler, two or more seemingly independent melodic lines intertwine in a manner that, nonetheless, manifests an undeniable unity. In the same way, in the hands of the Creator, the created order in all of its variety and the biblical story line seem to follow their own trajectories independent of each other. Nonetheless, nature and Scripture intertwine in a unity that unfolds only with continued study.

the different Gospels. In a few cases, it is difficult to reconcile the differences in detail between the accounts. Anyone who has attempted to work out a perfectly consistent and coherent sequence of the events that occurred on the first Easter morning from the accounts of the resurrection in the four Gospels will experience some puzzlement. No one has yet been able to work out a completely satisfactory chronology. Nevertheless, the Christian church has rightly proclaimed the bodily resurrection of Jesus Christ for two millennia because the Bible teaches it and witnesses to it (despite claims by some scholars to the contrary), the corporate faith of the church and the personal faith of individual believers utterly depend upon it, and the historical evidence strongly favors it.[4]

There also are difficulties associated with the doctrines of the Trinity, the person of Jesus or the sovereignty of God in relation to human freedom. The Scriptural data point to the divinity of God the Father, Jesus Christ and the Holy Spirit. The Bible treats these three as distinct persons and yet as only one God. Scripture everywhere condemns polytheism. Because no human being can comprehend logically how one God can exist as three distinct coequal persons, some groups, such as Jehovah's Witnesses, refuse to accept the biblical data by demoting Jesus to the status of a lesser god who was created before anything else, and by reducing the Holy Spirit to an impersonal divine influence. Rather than reject the doctrine of the Trinity, however, the Christian church has generally held to the biblical data despite the puzzles posed. Most believers confess their belief in the Trinity in the Apostles' Creed and joyously sing what is arguably the greatest hymn ever composed: "Holy, holy, holy! Lord God Almighty! Early in the morning, our song shall rise to thee; Holy, holy, holy! Merciful and Mighty! God in three Persons, blessed Trinity!" They confess and sing those words without agonizing over how Jesus and the Holy Spirit can be God along with the Father without there being three gods.

In regard to Jesus, no one can comprehend how a person can be both God and man with both a divine and a human nature at the same time. How can Jesus at one and the same time know everything as God and yet know some things only partially or not at all as a man? No one knows how that works except Jesus. As with the Trinity, many people reject the biblical data because they cannot logically explain how a human being can also be God Almighty, the Creator of heaven and earth. During early church history, the Docetists denied that Jesus had a real physical body, considering it a kind of specter;

---

[4]For excellent defenses of the historical veracity of the resurrection of Jesus, see William Lane Craig, *Assessing the New Testament Evidence for the Historicity of the Resurrection of Jesus* (Lewiston, N.Y.: E. Mellen, 1989); and N. T. Wright, *The Resurrection of the Son of God* (Minneapolis: Fortress, 2003).

while in our modern age, some liberal theologians have stripped Jesus of his divinity and reduced him to the status of a great religious teacher who was exceptionally close to God, but, although he died like all the rest of us, was, through his high ethical teachings and example, "resurrected" in the hearts and memories of his followers. But the Christian church, guided by the Holy Spirit of truth as promised by Jesus himself, maintains its confession of both the divine and human natures of Jesus despite the tension this may cause. Moreover, most believers readily confess without mental anguish that Jesus the man is himself God, though they don't know how that works.

Mindful of their creatureliness and fallibility, Christians confess that they "know in part" and "see through a glass, darkly" (1 Corinthians 13:9, 12 KJV). Acknowledging that there is much in Scripture that they do not comprehend and that is difficult, Christians nonetheless remain persuaded of the full authority of Scripture through the teaching of Scripture about itself and the internal testimony of the Holy Spirit, with the encouragement of the tradition of the church through two millennia. Christians accept biblical statements and teachings, properly interpreted, as correct and normative, and then live with any resulting tensions. The Christian church should continue to seek out the underlying unity of the Bible and to resolve difficulties without doing so in an artificial manner that forces the biblical data to fit our harmonizing theories.

THE UNITY OF CREATION (OR NATURE)

Many non-Christians, and particularly non-Christian scientists, reject the historical reliability of the Bible because of its textual difficulties, both real and imagined. And many reject the message of the Bible because of its doctrines. The presence of difficulties, however, does not provide adequate grounds for rejection of the authority, infallibility and historical reliability of the Bible. Nor does the existence of alleged discrepancies between Scriptural statements and the details of nature provide adequate grounds for such rejection by unbelievers. In the case of non-Christians, the difficulties very frequently arise from insufficient acquaintance with the Bible and from superficial and deficient interpretations of biblical texts. Moreover, the theological understanding of most non-Christians generally needs considerable correction.

Beyond that, the same sorts of problems exist in our attempts to understand the created world. There are always loose ends, puzzles, difficulties, discrepancies and apparent contradictions in the natural (and social) sciences owing to the fact that scientists are human, their understanding of nature is limited, and they don't always have access to all the information that is pertinent to a situation that they might be investigating. Even though scientists may not have all the information they would like or may not have apprehended all the facts of nature, they still proceed to develop natural science on the basis of the data that they do have. In fact they thrive on the existence of the puzzles. It is precisely the problems and loose ends that lead to progress as scientists seek to uncover further knowledge and understanding. A science without any tensions and seemingly intractable problems would not be any fun at all! One should be very suspicious of a science devoid of perplexities. Some theories like quantum mechanics, chaos theory and the theory that postulates the existence of dark matter even entail an open-endedness that provides room for mystery and apparent contradictions. Scientists will undoubtedly clear up some of the loose ends as our knowledge increases and new discoveries are made, but there are always going to be puzzlements in our interpretation of God's created works.

In chapter four we alluded to one very prominent illustration of such a scientific difficulty in which a number of late nineteenth-century geologists believed that the Earth is several hundred million years old, whereas, in contrast, Lord Kelvin suggested from physical evidence that Earth is closer to 20 million years old, bringing him into sharp conflict with the views of these geologists. At that time, two bodies of evidence drawn from different parts of nature, when interpreted by competent scientists, indicated two radically divergent ideas regarding Earth's age. Consequently, there was a serious conflict in scientific thinking during the late nineteenth century.

What happened within the scientific community as a result of this situation? There was considerable controversy, especially between some geologists and some physicists. Each group was convinced that the other group was missing some key piece of evidence. At the same time, both physicists and geologists remained deeply persuaded that nature is a unity and that nature has only one "right" answer regarding Earth's antiquity. Scientists attempted to work out plausible "harmonizations" of the discrepancies between the data of physics and the data of geology. Toward the end of his career, Kelvin was greatly encouraged when Clarence King proposed on geological and geophysical grounds an age of 24 million years. Many other geologists had also estimated an age on the order of only a few tens of millions of years. In the view of many scientists, some degree of resolution of the situation had been achieved.

Let's note what did not happen during the controversy. Geologists did not abandon the known data of nature, lose interest in nature or turn their back on nature. Nor did the physicists. No one accused nature of contradicting itself. No one accused nature of being unreliable or deceptive. No physicist repudiated the field of geology. No geologist repudiated the field of physics. All participants in the discussion acknowledged the difficulty and tried to resolve it, fully persuaded of the unity of nature. By the end of the nineteenth century, many physicists and geologists were modestly confident that they had uncovered the underlying unity of nature: the evidence roughly pointed to an Earth that is only a few tens of millions of years old. The tension greatly eased—but not for long.

In reality, geologists and physicists had neither resolved the problem nor discovered the underlying harmony and unity of nature. The claim of geologists who thought that Earth was much older than 100 million years eventually was substantiated, and the claim of Kelvin and his followers was rejected, but not because either group had badly misinterpreted the available evidence or distorted data. An important new piece of the puzzle, the phenomenon of radioactivity, was discovered. Radioactivity provided a source of heat that was wholly unknown to Lord Kelvin as well as a method for calculating the ages of minerals. The old consensus was abandoned, and the scientific difficulty was quickly resolved more satisfactorily because of the discovery of new data followed by new and improved interpretations of geophysical evidence.

The fact that the study of creation entails the solution of puzzles should alert everyone to the fact that the study of God's Word, the Bible, might also place a few difficulties in our way. If a non-Christian wishes to repudiate the Bible on the grounds of its alleged unreliability, then he or she must do better than simply point to "difficulties." Well, of course there are difficulties. The whole of our existence confronts us with "difficulties."

THE UNITY OF BOTH SCRIPTURE AND NATURE AGAIN

Why, then, should biblical studies and the natural sciences be in perfect agreement if each by itself has plenty of internal tensions? Both Christians and non-Christians should never expect to have a unified knowledge of both the Bible and Earth devoid of data that lack adequate explanation and theories without difficulties. Let's keep reminding ourselves that the difficulties do not so much exist in and between the created order and the Bible as they do in and between our theoretical reconstructions of these two very different revelations from God. No Christian should be threatened by discrepancies between science and biblical studies any more than they should be threatened by discrepancies within biblical studies or within geology. It should not be a matter of great distress to a Christian if the data of nature do not seem to dovetail with the data of Genesis 6—9 regarding the flood or the creation account of Genesis 1.

The fact that many people incorrectly perceive a serious disagreement between the Bible and nature should not cause rejection of Genesis 1 or the reality of God's work of creation, because the disagreements pertain to human theoretical understanding, not to reality. Nor should such a situation cause anyone to dismiss nature as a reliable and legitimate object of theoretical study. Nor should anyone knowingly distort or falsify evidence to force it into conformity with what we think Scripture teaches. It is a serious mistake for Christians who believe that Genesis 1 teaches creation of the world in six twenty-four-hour days to reject or distort geological evidence for Earth's vast antiquity in order to achieve a harmony between Scripture and nature, because such a harmony would be imaginary. It is healthier to maintain a belief in a six twenty-four-hour-day creation and belief in an ancient Earth in tension (obviously a very serious tension!) than to deny realities.

On the other hand, it is also a serious mistake for Christians who accept scientific evidence for an ancient Earth to reject the biblical teaching that God created the world or to knowingly distort the biblical text in order to achieve a harmony. Proponents of an ancient Earth need to be more careful in interpreting Genesis 1. All too often, Christian advocates of an ancient Earth have adopted well-intended, but ultimately unsatisfactory exegeses of Genesis 1 such as the restitution or day-age interpretations in hopes of achieving harmony between Scripture and nature.[5] Too many of us have arrived at deficient harmonizations, and we need to correct our faulty biblical interpretations.[6] It is healthier to maintain a belief in an old Earth in tension with the raw data of Genesis 1 than to persist in distorting the biblical text simply to achieve harmony. We should be content to let both bodies of revelation speak for themselves and listen as carefully as we can.

This is not to say, of course, that Christians may never offer new interpretations of Genesis 1 or other biblical texts for fear of making mistakes, but if the consensus of biblical scholars (if there is such a thing!) acknowledges serious flaws in those interpretations, then such interpretations eventually ought to be withdrawn. For example, the restitution hypothesis, so popular in the nineteenth century, has largely been abandoned because biblical scholars have repeatedly shown that Genesis 1:2, the text on which the entire hypothesis rests, cannot properly be translated as proposed by its proponents.

Too often, those of us who have been involved in the discussion of the interrelationship between biblical interpretation and natural scientific investigation have been guilty of accepting or rejecting hypotheses on the basis of how well or how poorly the hypothesis fits a set of data to which it was not particularly applicable. For example, some Christians have at times been guilty of interpreting Scripture more on the basis of data from nature than on the data

---

[5]For a critique of concordism, see Paul Seely, "The First Four Days of Genesis in Concordist Theory and in Biblical Context," *Perspectives on Science and Christian Faith* 49 (1997): 85-95; and Davis A. Young, "Scripture in the Hands of Geologists (Part Two)," *Westminster Theological Journal* 49 (1987): 257-304.

[6]For example, the lead author of this book formerly adopted a version of the day-age interpretation that he has abandoned. See Davis A. Young, *Creation and the Flood: An Alternative to Flood Geology and Theistic Evolution* (Grand Rapids: Baker, 1977).

of Scripture. This practice needs to be discontinued.

Henry Morris, for example, rejected the classic day-age hypothesis of interpreting Genesis 1 on textual grounds, a perfectly legitimate procedure whether one agrees with his literal six-day interpretation or not.[7] But Morris also rejected the validity of the day-age interpretation on the grounds that the order of events suggested by geologic investigation does not correspond with the order of events listed in Genesis 1. Ironically, some nineteenth-century advocates of the day-age hypothesis adopted that view because they thought that the order of events disclosed by geology *does* correspond to the order of events in Genesis 1! Morris has listed a number of points at which he believes there is poor or contradictory correspondence between the sequence in Genesis 1 and the sequence of geology. In other words, he assumed that the geological evidence of the supposed sequence of events in Earth history could be used to negate the idea that the "days" of Genesis 1 could have been long periods of time.

But such a comparison between the Bible and nature does not provide legitimate grounds for rejecting the day-age hypothesis. That hypothesis is an interpretation of Genesis 1 that should be defended or rejected solely on the grounds that Scripture affirms or denies it. Evidence from nature may suggest the day-age or other interpretations that may be tested against the biblical text, but the evidence from nature cannot determine what the Bible says about any particular subject. Therefore, simply because the order of events in nature is somewhat different from the order in Genesis 1 does not prove that the days of Genesis 1 were not long periods of time in which much geologic activity may have taken place. Nor does it prove that they were ordinary solar days. That question must be decided on the basis of the internal textual data of Scripture. Viewed from another angle, if scientific evidence demanded a young Earth and the textual evidence demanded that the six "days" were long periods of time, we should have to adopt a day-age interpretation in spite of the putative scientific data!

On the other hand, some opponents of young-Earth creationism have mistakenly claimed that the days of Genesis 1 cannot be twenty-four-hour days on the grounds that geologic evidence is so overwhelmingly convincing that the Earth is extremely old. Christians must, therefore, they think, adopt another understanding of the days of Genesis 1 that spares them the embarrassment of maintaining that the Bible teaches that the days are twenty-four hours long. Although sympathetic toward that point of view, we believe that it is flawed. The data obtained from creation (or from archeological sites) is

---

[7]Henry M. Morris, *Scientific Creationism*, 1st ed. (El Cajon, Calif.: Master Books, 1974), pp. 221-30.

a valuable God-given tool that may raise questions to ask of the biblical text and that might encourage another hard look at the data of the Bible to see if the text has previously been interpreted correctly. Creation, however, cannot force on us the proper interpretation of the Bible any more than the biblical text can force on a geologist the proper interpretation of the geology of the Rocky Mountains.

The biblical text must finally be interpreted in terms of its own claims and assertions, just as geology must be interpreted in terms of its own principles. That said, biblical interpretation that is divorced from information obtained from other sources, such as literature, history, anthropology, linguistics, geology, archeology and the like is bound to be seriously defective. Biblical interpretation carried out in an intellectual vacuum without regard to insights derived from extrabiblical sources will end up being controlled, shaped, and limited by other influences such as the training, experience, biases, breadth of knowledge and culture of the interpreter, influences of which he or she may often be unaware.

In the end, the question of the character of the days of Genesis 1 must be decided by the text and context of that passage and by the analogy of Scripture. It cannot finally be decided by information from the created world and archeological data, but the created world and archeological data are bound to provide new data that clarify textual data in striking ways. To be specific, scientific data may indicate an ancient Earth, but those data do not determine that we should adopt the day-age interpretation any more than they determine that we should adopt the restitution interpretation, the prophetic-day interpretation or some other interpretation. The text itself has to lead us to the proper interpretation, but only when considered in the light of other evidence.

An interpreter who ignores such inputs may end up reading the text as if it had been written in the modern era. Hence, the "earth" is unwittingly assumed to be a planetary globe, and a sequence of days is assumed to be a week. However, an interpreter who takes knowledge of the ancient Near East into account may recognize that "earth" to the original readers was a flat disk and that seven days often functioned as a literary device. Thus, what the text says or teaches may well become clearer to the modern reader if it is read through the proper filter or, to use Calvin's old metaphor, read through the proper spectacles. In other words, one should approach the text with as much knowledge as possible from all relevant sources.

Sometimes the second look at Scripture that is prompted by the investigation of creation has led scholars to change their interpretation of Scripture, as was certainly the case in the days of the Copernican controversy. The new biblical interpretation in that case is one that is consistent with the data of

Scripture itself, and not one that was demanded for the Bible by the evidence from creation.[8] One might argue, as most commentators have, in response to the Copernican theory, that verses like Psalm 93:1 or Matthew 5:45 simply use phenomenal language in which the sun and the moon appear to the ordinary terrestrial observer to be moving around the earth. That interpretation, however, is not imposed on Scripture by the "truth" of the heliocentric theory. One might equally well interpret these texts literally when they speak of the sun rising or the earth not moving while also asserting that the divine author accommodated himself to the geocentric world picture of the original readers and hearers. On this view, neither the divine nor human author intended to teach absolute physical truth that is binding of all believers of all ages. Neither is that interpretation imposed on the Bible by the "truth" of the Copernican theory.

Tayler Lewis was right to critique many of the nineteenth-century harmonists for often allowing their science to do their interpreting of Scripture.[9] They thought that, because science argued for the vast antiquity of Earth, therefore, the restitution hypothesis or the day-age hypothesis, depending upon which hypothesis one favored, is the teaching of Scripture. Christians who accept an old Earth cannot reject the traditional twenty-four-hour-day hypothesis simply because it doesn't agree with science. The character of the days of creation is a textual question.

So, take heart, those of you who are committed to a literal interpretation of Genesis 1 with its six, historical, sequential, twenty-four-hour days only a few thousand years ago. Don't be overly distressed by a disjunction (although a huge one!) between your belief and the scientific consensus of an old Earth. Consider the disjunction an issue to be resolved, not an occasion for a crisis of your faith. But do not seek a solution to the disjunction in the pseudo-science of young-Earth creationism.

But are those of you who fall into this category doomed to live with this huge disjunction for the rest of your life? We certainly hope not. Rather, why not use this scientific evidence in a positive manner as a God-given tool that provides a marvelous occasion for you to rethink your interpretation of Genesis 1? No, we are not asking that you abandon the truthfulness, reliability or historical character of Genesis. Far from it. We are not asking you to abandon the undeniable biblical teaching that God made the entire cosmos and everything in it. We are simply asking that you reconsider your interpretation that

---

[8]On the Copernican controversy, see Kenneth J. Howell, *God's Two Books: Copernican Cosmology and Biblical Interpretation in Early Modern Science* (Notre Dame, Ind.: University of Notre Dame Press, 2002).

[9]Tayler Lewis, *The Six Days of Creation* (Schenectady, N.Y.: Van Debogert, 1855).

God created the cosmos in six solar days a few thousand years ago in a series of instantaneous miracles devoid of any process, and to do that reconsideration in light of our ever-growing knowledge of the ancient Near Eastern context in which the Old Testament was written as well as in light of geological and astronomical discoveries.

In chapters six and seven, we pointed to textual evidence that, although the days of Genesis 1 may be ordinary days, they are used as a literary device symbolizing the completeness of God's handiwork. The days were not used to say that God literally took 144 hours to construct the universe.

Why, then, was this textual evidence dismissed or missed for centuries? Why is it only in recent decades that interpreters have begun to make assertions about the pattern of six days plus one day as a literary device? The situation involving the days of Genesis 1 is analogous to the situation involving the age of Earth in the late nineteenth century. That debate took on a completely different character with the discovery of relevant new information pertaining to the phenomenon of radioactivity. Similarly, relevant new information has also come into play in the interpretation of Genesis 1 relatively recently. The languages, literatures, cultures and histories of the ancient Mesopotamian, Canaanite, Hittite and Egyptian worlds had largely been lost to the early Christian church. Apart from what was preserved in the Old Testament and a few Greek writings, knowledge of the Sumerians, Assyrians, Babylonians, Persians, Egyptians and many other Near Eastern groups was scant. Not until the eighteenth century did archeological investigation begin to uncover those worlds. Decipherment of ancient languages and translation of texts from Iraq and Egypt progressed through the nineteenth century. In the twentieth century, there was an enormous explosion in knowledge of the languages, cultures, thought-forms, religions, governments, international relations, sciences, arts and histories of much of the ancient Near East.[10]

Christians can no longer ignore that vast wealth of extrabiblical information as they undertake interpretation of Genesis 1. Written in the milieu of the ancient Near Eastern world, Genesis 1 must be studied in the light of the literary conventions, symbols, images and other creation accounts of that world. That extrabiblical information does not determine the meaning of the text of Genesis 1, but it certainly opens up new questions and suggests possible answers to what the biblical creation story is saying and how it says it. The Christian church ignores the relevant ancient Near Eastern context of the Old Testament to its detriment.

---

[10]See, for example, James B. Pritchard, *Ancient Near Eastern Texts Relating to the Old Testament* (Princeton, N.J.: Princeton University Press, 1950).

SOME FINAL THOUGHTS

As Christians present the gospel and defend Scripture as God's truth, particularly to scientifically trained unbelievers, they may be confronted with questions about the relationship between the Bible and the natural sciences. Nonbelievers may charge that the Bible is logically inconsistent, scientifically naive or full of mistakes. The unbeliever may find it difficult to cross intellectual barriers posed by real and imagined difficulties with the Bible. The strong temptation for Christians in some instances may be to sweep difficulties under the rug by twisting Scripture or twisting nature to eliminate perceived problems. That temptation must be resisted. In dealing with non-Christians, believers must freely admit that there are puzzles that cannot be solved to their own satisfaction. The non-Christian is likely to respect a presentation of Christianity far more if Christians present the truth of the gospel with love, humility and a willingness to say, "I don't know the answer," when indeed they do not know the answer.

Sadly, too many Christians have distorted the content of the natural sciences in order to gain an accommodation with what they perceive to be the natural interpretation of Scripture. This is, in fact, what has happened with the modern young-Earth creationist movement. Having locked themselves into fixed interpretations of the creation and flood accounts, they find themselves in profound and widening disagreement with the results of modern geology and other sciences. Unwilling to allow conflict to exist, they have sought harmonization with science, not by reevaluating their biblical exegesis but by wholesale distortion of science and the data of nature. They have tried to force nature to say things it does not say.

In this book we have documented that young-Earth creationists have ignored data when convenient. They have misinterpreted other data. They have often misrepresented the views of mainstream geology. They have typically failed to attend to larger geological contexts in focusing on isolated details that seem to support their theories. They have attempted to develop an alternative science that lacks a solid empirical foundation and that cannot duplicate the successes of mainstream geology. They have all too often supported their alternative science with quotations from mainstream geologic literature taken out of context. Their alternative science does agree with their biblical interpretations, but their approach provides no legitimate solutions for biblical studies, theology or geology, because it leads to an illusory harmony between theology and science.

Their wholly new science has nothing to do with the real world in which God has placed us, the real world that God himself made, shapes, sustains and controls and whose factuality has been determined and given by the sovereign

God of the universe. Young-Earth creationism and Flood geology have almost nothing to do with the totality of evidence from Earth, whether chemical element distribution, sedimentary strata, fossils, magmatic activity, glaciation or metamorphism, except in the most superficial way.

We do applaud young-Earth creationists for their desire to provide an alternative to the thoroughgoing materialistic worldview that pervades the popular writings of several world-class scientists whose names are well known to the public. We wholeheartedly concur with them in that effort. Young-Earth creationism, however, fails to provide an appropriate antidote to such materialism. Rather than combat materialism with pseudo-science, let's challenge that false philosophy with a robust biblical theism that lays bare the inadequacy of materialism to account for the origin and ongoing existence of anything—whether it is matter, order, coherence, law, consciousness, good and evil, or purpose and meaning.

The Bible is indeed the infallible Word of God. But the created order is also from God, and the created order leads us to believe that Earth is extremely old. Genuine scientific investigation of God's world is an exciting enterprise that God wants his people to engage in. Instead of the flight-from-reality "science" of young-Earth creationism, the church and the world need more of the vigorous approach to both the creation and Scripture that we found in men of the nineteenth century like William Buckland, Hugh Miller, Thomas Chalmers, John Fleming, Edward Hitchcock, James Dana, William Dawson, Arnold Guyot and Alexander Winchell.

Wouldn't it be thrilling if Christians would stop expending energy espousing and defending this false conception of the biblical doctrine of creation? How much more satisfying it would be to expend our time in properly interpreting and applying the Bible in light of all the knowledge that God has given us, rather than in a vacuum. How much more satisfying it would be to expend our time in light of all the knowledge God has given us, and in uncovering the awesome mysteries of an Earth that God in his overabounding love has actually given us! Vigorous Christian theology, biblical studies, geology and other natural sciences, unafraid of truth even when it surprises, are far more serviceable in effective evangelism and apologetics than the dream world of young-Earth creationism. In the end, Christian devotion to God's truth, not enslavement to tradition, leads to the greater glory of our Creator—our Maker, Defender, Redeemer and Friend.

Name Index

Subject Index

Scripture Index